Simplified Algebra, Differential Calculus, Statistics and Probability

A Well Simplified Math Book for high Schools and Colleges

Kingsley Augustine

Printed by Amazon KDP

TABLE OF CONTENTS

PREFACE

Simplified Algebra, Differential Calculus, Statistics and Probability serves as a useful companion for students in high schools and higher institutions of learning. It is a valuable textbook for students who want to write entrance test or examination into colleges and universities. This book consists of step-by-step explanation of topics presented in a way that is easy for students to understand. It contains very many worked examples and many self-assessment exercises to ensure that students get a mastery of each topic covered. The answers to the exercises are provided at the end of the book.

What makes this book a unique mathematical asset, is its detailed step by step approach in explaining the topics covered in these branch of mathematics. Instead of solving questions by going straight to the point, leaving you confused and frustrated, this textbook teaches you in simple English, explaining each step taken at a time. Thus, allowing anyone, regardless of their experience in algebra, differential calculus, statistics and probability, to understand each topic with ease, and hence make mathematics more interesting.

I give all thanks and Glory to God Almighty, for giving me the grace to write this book. I also wish to express my deep appreciation to my wife Mrs. Mercy Augustine for her patience, understanding and encouragement when I was writing this book. I also thank my children, Dora, Merit and Elvis for their moral support.

Kingsley Augustine.

CHAPTER 1
BASIC ARITHMETIC OPERATIONS

In order to fully understand the content of this book, it is important to be familiar with basic arithmetic operations. The following topics should remind us of these basics.

Addition and Subtraction of Integers (Directed Numbers)

a. -11 + 7 = -4
Since the two numbers have different signs, one way of solving this problem is to subtract the smaller number from the larger number. Then take the sign of the higher number in the question as the sign of the answer. In this example, 7 is the smaller number while 11 is the larger number. Hence, subtract 7 from 11 to get 4. The sign of the higher number (i.e. 11) is minus, so the answer 4 will carry a minus sign. This will finally give us -4

b. -6 + 15 = 9
Since their signs are different, subtract the smaller number from the larger number. Finally, take the sign of the higher number as the sign of the answer. In this example, 6 subtracted from 15 is 9. The sign of the higher number (i.e. 15) is plus, so the answer 9 will carry a plus sign. This will finally give us +9 or 9 as the answer as shown above.

c. 3 - 10 = -7
Their signs are different. So, subtract 3 from 10. This gives 7. The sign of the larger number (i.e. 10) is minus, so the answer 7 will carry a minus sign. This will finally give us –7.

d. -4 - 8= -12
In this case, their signs are the same. So, add the two numbers, then give the sign that they have, to the answer. In this example, 4 added to 8 gives 12. Their sign is minus, so the answer 12 will carry a minus sign. This will finally give us -12.

e. 5 - (-6) = 5 + 6 = 11
Two negative signs that are closed to each other multiply to become positive.

f. -6 - (-2) = -6 + 2 = -4.   (This is similar to (a) above where the larger number has a negative sign.

g. -14 - 17 = -31
Since their signs are the same, add the two numbers to get the answer, and remember to give the sign that they have to the answer. In this example, 14 added to 17 gives 31. Their sign is minus, so the answer 31 will carry a minus sign. This will finally give us -31.

Multiplication and Division of Integers (Directed Numbers)
a. -2 x 3 = -6

Carry out multiplication and division as the usual way of multiplying and dividing numbers. However, when one of the numbers is negative, the answer will be negative. When the two numbers are negative, the answer will be positive.

b. $5 \times -4 = -20$

c. $-2 \times -6 = 12$

d. $-7 \times -3 = 21$

e. $-10 \div 2 = \dfrac{-10}{2} = -5$

f. $18 \div -6 = \dfrac{18}{-6} = -3$

g. $-48 \div -12 = \dfrac{-48}{-12} = 4$

h. $-24 \div -6 = \dfrac{-24}{-8} = 3$

Precedence (Rule of PEMDAS or BODMAS)

Precedence is the order which arithmetic operations are followed during simplifications. The order of operation is easily remembered by using the acronym below:

PEMDAS: Parenthesis, Exponent, Multiplication, Division, Addition, Subtraction.

This means that parenthesis is done first before exponents (powers and roots), then multiplication or division, and addition or subtraction. Multiplication and division are at the same level, hence deal with anyone that comes first from left to right. Similarly, addition and subtraction are at the same level, hence deal with anyone that comes first from left to right. Another acronym that can be used for precedence is:

BODMAS: Bracket, Order, Division, Multiplication, Addition, Subtraction.

This means that bracket is done first before order (powers and roots), then multiplication or division, and addition or subtraction.

Note that the 'O' in BODMAS does not mean 'Of' as popularly used in some books. However, the operation, 'Of', which can be used in some arithmetic processes, should be treated as multiplication.

Between the two acronyms i.e. PEMDAS and BODMAS, I prefer the use of BODMAS because I think it is easier to remember.

Examples

1. $8 \div 2 \times 5$

Here, the rule of BODMAS has to be applied. We have to carry out the division first before the multiplication. This is done as follows:

$\qquad 8 \div 2 \times 5 = (8 \div 2) \times 5 \qquad$ (The brackets show the operation that will be done first)

$$= 4 \times 5 = 20$$

2. $15 \times 2 \div 5$

Since division and multiplication are at the same level, we simplify anyone that comes first. We can work from left to right and carry out the multiplication first before division. This is done as follows:

$$15 \times 2 \div 5 = (15 \times 2) \div 5$$
$$= 30 \div 5$$
$$= 6$$

3. $6 + 24 \div 3 \times (5 - 3) \text{ of } \dfrac{3}{4}$

By applying the rule of BODMAS, the part in the bracket will be handled first. This gives:

$$6 + 24 \div 3 \times (5 - 3) \text{ of } \dfrac{3}{4} = 6 + 24 \div 3 \times 2 \text{ of } \dfrac{3}{4}$$

The next step will be division. 24 divided by 3 gives 8. This is written as follows:

$$6 + 8 \times 2 \text{ of } \dfrac{3}{4}$$

The 'Of' is regarded as multiplication. This now simplifies to give

$$6 + 8 \times 2 \times \dfrac{3}{4} = 6 + 16 \times \dfrac{3}{4}$$
$$= 6 + 4 \times 3 \qquad \text{(Note that 4 divides into 16 to give 4)}$$
$$= 6 + 12$$
$$= 18$$

4. $4 \times 5 - 9 \div [3 + 2^3 - (3 \times 2)]$

The inner bracket simplifies to give 6. This is given below.

$$4 \times 5 - 9 \div (3 + 2^3 - 6)$$

The exponent in the bracket gives 8 (i.e. $2^3 = 8$) as shown below.

$$4 \times 5 - 9 \div (3 + 8 - 6)$$

The addition and subtraction in the bracket simplifies to give $3 + 8 - 6 = 5$. This gives:

$$4 \times 5 - 9 \div 5$$

The multiplication and division simplify to give:

$$20 - \dfrac{9}{5} = \dfrac{100 - 9}{5}$$
$$= \dfrac{91}{5} = 18\dfrac{1}{5}$$

5. $4 + 6 - 3^2 \times (2 \times \sqrt{16})7 + 1$

The root in the bracket simplifies to give 4 as shown below:

$$4 + 6 - 3^2 \times (2 \times 4)7 + 1$$

The term in the bracket gives 8 as follows:

$$4 + 6 - 3^2 \times (8)7 + 1$$

The exponent or order or power simplifies to give 9 as shown below:

$$4 + 6 - 9 \times (8)7 + 1$$

The term outside the bracket multiplies the term in the bracket to give:

$$4 + 6 - 9 \times 56 + 1$$

We simplify the multiplication operation to give:

4 + 6 – 504 + 1

Starting from left to right we carry out the addition and subtraction as follows:

10 – 504 + 1 = – 494 + 1

= – 493

Exercise 1

1. Evaluate the following:

a. −20 + 5
b. −32 + 21
c. 2 − 7
d. −3 − 9
e. 8 − (−11)
f. −11 − (−17)
g. −56 − 26

2. Find the values of the following:
a. −5 x 2
b. 2 x −11
c. −4 x −9
d. −6 x −5
e. −24 ÷ 6
f. 42 ÷ −7
g. −72 ÷ − 9
h. −48 ÷ −12

3. Apply precedence in evaluating the following:
a. 22 ÷ 11 x 6
b. 8 x 15 ÷ 5
c. $13 + 30 \div 15 \times (16 - 8) \text{ of } \frac{2}{3}$
d. $12 \times \frac{5}{24} - 9 \div [1 + 3^4 - (2 \times 5)]$
e. $2 + 11 - 4^3 \times (3 \times \sqrt{4})5 - 3$

CHAPTER 2
SIMPLIFICATION, FACTORIZATION AND SUBSTITUTION IN ALGEBRA

The use of letters to represent numbers is known as algebra. In algebra, we imagine the letters to stand for numbers in order for operations to be easy.

Examples
1. What is y kilograms in grams

<u>Solution</u>
 Since 1kg = 1000g
 and 2kg = 2 x 1000g
 Then ykg = y x 1000g
     = 1000yg

Note that y x 1000 = 1000 x y = 1000y. In algebra, when a number is multiplied by a letter, the number is usually written before the letter even if the letter comes first in the multiplication. Also, when two or more letters are multiplied together, the letters are usually written in alphabetical order. For example x x d x f is written as dfx (in alphabetical order).
Also note that m x n is written as mn, m÷ n is written as $\frac{m}{n}$, m + n remains m + n, while m − n remains m − n.

2. A boy was 13 years old y years ago. How old is he now? How old will he be in c years time?

<u>Solution</u>
It is easy to evaluate this question if the letters are imagined to be numbers. For example, if the boy was 13 years old 4 years ago, then he will be 17 years old now (i.e. 13 + 4 = 17).
Therefore, the boy's present age is obtained by addition. So, his present age is (13 + y) years.
Similarly, in c years time, his age will be obtained by addition. So, in c years time, the boy's age will be his present age + c. This gives:
(13 + y + c)years (Note that his present age is 13 + y)

3. A man is x years old. How old was he p years ago? How old will he be in 5 years time?

<u>Solution</u>
His present age is x years. His age p years ago will be obtained by subtracting p from his present age.
So, p years ago, the man's age was (x − p) years
Also, in 5 years time, his age will be:
(x + 5)years

4(a) What is $a in cents?

(b) Express x minutes in hours.

Solution

(a) Since $1 = 100 cents

and $2 = 2 x 100 cents

Then, $a = a x 100 cents

= 100a cents

(b) Since 60 minutes = 1 hour

1 minute = $\dfrac{1}{60}$ x 1 = $\dfrac{1}{60}$ hours

2 minutes = $\dfrac{1}{60}$ x 2 hours

10 minutes = $\dfrac{1}{60}$ x 10 hours

Then, x minutes = $\dfrac{1}{60}$ x x hours

$\therefore$ x minutes = $\dfrac{x}{60}$ hours

5. The perimeter of a rectangle is 50cm and the breadth is $y$cm. Find the area of the rectangle in terms of y.

Solution

Perimeter of a rectangle = 2(l + b), where l = length and b = breadth.

50 = 2(l + y)   (since breadth = y)

$\dfrac{50}{2} = \dfrac{2(l+y)}{2}$ (When you divide both sides by 2 in order to eliminate the 2)

$\therefore$ 25 = l + y

$\therefore$ 25 − y = l (Note that when y is taken to the other side of the equation, its sign will change. Hence +y becomes −y).

$\therefore$ length, l = 25 − y

Area of rectangle = l x b

$\therefore$ Area = (25 − y) x y

= 25y − y^2

Note that y x y = y^2. Also, when a number or letter multiplies an expression in a bracket, the letter or number is used to multiply each term in the bracket.

6. One side of a rectangle is ($2x$ + 6)cm. If the perimeter of the rectangle is ($6x$ + 8)cm, find the area of the rectangle in terms of x.

Solution

Let l = ($2x$ + 6)

$\therefore$ perimeter = 2(l + b)

$6x$ + 8 = 2($2x$ + 6 + b)

15

Divide both sides by 2. This gives:

$$\frac{6x + 8}{2} = \frac{2(2x + 6 + b)}{2}$$

$$\frac{6x}{2} + \frac{8}{2} = 2x + 6 + b \qquad \text{(2 has canceled out)}$$

$$3x + 4 = 2x + 6 + b$$

Collecting terms in x and constant terms on the same side of the equation gives b as follows:

$$3x + 4 - 2x - 6 = b$$

$$3x - 2x + 4 - 6 = b$$

$$x - 2 = b$$

∴ Breadth, b = $x - 2$

The area of the rectangle is given by:

Area = l x b

$$= (2x + 6) \times (x - 2)$$

$$= (2x + 6)(x - 2)$$

Area = $2x^2 - 4x + 6x - 12$

Area = $2x^2 + 2x - 12$

Note that in expanding the bracket above, each term in the first bracket is used to multiply each term in the second bracket. Ensure you carry the sing of each term along with the term when carrying out the expansion.

Expansion of Brackets

The expression a($x + y$) means a x ($x + y$), and it is expanded as follows:

a($x + y$) = ax + ay

The expression (a + b)($x + y$) means (a + b) x ($x + y$), and it is expanded as follows:

(a + b)($x + y$) = ax + ay + bx + by. Each term in the first bracket is used to multiply each term in the second bracket.

Examples

1. Expand 2($x - $3y)

Solution

2($x - $3y) = 2($x$) +2(-3y)

= 2$x - $6y

2. Expand and simplify the expression, 2$x - $[3 − (a + b)]

Solution

2$x - $[3 − (a + b)] = 2$x - $(3 − a − b)

= 2$x - $3 + a + b

Note that the inner bracket was expanded first before the outer bracket. Also, a negative sign outside a bracket is taken as −1 outside the bracket. For example

−(a+b) = −1(a + b) = (−1 x a) + (−1 x b) = −a − b as expressed above.

3. Remove bracket and simplify $a + 2(a - b) - (a + 3b)$

Solution
$a + 2(a - b) - (a + 3b) = a + 2a - 2b - a - 3b$
Collecting like terms together gives:
$a + 2a - a - 2b - 3b$
$= 2a - 5b$
Note that terms that have the same letters are like terms. Numbers without letters are also like terms.

4. Simplify $2x(x - 3) - 2(x^2 + 2x)$

Solution
$2x(x - 3) - 2(x^2 + 2x) = 2x(x) + 2x(-3) - 2(x^2) - 2(+2x)$
$= 2x^2 - 6x - 2x^2 - 4x$
Collecting like terms together gives:
$2x^2 - 2x^2 - 6x - 4x$
$= -10x$ (since $2x^2 - 2x^2 = 0$)

5. Expand $(a - b)(a - c)$

Solution
$(a - b)(a - c) = a \times a + (a \times -c) + (-b \times a) + (-b \times -c)$
$= a^2 - ac - ab + bc$ (This cannot simplify further)
Take note of how the negative sign was used in the expansion.

6. Expand and simplify $(2x + 5y)(3x - 2y)$

Solution
$(2x + 5y)(3x - 2y) = (2x \times 3x) + (2x \times -2y) + (5y \times 3x) + (5y \times -2y)$
$= 6x^2 - 4xy + 15xy - 10y^2$
$= 6x^2 + 11xy - 10y^2$
-Note that $-4xy$ and $+15xy$ are like terms which simplify to $+11xy$. Note that $x \times x = x^2$. Similarly, $x^2 \times x = x^3$. This is done by adding their powers. Note that x has a power of 1 which is not usually written.

7. Expand and simplify $(3a - 3)^2$

Solution
$(3a - 3)^2 = (3a - 3)(3a - 3)$
$= (3a \times 3a) + (3a \times -3) + (-3 \times 3a) + (-3 \times -3)$
$= 9a^2 - 9a - 9a + 9$
$= 9a^2 - 18a + 9$

Note : that $-x- = +$ (i.e. minus multiplied by minus gives plus), $-x+ = -$, $+x- = -$. For example $-2a \times -a = +2a^2$, and $2 \times -3 = -6$

8. Find the coefficient of ab in the expansion and simplification of: $(2a - 3b)^2 - (a - b)^2$

Solution

$(2a-3b)^2 - (a - b)^2 = (2a -3b)(2a -3b) - [(a-b)(a-b)]$
$= 4a^2 - 6ab - 6ab + 9b^2 - (a^2 - ab - ab + b^2)$
$= 4a^2 - 6ab - 6ab + 9b^2 - a^2 + ab + ab - b^2$

Collecting like terms together gives
$4a^2 - a^2 - 6ab - 6ab + ab + ab + 9b^2 - b^2$
$= 3a^2 - 10ab + 8b^2$

$\therefore$ The coefficient of ab $= -10$

Note that coefficient is a number multiplying a particular variable or letter/letters

9. Simplify and find the coefficient of d^2 in $3d(2d + 3) - (3d + 1)(2d + 1)$

Solution

$3d(2d +3) - (3d +1)(2d+1) = 3d(2d +3) - [(3d +1)(2d +1)]$

Note that the outer bracket is introduced due to the negative sign multiplying the bracket
$= 6d^2 + 9d - (6d^2 + 3d + 2d + 1)$
$= 6d^2 + 9d - 6d^2 - 3d - 2d - 1$
$= 6d^2 - 6d^2 + 9d - 3d - 2d - 1$
$= 0d^2 + 4d - 1$
$= 4d -1$

$\therefore$ The coefficient of $d^2 = 0$

Lowest Common Multiple (L.C.M) of Algebraic Terms

The LCM of a given set of numbers is the first multiple of the largest number which can divide all the other numbers without a remainder .

For letters, the LCM is obtained by multiplying all the letter together. However, when a particular letter is repeated in the various terms, the LCM is the letter that has the highest power.

Examples
Find the LCM of the following:
1. x, y, z
2. $2ab, 3ax, ac$
3. $5ax^2, 3a^2y, 3xy$
4. 6a and 3b
5. $(x - y)$ and $(2x + y)$
6. $(a - 2b)$ and $(4a - 8b)$

1. The LCM of x, y, $z = xyz$ (Their multiplication)

2. 2ab, 3ax, ac. Here, first find the LCM of 2 and 3. The larger number is 3. The multiples of 3 are 3, 6, 9, 12, … e.t.c. Out of these multiples the first one that can divide 2 and 3 without a reminder is 6. So the LCM of 2 and 3 is 6.
For the letters, the highest power of a is 1, so the LCM of a in all the terms is a. Similarly, the LCM of b, c, and x are b, c, and x, since they all have power of 1. We now multiply all these LCMs to obtain the overall LCM. Therefore the LCM of 2ab, 3ax and ac is 6abc$x$.

3. 5ax^2, 3a^{2}y, 3xy. For the numbers, the largest is 5. The multiples of 5 are 5, 10, 15, 20 … e.t.c. The first multiple that can divide 3, 3 and 5 is 15. Hence, the LCM of 5, 3, 3 is 15.
For the letters, a in its highest power is a^2, for x is x^2 and for y is y. So the LCM of 5ax^2, 3a^{2}y and 3xy is 15a$^2x^2$y.

4. 6a and 3b. For the numbers, the larger value is 6. The multiples of 6 are 6, 12, 18, … e.t.c. The first multiple that can divide 6 and 3 is 6. The letters are single in the two terms. So the LCM of 6a and 3b is 6ab.

5. $(x - y)$ and $(2x + y)$. Here, the terms in the bracket are not the same and they cannot be factorized to look alike, so the LCM is obtained by multiplying them together.
∴ The LCM of $(x - y)$ and $(2x + y)$ is $(x - y)(2x + y)$

6. (a – 2b) and (4a – 8b). If we look at the second expression carefully, we will observe that it can be factorized to have an expression like the first one. If we factorize the second expression it will give 4(a – 2b). Hence, we can say that we are looking for the LCM of (a – 2b) and 4(a – 2b). Since (a – 2b) is common, we simply pick one as an LCM and then we take 4 as an LCM since it is the only number present. When we multiply 4 and (a – 2b) (i.e. one of the common expression) it gives 4(a – 2b).
∴ The LCM of (a – 2b) and (4a – 8b) is 4(a – 2b).

Addition and Subtraction of Algebraic Fractions

Examples
1. Simplify the following:

(a) $\dfrac{3x}{8} + \dfrac{x}{8}$

(b) $\dfrac{1}{5x} + \dfrac{3}{4x}$

(c) $\dfrac{2a}{3b} - \dfrac{3}{4x}$

(d) $\dfrac{3x+2}{5a} + \dfrac{2b-1}{10b}$

(e) $\dfrac{a+3b}{a} - \dfrac{2a-b}{4b} - 3$

Solutions

(a) $\dfrac{3x}{8} + \dfrac{x}{8}$

The LCM of 8 and 8 is 8. So, divide 8 (i.e. LCM) by each of the denominators and multiply the answer by the corresponding numerator. This gives:

$$\dfrac{3x}{8} + \dfrac{x}{8} = \dfrac{(1 \times 3x)+(1 \times x)}{8} = \dfrac{3x+x)}{8} = \dfrac{4x}{8}$$

$$= \dfrac{x}{2} \quad \text{(in its lowest term)}$$

(b). $\dfrac{1}{5x} + \dfrac{3}{4x}$

The LCM of $5x$ and $4x$ is $20x$. Divide $20x$ by each denominator and multiply the value obtained by the corresponding numerator. This gives:

$$\dfrac{1}{5x} + \dfrac{3}{4x} = \dfrac{(4 \times 1)+(5 \times 3)}{20x} = \dfrac{4+15}{20x} = \dfrac{19}{20x}$$

Note that $\dfrac{20x}{5x} = 4$, while $\dfrac{20x}{4x} = 5$. These values are used to multiply their corresponding numerators as shown in the solution above.

(c) $\dfrac{2a}{3b} - \dfrac{3}{4x} = \dfrac{(4x \times 2a) - (3b \times 3)}{12bx} = \dfrac{8ax - 9b}{12bx}$

Note that the answer above cannot be simplified further. Also, $\dfrac{12bx}{3b} = 4x$, while $\dfrac{12bx}{4x} = 3b$, as b and x cancels out respectively.

(d) $\dfrac{3x+2}{5a} + \dfrac{2b-1}{10b} = \dfrac{2b(3x+2) + a(2b-1)}{10ab} = \dfrac{6bx + 4b + 2ab - a}{10ab}$ (This cannot be simplified further)

(e) $\dfrac{a+3b}{a} - \dfrac{2a-b}{4b} - \dfrac{3}{1}$ (The 3 is expressed as $\dfrac{3}{1}$)

This gives: $\dfrac{a+3b}{a} - \dfrac{2a-b}{4b} - \dfrac{3}{1} = \dfrac{4b(a+3b) - a(2a-b) - 4ab(3)}{4ab}$

$$= \dfrac{4ab + 12b^2 - 2a^2 + ab - 12ab}{4ab}$$

$$= \dfrac{4ab + ab - 12ab + 12b^2 - 2a^2}{4ab}$$

$$= \dfrac{-7ab + 12b^2 - 2a^2}{4ab}$$

$$= \frac{12b^2 - 7ab - 2a^2}{4ab}$$

2. Express each of the following as a single fraction in its simplest form.

(a) $\dfrac{3x + 2}{3} - \dfrac{x - 1}{4} - \dfrac{5}{12}$

(b) $\dfrac{5}{a + 4} - \dfrac{2}{a - 2}$

(c) $\dfrac{a - 3}{a + 2} - \dfrac{a - 2}{a - 5}$

<u>Solutions</u>

(a) $\dfrac{3x + 2}{3} - \dfrac{x - 1}{4} - \dfrac{5}{12} = \dfrac{4(3x+2) - 3(x - 1) - (1 \times 5)}{12}$

$$= \frac{12x + 8 - 3x + 3 - 5}{12}$$

$$= \frac{12x - 3x + 8 + 3 - 5}{12}$$

$$= \frac{9x + 6}{12}$$

$$= \frac{3(3x + 2)}{12} \quad \text{(After factorizing the numerator)}$$

$$= \frac{3x + 2}{4} \quad \text{(After equal division by 3)}$$

(b) $\dfrac{5}{a + 4} - \dfrac{2}{a - 2}$

The LCM of a + 4 and a − 2 is (a + 4)(a − 2)

Also, if we use this LCM to divide the first denominator, it gives: $\dfrac{(a + 4)(a - 2)}{a + 4} = a - 2$ (Since a + 4 will cancel out).

Similarly, the LCM divided by the second denominator gives: $\dfrac{(a + 4)(a - 2)}{a - 2} = a + 4$ (Since a − 2 will cancel out)

$\therefore \dfrac{5}{a + 4} - \dfrac{2}{a - 2} = \dfrac{(a-2)5 - (a+4)2}{(a + 4)(a - 2)} = \dfrac{5a - 10 - (2a + 8)}{(a + 4)(a - 2)}$

$$= \frac{5a - 10 - 2a - 8}{(a + 4)(a - 2)}$$

$$= \frac{3a - 18}{(a + 4)(a - 2)}$$

$$= \frac{3(a-6)}{(a+4)(a-2)}$$

(c) $\dfrac{a-3}{a+2} - \dfrac{a-2}{a-5}$

The LCM of a + 2 and a − 5 is (a + 2)(a − 5)

$$\therefore \frac{a-3}{a+2} - \frac{a-2}{a-5} = \frac{(a-3)(a-5)-(a-2)(a+2)}{(a+2)(a-5)}$$

$$= \frac{a^2-5a-3a+15-(a^2+2a-2a-4)}{(a+2)(a-5)}$$

$$= \frac{a^2-8a+15-(a^2-4)}{(a+2)(a-5)} \qquad \text{(Note that +2a − 2a = 0)}$$

$$= \frac{a^2-8a+15-a^2+4)}{(a+2)(a-5)}$$

$$= \frac{a^2-a^2-8a+15+4}{(a+2)(a-5)}$$

$$= \frac{-8a+19}{(a+2)(a-5)}$$

Highest Common Factor (H.C.F) of Algebraic Terms

The HCF of a given set of numbers is the number which is just lower than or equal to the smallest number in the given set, and which all the numbers in the given set can divide without a remainder.

For letters, the HCF of a particular letter is obtained by taking the letter with the lowest power in the given set.

Examples
Find the HCF of the following:
1. 4, 6, 8
2. 3, 9, 15
3. $8a^2b$, $20ab^2$
4. a^2b^4c, $a^3b^2c^3$
5. $36xy^3z^5$, $54xy^5z^3$, $72xy^4z^5$

Solutions
1. The H.C.F of 4, 6, 8 is 2. The smallest number in the set is 4. So, a number that is just lower than or equal to 4 which 4, 6, and 8 can divide without a remainder is 2.

So, their H.C.F is 2.

2. The H.C.F of 3, 9, and 15 is 3. This is because the lowest number is 3, and all the numbers can divide 3 without remainder.

3. For the terms, $8a^2b$ and $20ab^2$, the H.C.F of 8 and 20 is 4. The lower number is 8, and a number which is just lower than 8 or equal to 8, and which 8 and 20 can divide without remainder is 4. Concerning the letters, between a^2 and a, we take "a", since it has the lower power. Similarly, between b and b^2 we take b since it has the lower power. Therefore, combining all these answers together gives the H.C.F of $8a^2b$ and $20ab^2$ as 4ab.

4. a^2b^4c, $a^3b^2c^3$
For the letter a, we take a^2 since it has the smaller power. For the letter b, we take b^2, while for the letter c, we take c since it has the lower power.
Therefore, the H.C.F of a^2b^4c and $a^3b^2c^3$ is a^2b^2c.

5. $36xy^3z$, $54xy^5z^3$, $72xy^4z^5$
For the numbers, 36, 54 and 72, the smallest is 36. All the numbers can divide 2, 3, 6, 9 and 18. Note that these factors must not be greater than 36 which is the smallest in the set. So, among 2, 3, 6, 9 and 18, the highest is 18. So, 18 is the number which is just lower than or equal to 36, and which all the numbers (i.e. 36, 54, 72) can divide without remainder.
For the letters, the lowest power of x is 1 (i.e. x), the lowest power of y is 3 (i.e. y^3), while the lowest power of z is 3 (i.e. z^3). So, the H.C.F of the letters is xy^3z^3
∴ The HCF of $36xy^3z^5$, $54xy^5z^3$, and $72xy^4z^5$ = $18xy^3z^3$.

Factorization of Algebraic Expression

An algebraic expression can be expressed as a product of its factors. When factorizing expressions we simply take the HCF of the terms in the expression and put it outside a bracket. Then we divide each individual term in the expression by the HCF in order to obtain the new terms that will be inside the bracket.

Examples
1. Factories 3a – 12b

Solution
The HCF of 3 and 12 is 3. For the letters, since the letters are not repeated, then their HCF which is 1 will not be used here. When there is no common letter or number to a set of terms, then their HCF is taken as 1.
So, factorize the expression by taking the HCF and then introducing a bracket. Then, divide each term of the expression by the HCF. This is done as follows:
$3a - 12b = 3(\frac{3a}{3} - \frac{12b}{3})$ (The HCF of 3a and 12b is 3, which is used outside the bracket, and also used to divide each term in the expression)

= 3(a – 4b)   (After simplifying the division above)

∴ 3a – 12b = 3(a – 4b)

(Note that the expansion of the factorized expression will give the original expression)

2. Factorize 15y + 10.

<u>Solution</u>

15y + 10

The HCF of 15y and 10 is 5. The letter is not considered since it appears in only one of the terms

∴ $15y + 10 = 5(\dfrac{15y}{5} + \dfrac{10}{5})$

= 5(3y + 2)   (This is obtained after the division of 15 by 5, and 10 by 5)

∴ 15y + 10 = 5(3y + 2)

3. Factorize: –6m + 15mn

<u>Solution</u>

–6m + 15mn

The HCF of –6m and 15m is –3m. Always take the sign of the first term.

Note that n is not present in both terms, so it is not considered in the HCF.

∴ $-6m + 15mn = -3m(\dfrac{-6m}{-3m} + \dfrac{15mn}{-3m})$

= –3m(2 – 5n)

(Note that –6m divided by –3m is 2, while 15mn divided by –3m is –5n)

4. Factorize $-4a^5 + 2a^3b - 10a^2b^2$

<u>Solution</u>

$-4a^5 + 2a^3b - 10a^2b$

The HCF of $-4a^5$, $2a^3b$ and $10a^2b$ is $-2a^2$. Note that b is not present in all the three terms, so it is not considered in the HCF. Since $-2a^2$ is the HCF of the terms, it also means that $-2a^2$ is what is common to the three terms in the expression.

∴ $-4a^5 + 2a^3b - 10a^2b^2 = -2a^2(\dfrac{-4a^5}{-2a^2} + \dfrac{2a^3b}{-2a^2} - \dfrac{10a^2b^2}{-2a^2})$

$= -2a^2(2a^3 - ab + 5b^2)$

Note that $\dfrac{a^5}{a^2} = a^{5-2} = a^3$ (simply subtract the powers). Also note that a negative sign outside a bracket changes the sign of every term in the bracket.

5. Factorize $3m^3 - 2m^2 + m$

<u>Solution</u>

$3m^3 - 2m^2 + m$

The terms m^3, m^2 and m have m in common. This means that m is their HCF. 3 and 2 have nothing but 1 in common. So, their HCF is not considered here.

$$\therefore \quad 3m^3 - 2m^2 + m = m\left(\frac{3m^3}{m} - \frac{2m^2}{m} + \frac{m}{m}\right)$$
$$= m(3m^2 - 2m + 1)$$

Factorization by Grouping

In factorization by grouping four terms are usually given in the expression. The first two of the terms will be factorized before the other two terms are factorized. After this, the common term obtained is used to carry out the final factorization of the terms.

Examples

1. Factorize $ax + ay + 3bx + 3by$

Solution

The terms ax and ay have 'a' in common. This also means that their HCF is 'a'. The terms $3bx$ and $3by$ have $3b$ in common. This also means that their HCF is $3b$. We now factorize each pair as follows:

$ax + ay + 3bx + 3by = a(x + y) + 3b(x + y)$

The final step is to take the factors outside each of the bracket (i.e. a and 3b) and enclose them in a bracket. Then, one of the two equal bracket terms is also taken along. This gives:

$a(x + y) + 3b(x + y) = (a + 3b)(x + y)$

This also means that $(x + y)$ is common to the terms above. So, it can be factorized as follows:

$$a(x + y) + 3b(x + y) = (x + y)\left[\frac{a(x + y)}{(x + y)} + \frac{3b(x + y)}{(x + y)}\right]$$

$(x + y)$ cancels out to give: $(x + y)(a + 3b)$ as obtained above.

Hence, $ax + ay + 3bx + 3by = (x + y)(a + 3b)$

2. Factorize $2ab - 5a + 2b - 5$

Solution

'a' is common to the first two terms. The next two terms have nothing in common, so we can take 1 to be the common term between them.

$\therefore \quad 2ab - 5a + 2b - 5 = a(2b - 5) + 1(2b - 5)$

Taking one of the bracket (since they are equal) along with the factors outside each of the bracket gives:

$a(2b - 5) + 1(2b - 5) = (2b - 5)(a + 1)$

3. Factorize $2ax - 2ay - 3bx + 3by$

Solution

2a is common to the first two terms, while $-3b$ is common to the last two terms. Note that the sign of the first of the two terms must be taken. For example, $-ab + 2ax$ has $-a$ as the common term. Also, when a negative term is used as a common term, the inner sign of the bracket formed will change. For example, $-2a - 3ax$ will give $-a(2 + 3x)$. Notice the change of sign when the bracket is formed, especially as $-3ax$ becomes $+3x$ in the bracket.

$$\therefore \quad 2ax - 2ay - 3bx + 3by = 2a(x - y) - 3b(x - y)$$
$$= (x - y)(2a - 3b)$$

Notice the change in sign of $+3by$ to $-y$ after factorizing.

4. Factorize $15 - xy + 5y - 3x$

<u>Solution</u>

Since the first two terms have nothing in common, and the last two terms have nothing in common, then we regroup the expression. To regroup the expression means to rearrange the terms such that the first two terms will have a common factor and the last two terms will have a common factor. This gives:

$15 - xy + 5y - 3x = 15 + 5y - xy - 3x$

$\therefore$ The expression to be factorized is:

$15 + 5y - xy - 3x$

The first two terms have 5 as a factor while the last two terms have $-x$ as a factor. The expression is now factorized as follows:

$$15 + 5y - xy - 3x = 5(3 + y) - x(y + 3)$$
$$= (3 + y)(5 - x)$$

Note that $(3 + y)$ is also equal to $(y + 3)$.

Difference of Two Squares

$$(a + b)(a - b) = a^2 - ab + ab - b^2$$
$$= a^2 - b^2 \quad \text{(since } -ab + ab = 0\text{)}$$
$$\therefore \quad a^2 - b^2 = (a + b)(a - b)$$

This can be used to factorize the difference of the squares of two quantities.

Examples

1. Factorize $81 - w^2$

<u>Solution</u>

$$81 - w^2$$

We can express 81 as a square in order to make the expression a difference of two squares. This gives $81 - w^2 = 9^2 - w^2$

$$= (9 + w)(9 - w) \quad \text{[This is similar to } a^2 - b^2 = (a + b)(a - b) \text{ as explained above]}$$

2. Factorize $16y^2 - 4z^2$

<u>Solution</u>

$$16y^2 - 4z^2 = 4^2y^2 - 2^2z^2 \quad \text{(Since } 16 = 4^2 \text{ and } 4 = 2^2\text{)}$$
$$= (4y)^2 - (2z)^2 \quad \text{(This is a difference of two squares)}$$

$$= (4y + 2z)(4y - 2z)$$

3. Factorize $5 - 5m^2$

Solution
The common factor is 5. This is used to factorize the expression as follows:
$$5 - 5m^2 = 5(1 - m^2)$$
But, $1 - m^2$ is a difference of two squares because it can be expressed as
$$1^2 - m^2 \quad \text{(Since } 1^2 = 1\text{)}$$
$$\therefore \; 5(1 - m^2) = 5(1^2 - m^2)$$
We now factorize the terms in the bracket (i.e. difference of two squares) as follows:
$$5(1^2 - m^2) = 5(1 + m)(1 - m)$$
$$\therefore \; 5 - 5m^2 = 5(1 + m)(1 - m)$$

4. Factorize $25x^2 - 9$

Solution
$$25x^2 - 9 = 5^2x^2 - 3^2$$
$$= (5x)^2 - 3^2$$
$$= (5x + 3)(5x - 3)$$

5. Find the value of $62^2 - 38^2$ without using calculator.

Solution
$$62^2 - 38^2 = (62 + 38)(62 - 38)$$
$$= 100 \text{ x } 24 = 2400$$
Notice that what was obtained from one bracket was used to multiply what was obtained from the other bracket.

6. Find the value of $84^2 - 16$ without using calculator.

Solution
$84^2 - 16 = 84^2 - 4^2$ (Both terms must be expressed in squares).
$$\therefore \quad 84^2 - 4^2 = (84 + 4)(84 - 4)$$
$$= 88 \text{ x } 80$$
$$= 7040$$

Factorization of Simple Quadratic Expression
A quadratic expression is an expression in which 2 is the highest power of the unknown. For example, $2x^2 - 5x + 8$ is a quadratic expression.

Examples

1. Factorize $x^2 + 8x + 12$

<u>Solution</u>

Since the co–efficient of x^2 is 1 (i.e. $1x^2$ also means x^2), then we can apply a short method as follows:

$$x^2 + 8x + 12 = (x \quad)(x \quad)$$

Find two numbers such that their product is +12 and their sum is +8. These two numbers must be factors of 12 since their product has to give 12. An easy way to find the factors of a number is represented below:

2	12
2	6
3	3
	1

∴ The factors are 12 and 1, 6 and 2, 3 and 4. These are the numbers (or multiplication of numbers) on the right side and left side of the line above, along with a number that must be used to multiply them in order to obtain 12.

∴ Among the three pairs of factors of 12 above, the factors whose product is +12, and sum is +8 are +6 and +2. Note that the correct sign must be taken along with the factors. These two numbers are now used to complete the two brackets above. This gives:

$x^2 + 8x + 12 = (x + 6)(x + 2)$

2. Factorize $c^2 - 8c - 20$

<u>Solution</u>

The co–efficient of c^2 is 1. So let's apply the direct method.

$$c^2 - 8c - 20 = (c \quad)(c \quad)$$

Find two numbers such that their product is –20 (i.e. the last term) and their sum is –8 (i.e. the middle term). These two numbers must be factors of 20. We can obtain the factors as follows:

2	20
2	10
5	5
	1

∴ The factors are 20 and 1 (since 20 x 1 = 20). The second factors are 10 and 2 (since 10 x 2 = 20). The third factors are 5 and 4 (since 5 x 4 = 20). Note that 4 is obtained from 2 x 2. Among these factors, it is –10 and +2 that will give a product of –20 and a sum of –8. Note that you have to try different signs with the factors in order to get the right sign. The two numbers are

now used to factorize the expression by putting them in the bracket above. This gives:
$$c^2 - 8c - 20 = (c - 10)(c + 2)$$

3. Factorize $m^2 - 11m + 24$

Solution
$$m^2 - 11m + 24 = (m \qquad)(m \qquad)$$
Two numbers whose product is +24 and sum is –11 are –8 and –3. This is because –8 x (–3) = +24,
and –8 + (–3) = –11. These two numbers are now used to complete the brackets above.
$\therefore$ $m^2 - 11m + 24 = (m - 8)(m - 3)$
Note that when these brackets are expanded, the original quadratic expression will be obtained.

4. Factorize $2a^2 - 10a + 12$

Solution
The co–efficient of a^2 is 2. It is not 1, so the direct method used above cannot be applied here. Hence the first step in this case is to multiply the first and last terms (i.e. $2a^2$ and +12). This gives: $2a^2 \times 12 = 24a^2$
We now find two numbers as coefficients of 'a' such that their product is $24a^2$ and their sum is –10a (i.e. the middle term in the original expression). These two numbers must be factors of $24a^2$. They are –6a and –4a. We now substitute these two terms for –10a in the original expression. This gives:
$$2a^2 - 10a + 12 = 2a^2 - 6a - 4a + 12$$
We now factorize by grouping. Between the first two terms 2a is common while between the last two terms –4a is common. This now factorizes as follows:
$$\begin{aligned} 2a^2 - 10a + 12 &= 2a^2 - 6a - 4a + 12 \\ &= 2a(a - 3) - 4(a - 3) \\ &= (a - 3)(2a - 4) \end{aligned}$$

Substitution
Examples
1. Evaluate the following if $x = 2$, and y = 3
(a) $2x - 3y$
(b) $x + 2(3x - y)$

Solutions
(a) $2x - 3y$
Substitute 2 for x and 3 for y. Note that $2x$ means 2 x x. Similarly, 3y means 3 x y
$\therefore$ $2x - 3y$
    $= (2 \times 2) - (3 \times 3)$

$= 4 - 9$

$= -5$

(b) $x + 2(3x - y)$ (Substitute 2 for x and 3 for y)

$= 2 + 2[(3 \times 2) - 3]$

$= 2 + 2(6 - 3)$

$= 2 + 2(3)$

$= 2 + 6 = 8$

Note that the term in the bracket is first evaluated. Also, a term outside a bracket is multiplying the contents of the bracket.

2. Evaluate the following given that a = –1, b = 3 and c = –5

(a) $2(a - 2c) - 4b$

(b) $(b - a)(c + 4a)$

(c) $\dfrac{a + bc}{b - ac}$

Solutions

(a) $2(a - 2c) - 4b$

$= 2[-1 - (2(-5))] - (4 \times 3)$ (Substitute the values of the letters as given in the question)

$= 2[-1 - (-10)] - 12$

$= 2(-1 + 10) - 12$

$= 2(9) - 12$

$= 18 - 12$

$= 6$

b. $(b - a)(c + 4a)$

$= [3 - (-1)][-5 + (4(-1))]$

$= (3 + 1)[-5 + (-4)]$

$= 4(-5 - 4)$

$= 4 \times (-9)$

$= -36$

Note that it is advisable to enclose each multiplication in a bracket. For example 4b is expressed as (4 x 3), and 4a is expressed as [4 x (–1)]. Always evaluate the contents of inner brackets before the outer brackets and then outside the brackets.

(c) $\dfrac{a + bc}{b - ac}$

$= \dfrac{-1 + [3 \times (-5)]}{3 - [(-1) \times (-5)]}$

$= \dfrac{-1 + (-15)}{3 - (5)}$

$$= \frac{-1 - 15}{3 - 5}$$

$$= \frac{-16}{-2}$$

= 8   (The negative sign cancels out)

Note that the numerator and denominator have to be reduced to a single value before division is possible.

3. Given that p = 3, and q = –1, find the value of x when $x = p^2q - q^2p$.

Solution

$x = p^2q - q^2p$

$x = [3^2 \times (-1)] - [(-1)^2 \times 3]$ (Simply substitute the given values of p and q)

$= [9 \times (-1)] - (1 \times 3)$

$= -9 - 3 = -12$

Note that the terms having powers must be evaluated first before further evaluations.

4. Evaluate: $ab\sqrt{c^2 + b^2}$, given that a = 2, b = –3, c = 4

Solution

$ab\sqrt{c^2 + b^2}$

$= [2 \times (-3)] \times \sqrt{4^2 + (-3)^2}$

$= (-6) \times \sqrt{16 + 9}$

$= -6 \times \sqrt{25}$

$= -6 \times 5 = -30$

5. If x = –7 and y = 3, calculate the values of:

(a) $(\frac{x + y}{x - y})^2$

(b) $2x^2y + y^2x$

Solutions

(a) $(\frac{x + y}{x - y})^2$

$= (\frac{-7 + 3}{-7 - 3})^2$

$= (\frac{-4}{-10})^2$

$= (\frac{2}{5})^2$ (After simplifying the term in the bracket)

$$= \frac{2^2}{5^2}$$

$$= \frac{4}{25}$$

Note that the contents of the bracket must be evaluated to a single digit before the use of the power. The power is used for both the numerator and denominator.

(b) $2x^2y + y^2x$
   $= [2(-7)^2(3)] + [(3^2)(-7)]$
   $= (2 \times 49 \times 3) + [9 \times (-7)]$
   $= 294 + (-63)$
   $= 294 - 63 = 231$

Exercise 2

1. What is y kilometers in meters?
2. A man was 38 years old m years ago. How old is he now? How old will he be in p years time?
3. A man is b years old. How old was he c years ago? How old will he be in 12 years time?
4.(a) What is $y in cents?
(b) Express n seconds in minutes. What is n seconds in hours?
5. The perimeter of a rectangle is 44cm and the breadth is $x$cm. Find the area of the rectangle in terms of x.
6. One side of a rectangle is $(x + 10)$cm. If the perimeter of the rectangle is $(4x + 18)$cm, find the area of the rectangle in terms of x.
7. Expand $5(2x - y)$
8. Expand and simplify the expression, $11x - [5 - 3(x + y)]$
9. Remove bracket and simplify $3a + (a - 2b) - (2a + b)$
10. Simplify $3x(7x + 2) - (10x^2 - 9x)$
11. Expand $(2a + 3b)(a - 5b)$
12. Expand and simplify $(3x - 2y)(5x - y)$
13. Expand and simplify $(4a - 5)^2$
14. Find the coefficient of xy in the expansion and simplification of: $(3x + 2y)^2 - (2x - 3y)^2$
15. Simplify and find the coefficient of y^2 in $5y(y - 2) - (2y - 3)(y - 4)$
16. Find the LCM of the following:
a. l, m, n
b. 6cd, 2ed, ec
c. $2a^3x$, $5ay^2$, $6x^2y$
d. 20m and 10n
e. $(x + y)$ and $(3x - y)$
17. Simplify the following:
a. $\dfrac{2x}{5} + \dfrac{x}{5}$

b. $\dfrac{4}{9x} + \dfrac{1}{3x^2}$

c. $\dfrac{a}{10b} - \dfrac{7}{15x}$

d. $\dfrac{2x+5}{8a} + \dfrac{b-3}{5b}$

e. $\dfrac{5a-b}{3} - \dfrac{a-5b}{5a^2} - 1$

18. Express each of the following as a single fraction in its simplest form.

a. $\dfrac{x-4}{10} - \dfrac{3x-2}{5} + \dfrac{4}{15}$

b. $\dfrac{2}{3a+7} - \dfrac{1}{6a+14}$

c. $\dfrac{2a-1}{3a-1} - \dfrac{3a+4}{a-2}$

19. Find the HCF of the following:
a. 5, 20, 30
b. 6, 14, 24
c. $12a^3b^2$, $30a^2b^3$
d. ab^2c^3, ab^5c^2
e. $24x^2yz^3$, $72x^2y^2z^4$, $36x^3yz^2$
20. Factorize $4m - 8n$
21. Factorize $12y - 8$.
22. Factorize: $-20y - 50yz$
23. Factorize $5m^4 - 10m^2 - m^3$
24. Factorize $2x + 6y + 5mx + 15my$
25. Factorize $3ab + 7a - 3b - 7$
26. Factorize $5bx - 5by + nx - ny$
27. Factorize $15 - xy + 3y - 5x$
28. Factorize $100 - m^2$
29. Factorize $25x^2 - 9y^2$
30. Factorize $7 - 7b^2$
31. Factorize $36p^2 - 4$
32. Find the value of $63^2 - 37^2$ without using calculator.
33. Find the value of $92^2 - 64$ without using calculator.
34. Factorize $x^2 + 6x + 9$
35. Factorize $b^2 - b - 20$
36. Factorize $n^2 - 14n + 48$
37. Factorize $5a^2 - 11a - 12$
38. Evaluate the following if $x = -3$, and $y = -2$

a. $x^2y - y^2x$

b. $(y^2 - x^2)(y - x)$

39. Evaluate the following given that a = −1, b = 3 and c = −5

a. 2(4a − c) + 7b

b. (3a − b)(b − 2c)

c. $\dfrac{ab - bc}{a - bc}$

40. Given that p = 1, and q = −3, find the value of x when $x = pq - q^2q$.

41. If x = −4 and y = 2, calculate the values of:

a. $(\dfrac{x + 2y}{x - 6y})^2$

b. $2y^2 - x^2y$

CHAPTER 3
LAWS OF INDICES

The following are the laws of indices. They are true for all values of a, b and $x \neq 0$

Law 1: $x^a \times x^b = x^{a+b}$

Law 2: $x^a \div x^b = x^{a-b}$

Law 3: $x^0 = 1$

Law 4: $x^{-a} = \dfrac{1}{x^a}$ or $bx^{-a} = \dfrac{b}{x^a}$ or $\left(\dfrac{b}{x}\right)^{-a} = \left(\dfrac{x}{b}\right)^a$

Examples
Simplify the following:

1. $10^5 \times 10^4$

2. $m^8 \div m^5$

3. $\dfrac{a^{-8}}{a^3}$

4. $5x^2 \times 4x^0 \times 2x^{-6}$

5. $y^{-5} \div b^0$

Solution

1. $10^5 \times 10^4 = 10^{5+4} = 10^9$

2. $m^8 \div m^5 = m^{8-5} = m^3$

3. $\dfrac{a^{-8}}{a^3} = a^{-8-3} = a^{-11} = \dfrac{1}{a^{11}}$

4. $5x^2 \times 4x^0 \times 2x^{-6} = (5 \times 4 \times 2)x^{2+0+(-6)} = 40x^{2-6} = 40x^{-4} = \dfrac{40}{x^4}$

5. $y^{-5} \div b^0 = y^{-5} \div 1 = y^{-5} = \dfrac{1}{y^5}$

Product of Indices
In applying product of indices, the following are true:

$(x^a)^b = x^{ab}$

Similarly, $(x^a y^b)^c = x^{ac} y^{bc}$ and $\left(\dfrac{x}{y}\right)^a = \dfrac{x^a}{y^a}$

Examples

Simplify the following:

1. $(h^4)^{-5}$
2. $(2^{-3})^2$
3. $(-c^3)^2$
4. $(-4u^2 v)^3$

<u>Solution</u>

1. $(h^4)^{-5} = h^{4\times(-5)} = h^{-20} = \dfrac{1}{h^{20}}$

2. $(2^{-3})^2 = 2^{-3\times2} = 2^{-6} = \dfrac{1}{2^6} = \dfrac{1}{64}$

3. $(-c^3)^2 = -c^{3\times2} = -c^6 = c^6$ (A negative number that is raised to an even number power will give a positive value).

4. $(-4u^2 v)^3 = -4^{1\times3} u^{2\times3} v^{1\times3} = -4^3 u^6 v^3 = -64u^6 v^3$

Fractional Indices

In applying fractional indices, the following are true:

$$x^{1/a} = \sqrt[a]{x} \quad \text{and} \quad x^{a/b} = \sqrt[b]{x^a} \quad \text{or} \quad x^{a/b} = (\sqrt[b]{x})^a$$

In all cases, $x \neq 0$

Examples

Simplify the following:

1. $27^{\frac{1}{3}}$

2. $9^{-\frac{1}{2}}$

3. $(25a^2)^{\frac{1}{2}}$

4. $\sqrt{1\dfrac{9}{16}}$

5. $\left(\dfrac{16}{54}\right)^{-\frac{2}{3}}$

<u>Solutions</u>

1. $27^{\frac{1}{3}} = \sqrt[3]{27} = 3$

2. $9^{-\frac{1}{2}} = \dfrac{1}{9^{\frac{1}{2}}} = \dfrac{1}{\sqrt{9}} = \dfrac{1}{3}$ (Note that $\sqrt[2]{9}$ is $\sqrt{9}$ since 2 is not usually written with the square root sign).

3. $(25a^2)^{\frac{1}{2}} = 25^{\frac{1}{2}}a^{(2 \times \frac{1}{2})} = 25^{\frac{1}{2}}a^1 = (\sqrt{25}) \times a = 5a$

4. $\sqrt{1\frac{9}{16}} = \sqrt{\frac{25}{16}} = \frac{5}{4}$

5. $\left(\frac{16}{54}\right)^{-\frac{2}{3}} = \left(\frac{8}{27}\right)^{-\frac{2}{3}}$ (When the fraction is expressed in its lowest term)

$\left(\frac{8}{27}\right)^{-\frac{2}{3}} = \left(\frac{27}{8}\right)^{\frac{2}{3}} = \frac{27^{2/3}}{8^{2/3}} = \frac{(\sqrt[3]{27})^2}{(\sqrt[3]{8})^2} = \frac{3^2}{2^2} = \frac{9}{4}$

Note that by taking the inverse of the term in the bracket, the negative power becomes positive.

Equations in Indices

Examples

Solve the following equations:

1. $4^{x-1} = 64$
2. $n^{-\frac{2}{3}} = 9$
3. $2a^{-3} = -16$
4. $9^x = 27$
5. $5x = 40x^{-\frac{1}{2}}$

Solutions

1. $4^{x-1} = 64$

This is solved by expressing both sides of the equation in the same base and then equating the powers. This gives:

$4^{x-1} = 4^3$ (Note that $64 = 4^3$)

Equating the powers gives:

$x - 1 = 3$

$x = 3 + 1$

$x = 4$

2. $n^{-\frac{2}{3}} = 9$

In this case, the unknown letter is the base. To solve this, make the power of n to 1 by multiplying this power by its inverse and using the same sign of the power. The other side of the equation should also be raised to the same power. This gives:

$(n^{-\frac{2}{3}})^{-3/2} = 9^{-3/2}$

$n^{(-2/3 \times -3/2)} = \frac{1}{9^{3/2}}$ (The inverse of $\frac{2}{3}$ is $\frac{3}{2}$. The sign of the original power is also used with the inverse)

$n^1 = \frac{1}{(\sqrt{9})^3} = \frac{1}{3^3} = \frac{1}{27}$ (Note that $-\frac{2}{3} \times -\frac{3}{2} = 1$. This gives n^1)

$n = \frac{1}{27}$

3. $2a^{-3} = -16$

Divide both sides by 2.

$$a^{-3} = \frac{-16}{2}$$

$$a^{-3} = -8$$

Now make the power of 'a' to be 1 by multiplying this power by its inverse. This gives:

$$(a^{-3})^{-\frac{1}{3}} = (-8)^{-\frac{1}{3}}$$

$$a^1 = \frac{1}{(-8)^{\frac{1}{3}}} = \frac{1}{\sqrt[3]{-8}} = \frac{1}{-2}$$

$$a = -\frac{1}{2}$$

4. $9^x = 27$

Expressing both sides of the equation in the same base gives:

$$(3^2)^x = 3^3$$

$$3^{2x} = 3^3$$

Equating the powers gives

$$2x = 3$$

$$x = \frac{3}{2}$$

5. $5x = 40x^{-\frac{1}{2}}$

Divide both sides by 5.

$$x = \frac{40x^{-\frac{1}{2}}}{5}$$

$$x = 8x^{-\frac{1}{2}} \quad \text{(Since } 40 \div 5 \text{ gives 8)}$$

Divide both sides by $x^{-\frac{1}{2}}$

$$\frac{x}{x^{-\frac{1}{2}}} = \frac{8x^{-\frac{1}{2}}}{x^{-\frac{1}{2}}}$$

Cancelling out $x^{-\frac{1}{2}}$ on the right hand side gives,

$x^{1-(-\frac{1}{2})} = 8$ (Note that x can be expressed as x^1. Also, from the law of indices, $x \div x^{-\frac{1}{2}} = x^{1-(-\frac{1}{2})}$)

$$\therefore x^{1+\frac{1}{2}} = 8$$

$$x^{3/2} = 8$$

Make the power of x to be 1 by multiplying it by its inverse. Also raise the power of 8 to the same inverse. This gives:

$$(x^{3/2})^{\frac{2}{3}} = 8^{\frac{2}{3}}$$

$$x^1 = 8^{\frac{2}{3}}$$

$$x = (\sqrt[3]{8})^2 = (2)^2$$

$$x = 4$$

Exercise 3

1. Simplify the following:
   a. $-3(te^3)^4$
   b. $(4ab^3)^3$
   c. $\dfrac{(-a)^2 \times a^7}{(-a)^5}$
   d. $(-g^4)^5$
   e. $\dfrac{(m^2)^3}{m^4 \times (-m)}$

2. Simplify the following:
   a. $(3a)^{-1}$
   b. $(a^2)^{-\frac{1}{2}}$
   c. $(49x^3)^{\frac{1}{2}}$
   d. $(27x^{3/2})^{\frac{2}{3}}$

3. Solve the following equations:
   a. $x^{-\frac{1}{2}} = 5$
   b. $a^{-2} = 9$
   c. $9^{x-2} = 27$
   d. $\dfrac{4^{2x-1}}{16^2} = 64$

CHAPTER 4
LINEAR EQUATIONS AND CHANGE OF SUBJECT OF FORMULAE

Linear Equations

Linear equations can sometimes be expressed with brackets, with fractions or both. When solving linear equations with fractions, clear the fractions by multiplying each term of the equation by the L.C.M of the denominators of the equation.

Examples

1. Solve the following equations:
a. $2x + 4(3 - x) = 11$
b. $6(a - 3) - 2(5a - 8) = -4$
c. $a - 5(2 + a) - (3a - 4) = 2(2a - 1) - 7$

<u>Solutions</u>

a. $2x + 4(3 - x) = 11$

Expanding the bracket gives:

$\quad 2x + 12 - 4x = 11$

Collect like terms. This means to collect terms in x on one side of the equation and constant terms on the other side of the equation.

$\quad 12 - 11 = 4x - 2x$ (When a number crosses the equality sign, its sign will change)

$\quad 1 = 2x$

Divide both sides by 2

$\quad \therefore \quad x = \dfrac{1}{2}$

b. $6(a - 3) - 2(5a - 8) = -4$

Expanding the brackets gives:

$6a - 18 - 10a + 16 = -4$ (Note that $-2 \times -8 = +16$)

Collect like terms in 'a' on one side of the equation.

$6a - 10a = -4 + 18 - 16$

$-4a = -2$

Divide both sides by -4

$\dfrac{-4a}{-4} = \dfrac{-2}{-4}$

$a = \dfrac{-2}{-4}$

$a = \dfrac{1}{2}$

c. $a - 5(2 + a) - (3a - 4) = 2(2a - 1) - 7$

Expanding the brackets gives:

$a - 10 - 5a - 3a + 4 = 4a - 2 - 7$ (Note that $-5 \times (+a) = -5a$, and $- x -4 = -1 \times -4 = +4$)

40

Collect like terms on one side of the equation.

$7 + 2 + 4 - 10 = 5a - a + 3a + 4a$

$3 = 11a$

Divide both sides by 11

$$\frac{3}{11} = \frac{11a}{11}$$

$$a = \frac{3}{11}$$

2. Solve the following equations:

a. $\frac{3}{4}x - \frac{1}{3}(x-2) = \frac{5}{6} - (2x-1)$

b. $\frac{1}{6}(5x-2) - \frac{2}{3}(4-x) = 1$

c. $\frac{5}{2m-3} - \frac{3}{4} = \frac{1}{6} + 7$

Solutions

a. $\frac{3}{4}x - \frac{1}{3}(x-2) = \frac{5}{6} - (2x-1)$

In order to clear the fractions, multiply each term in the equation by 12 which is the L.C.M of the denominators, i.e. 4, 3 and 6. This gives:

$12(\frac{3x}{4}) - 12 \times \frac{1}{3}(x-2) = 12(\frac{5}{6}) - 12(2x-1)$

Cancelling out by using the 12 to divide the various denominators gives:

$3(3x) - 4(x-2) = 2(5) - 12(2x-1)$

$9x - 4x + 8 = 10 - 24x + 12$

Collect like terms:

$9x - 4x + 24x = 10 + 12 - 8$

$29x = 14$

$$x = \frac{14}{29}$$

b. $\frac{1}{6}(5x-2) - \frac{2}{3}(4-x) = 1$

In order to clear the fractions, multiply each term in the equation by 6 which is the L.C.M of the denominators, i.e. 6 and 3. This gives:

$6 \times \frac{1}{6}(5x-2) - 6 \times \frac{2}{3}(4-x) = 6 \times 1$

Cancelling out by using the 6 to divide the various denominators gives:

$(5x-2) - 2 \times 2(4-x) = 6$

41

$$5x - 2 - 4(4 - x) = 6$$
$$5x - 2 - 14 + 4x = 6$$
$$5x + 4x = 6 + 2 + 14$$
$$9x = 22$$
$$\therefore \quad x = \frac{22}{9}$$

c. $\quad \dfrac{5}{2m-3} - \dfrac{3}{4} = \dfrac{1}{6} + 7$

Multiply each term in the equation by $12(2m - 3)$ which is the L.C.M of the denominators, i.e. $(2m - 3)$, 4 and 6. This gives:

$$12(2m - 3)\frac{5}{2m-3} - 12(2m - 3)\frac{3}{4} = 12(2m - 3)\frac{1}{6} + 12(2m - 3) \times 7$$

Cancelling out gives:

$$12(5) - 3(2m - 3)3 = 2(2m - 3) + 12(2m - 3)7$$
$$60 - 9(2m - 3) = 4m - 6 + 84(2m - 3)$$
$$60 - 18m + 27 = 4m - 6 + 168m - 252$$
$$60 + 27 + 252 + 6 = 4m + 168m + 18m$$
$$345 = 190m$$
$$m = \frac{345}{190}$$

$$\therefore \quad m = \frac{69}{38} \qquad \text{(After equal division by 5)}$$

Change of Subject of Formulae

If $m = b + c$, then m is the subject of the formula. If it is rearranged to give $b = m - c$, then b is now the new subject of the formula.

In changing the subject of a formula, simply solve the equation for the letter which is to become the new subject.

Examples

1. Make h the subject of the formula: $s = \dfrac{wd}{h}\left(h - \dfrac{d}{2}\right)$

Solution

$$s = \frac{wd}{h}\left(h - \frac{d}{2}\right)$$

Expanding the bracket gives:

$$s = \frac{wd}{h}(h) - \frac{wd}{h}\left(\frac{d}{2}\right)$$

Canceling out the h gives:

$$s = wd - \frac{wd^2}{2h}$$

To clear fractions, multiply throughout by 2h (LCM)

$$2h(s) = 2h(wd) - 2h\left(\frac{wd^2}{2h}\right)$$

$2hs = 2hwd - wd^2$ (Note that the 2h at the end on the right side has cancelled out).

Collect terms in h on one side.

$wd^2 = 2hwd - 2hs$

Factorizing the right hand side gives:

$wd^2 = h(2wd - 2s)$

Divide both sides by $(2wd - 2s)$

$$\frac{wd^2}{2wd - 2s} = \frac{h(2wd - 2s)}{2wd - 2s}$$

Cancelling out the $2wd - 2s$ on the right hand side gives:

$$h = \frac{wd^2}{2wd - 2s}$$

2. Given that $I = \dfrac{E}{\sqrt{R^2 + W^2 L^2}}$, express R in terms of I, E, W and L.

Solution

$$I = \frac{E}{\sqrt{R^2 + W^2 L^2}}$$

Cross multiply to obtain:

$$I\sqrt{R^2 + W^2 L^2} = E$$

Square both sides to remove the square root sign.

$$(I\sqrt{R^2 + W^2 L^2})^2 = E^2$$

$I^2(R^2 + W^2 L^2) = E^2$ (Note that the square also affect I in the bracket)

$I^2 R^2 + I^2 W^2 L^2 = E^2$

$I^2 R^2 = E^2 - I^2 W^2 L^2$

Divide both sides by I^2

$$R^2 = \frac{E^2 - W^2 I^2 L^2}{I^2}$$

This can also be simplified as follows:

$$R^2 = \frac{E^2}{I^2} - \frac{I^2 W^2 L^2}{I^2}$$

Canceling out I^2 on the right side gives:

$$R^2 = \frac{E^2}{I^2} - W^2 L^2$$

Take the square root of both sides in order to remove the square on R^2.

$$\therefore \quad R = \sqrt{\frac{E^2}{I^2} - W^2 L^2}$$

Note that when a term/terms in a square root sign are squared, it gives only the terms in the root sign. For example, $(\sqrt{m^2 b^2})^2 = m^2 b^2$. Take note of the absence of the root sign

3. Make x the subject of the formula $R = \sqrt{\dfrac{ax - P}{Q + bx}}$

Solution

$$R = \sqrt{\frac{ax - P}{Q + bx}}$$

Square both sides to remove the square root sign
$$R^2 = \frac{ax - P}{Q + bx}$$
By cross multiplication it gives:
$R^2(Q + bx) = ax - P$
$R^2 Q + R^2 bx = ax - P$
Collecting terms in x gives
$ax - R^2 bx = R^2 Q + P$
Factorizing the left hand side gives:
$x(a - R^2 b) = R^2 Q + P$
Divide both sides by $(a - R^2 b)$
$$\frac{x(a - R^2 b)}{(a - R^2 b)} = \frac{R^2 Q + P}{a - R^2 b}$$

$$x = \frac{R^2 Q + P}{a - R^2 b}$$

Exercise 4

1. Solve the following equations:
a. $5x + 2(3 - x) = 10$
b. $2(2a - 3) - 5(4a - 1) = -6$
c. $b - 4(1 + b) - (5b - 1) = -(b - 3) - 2$

2. Solve the following equations:
a. $\dfrac{1}{4}x - \dfrac{2}{3}(x - 1) = \dfrac{3}{4} - (5x - 2)$

b. $\frac{1}{6}(5x-2) - \frac{5}{12}(3-2x) = 1$

c. $\frac{2}{2n-3} - \frac{3}{5} = 2\frac{1}{2}$

3. Make p the subject of the formula: $tp = md(p - \frac{d}{3})$

4. Given that $V = \frac{P}{\sqrt{E^2 + I^2 C^2}}$, express C in terms of I, V, E and P.

5. Make x the subject of the formula $R = \sqrt{\frac{bx - S}{T + ax}}$

6. Make m the subject of the formula: $Amq = md(\frac{q}{5} - \frac{dA}{5})$

7. Given that $L = \frac{TC^2}{Y^3 - B}$, express Y in terms of L, T, B and C.

8. Make x the subject of the formula $\sqrt{\frac{Q^2}{x}} = \sqrt{\frac{M}{2x^3}}$

CHAPTER 5
LINEAR EQUATIONS FROM WORD PROBLEMS

Examples

1. The sum of 6 and one-third of a certain number is one more than twice the number. Find the number.

<u>Solution</u>

Let the number be x

One-third of the number $= \frac{1}{3}x$

The sum of 6 and one-third of the number $= 6 + \frac{1}{3}x$.

Twice the number $= 2x$

$6 + \frac{1}{3}x$ is one more than $2x$. This means that:

$6 + \frac{x}{3} - 2x = 1$

Multiply each term by 3 which is the LCM of all the denominators. Note that all the whole numbers have 1 as their denominator. This multiplication by the LCM of their denominators is done in order to clear the fraction. Each term is now multiplied by 3 as follows:

$(6 \times 3) + 3(\frac{x}{3}) - (2x \times 3) = 1 \times 3$

$18 + x - 6x = 3$ (Note that in $3(\frac{x}{3})$, the 3 cancels out to give x)

Collect terms in x on one side and the other terms on the other side of the equation. This gives:

$18 - 3 = 6x - x$ (Note the change in the signs of the terms that cross the equality sign).

$15 = 5x$

Divide both sides of the equation by 5 (i.e. the co-efficient of x which is the unknown).

$\frac{15}{5} = \frac{5x}{5}$

$\therefore \quad 3 = x$ (The 5 cancels out)

$\therefore \quad x = 3$

2. When 10 is subtracted from the product of 6 and a certain number, and the result is divided by 4, the answer is equal to the number. What is the number?

<u>Solution</u>

Let the number be a

The product of 6 and the number = 6a

When 10 is subtracted from this product it gives: 6a − 10

This result divided by 4 is $\frac{6a-10}{4}$

The answer is equal to the number means that:

$\frac{6a-10}{4} = a$

Cross multiply to solve this equation. This gives:

6a − 10 = 4a

collect like terms in a on the left hand side

$6a - 4a = 10$

$2a = 10$

Divide both sides by 2

$$\frac{2a}{2} = \frac{10}{2}$$

$\therefore \quad a = 5$

3. A man is three times as old as his son. Five years ago, the sum of their ages was 38. Find the age of the man and his son.

<u>Solution</u>

Let the son's age be x

$\therefore$ The man's age $= 3x$ (i.e. three times his son's age)

Five years ago, the son's age was: $x - 5$

Similarly, five years ago the man's age was: $3x - 5$

The sum of their age five years ago, was:

$x - 5 + 3x - 5$

Since this sum is 38, we express it as follows:

$x - 5 + 3x - 5 = 38$

$x + 3x = 38 + 5 + 5$

$4x = 48$

$\therefore \qquad x = \dfrac{48}{4}$

$x = 12$

$\therefore$ The son's age $= 12$ years

The man's age $= 3x = 3 \times 12 = 36$ years

4. A girl is 5 years old and her mother is 32 years old. In how many years time will the mother be twice as old as the girl?

<u>Solution</u>

Let the number of years time be y.

At that time, the girl's age will be $= 5 + y$

At that time the mother's age will be $= 32 + y$

Since the mother's age will be twice the girl's age, it means that:

$2(5 + y) = 32 + y$ (Note that $2(5 + y)$ is twice the girl's age)

$\therefore \quad 10 + 2y = 32 + y$

$2y - y = 32 - 10$ (After collecting like terms)

$y = 22$

$\therefore$ In 22 years time, the mother will be twice as old as the girl.

5. A cyclist takes 5hours to travel from a town A to a town B. When coming back, the journey took him 1hour less because he increased his speed by 15km/h. What was his speed when he

was going from A to B.

<u>Solution</u>
Let his speed from A to B be x

$$\text{Speed} = \frac{\text{Distance}}{\text{Time}}$$

∴   Distance = speed x time

$= x \times 5 = 5x$ (When going from A to B)

When coming from B to A (i.e. the return journey), his speed was $x + 15$ (since he increased His speed by 15km/h), while his time was: $5 - 1 = 4$ (Since he took 1 hour less)

∴   Distance = speed x time

$= (x + 15) \times 4$

$= 4(x + 15)$

Since the distance from A to B is equal to the distance from B to A, then it follows that:

$5x = 4(x + 15)$ (Distance A to B is $5x$, while distance B to A is $4(x + 15)$. Both are equal)

$5x = 4x + 60$

$5x - 4x = 60$

$x = 60$

His speed when going from A to B was 60km/h

6. A man walked for 3hours at 8km/h. He then cycled at 12km/h for a certain period of time. If his total distance was 54km, for how many hours did he cycle?

<u>Solution</u>

$$\text{Speed} = \frac{\text{Distance}}{\text{Time}}$$

When he walked, his total distance was:

Distance = speed x time

$= 8 \times 3$

$= 24$km

When he cycled, his total distance was:

Distance = speed x time

Let the time taken to cycle be x

∴   Distance cycled = 12 x x

$= 12x$

His total distance is given by:

$24 + 12x$

Since this total distance was 54km, then:

$24 + 12x = 54$

$12x = 54 - 24$

$12x = 30$

∴ $x = \dfrac{30}{12}$

$x = \dfrac{5}{2}$ (After equal division by 6)

$$x = 2\frac{1}{2}$$

∴   The man cycled for $2\frac{1}{2}$ hours

7. A sum of $324 is made up of $10 notes and $1 notes. If there are eight times as many $1 notes as there are $10 notes, find the number of each note.

<u>Solution</u>
Let there be x $1 notes.
∴ The number of $10 notes = 8$x$ (i.e. eight times as many $1 notes)
∴ Total amount of $1 notes = x x 1 = x (This means $$x$)
Similarly, total amount of $10 notes = 8$x$ x 10 = 80x (This means 80x$)
Since this total gives the sum of $324, it means that:
     x + 80x = 324
       81x = 324   (Note that x means 1x)
∴ $x = \dfrac{324}{81}$
      $x = 4$
∴ Number of $1 = 4
Number of $10 notes = 8$x$ = 8 x 4 = 32

8. Chloe has $30 and Peter has $186. If Chloe saves $5 a day and Peter spends $7 a day, after how many days will they have equal amounts?

<u>Solution</u>
Let the number of days for both of them to have equal amount of money be y.
If Chloe saves $5 a day, then after y days, she must have saved $5y (i.e. 5 x y)
If Peter spends $7 a day, then after y days he must have spent $7y (i.e. 7 x y)
So, total amount that Chloe will have = 30 + 5y   (i.e. her initial money plus her savings)
Total amount that Peter will have = 186 – 7y   (i.e. his initial money minus his spending)
Since they will both have equal amount of money, then:
     30 + 5y = 186 – 7y
∴   5y + 7y = 186 – 30
      12y = 156
∴ $y = \dfrac{156}{12}$
      y = 13
∴ They will have equal amounts of money after 13 days.

9. A total of x litres of water is needed to fill 30 tanks with the same amount of water in each tank. If the size of each tank is 3 litres less, there will be enough water to fill 32 tanks. What is the value of x.

<u>Solution</u>

Amount of water in each of the 30 tank is: $\dfrac{x}{30}$

When the size of each of these 30 tanks is 3 litres less than its original size, then each tank will contain: $\dfrac{x}{30} - 3$

Since this amount of water can fill each of the 32 tanks, then the total amount of water in the 32 tanks is:

$$32(\dfrac{x}{30} - 3)$$

This total amount of water has not changed. It is also equal to x litres (i.e. the original quantity of water). This means that:

$$32(\dfrac{x}{30} - 3) = x$$

$$\therefore \quad \dfrac{32x}{30} - 96 = x$$

Multiply each term by 30 in order to clear fraction. This gives:

$$30(\dfrac{32x}{30}) - 30(96) = 30(x)$$

$$32x - 2880 = 30x$$

$$32x - 30x = 2880$$

$$2x = 2880$$

$$\therefore \quad x = \dfrac{2880}{2}$$

$$x = 1440$$

$$x = 1440 \text{ litres}$$

10. When a plane travels a certain journey at an average speed of 200km/h, it arrives an airport at 5am. When it travels at an average speed of 240km/h, it arrives at the airport at 3am. What is the length of the journey?

<u>Solution</u>

Let the duration of the journey when it travels at 200km/h be m hours.

∴ The duration when it travels at 240km/h will be:

    m – 2   (i.e. it arrives 2 hours earlier, from 5am to 3am. This means it takes lesser time)

But, average speed = $\dfrac{\text{Total distance}}{\text{Total time}}$

∴ At 200km/h: $200 = \dfrac{\text{Total distance}}{m}$

∴ Total distance = 200 x m

                = 200m

Similarly, at 240km/h: Total distance = 240(m – 2)

Since the distance is the same, then:

    200m = 240(m – 2)

    200m = 240m – 480

     480 = 240m – 200m

     480 = 40m

$$\therefore \quad m = \frac{480}{40}$$
$$m = 12$$

Recall that: Total distance = 200m or 240(m – 2)

$\therefore$ Total distance = 200 x 12 = 2400

$\therefore$ The length of the journey is 2400km

11. A man walks to a village at 18km/h and returns at 12km/h. If the whole journey takes $6\frac{1}{4}$ hours, what is the total distance walked?

Solution

Let x be the time taken when he walked at 18km/h

$\therefore$ The time taken for the return journey when he walked at 12km/h will be:

$6\frac{1}{4} - x$ (since total time for the two journeys is $6\frac{1}{4}$)

Also, speed $= \dfrac{\text{Distance}}{\text{Time}}$

$\therefore$ At 18km/h: $18 = \dfrac{\text{Distance}}{x}$

$\therefore$ Distance = $18x$

Similarly,

At 12km/h: $12 = \dfrac{\text{Distance}}{6\frac{1}{4} - x}$

$\therefore$ Distance = $12(6\frac{1}{4} - x)$

Since the to and fro distances are the same, then:

$$18x = 12(6\frac{1}{4} - x)$$
$$18x = 12(\frac{25}{4} - x)$$
$$18x = 12(\frac{25}{4}) - 12(x)$$
$$18x = 3(25) - 12x$$
$$18x + 12x = 75$$
$$30x = 75$$
$$\therefore \qquad x = \frac{75}{30}$$
$$x = 2\frac{1}{2}\text{hours}$$

$\therefore$ Distance = $18x$ or $12(6¼ - x)$

$\therefore$ Distance = 18 x 2½

$$= 18 \times \frac{5}{2} = 45\text{km}$$

This distance is for one part of the journey

$\therefore$ Total distance for the to and fro journey is:

45 x 2 = 90

$\therefore$ The man walked a total distance of 90km.

12. The result of taking 3 from a number and multiplying the answer by 4 is the same as taking 3 from five times the number. Find the number.

Solution
Let the number be c
3 taken from c means: c – 3
Multiplying this by 4 gives: 4(c – 3)   (This is the first part of the statement)
Five times the number is 5c
3 taken from 5c is 5c – 3   (This is the second part of the statement)
Since the first part = the second part, then:

$$4(c – 3 = 5c – 3$$
$$4c – 12 = 5c – 3$$
$$4c – 5c = 12 – 3$$
$$–c = 9$$
$$∴ \quad –9 = c$$
$$∴ \quad c = –9$$

Hence, the number is –9.

13. The sum of three consecutive numbers is 63. Find the numbers.

Solution
Let the first number be x.
∴   The second number will be $x + 1$, while the third number will be $x + 2$. For example, if the first number is 10, then the second number will be 11 which is (10 + 1), while the third number will be 12 which is (10 + 2).
So, the numbers are x, $x + 1$ and $x + 2$.
Since their sum is 63, then it follows that:

$$x + (x + 1) + (x + 2) = 63$$
$$3x + 3 = 63$$
$$3x = 63 – 3$$
$$3x = 60$$
$$∴ \quad x = \frac{60}{3}$$
$$x = 20$$

∴ The numbers are 20, 21 and 22 from x, $x + 1$ and $x + 2$.

14. The sum of two numbers is 21. Five times the first number added to two times the second number is 66. Find the two numbers.

Solution
Let the first number be x.
∴ The second number is $21 – x$. Note that the value of the sum of two numbers minus the first number gives the second number. For example, if 8 + 6 = 14, then 14 – 8 = 6 or 14 – 6 = 8. This shows that

Sum − one of the numbers = the other number

∴ five times the first number is : 5 x x = $5x$

2 times the second number is: 2 x $(21 − x)$ = $2(21 − x)$

When these two expressions are added together, it gives 66 as follows:

$5x + 2(21 − x) = 66$

$5x + 42 − 2x = 66$

$5x − 2x = 66 − 42$

$3x = 24$

∴ $x = \dfrac{24}{3}$

$x = 8$

∴ The first number is 8 and the second number is 21 − 8 = 13, i.e. 8 and 13.

15. 2 is added to twice a certain number and the sum is doubled. The result is 10 less than 5 times the original number. Find the number.

Solution

Let the number be y.

Twice the number is 2 x y = $2y$

2 added to twice the number = $2 + 2y$

When this sum is doubled, it gives: 2 x $(2 + 2y)$ = $2(2 + 2y)$

From second sentence, 5 times the number = 5 x y = $5y$

10 less than 5 times the number is: $5y − 10$

Since the result from first sentence = the result from second sentence, then it follows that:

$2(2 + 2y) = 5y − 10$

$4 + 4y = 5y − 10$

$4 + 10 = 5y − 4y$

$14 = y$

∴ $y = 14$

The number is 14.

16. The sum of two numbers is 38. When 8 is added to twice one of the numbers, the result is 5 times the other number. Find the two numbers.

Solution

Let one of the numbers be n.

∴ The other number = 38 − n (Since their sum is 38)

Twice one of the numbers is: 2 x n = $2n$

8 added to twice one of the numbers is: $8 + 2n$

5 times the other number is: 5 x $(38 − n)$ = $5(38 − n)$

Since these two results are equal according to the question, then we equate them as follows:

$8 + 2n = 5(38 − n)$

$8 + 2n = 190 − 5n$

$2n + 5n = 190 − 8$

$7n = 182$

$$n = \frac{182}{7}$$

$n = 26$

∴ The two numbers are n = 26 and 38 − n = 38 − 26 = 12.

They are 26 and 12

17. A woman's age and her son's age add up to 45 years. Five years ago, the woman was 6 times as old as her son. How old was the woman when the son was born?

Solution

Let the woman's age be x

∴ The son's age is = $45 − x$ (Since their sum is 45)

Five years ago, the woman's age = $x − 5$

Five years ago, the son's age was: $45 − x − 5 = 40 − x$

Since five years ago, the woman's age was 6 times that of her son, then this gives:

$x − 5 = 6(40 − x)$

$x − 5 = 240 − 6x$

$x + 6x = 240 + 5$

$7x = 245$

$$x = \frac{245}{7}$$

$x = 35$

∴ The woman's age is x = 35years.

The son's age is $45 − x = 45 − 35 = 10$years.

∴ The woman's age when the son was born is given by the difference between their ages (i.e. how many years older is the woman than her son). This gives

$35 − 10 = 25$

∴ The woman was 25 years old when the son was born.

18. The numerator of a fraction is 3. If 9 is added to the numerator and 4 is added to the denominator, the fraction is doubled. What is the fraction?

Solution

Let the denominator be y

∴ The fraction is $\frac{3}{y}$

6 added to the numerator = 3 + 6 = 9

4 added to the denominator = y + 4

∴ The new fraction is: $\frac{9}{y + 4}$

This new fraction is double of the initial fraction. This gives

$$\frac{9}{y + 4} = 2\left(\frac{3}{y}\right)$$

$$\frac{9}{y + 4} = \frac{6}{y}$$

Cross multiply to give:

$$9y = 6(y + 4)$$
$$9y = 6y + 24$$
$$9y - 6y = 24$$
$$3y = 24$$
$$\therefore \quad y = \frac{24}{3}$$
$$y = 8$$

$\therefore$ The fraction is: $\frac{3}{y}$, which is $\frac{3}{8}$

Exercise 5

1. The sum of 9 and one-fourth of a certain number is five less than twice the number. Find the number.

2. When 2 is subtracted from the product of 5 and a certain number, and the result is divided by 3, the answer is equal to one and a half times the number. What is the number?

3. A man is six times as old as his daughter. Four years ago, the man was four times as old as his daughter. Find the age of the man and his daughter.

4. A boy is 16 years old and his mother is 42 years old. In how many years time will the mother be twice as old as the boy?

5. A cyclist takes 10hours to travel from a town A to a town B. When coming back, the journey took him 2hours more because he reduced his speed by 30km/h. What was his speed when he was coming from B to A.

6. A man runs for 2hours at 6km/h. He then cycled at 16km/h for a certain period of time. If his total distance was 72km, for how many hours did he cycle?

7. A sum of $186 is made up of $10 notes and $1 notes. If there are three times as many $10 notes as there are $1 notes, find the number of each note.

8. John has $20 and Rose has $200. If John saves $3 a day and Rose spends $9 a day, after how many days will they have equal amount of money?

9. A total of x litres of wine is needed to fill 21 tanks with the same amount of wine in each tank. If the size of each tank is 55 litres more, the wine will fill only 10 tanks. What is the value of x.

10. When a plane travels a certain journey at an average speed of 180km/h, it arrives an airport at 2pm. When it travels at an average speed of 150km/h, it arrives at the airport at 4.30pm. What is the length of the journey?

11. A man walks to a village at 12km/h and returns at 15km/h. If the whole journey takes 4hours, what is the total distance walked?

12. The result of taking 5 from a number and multiplying the answer by 8 is the same as adding 8 to four times the number. Find the number.

13. The sum of three consecutive odd numbers is 45. Find the numbers.

14. The sum of two numbers is 29. Two times the first number added to six times the second number is 78. Find the two numbers.

15. 6 is added to thrice a certain number and the sum is halved. The result is 3 less than 2 times the original number. Find the number.

16. The sum of two numbers is 40. When 10 is added to twice one of the numbers, the result is 4 times the other number. Find the two numbers.

17. A woman's age and her son's age add up to 52 years. 8 years ago, the woman was 17 times as old as her son. How old was the woman when the son was born?

18. The numerator of a fraction is 5. If 2.5 is added to the numerator and 8 is subtracted from the denominator, the fraction is halved. What is the fraction?

CHAPTER 6
SIMULTANEOUS LINEAR EQUATIONS

In simultaneous equations, we solve a pair of equations and determine the values of two variables. Apart from graphical method, the two methods of solving simultaneous equations are:
1. Substitution Method
2. Elimination Method

Substitution Method

In this method, we make one of the variables a subject of formula from one of the equations, and substitute it into the other equation. It is easier to use this method when one of the variables has a coefficient of 1.

Examples

1. Solve simultaneously the equations: $x + y = 4$ and $2x - y = 5$

<u>Solution</u>
Let us rewrite this equation and label them equation 1 and 2 for easy identification. This gives:

$x + y = 4$ ....Equation (1)

$2x - y = 5$ ....Equation (2)

A careful look at the two equations shows that x and y in equation 1 have coefficients of 1, while y in equation 2, has a coefficient of -1. Any of these three variables can easily be made the subject of formula in their respective equation. So, let us make x in equation 1 the subject of formula of the equation. This is done as follows:

$x + y = 4$ ....Equation (1)

$x = 4 - y$

Let us represent this equation as equation 3 as follows:

$x = 4 - y$ ....Equation (3)

Since $x = 4 - y$, let us substitute $4 - y$ for x in equation 2. Note that we made x the subject of formula from equation 1, so the substitution of the expression for x should be done in the other equation (i.e. equation 2), and not in the same equation from which x was made the subject of the formula. Hence we substitute $4 - y$ for x in equation 2 as follows:

$2x - y = 5$ ....Equation (2)

$2(4 - y) - y = 5$ (Note that x has been replaced with $4 - y$ as stated above)

$8 - 2y - y = 5$

$-3y = 5 - 8$

$-3y = -3$

$y = \dfrac{-3}{-3}$ (When we divide both sides by -3)

$y = 1$

Now, substitute 1 for y in equation 3 in order to obtain x. Note that we can use any of the three equations, but equation 3 is the best to use since it directly gives us the value of x. This gives:

$x = 4 - y$ ....Equation (3)

$x = 4 - 1$

$x = 3$

Therefore, $x = 3$ and $y = 1$

2. Solve the simultaneous equations: $3x + 2y = 10$ and $4x - y = 6$

Solution

Let us rewrite this equation and label them equation 1 and 2 for easy identification. This gives:

$3x + 2y = 10$ ....Equation (1)

$4x - y = 6$ ....Equation (2)

A careful look at the two equations shows that only y in equation 2 has a coefficients of −1. So, let us make y in equation 2 the subject of formula of the equation. This is done as follows:

$4x - y = 6$ ....Equation (2)

Hence, $4x - 6 = y$

Or, $y = 4x - 6$ (Note that p = q also means that q = p)

Let us label this equation as equation 3 as shown below.

$y = 4x - 6$ ....Equation (3)

Since $y = 4x - 6$, let us substitute $4x - 6$ for y in equation 1. Note that we made y the subject of formula from equation 2, so the substitution of the expression for y should be done in the other equation (i.e. equation 1), and not in the same equation from which y was made the subject of the formula. Hence we substitute $4x - 6$ for y in equation 1 as follows:

$3x + 2y = 10$ ....Equation (1)

$3x + 2(4x - 6) = 10$ (Note that y has been replaced (substituted) with $4x - 6$)

$3x + 8x - 12 = 10$

$11x = 10 + 12$

$11x = 22$

$x = \dfrac{22}{11}$ (When we divide both sides by 11)

$x = 2$

Now, substitute 2 for x in equation 3 in order to obtain y. Note that we can use any of the three equations, but equation 3 is the best to use since it directly gives us the value of y. This gives:

$y = 4x - 6$ ....Equation (3)

$y = 4(2) - 6$

$y = 8 - 6$

$y = 2$

Therefore, $x = 2$ and $y = 2$

3. Solve simultaneously, the equations $m - 6n = 1$ and $9m + n = -46$

<u>Solution</u>

$m - 6n = 1$ ....Equation (1)

$9m + n = -46$ ....Equation (2)

From equation 2:

$n = -46 - 9m$ ....Equation (3)

Note that we can also easily make m the subject of the formula from equation 1. Since, n = -46 - 9m, let us substitute -46 - 9m for n in equation 1. This gives:

$m - 6n = 1$ ....Equation (1)

$m - 6(-46 - 9m) = 1$

$m + 276 + 54m = 1$

$55m = 1 - 276$

$55m = -275$

$m = \dfrac{-275}{155}$

$m = -5$

Now, substitute -5 for m in equation 3 in order to find n. This gives:

$n = -46 - 9m$ ....Equation (3)

$n = -46 - 9(-5)$

$n = -46 + 45$

$n = -1$

Therefore, n = -1 and m = -5

4. Solve simultaneously, the equations 3a - 5b = 25 and 7a + 2b = 31

<u>Solution</u>

$3a - 5b = 25$ ....Equation (1)

$7a + 2b = 31$ ....Equation (2)

A careful look at the two equations shows that there is no variable with a coefficient of 1. However, we can still make any of the variable a subject of its equation, and substitute it into the other equation. Hence, from equation 1, let as make b the subject of the formula as follows:

$3a - 5b = 25$ ....Equation (1)

$3a - 25 = 5b$

Or, $5b = 3a - 25$ (Note that m = n also means that n = m)

Dividing both sides by 5 gives:

$b = \dfrac{3a - 25}{5}$ ....Equation (3)

Since $b = \dfrac{3a - 25}{5}$, let us substitute $\dfrac{3a - 25}{5}$ for b in equation 2. This gives:

$7a + 2b = 31$ ....Equation (2)

$7a + 2(\dfrac{3a - 25}{5}) = 31$

$$7a + \frac{6a - 50}{5} = 31$$

Multiply each term by 5 in order to clear out the fractions. This gives:

$$5(7a) + 5(\frac{6a - 50}{5}) = 5(31)$$

$$35a + 6a - 50 = 155$$
$$41a = 155 + 50$$
$$41a = 205$$
$$a = \frac{205}{41}$$
$$a = 5$$

Now, substitute 5 for a in equation 3. This gives:

$$b = \frac{3a - 25}{5} \quad \dots\dots\dots\dots\dots\text{Equation (3)}$$

$$= \frac{3(5) - 25}{5}$$

$$= \frac{15 - 25}{5}$$

$$= \frac{-10}{5}$$
$$b = -2$$

Therefore, a = 5 and b = −2

5. Solve the equations 8p − 3q = −5 and 2p − 11q = −32

Solution
$$8p - 3q = -5 \quad \dots\dots\dots\dots\dots\text{Equation (1)}$$
$$2p - 11q = -32 \quad \dots\dots\dots\dots\text{Equation (2)}$$
A careful look at the two equations shows that there is no variable with a coefficient of 1. However, let us make p the subject of formula in equation 2. This is done as follows:
$$2p - 11q = -32 \quad \dots\dots\dots\dots\text{Equation (2)}$$
$$2p = -32 + 11q$$
Dividing both sides by 2 gives:

$$p = \frac{-32 + 11q}{2} \quad \dots\dots\dots\dots\dots\text{Equation (3)}$$

Substitute $\frac{-32 + 11q}{2}$ for p in equation 1. This gives:

$$8p - 3q = -5 \quad \dots\dots\dots\dots\dots\text{Equation (1)}$$

$$8(\frac{-32 + 11q}{2}) - 3q = -5$$

$$4(-32 + 11q) - 3q = -5$$
$$-128 + 44q - 3q = -5$$
$$41q = -5 + 128$$
$$41q = 123$$
$$q = \frac{123}{41}$$
$$q = 3$$

Substitute 3 for q in equation 3 in order to find p. This gives:

$$p = \frac{-32 + 11q}{2} \quad \text{......................Equation (3)}$$

$$= \frac{-32 + 11(3)}{2}$$

$$= \frac{-32 + 33}{2}$$

$$p = \frac{1}{2}$$

Hence, $p = \frac{1}{2}$ and q = 3

Elimination Method

Elimination method is used in solving simultaneous equations, usually when there is no variable with a coefficient of 1. In elimination method, we get rid of (eliminate) one of the variables by making its coefficient the same in both equations.

Examples

1. Solve the equation: $4x - 5y = 5$ and $2x - 3y = 2$.

<u>Solution</u>
$$4x - 5y = 5 \quad \text{....................Equation (1)}$$
$$2x - 3y = 2 \quad \text{....................Equation (2)}$$
When using elimination method, any operation of multiplication and division can be used to manipulate any of the equations in order to make a particular variable to have equal coefficient.

A careful look at the two equations shows that we can make the coefficient of x in equation (2) to become 4 so that the coefficient of x in the two equations will be 4. In order to make the coefficient of x to be 4 in equation (2), we multiply equation (2) by 2. This gives:
$$2(2x) - 2(3y) = 2(2) \quad \text{(Note that each term in equation (2) should be multiplied by 2)}$$
$$4x - 6y = 4 \quad \text{....................Equation (3)} \quad \text{(Note that our result is labelled equation 3)}$$
Now, equation 1 and 3 have the same coefficient of x. This means that x can now be eliminated by subtracting one equation from the other. Note that the coefficient of x which is 4 in equation 1 and 3 is a positive value. Hence one equation will be subtracted from the other in order to eliminate x. However, when the coefficients of a variable have opposite signs (i.e.

positive and negative), we will have to add the two equations in order to eliminate that variable.

In this case, for us to eliminate x, let us subtract equation 3 from equation 1 (we can also subtract equation 1 from equation 3) as follows:

$4x - 5y - (4x - 6y) = 5 - 4$ (Take note of the use of bracket in the subtraction)

Note that when subtracting, you subtract the left hand side of both equations separately, and subtract the right hand side of both equations separately as shown above. This simplifies to give:

$4x - 5y - 4x + 6y = 1$ (Note the changes in sign of the terms in the bracket)

$4x - 4x - 5y + 6y = 1$

$y = 1$ (Note that $4x - 4x = 0$, and $-5y + 6y = y$)

It is important to note that the terms in equation 1 come first on both sides of the equation during subtraction since we are carrying out: Equation 1 minus Equation 3.

Now that we have obtained y = 1, let us substitute 1 for y in equation 2 in order to find x. Note that the substitution can be done in any of the three equations.

$2x - 3y = 2$ ....Equation (2)

$2x - 3(1) = 2$ (Note that 1 has been substituted for y)

$2x - 3 = 2$

$2x = 2 + 3$

$2x = 5$

$x = \dfrac{5}{2}$

Hence, $x = \dfrac{5}{2}$ and y = 1

2. Solve simultaneously, the equations $5x - 4y = -2$ and $4x + 5y = 23$.

Solution

$5x - 4y = -2$ ....Equation (1)

$4x + 5y = 23$ ....Equation (2)

An easy way of making the coefficient of x to be the same in both equations is to multiply one equation by the coefficient of x in the other equation. This means that we multiply equation (1) by 4 (i.e. the coefficient of x in equation 2), and multiply equation (2) by 5 (i.e. the coefficient of x in equation 1). This will make the coefficient of x to be 20 in the two equations. Hence, let us multiply equation (1) by 4 and equation (2) by 5 in order to obtain equation (3) and equation (4) respectively as shown below:

$4(5x) - 4(4y) = 4(-2)$ ....Equation (3)

$5(4x) + 5(5y) = 5(23)$ ....Equation (4)

These two equations simplifies to give:

$20x - 16y = -8$ ....Equation (3)

$20x + 25y = 115$ ....Equation (4)

Now, equation 3 and 4 have the same coefficient of x. This means that x can now be eliminated by subtracting one equation from the other. Hence, let us subtract equation 3 from equation 4.

Note that any order can be followed, i.e. we can subtract equation 4 from equation 3. Therefore, equation 4 minus equation 3 gives:

$20x + 25y - (20x - 16y) = 115 - (-8)$ (Note the subtraction of left and right side separately)

$20x + 25y - 20x + 16y = 115 + 8$ (Note the changes in sign of the terms in the bracket)

$20x - 20x + 25y + 16y = 123$

$41y = 123$ (Note that $20x - 20x = 0$, and $25y + 16y = 41y$)

$y = \dfrac{123}{41}$

$y = 3$

Now that we have obtained y = 3, let us substitute 3 for y in equation 1 in order to find x. Note that the substitution can be done in any of the four equations.

$5x - 4y = -2$ ....Equation (1)

$5x - 4(3) = -2$ (Note that 3 has been substituted for y)

$5x - 12 = -2$

$5x = 12 - 2$

$5x = 10$

$x = \dfrac{10}{5}$

$x = 2$

Hence, $x = 2$ and y = 3

3. Solve simultaneously, the equations $11x + 3y = 4$ and $7x - 2y = -17$.

Solution

$11x + 3y = 4$ ....Equation (1)

$7x - 2y = -17$ ....Equation (2)

In our last two examples we made the coefficient of x to be equal. In this example let us make the coefficient of y to be equal. An easy way of making the coefficient of y to be the same in both equations is to multiply equation (1) by 2 (i.e. the coefficient of y in equation 2), and multiply equation (2) by 3 (i.e. the coefficient of y in equation 1). This will make the coefficient of y to be 6 (we are not taking the signs into consideration) in the two equations. Hence, let us multiply equation (1) by 2 and equation (2) by 3 in order to obtain equation (3) and equation (4) respectively as shown below:

$2(11x) + 2(3y) = 2(4)$ ....Equation (3)

$3(7x) - 3(2y) = 3(-17)$ ....Equation (4)

These two equations simplifies to give:

$22x + 6y = 8$ ....Equation (3)

$21x - 6y = -51$ ....Equation (4)

Now, equations 3 and 4 have the same coefficient of y. This means that x can now be eliminated by adding the two equations together. Hence, let us add equation 3 with equation 4. Note that in example 1 and 2 above, we subtracted one equation from the other. This was because the signs of the coefficients of x were the same. However, in this example (example 3), the signs of the coefficients of y (i.e. the variable to be eliminated) are not the same. Hence, we have to add them together in order to eliminate y. This summarizes the rule that when the signs of the coefficient of the variable to be eliminated are the same, we subtract one equation from the other in order to eliminate that variable. However, when the coefficients of a variable to be eliminated have opposite signs (i.e. positive and negative), we will have to add the two equations in order to eliminate that variable.

Therefore, equation 4 plus equation 3 gives:

$22x + 6y + (21x - 6y) = 8 + (-51)$ (Take note of the addition of left and right side separately)

$22x + 6y + 21x - 6y = 8 - 51$

$22x + 21x + 6y - 6y = -43$

$43x = -43$

$x = \dfrac{-43}{43}$

$x = -1$

Now that we have obtained $x = -1$, let us substitute -1 for x in equation 1 in order to find y.

$11x + 3y = 4$ ....Equation (1)

$11(-1) + 3y = 4$

$-11 + 3y = 4$

$3y = 4 + 11$

$3y = 15$

$y = \dfrac{15}{3}$

$y = 3$

Hence, $x = -1$ and $y = 3$

4. Solve simultaneously, the equations: 6m − 5n − 3 = 0 and 4m + 3n + 17 = 0

Solution

Each of the equations above can be rearranged by taking the constant term to the right hand side of the equations. This gives:

6m − 5n = 3 ....Equation (1)

4m + 3n = −17 ....Equation (2)

Let us make the coefficient of n to be equal in the two equations. An easy way of doing this is to multiply equation (1) by 3 (i.e. the coefficient of n in equation 2), and multiply equation (2) by 5 (i.e. the coefficient of n in equation (1). This will make the coefficient of n to be 15 in the two equations. Hence, equation (1) multiplied by 3 and equation (2) multiplied by 5 will give equation (3) and equation (4) respectively as shown below:

3(6m) − 3(5n) = 3(3) ....Equation (3)

5(4m) + 5(3n) = 5(−17) ....Equation (4)

These two equations simplify to give:

18m − 15n = 9 ....Equation (3)

20m + 15n = − 85 ....Equation (4)

Now, equations 3 and 4 have the same coefficient of n. This means that n can now be eliminated by adding the two equations together. We have to add the two equations since the signs of the coefficients of n are different.

Another method of carrying out our operation (either addition or subtraction) is to arrange the two equations to be one above the other and then add or subtract the corresponding like terms. Hence, equation 3 plus equation 4 can be used to describe this method as shown below.

$$18m - 15n = 9 \ldots\ldots\ldots\ldots\ldots\ldots\text{Equation (3)}$$
$$20m + 15n = -85 \ldots\ldots\ldots\ldots\ldots\text{Equation (4)}$$

Equation (3) + equation (4): $38m \qquad = -76$

In the addition above, the followings were carried out in each column:

$$18m + 20m = 38m$$
$$-15n + 15n = 0,$$

and $9 + (-85) = 9 - 85 = -76.$

Also, zero n was not written since n has been eliminated.

We now continue from above as follows:

$$38m = -76$$
$$m = \frac{-76}{38}$$
$$m = -2$$

Substitute −2 for m in equation (1) in order to obtain n. This gives:

$$6m - 5n = 3 \ldots\ldots\ldots\ldots\ldots\text{Equation (1)}$$
$$6(-2) - 5n = 3$$
$$-12 - 5n = 3$$
$$-5n = 3 + 12$$
$$-5n = 15$$
$$n = \frac{15}{-5}$$
$$n = -3$$

Therefore, m = −2 and n = −3

5. Solve the simultaneous equations $5p - 8q = 16$ and $12q - p = 24$

Solution

The second equation above has been arranged differently from the first one. In the first equation the term in p is written first, but in the second equation the term in q has been written first. Therefore in order to avoid making mistake, it is advisable to have the same pattern of arrangement in the two equations.

Hence, let us now label the two equations and rearrange the second equation as shown below.

$$5p - 8q = 16 \ldots\ldots\ldots\ldots\ldots\text{Equation (1)}$$
$$- p + 12q = 24 \ldots\ldots\ldots\ldots\ldots\text{Equation (2)}$$

Let us make the coefficient of p to be equal in the two equations. In order to do this we simply multiply equation (2) by 5 so that the coefficient of p in the two equations will be 5 (we have ignored the signs). Hence, if we multiply equation (2) by 5, it will give us equation (3) as follows:

$$5(- p) + 5(12q) = 5(24) \ldots\ldots\ldots\ldots\text{Equation (3)}$$

This simplifies to give:

$$-5p + 60q = 120 \ldots\ldots\ldots\ldots\text{Equation (3)}$$

Now, equations 1 and 3 have the same coefficient of p. This means that p can now be eliminated by adding the two equations together. We have to add the two equations since the signs of the coefficients of p are different.

Hence, equation (1) and (3) can be brought together and added to eliminate p as shown below.

$$5p - 8q = 16 \dots\dots\dots\dots\text{Equation (1)}$$

$$-5p + 60q = 120 \dots\dots\dots\text{Equation (3)}$$

Equation (1) + equation (3): $52q = 136$

Note that 5p + (−5p) = 0, −8q + 60q = 52q, and 16 + 120 = 136, as all obtained above when the corresponding like terms are added. Also, zero p was not written since it has been eliminated. We now continue from above as follows:

$$52q = 136$$
$$q = \frac{136}{52}$$
$$q = \frac{34}{13} \qquad \text{(After equal division by 4)}$$

Substitute $\frac{34}{13}$ for q in equation (2) in order to obtain p. This gives:

$$-p + 12q = 24 \dots\dots\dots\dots\text{Equation (2)}$$
$$-p + 12(\frac{34}{13}) = 24$$
$$-p + \frac{408}{13} = 24$$
$$-p = 24 - \frac{408}{13}$$
$$-p = \frac{312 - 408}{13}$$
$$-p = \frac{-96}{13}$$

$$p = \frac{96}{13} \qquad \text{(After dividing both sides by −1 in order to make p positive)}$$

Therefore, $p = \frac{96}{13}$, and $q = \frac{34}{13}$

6. Solve simultaneously, the following equations:
$$2c + 5d = 0$$
$$3c - 2d = 19$$

Solution
$$2c + 5d = 0 \dots\dots\dots\text{Equation (1)}$$
$$3c - 2d = 19 \dots\dots\dots\text{Equation (2)}$$
Let us make the coefficient of c to be equal in the two equations. In order to do this we multiply equation (1) by 3 (i.e. the coefficient of c in equation 2), and multiply equation (2) by 2 (i.e. the

coefficient of c in equation 1). This will make the coefficient of c to be 6 in the two equations. Hence, equation (1) multiplied by 3 and equation (2) multiplied by 2 will give equation (3) and equation (4) respectively as shown below:

$$3(2c) + 3(5d) = 3(0) \text{ ....Equation (3)}$$
$$2(3c) - 2(2d) = 2(19) \text{ ....Equation (4)}$$

These two equations simplify to give:

$$6c + 15d = 0 \text{ ....Equation (3)}$$
$$6c - 4d = 38 \text{ ....Equation (4)}$$

We now eliminate c by subtracting equation (4) from equation (3). We have to subtract the two equations since the signs of the coefficients of c are the same.

Hence, equation 3 minus equation 4 is evaluated as shown below.

$$6c + 15d = 0 \text{ ....Equation (3)}$$
$$6c - 4d = 38 \text{ ....Equation (4)}$$

Equation (3) – equation (4): $19d = -38$

Note that $6c - 6c = 0$, $15d - (-4d) = 19d$, and $0 - 38 = -38$, as all obtained above.

We now continue from above as follows:

$$19d = -38$$
$$d = \frac{-38}{19}$$
$$d = -2$$

Substitute –2 for d in equation (1) in order to obtain c. This gives:

$$2c + 5d = 0 \text{ ....Equation (1)}$$
$$2c + 5(-2) = 0$$
$$2c - 10 = 0$$
$$2c = 10$$
$$c = \frac{10}{2}$$
$$c = 5$$

Therefore, c = 5 and d = –2

7. Solve simultaneously, the following equations:
$$11x - 7y = -11$$
$$5x + 3y = 29$$

Solution
$$11x - 7y = -11 \text{ ....Equation (1)}$$
$$5x + 3y = 29 \text{ ....Equation (2)}$$

Let us make the coefficient of x to be equal in the two equations. In order to do this, multiply equation (1) by 5 (i.e. the coefficient of x in equation 2), and multiply equation (2) by 11 (i.e. the coefficient of x in equation 1). This will give equation (3) and equation (4) respectively as shown below:

$$5(11x) - 5(7y) = 5(-11) \text{ ....Equation (3)}$$
$$11(5x) + 11(3y) = 11(29) \text{ ....Equation (4)}$$

These two equations simplifies to give:

$55x - 35y = -55$ ....Equation (3)

$55x + 33y = 319$ ....Equation (4)

We now eliminate x by subtracting equation (4) from equation (3) (note that we can also subtract equation 3 from equation 4). We have to subtract the two equations since the signs of the coefficients of x are the same.

Hence, equation 3 minus equation 4 is evaluated as shown below.

$55x - 35y = -55$ ....Equation (3)

$55x + 33y = 319$ ....Equation (4)

Equation (3) – equation (4): $-68y = -374$

Note that $55x - 55x = 0$, $-35y - (+33y) = -68y$, and $-55 - 319 = -374$, as all obtained above.

Hence: $-68y = -374$ (From above)

$$y = \frac{-374}{-68}$$

$$y = \frac{11}{2}$$ (When expressed in its lowest term)

$$y = 5\frac{1}{2}$$

Substitute $\frac{11}{2}$ for y in equation (2) (we can use any equation) in order to obtain x. This gives:

$5x + 3y = 29$ ....Equation (2)

$$5x + 3(\frac{11}{2}) = 29$$

$$5x + \frac{33}{2} = 29$$

$$5x = 29 - \frac{33}{2}$$

$$5x = \frac{58 - 33}{2}$$

$$5x = \frac{25}{2}$$

$$x = \frac{\frac{25}{2}}{5}$$

$$x = \frac{25}{2} \times \frac{1}{5}$$

$$= \frac{5}{2}$$

$$x = 2\frac{1}{2}$$

Therefore, $x = 2\frac{1}{2}$ and $y = 5\frac{1}{2}$

Exercise 6

1. Solve simultaneously the equations: $2x - y = -3$ and $5x + y = -4$
2. Solve the simultaneous equations: $x + y = 5$ and $2x - y = 13$
3. Solve simultaneously, the equations $5m - n = 8$ and $3m - n = 4$
4. Solve simultaneously, the equations: $a - 3b = -5$ and $4a - 3b = 25$
5. Solve the equations: $p - q = 7$ and $p - 3q = 13$
6. Solve the equation: $10x - 5y = -30$ and $20x + 3y = -34$
7. Solve simultaneously, the equations $4x - y = 6$ and $3x + y = 15$.
8. Solve simultaneously, the equations $x + 5y = 2$ and $9x - y = 12$.
9. Solve simultaneously, the equations: $11m - 3n - 11 = 3$ and $8m + 5n - 3 = 0$
10. Solve the simultaneous equations $2p - 15q = 1$ and $2q - 11p = 5$
11. Solve simultaneously, the following equations:

$$5c + 5d = 10$$
$$c - d = 5$$

12. Solve simultaneously, the following equations:

$$x - 4y = -20$$
$$12x + 5y = 10$$

CHAPTER 7
WORD PROBLEMS LEADING TO SIMULTANEOUS LINEAR EQUATIONS

Examples

1. Three spoons and five knives cost $46. Five spoons and ten knives cost $90. Find the cost of a spoon and a knife.

Solution

Let the cost of a spoon be x, and the cost of a knife be $y.

∴   From the first sentence, we have:

   $3x + 5y = 46$ ....Equation (1)

Similarly, from the second sentence, we have:

   $5x + 10y = 90$ ....Equation (2)

Bringing equation (1) and (2) together gives:

   $3x + 5y = 46$ ....Equation (1)

   $5x + 10y = 90$ ....Equation (2)

In order to make the coefficient of y to be the same in equation (1) and (2) (so as to solve by elimination method), lets simply divide equation (2) by 2. This gives:

$$\frac{5x + 10y = 90}{2}$$

This is simplified by dividing each term by 2 to give: $2.5x + 5y = 45$ .... equation (3)
Note that the coefficient of y is made to be the same in order to eliminate y (elimination method).

Therefore bringing equations (1) and (3) together gives:

   $3x + 5y = 46$ ....Equation (1)

   $2.5x + 5y = 45$ ....Equation (3)

Equation (1) – equation (3): $0.5x \quad = 1$

$$x = \frac{1}{0.5}$$

$$x = 2$$

Substitute 2 for x in any of the equations. Let us use equation (1)

∴ $3x + 5y = 46$ ....Equation (1)

   $3(2) + 5y = 46$

   $6 + 5y = 46$

∴ $5y = 46 – 6$

   $5y = 40$

∴ $y = \frac{40}{5}$

   $y = 8$

∴   A spoon cost $2, while a knife cost $8.

2. One-quarter of a boy's score in English plus one-fifth of his score in Mathematics makes 31. Half of his score in mathematics added to two-third of his score in English gives 80. What was his score in each subject?

<u>Solution</u>

Let his score in English be 'a' and his score in Mathematics be 'b'.

$\therefore$ From the first sentence, we have:

$$\frac{1}{4}a + \frac{1}{5}b = 31$$

Multiply each term by 20 (L.C.M of 4 and 5) in order to remove the fractions. This gives:

$$20\left(\frac{a}{4}\right) + 20\left(\frac{b}{5}\right) = 20 \times 31$$

$\therefore$ $5a + 4b = 620$ ....Equation (1)

From the second sentence of the question, we have

$$\frac{2}{3}a + \frac{1}{2}b = 80$$

Multiply each term by 6 (i.e. L.C.M of 3 and 2). This gives:

$$6\left(\frac{2a}{3}\right) + 6\left(\frac{b}{2}\right) = 6 \times 80$$

$2(2a) + 3(b) = 480$ (After 6 has been used to divide each of the denominator)

$\therefore$ $4a + 3b = 480$ ....Equation (2)

Bringing equations (1) and (2) together in order to solve them simultaneously gives:

$5a + 4b = 620$ ....Equation (1)

$4a + 3b = 480$ ....Equation (2)

Multiply equation (1) by 3 and equation (2) by 4 in order to make the coefficients of b to be the same in the two equations. This gives:

$3(5a + 4b = 620)$

$4(4a + 3b = 480)$

These simplify to:

$15a + 12b = 1860$ .... Equation (3)

$\underline{16a + 12b = 1920}$ .... Equation (4)

Equation (4) – Equation (3): a $= 60$

$\therefore$ $a = 60$

Substitute 60 for a in equation (2). This gives:

$4a + 3b = 480$ ....Equation (2)

$4(60) + 3b = 480$

$240 + 3b = 480$

$3b = 480 - 240$

$\therefore$ $3b = 240$

$b = \dfrac{240}{3}$

$\therefore$ $b = 80$

$\therefore$ He scored 60 in English and 80 in Mathematics

3. Divide 63 into two parts such that one part is two-fifth of the other.

Let the two parts be x and y.

$\therefore$ $x + y = 63$ ....Equation (1)   (Note that the two parts sum up to 63)

$\dfrac{2}{5}x = y$ ....Equation (2)   (Note that y has been made $\frac{2}{5}$ of x as stated above)

71

Substitute $\dfrac{2x}{5}$ for y in equation (1)

$\qquad x + y = 63$ ....Equation (1)

$\qquad x + \dfrac{2x}{5} = 63$

Multiply each term by 5 in order to remove the fraction. This gives:

$\qquad 5x + 5\dfrac{2x}{5} = 5 \times 63$

$\qquad 5x + 2x = 315$

$\qquad\qquad 7x = 315$

$\therefore \qquad\qquad x = \dfrac{315}{7} \qquad\qquad 7$

$\qquad\qquad x = 45$

From equation (1), y = 63 − x

$\qquad\qquad\qquad y = 63 − 45$

$\qquad\qquad\qquad\qquad y = 18$

$\therefore$ The two parts are 45 and 18.

4. A bicycle moves for p hours at 10km/h and q hours at 15km/h. If his total journey is 80km in 7 hours, find the values of p and q.

Solution

Recall that: Speed $= \dfrac{\text{Distance}}{\text{Time}}$

$\therefore$ At 10km/h, we have: $10 = \dfrac{\text{Distance}}{p}$ (Since time = p)

$\therefore$ Distance = 10p   (After cross multiplication)

Similarly, at 15km/hr, we have:

$\qquad$ Distance = 15q   (Since time is q in this case)

$\therefore$ Total distance is given by:

$\qquad$ 10p + 15q = 80 ....Equation (1)

The total time is: p + q = 7 ....Equation (2)   (Note that the journey took 7 hours)

Multiply equation (2) by 10 in order to make the coefficient of p to be 10 in the two equations. This gives:

$\qquad$ 10(p + q = 7)

$\qquad$ 10p + 10q = 70 ....Equation (3)

Bringing equations (1) and (3) together in order to solve them simultaneously gives:

$\qquad\qquad\qquad$ 10p + 15q = 80 ....Equation (1)

$\qquad\qquad\qquad$ <u>10p + 10q = 70</u> ....Equation (3)

Equation (1) − equation (3):$\qquad$ 5q = 10

$\qquad\qquad\qquad\qquad q = \dfrac{10}{5}$

$\qquad\qquad\qquad\qquad$ q = 2

From equation (2), p can be expressed as:

$\qquad$ p = 7 − q

$\qquad$ p = 7 − 2   (Since q = 2)

p = 5

∴ p = 5 hours and q = 2 hours.

5. The perimeter of an isosceles triangle is 56cm. If the two equal sides are (8y)cm and (12y − 4x + 2)cm, while the third side is (2x + 4y)cm, find the values of x and y.

Solution

Equating the two equal sides since they are equal gives:

8y = 12y − 4x + 2

∴ 4x + 8y − 12y = 2   (By collecting the unknown terms on the left hand side)

4x − 4y = 2

Dividing each term by 2 in order to simplify the equation gives:

2x − 2y = 1 ....Equation (1)

Adding the three sides of the triangle gives the perimeter as follows:

8y + 12y − 4x + 2 + 2x + 4y = 56

24y − 2x = 56 − 2

−2x + 24y = 54 ....Equation (2)

Bringing equations (1) and (2) together gives:

$$2x - 2y = 1 \text{ ....Equation (1)}$$
$$-2x + 24y = 54 \text{ ....Equation (2)}$$

In order to eliminate x, we add equation (1) and (2):   22y = 55

$$y = \frac{55}{22}$$

$$y = \frac{5}{2}$$ (After equal division by 11)

$$y = 2\frac{1}{2}$$

Substitute $\frac{5}{2}$ for y in equation (1)

2x − 2y = 1 ....(Equation (1)

$$2x - 2(\frac{5}{2}) = 1$$

2x − 5 = 1

2x = 1 + 5

2x = 6

∴ $$x = \frac{6}{2}$$

x = 3

∴ x = 3 and y = $2\frac{1}{2}$

6. The three sides of an equilateral triangle are (3x)cm, (4y)cm and (x + y + 3)cm. Find:

a. The values of x and y.

b. The sides of the triangle.

Solutions

a. Since the three sides of an equilateral triangle are equal, then it follows that: $3x = 4y = x + y + 3$

Equating any two of the terms above can be used to obtain two equations as follows:

$3x = 4y$

$\therefore$ $3x - 4y = 0$ ....Equation (1)

and $3x = x + y + 3$

$\therefore$ $3x - x - y = 3$

$2x - y = 3$ ....Equation (2)

From equation (2), $y = 2x - 3$ ....Equation (3)

Substitute $2x - 3$ for y in equation (1)

$\therefore$ $3x - 4y = 0$ ....Equation (1)

$3x - 4(2x - 3) = 0$

$3x - 8x + 12 = 0$ (Note that $- 4 \times - 3 = +12$)

$\therefore$ $-5x = -12$

$x = \dfrac{-12}{-5}$

$x = \dfrac{12}{5}$ (The negative sign has cancelled out)

$\therefore$ $x = 2\dfrac{2}{5}$

From equation (3) substitute: $x = \dfrac{12}{5}$

$\therefore$ $y = 2x - 3$ ....Equation (3)

$y = 2(\dfrac{12}{5}) - 3$

$= \dfrac{24}{5} - 3$

$= \dfrac{24 - 15}{5}$

$y = \dfrac{9}{5}$

$y = 1\dfrac{4}{5}$

Therefore, $x = 2\dfrac{2}{5}$ and $y = 1\dfrac{4}{5}$

b. Each side of the triangle is given by:

$3x$ or $4y$ or $x + y + 3$

$\therefore$ $3x$ will give:

$3 \times \dfrac{12}{5} = \dfrac{36}{5}$

$= 7\dfrac{1}{5}$cm

$\therefore$ Each side of the triangle is $7\dfrac{1}{5}$cm

7.   The sum of the digits of a two digits number is 13. If the digits are interchanged, the number is decreased by 45. Find the number.

Let the two digits number be xy. Therefore, the number itself is obtained as follows:

$10x + y$ (For example 24 = (2 x 10) + 4 = 20 + 4 = 24)

Note that x is the tens digit while y is the unit digit. When the digits are interchanged to obtain yx, then the new number is: $10y + x$

Since the original number is decreased by 45 after the digits are interchanged, it means that the first number is greater than the second number by 45.

$\therefore$ $(10x + y) - (10y + x) = 45$

$\qquad 10x + y - 10y - x = 45$

$\qquad\qquad 9x - 9y = 45$

$\therefore$ $\qquad\qquad\qquad x - y = 5$ (After dividing each term by 9 in order to simplify the equation)

$\therefore$ $\quad x - y = 5$ ....Equation (1)

Recall from the first sentence that:

$\qquad\qquad\qquad x + y = 13$ ....Equation (2)

$\qquad\qquad\qquad \underline{x - y = 5}$ ....Equation (1) (This is simply equation 1 brought down)

Equation (2) – (1): $\quad 2y = 8$ (x has been eliminated by the subtraction. Also $+y - (-y) = 2y$)

$\therefore$ $\qquad\qquad\qquad\qquad y = \dfrac{8}{2}$

$\qquad\qquad\qquad\qquad y = 4$

From equation (1), substitute 4 for y. This gives:

$\quad x - y = 5$ ....Equation (1)

$\quad x - 4 = 5$

$\qquad x = 5 + 4$

$\qquad x = 9$

$\therefore$ The number is 94 (i.e. xy)

Check: The number is 94. When the digits are interchanged it becomes 49. Their difference is 94 – 49 = 45. Hence it is correct as given in the question.

8. In a two digits number, the difference between the digits is 1. The number is 1 more than 5 times the sum of the digits. Find the number if the tens digit is less than the unit digit.

Let the number be xy (x is the tens digit)

$\quad y - x = 1$ ....Equation (1) (Note that the unit digit is greater than the tens digit). This equation can also be expressed as:

$\quad -x + y = 1$

The sum of the digits is $x + y$. And 5 times the sum of the digits is $5(x + y)$

The number itself is:

$\quad 10x + y$

Since the number is 1 more than 5 times the sum of the digits, then it follows that:

$\quad 10x + y - 5(x + y) = 1$

$\quad 10x + y - 5x - 5y = 1$

$\quad 5x - 4y = 1$ ....Equation (2)

Bringing equation (1) and (2) together in order to solve them simultaneously gives:

$-x + y = 1$ ....Equation (1)

$5x - 4y = 1$ .... equation (2)

Multiply equation (1) by 5. This is done in order to make the coefficient of x in the two equations to be 5. This gives:

$-5x + 5y = 5$ ....Equation (3)

Bringing equation (2) and (3) together gives:

$5x - 4y = 1$ ....Equation (2)

$-5x + 5y = 5$ ....Equation (3)

Equation (2) + equation (3): $y = 6$

Substitute 6 for y in equation (1)

$\therefore$ $-x + y = 1$ ....Equation (1)

$-x + 6 = 1$

$\therefore$ $x = 6 - 1$

$x = 5$

$\therefore$ The number is 56 (i.e. xy)

9.   The sum of a man's age and his son's age is 45 years. Five years ago, the man was 6 times as old as his son. How old was the man when his son was born?

Solution

Let the man's present age be x, and his son's present age be y.

$\therefore$ $x + y = 45$ ....Equation (1)

Five years ago, the man's age was $x - 5$, while his son's age was y $-$ 5. At that time, his age was 6 times that of his son. This means that:

$x - 5 = 6(y - 5)$

$x - 5 = 6y - 30$

$x - 6y = -30 + 5$

$x - 6y = -25$ ....Equation (2)

Bringing equations (1) and (2) together gives:

$x + y = 45$ ....Equation (1)

$x - 6y = -25$ ....Equation (2)

Equation (1) $-$ (2) $7y = 70$ (Note that $+y -(-6y) = 7y$, and $45 -(-25) = 70$)

$$y = \frac{70}{7}$$

$\therefore$ $y = 10$

Substitute 10 for y in equation (1). This gives:

$x + y = 45$ ....Equation (1)

$x + 10 = 45$

$x = 45 - 10$

$x = 35$

$\therefore$ The man's present age is 35 years while his son's present age is 10 years. This means that the son was born 10 years ago. Therefore, 10 years ago, the man's age was $35 - 10 = 25$ years.

$\therefore$ The man was 25 years when his son was born.

10. If 1 is added to the numerator of a fraction, and 2 is added to the denominator, the fraction becomes $\frac{1}{3}$. If 3 is added to both the numerator and denominator of the fraction, the fraction becomes $\frac{1}{2}$. Find the fraction.

Solution

Let the numerator be x and the denominator be y

∴ The fraction is $\frac{x}{y}$

From the first sentence:

$$\frac{x+1}{y+2} = \frac{1}{3}$$

∴ 3(x + 1) = 1(y + 2)   (When we cross multiply)

3x + 3 = y + 2

3x − y = 2 − 3

∴ 3x − y = −1 ....Equation (1)

From the second sentence:

$$\frac{x+3}{y+3} = \frac{1}{2}$$

∴ 2(x + 3) = 1(y + 3)

2x + 6 = y + 3

2x − y = 3 − 6

∴ 2x − y = −3 ....Equation (2)

Bringing equations (1) and (2) together gives:

3x − y = −1 ....Equation (1)

2x − y = −3 ....Equation (2)

Equation (1) − (2)   x   = 2

Hence $x = 2$

Substitute 2 for x in equation (1)

∴ 3x − y = −1 ....Equation (1)

3(2) − y = −1

6 − y = −1

6 + 1 = y

∴ y = 7

∴ The fraction is $\frac{2}{7}$ (i.e. $\frac{x}{y}$)

Exercise 7

1. Five books and eight pens cost $90. Six books and four pens cost $80. Find the cost of a book and a pen.

2. One-third of a girl's score in Mathematics plus two-seventh of her score in Biology makes 30. One-fifth of her score in Biology added to half of her score in Mathematics gives 29. What was her score in each subject?

3. Divide 75 into two parts such that one part is one-third of the other.

4. A man walks for m hours at 12km/h and n hours at 18km/h. If his total journey is 114km in 8 hours, find the values of m and n.

5. The perimeter of an isosceles triangle is 50cm. If the two equal sides are (6y)cm and $(2y - 4x + 4)$cm, while the third side is $(2x + 6y)$cm, find the values of x and y. Hence, determine the length of each side of the triangle.

6. The three sides of an equilateral triangle are $(5x)$cm, $(4y)$cm and $(x + 2y + 9)$cm. Find:
a. The values of x and y.
b. Each side of the triangle.

7. The sum of the digits of a two digits number is 11. If the digits are interchanged, the number is decreased by 27. Find the number.

8. In a two digits number, the difference between the digits is 7. The number is 4 more than 8 times the sum of the digits. Find the number if the tens digit is greater than the unit digit.

9. The sum of a woman's age and her daughter's age is 65 years. Ten years ago, the woman was 8 times as old as her daughter. How old was the woman when her daughter was born?

10. If 5 is added to the numerator of a fraction, and 2 is added to the denominator, the fraction becomes $\frac{4}{5}$. If 2 is added to both the numerator and denominator of the fraction, the fraction becomes $\frac{1}{2}$. Find the fraction.

CHAPTER 8
QUADRATIC EQUATION

An equation such as $x^2 + 7x - 18 = 0$, in which 2 is the highest power of the unknown variable is called quadratic equation. There are four methods of solving quadratic equations. They are:
1. By factorization
2. By Completing the square
3. By use of quadratic equation formula
4. By graphical method

We are going to look at the first three methods.

Factorization of Quadratic Expression

Examples
1. Factorize the following quadratic expressions
a. $3a^2 + 5a$
b. $y^2 - 81$
c. $4m^2 - 49$
d. $18c^2 - 72$

Solutions
a. $3a^2 + 5a$
In this case the common factor (H.C.F) of $3a^2$ and $5a$ is 'a'. Hence we factorize directly as follows:
$3a^2 + 5a = a(3a + 5)$
The terms in the bracket are obtained by dividing each of $3a^2$ and $5a$ by 'a' which is the common factor.

b. $y^2 - 81$
$y^2 - 81$ is a difference (minus) of two squares. So, recall that the difference of two squares such as $a^2 - b^2$ is factorized as follows: $a^2 - b^2 = (a + b)(a - b)$
Therefore, $y^2 - 81 = (y)^2 - (9)^2$
$$= (y + 9)(y - 9)$$

c. $4m^2 - 49$
Each term in this expression is a square. The square root of $4m^2$ is 2m, while the square root of 49 is 7. Hence, $4m^2 - 49$ is a difference (minus) of two squares. So, recall that the difference of two squares such as $a^2 - b^2$ is factorized to obtain: $(a + b)(a - b)$
Therefore, $4m^2 - 49 = (2m)^2 - (7)^2$
$$= (2m + 7)(2m - 7) \quad \text{[In the same way as } a^2 - b^2 = (a + b)(a - b)]$$

d. $18c^2 - 72$
This is not exactly a difference of two squares, but it can be made a difference of two squares by first factorizing with 18 as a common factor. This gives:

$18c^2 - 72 = 18(c^2 - 4)$

The terms in the bracket are now difference of two squares. The square root of $c^2 = c$, while the square root of $4 = 2$. Hence, we now have:

$$18(c^2 - 4) = 18[(c)^2 - (2)^2]$$
$$= 18[(c + 2)(c - 2)]$$

Therefore, $18c^2 - 72 = 18[(c + 2)(c - 2)]$

In the four examples above, none of the quadratic expressions has a complete three terms of the usual quadratic expression such as $3x^2 + 7x - 4$. Hence it is easy to factorize such expression. However, when we have a complete quadratic expression such as $2x^2 - 11x - 30$, then we have to apply a different method of factorization. This method is explained in the examples below.

2. Factorize $2x^2 - 11x - 30$.

Solution

This is a complete quadratic expression with three terms. In order to factorize this expression, we first multiply the first and last terms. This gives:

$$2x^2(-30) = -60x^2$$

Then we look for two factors of $-60x^2$ (i.e. two numbers whose product is $-60x^2$) such that their sum will give $-11x$ (i.e. the middle term in the original quadratic expression).

Some simple rules that should be noted when looking for the two factors are as given below:

1. In a case such as $2x^2 + 7x + 5$, where the sign of the constant term (i.e. +5) and the middle term (i.e. +7) are both positive, then the two factors should be positive.

2. In a case such as $2x^2 - 7x + 5$, where the sign of the constant term is positive (i.e. +5) and the sign of the middle term is negative (i.e. −7), then the two factors should be negative.

3. In a case such as $2x^2 - 3x - 5$, where the sign of the constant term is negative (i.e. −5) and the sign of the middle term is negative (i.e. −3), then one factor will be positive and the other will be negative. However, the higher value of the two factors will take the negative sign (i.e. the sign of the middle term).

4. In a case such as $2x^2 + 3x - 5$, where the sign of the constant term is negative (i.e. −5) and the sign of the middle term is positive (i.e. +3), then one factor will be positive and the other will be negative. However, the higher value of the two factors will take the positive sign (i.e. from the middle term).

Let us summarize these four rules as shown below:

S/N	Sign of product	Sign of middle term	Sign of factors
1	+	+	+ and +
2	+	-	- and -
3	-	-	+ and - with the larger value having -
4	-	+	+ and - with the larger value having +

With these four rules in mind, finding the factors in a quadratic equation becomes easy.

Let us continue with the solution to the question above.

The question at hand which is $2x^2 - 11x - 30$, is applicable to rule 3 above. This means that our two factors will have different signs with the higher value having the negative sign.

From the $-60x^2$ obtained above, a simple way of finding the factors of 60 (first ignore the minus sign and x^2) is to break 60 into its prime factors as follows:

2	60
2	30
3	15
5	5
	1

The above division can help us obtain the twelve factors of 60 as follows: 1, 2, 3, 4 (from 2 x 2 in the division above), 5, 6 (from 2 x 3 in the division above), 10 (from 2 x 5 in the division above), 12 (from 2 x 2 x 3 in the division above), 15, 20 (from 2 x 2 x 5 in the division above), 30 and 60.

Note that only 1, 2, 3, 5, 15, 30 and 60 are visible from the division above. The other factors (4, 6, 10, 12 and 20) have to be obtained by multiplying any two or more of the visible factors. Let us rewrite the factors as follows: 1, 2, 3, 4, 5, 6, 10, 12, 15, 20, 30, and 60.

Now that we have our factors, we can pair them up to see which of them will give the required sum of -11. According to rule 3 above, the higher factor should have a negative sign. Let us now write out the pairs and assign a negative sign to the larger factor. Hence, the pair whose product will give us -60 are:

1 and -60 (This pair is out. Their sum cannot give -11)

2 and -30 (This pair is out)

3 and -20 (This pair is out)

5 and -12 (This pair is out)

6 and -10 (This pair is out)

4 and -15 (This is the required pair. Their sum gives -11)

Hence the only possible factors that can give a sum of -11 are 4 and -15. Hence we now include x on them so that their product will give $-60x^2$ and their sum will give $-11x$. Hence the numbers are $-15x$ and $+4x$. We now substitute these two numbers for $-11x$ in the original expression. This gives:

$2x^2 - 11x - 30 = 2x^2 + 4x - 15x - 30$ (Note that any of the two numbers can be written first).

We now factorize each pair of terms in this new expression (i.e. factorization by grouping). This gives:

$2x^2 + 4x - 15x - 30 = 2x(x + 2) - 15(x + 2)$

Note that $-15x - 30 = -15(x + 2)$. The negative signs from the question have change to + in the bracket. This is what happens when a negative number is used for the factorization. A positive sign in the question will also change to a negative sign in the bracket. We now have:

$2x^2 + 4x - 15x - 30 = 2x(x + 2) - 15(x + 2)$ [Here the common factor is $(x + 2)$]

$= (x + 2)(2x - 15)$ (Simply take one of the common brackets i.e. $(x + 2)$, and then the other two terms outside each of the bracket and enclose them in a bracket).

Therefore, $2x^2 - 11x - 30 = (x + 2)(2x - 15)$

3. Factorize $x^2 + 13x - 30$

Here, you will observe that the coefficient of x^2 is 1 (i.e. $1x^2$). This is not like the previous example, i.e. $2x^2 - 11x - 30$, where the coefficient of x^2 is 2. So, a short method of factorizing this expression is to go ahead and look for two numbers whose product is -30 and whose sum is $+13$, without any need of multiplying the first and last terms. This is explained as follows:

$x^2 + 13x - 30 = (x \quad)(x \quad)$ (simply put the variable x into the brackets)

In order to complete the bracket, we find two numbers whose product is -30 and whose sum is $+13$. This means that we have to look for the factors of 30.

The factors of 30 are: 1, 2, 3, 5, 6, 10, 15 and 30.

Now that we have our factors, we can pair them up to see which of them will give a possible sum of $+13$. According to rule 4 above, the higher factor should have a positive sign. Let us now write out the pairs and assign a positive sign to the larger factor. Hence the pair whose product will give us -30 are:

      -1 and $+30$ (This is out. Their sum cannot give $+13$)

      -3 and $+10$ (This is out. Their sum cannot give $+13$)

      -5 and $+6$ (This is out.)

      -2 and $+15$ (These are the required factors. Their sum will give $+13$)

Therefore, we now put them into the brackets above as follows:

$x^2 + 13x - 30 = (x + 15)(x - 2)$

4. Factorize $3x^2 + 11x + 6$

Solution

   $3x^2 + 11x + 6$

The first thing to do is to multiply the first and last term. This gives:

   $3x^2(6) = 18x^2$

We now look for two factors of 18 whose product is 18 and sum is $+11$. This is a case of rule 1 above where the constant term is positive and the middle term is also positive. Hence the two factors will be positive. The factors of 18 are 1, 2, 3 , 6, 9, 18. If we pair them up based on any two of the factors whose product is 18, we have:

(1 and 18), (2 and 9), (3 and 6).

Out of these pairs, only 2 and 9 can give us a sum of 11. Their signs must be positive based on rule 1 above. We now include x on them so that their product will give $18x^2$ and their sum will give $11x$. Hence the numbers are $+9x$ and $+2x$.

We now substitute these two numbers for $11x$ in the original expression. This gives:

$3x^2 + 11x + 6 = 3x^2 + 9x + 2x + 6$ (Note that any of the two numbers can be written first).

We now factorize by grouping. This gives:

$$3x^2 + 9x + 2x + 6 = 3x(x + 3) + 2(x + 3)$$
$$= (x + 3)(3x + 2)$$

Therefore, $3x^2 + 11x + 6 = (x + 3)(3x + 2)$

5. Factorize $5x^2 - 14x + 8$

<u>Solution</u>

$5x^2 - 14x + 8$

The first thing to do is to multiply the first and last terms. This gives:

$5x^2(8) = 40x^2$

We now find two factors of 40 whose product is 40 and sum is -14 (i.e. the middle term). This is a case of rule 2 above where the constant term is positive and the middle term is negative. Hence the two factors will be negative. The factors of 40 are 1, 2, 4 , 5, 8, 10, 20 and 40. According to rule 2 above, the factors will both have a negative sign. Let us now write out the pairs and assign a negative sign to each factor. Hence the pairs whose product will give us 40 are:

(-1 and -40), (-2 and -20), (-4 and -10), (-5 and -8). Note that the product of two negative numbers gives a positive value.

Out of these pairs, only -4 and -10 can give us a sum of -14. Hence we now include x on them so that their product will give $40x^2$ and their sum will give $-14x$. Hence the numbers are $-4x$ and $-10x$.

We now substitute these two numbers for $-14x$ in the original expression. This gives:

$5x^2 - 14x + 8 = 5x^2 - 4x - 10x + 8$ (Note that any of the two numbers can be written first).

We now factorize by grouping. This gives:

$5x^2 - 4x - 10x + 8 = x(5x - 4) - 2(5x - 4)$

Note that in the second pair (i.e. $-10x + 8$) the sign of the first number must be used. Hence the use of -2 to obtain $- 2(5x - 4)$.

Therefore with $(5x - 4)$ as a common factor we proceed with the factorization as follows:

$x(5x - 4) - 2(5x - 4) = (5x - 4)(x - 2)$

Therefore, $5x^2 - 14x + 8 = (5x - 4)(x - 2)$

Solving quadratic equations by by Factorization

In solving quadratic equation by factorization, we first factorize the equation and then equate each of the term or bracket to zero to solve the linear equation formed. The general form of a quadratic equation is given by: $ax^2 + bx + c = 0$. Hence if an equation is given by $3x^2 - 11x + 6$, we can compare it with $ax^2 + bx + c = 0$, and obtain:

a = 3, b = –11 and c = 6.

With these values of a, b and c, we can determine if a quadratic expression can be factorized or not. Hence, a quadratic equation can be factorized if:
$b^2 - 4ac$ gives a perfect square.
For example, let us determine which of the following can be factorized:

a. $x^2 + 5x + 6 = 0$

b. $4x^2 - 11x + 5 = 0$

c. $2x^2 + x - 6 = 0$

d. $5x^2 - 6x - 8 = 0$

e. $3x^2 - 13x + 8 = 0$

Now, let us see which of them will give $b^2 - 4ac$ as a perfect square.

a. $x^2 + 5x + 6 = 0$.
    a = 1, b = 5, c = 6
Hence, $b^2 - 4ac = 5^2 - (4 \times 1 \times 6)$
$$= 25 - 24$$
$$= 1$$
Therefore, 1 is a perfect square whose square root is 1. Hence, $x^2 + 5x + 6 = 0$ can be factorized.

b. $4x^2 - 11x + 5 = 0$.
In this case, a = 4, b = –11, c = 5
Hence, $b^2 - 4ac = (-11)^2 - (4 \times 4 \times 5)$
$$= 121 - 80$$
$$= 41$$
Therefore, 41 is not a perfect square. Its square root will not give us a whole number. Hence, $4x^2 - 11x + 5 = 0$ cannot be factorized.

c. $2x^2 + x - 6 = 0$.
    a = 2, b = 1, c = –6
Hence, $b^2 - 4ac = 1^2 - [4 \times 2 \times (-6)]$
$$= 1 - (-48)$$
$$= 1 + 48$$
$$= 49$$
Therefore, 49 is a perfect square whose square root is 7. Hence, $2x^2 + x - 6 = 0$ can be factorized.

d. $5x^2 - 6x - 8 = 0$.
    a = 5, b = –6, c = –8

84

Hence, $b^2 - 4ac = (-6)^2 - [4 \times 5 \times (-8)]$
$$= 36 - (-160)$$
$$= 36 + 160$$
$$= 196$$

Therefore, 196 is a perfect square whose square root is 14. Hence, $5x^2 - 6x - 8 = 0$ can be factorized.

e. $3x^2 - 13x + 8 = 0$.
   $a = 3, b = -13, c = 8$

Hence, $b^2 - 4ac = (-13)^2 - (4 \times 3 \times 8)$
$$= 169 - 96$$
$$= 73$$

Therefore, 73 is not a perfect square. Its square root will not give us a whole number. Hence, $3x^2 - 13x + 8 = 0$ cannot be factorized.

Examples

1. Solve the following equations by factorization:
a. $2a^2 + 8a = 0$
b. $r^2 - 64 = 0$
c. $9m^2 - 25 = 0$
d. $(3x + 2)^2 = 16$

Solutions
a. $2a^2 + 8a = 0$

Factorizing this quadratic equation by taking 2a as the common term gives:
   $2a(a + 4) = 0$

We now equate each term to zero and solve for 'a'. We equate each term to zero because the product of two or more terms is zero if at least one of the terms is equal to zero. Therefore equating each term above to zero and solving for 'a' gives:

Either,
   $2a = 0$

Dividing both sides by 2 gives:
   $a = \dfrac{0}{2}$
   $a = 0$

Or, $(a + 4) = 0$
   $a + 4 = 0$
   $a = -4$

Therefore, a = 0 or a = -4

b. $r^2 - 64 = 0$

This can be solved directly as follows:
   $r^2 - 64 = 0$
   $r^2 = 64$
   $r = \sqrt{64}$

$$r = \pm 8$$

Therefore, $r = +8$ or $r = -8$

Note that $\pm$ is used to obtain two values where one is given a positive value and the other is given a negative value.

Hence, $r = 8$ or $r = -8$

c. $9m^2 - 25 = 0$

$9m^2 = 25$

Divide both sides by 9. This gives:

$$m^2 = \frac{25}{9}$$

$$m = \sqrt{\frac{25}{9}}$$

$$m = \pm \frac{5}{3}$$

Therefore, $m = \frac{5}{3}$ or $m = -\frac{5}{3}$

d. $(3x + 2)^2 = 16$

Let us take the square root of both sides. This gives:

$$\sqrt{(3x + 2)^2} = \sqrt{16}$$

$$(3x + 2) = \pm 4$$

$$3x = \pm 4 - 2$$

$$x = \frac{+4 - 2}{3} \quad \text{or} \quad x = \frac{-4 - 2}{3}$$

$$x = \frac{2}{3} \quad \text{or} \quad x = \frac{-6}{3}$$

$$x = \frac{2}{3} \quad \text{or} \quad x = -2$$

2. Solve the equation $2x^2 - 5x - 3 = 0$

Solution

$2x^2 - 5x - 3 = 0$

Multiply the first and last term on the left hand side. This gives:

$2x^2(-3) = -6x^2$

Now look for two numbers in x such that their product is $-6x^2$ and their sum is $-5x$ (i.e. the middle term). For us to obtain the two factors let us recall our rules as shown below:

S/N	Sign of product	Sign of middle term	Sign of factors
1	+	+	+ and +
2	+	-	- and -
3	-	-	+ and - with the larger value having -
4	-	+	+ and - with the larger value having +

Based on the rules, since the product of the terms is negative (i.e. $-6x^2$) and their sum is also negative (i.e. $-5x$), as in rule 3 above, then the two factors of 6 that we need must have different signs, and with the higher of the factors having a negative sign. The pairs of factors of 6 whose product is 6 are (1 and 6) and (2 and 3). The larger factor should have a negative sign. If we assign a negative sign to the larger factors, then the pair can be written as: (+1 and −6), (2 and −3). Out of these two pairs, the one whose sum is −5 is (+1 and −6). Hence, we now include x on them so that their product will give $-6x^2$ and their sum will give $-5x$. Hence the numbers are $+x$ and $-6x$. Note that $+1x$ is also $+x$.

We now substitute these two numbers for $-5x$ in the original expression. This gives:

$2x^2 - 6x + x - 3 = 0$

We now factorize by grouping. This gives:

$2x(x - 3) + 1(x - 3) = 0$

With $(x - 3)$ as a common factor we proceed as follows:

$(x - 3)(2x + 1) = 0$

Now that we have factorized the equation, we equate each of the bracket to zero and solve for x in each of the linear equation formed. This gives:

Either, $x - 3 = 0$

$\qquad x = 3$

Or, $\quad 2x + 1 = 0$

$\qquad 2x = -1$

$\qquad x = -\dfrac{1}{2}$ (After dividing both sides by 2)

Therefore, $x = 3$ or $x = -\dfrac{1}{2}$

3. Solve the quadratic equation $y^2 + y - 12 = 0$

Solution

$y^2 + y - 12 = 0$

Since the coefficient of y^2 in this equation is 1, then we can factorize the equation by putting directly the appropriate factors into bracket.

$y^2 + y - 12 = 0$

$(y \quad)(y \quad) = 0$

Here we look for the factors of 12 whose product is −12 and whose sum is +1 (i.e. +1 from +y which also means +1y). Since the product is negative and the sum is positive, then the two factors must have different sign, and with the positive sign going to the larger factor. The factors of 12 whose product will give -12 and whose sum is +1 are +4 and −3. Note that the larger value has the positive sign (i.e. from +1). We now fill these two values into the bracket above. This gives:

$$y^2 + y - 12 = 0$$
$$(y + 4)(y - 3) = 0$$

Equating each bracket to zero gives:

$$y + 4 = 0$$
$$y = -4$$
$$y - 3 = 0$$
$$y = 3$$

Hence, $y = 3$ or $y = -4$

4. Solve the quadratic equation $m^2 - 9m + 18 = 0$

Solution

$$m^2 - 9m + 18 = 0$$

Since the coefficient of m^2 in this equation is 1, then we can factorize the equation by putting directly the appropriate factors into bracket.

$$m^2 - 9m + 18 = 0$$
$$(m\quad)(m\quad) = 0$$

Look for two factors of 18 whose product is +18 and whose sum is −9. Since the product is positive and the sum is negative, then the two factors must be negative. The factors of 18 whose product is +18 and whose sum is −9 are −6 and −3. We now fill these two values into the bracket above. This gives:

$$m^2 - 9m + 18 = 0$$
$$(m - 6)(m - 3) = 0$$

Equating each bracket to zero gives:

$$m - 6 = 0$$
$$m = 6$$
$$m - 3 = 0$$

$$m = 3$$

Hence, $m = 6$ or $m = 3$

5. Solve the equation $6n^2 = 5 - 13n$

Solution

$$6n^2 = 5 - 13n$$

Rearranging the terms in order to equate the equation to zero (i.e. make the right hand side zero) gives:

$$6n^2 + 13n - 5 = 0$$

Multiply the first and last term on the left hand side. This gives:

$$6n^2(-5) = -30n^2$$

Now look for two numbers in n such that their product is $-30n^2$ and their sum is $+13n$ (i.e. the middle term). Based on the rules given above for obtaining factors, and since the product of the terms is negative (i.e. $-30n^2$) and their sum is positive (i.e. $+13n$), then the two factors of 30 that we need must have different signs, and with the higher of the factors having a positive sign (i.e. from $+13$). The pairs of factors of 30 are: (1 and 30), (2 and 15), (3 and 10), (6 and 5). The larger factor should have a positive sign. If we assign a positive sign to the larger number of each pair of the factors, then the pairs can be written as: (-1 and 30), (-2 and 15), (-3 and 10), (-5 and 6)

Out of these pairs, the one whose sum is $+13$ is (-2 and 15).

Hence, we now include 'n' on them so that their product will give $-30n^2$ and their sum will give $+13n$. Hence the numbers are $-2n$ and $+15n$.

We now substitute these two numbers for $+13n$ in the original expression. This gives:

$$6n^2 + 15n - 2n - 5 = 0$$

We now factorize by grouping. This gives:

$$(6n^2 + 15n) - (2n - 5) = 0$$
$$3n(2n + 5) - 1(2n + 5) = 0$$

With $(2x + 5)$ as a common factor, we proceed as follows:

$$(2n + 5)(3n - 1) = 0$$

Now that we have factorized the equation, we equate each of the bracket to zero and solve for n in each of the linear equation formed. This gives:

Either, $2n + 5 = 0$

$$2n = -5$$
$$n = -\frac{5}{2}$$

Or, $3n - 1 = 0$

$$3n = 1$$
$$n = \frac{1}{3} \quad \text{(After dividing both sides by 3)}$$

Therefore, $n = -\dfrac{5}{2}$ or $n = \dfrac{1}{3}$

6. Solve the quadratic equation: $8 - 6x - 5x^2 = 0$

Solution

$$8 - 6x - 5x^2 = 0$$

Multiply the first and last term on the left hand side. This gives:

$$8(-5x^2) = -40x^2$$

Now look for two numbers in x such that their product is $-40x^2$ and their sum is $-6x$ (i.e. the middle term). Based on the rules given above for obtaining factors, and since the product of the terms is negative (i.e. $-40x^2$) and their sum is negative (i.e. $-6x$), then the two factors of 40 that we need must have different signs. The higher of the factors will have a negative sign (i.e. from -6, the middle number). The pairs of factors whose product is 40 are (1 and 40), (2 and 20), (4 and 10), (5 and 8). The larger factor should have a negative sign. If we assign a negative sign to the larger number of each pair of the factors, then the pairs can be written as: (1 and -40), (2 and -20), (4 and -10), (5 and -8)

Out of these pairs, the one whose sum is −6 is (4 and −10), since −10 + 4 = −6.

Hence, we now include 'x' on them so that their product will give −40x^2 and their sum will give −6x. Hence the numbers are +4x and −10x.

We now substitute these two numbers for −6x in the original equation. This gives:

$$8 + 4x - 10x - 5x^2 = 0$$

We now factorize by grouping. This gives:

$$(8 + 4x) - (10x - 5x^2) = 0$$

$$4(2 + x) - 5x(2 + x) = 0$$

With (2 + x) as a common factor, we proceed as follows:

$$(2 + x)(4 - 5x) = 0$$

Now that we have factorized the equation, we equate each of the brackets to zero and solve for x in each of the linear equation formed. This gives:

Either, $2 + x = 0$

$$x = -2$$

Or, $\quad 4 - 5x = 0$

$$4 = 5x$$

$$x = \frac{4}{5} \quad \text{(After dividing both sides by 5)}$$

Therefore, $x = -2$ or $x = \frac{4}{5}$

Construction of Quadratic Equation from Given Roots

The solutions of a quadratic equation are called roots. Since the roots are obtained from the factors, it follows that the factors can also be obtained from the roots. The multiplication of the factors gives the quadratic equation.

Consider the quadratic equation: $ax^2 + bx + c = 0$

If we divide each term by 'a', then we will obtain the quadratic equation in terms of its sum of roots and product of roots as follows:

$$x^2 + \frac{b}{a}x + \frac{c}{a} = 0$$

With this simplification, the sum of roots of the quadratic equation is:

$$-\frac{b}{a}$$

while the product of roots of the quadratic equation is:

$$\frac{c}{a}$$

Therefore, a quadratic equation can be generally expressed as follows:

$$x^2 - (\text{sum of roots})x + (\text{product of roots}) = 0$$

If the roots of a quadratic equation are given as α and β, then the quadratic equation can be represented as follows:

$$x^2 - (\alpha + \beta)x + (\alpha\beta) = 0$$

Let us now apply these principles with the following examples.

Examples

1. Find the quadratic equation whose roots are 4 and −7.

Solution

Method 1

Since the roots/solutions of the quadratic equation are 4 and –7, it follows that:

$x = 4$ or $x = -7$

Equating each solution to zero gives us the factors as follows:

$x - 4 = 0$ or $x + 7 = 0$ (Note that from $x = 4$, we have, $x - 4 = 0$, and from $x = -7$, $x + 7 = 0$)

The multiplication of these two factors gives us the equation as follows:

$(x - 4)(x + 7) = 0$

Expanding the brackets gives:

$x^2 + 7x - 4x - 28 = 0$

$x^2 + 3x - 28 = 0$ (Note that $7x - 4x = 3x$)

The quadratic equation is $x^2 + 3x - 28 = 0$

Method 2

Recall that a quadratic equation is given in term of its roots as follows:

$x^2 - (\text{sum of roots})x + (\text{product of roots}) = 0$

Hence, we can now use 4 and –7 to obtain the sum and product of roots as follows:

$x^2 - (\text{sum of roots})x + (\text{product of roots}) = 0$

$x^2 - (-7 + 4)x + [-7(4)] = 0$

$x^2 - (-3)x + (-28) = 0$

$x^2 + 3x - 28 = 0$

Hence the quadratic equation is $x^2 + 3x - 28 = 0$.

2. Find the quadratic equation whose roots are $\dfrac{2}{3}$ and 5.

Solution
Method 1

Since the roots/solutions of the quadratic equation are $\dfrac{2}{3}$ and 5, it follows that:

$x = \dfrac{2}{3}$ or $x = 5$

Equating each solution to zero gives us the factors as follows:

$x - \dfrac{2}{3} = 0$ or $x - 5 = 0$

In order to clear out the fraction in $x - \dfrac{2}{3} = 0$, we simply multiply each term by 3. This gives:

$3(x) - 3(\dfrac{2}{3}) = 3(0)$

$3x - 2 = 0$

Hence the two factors are:

$3x - 2 = 0$ and $x - 5 = 0$

The multiplication of these two factors gives us the equation as follows:

$(3x - 2)(x - 5) = 0$

Expanding the brackets gives:

$3x^2 - 15x - 2x + 10 = 0$

$3x^2 - 17x + 10 = 0$

Hence, the quadratic equation is $3x^2 - 17x + 10 = 0$

Method 2

Recall that a quadratic equation is given in term of its roots as follows:

$x^2 - $ (sum of roots)$x + $ (product of roots) $= 0$

Hence, we can now use $\frac{2}{3}$ and 5 to obtain the sum and product of roots as follows:

$x^2 - $ (sum of roots)$x + $ (product of roots) $= 0$

$x^2 - (\frac{2}{3} + 5)x + [\frac{2}{3}(5)] = 0$

$x^2 - (\frac{2 + 15}{3})x + (\frac{10}{3}) = 0$

$x^2 - (\frac{17}{3})x + (\frac{10}{3}) = 0$

In order to clear out the fractions, we multiply each term by 3 (i.e. the LCM of the denominators). This gives:

$3(x^2) - 3(\frac{17}{3})x + 3(\frac{10}{3}) = 0$

The 3 will cancel out each other in the fractions to give:

$3x^2 - 17x + 10 = 0$

Hence the quadratic equation is $3x^2 - 17x + 10 = 0$.

3. Find the quadratic equation whose roots are −2 and $-\frac{3}{4}$

Solution

Method 1

Since the roots of the quadratic equation are −2 and $-\frac{3}{4}$, it follows that:

$x = -2$ or $x = -\frac{3}{4}$

Equating each solution to zero gives us the factors as follows:

$x + 2 = 0$ or $x + \frac{3}{4} = 0$

In order to clear out the fraction in $x + \frac{3}{4} = 0$, we simply multiply each term by 4. This gives:

$4(x) + 4(\frac{3}{4}) = 4(0)$

$4x + 3 = 0$

Hence the two factors are:

$x + 2 = 0$ and $4x + 3 = 0$

The multiplication of these two factors gives us the equation as follows:

$(x + 2)(4x + 3) = 0$

Expanding the brackets gives:

$4x^2 + 3x + 8x + 6 = 0$

$4x^2 + 11x + 6 = 0$

Hence, the quadratic equation is $4x^2 + 11x + 6 = 0$

Method 2

Recall that a quadratic equation is given in term of its roots as follows:

$x^2 -$ (sum of roots)$x +$ (product of roots) $= 0$

Hence, we can now use -2 and $-\dfrac{3}{4}$ to obtain the sum and product of roots as follows:

$x^2 -$ (sum of roots)$x +$ (product of roots) $= 0$

$x^2 - (-2 - \dfrac{3}{4})x + [-2(-\dfrac{3}{4})] = 0$

$x^2 - (\dfrac{-8-3}{4})x + [-2(-\dfrac{3}{4}) = 0$

$x^2 - (-\dfrac{11}{4})x + (\dfrac{6}{4}) = 0$

$x^2 - (-\dfrac{11}{4})x + (\dfrac{3}{2}) = 0$ (Note that $\dfrac{6}{4} = \dfrac{3}{2}$ in its lowest term)

In order to clear out the fractions, we multiply each term by 4 (i.e. the LCM of the denominators). This gives:

$4(x^2) - 4(-\dfrac{11}{4})x + 4(\dfrac{3}{2}) = 0$

This will now give:

$4x^2 + 11x + 6 = 0$ (Note that $-4(-\dfrac{11}{4}) = +11$)

Hence the quadratic equation is $4x^2 + 11x + 6 = 0$.

4. A quadratic equation is given by: $x^2 - 7x - 10 = 0$. Find:

a. the sum of the roots of the equation

b. the product of the roots of the equation

Solution

a. $x^2 - 7x - 10 = 0$

Comparing this equation with $ax^2 + bx + c = 0$, shows that:

a = 1, b = -7, and c = -10   (Note that $x^2 = 1x^2$, hence a = 1)

Therefore, the sum of roots is given by:

$-(\dfrac{b}{a}) = -(-\dfrac{7}{1})$

$\qquad = 7$

Hence, the sum of the roots is 7

b. Also, the product of roots is given by:

$\dfrac{c}{a} = \dfrac{-10}{1} = -10$

93

Hence the product of the roots is –10

5. A quadratic equation is given by: $2x^2 + 13x - 26 = 0$. Find:
a. the sum of the roots of the equation
b. the product of the roots of the equation

Solution
a. $2x^2 + 13x - 26 = 0$
Comparing this equation to $ax^2 + bx + c = 0$, shows that:
   a = 2, b = 13, c = –26

Recall that the sum of roots of a quadratic equation is given by $-\dfrac{b}{a}$

Hence, sum of roots of the equation above is:
$$-\frac{b}{a} = -\frac{13}{2}$$
Therefore sum of roots of the equation is $-\dfrac{13}{2}$

b.   Recall that the product of the roots of a quadratic equation is given by: $\dfrac{c}{a}$

Hence product of roots of the equation above is:
$$= -\frac{26}{2}$$
$$= -13$$
Therefore, product of roots of the equation is –13.

6.   If 3 is a root of the equation $2x^2 + Kx - 21 = 0$, find the value of K. Hence find the other root.

Solution
      $2x^2 + Kx - 21 = 0$
In order to find K, substitute 3 for x in the equation. This gives:
      $2x^2 + Kx - 21 = 0$
      $2(3)^2 + K(3) - 21 = 0$
      $2(9) + 3K - 21 = 0$
      $18 + 3K - 21 = 0$
      $3K = 21 - 18$
      $3K = 3$
      $K = \dfrac{3}{3}$
      $K = 1$
Hence the value of K is 1.
The equation can now be written as $2x^2 + x - 21 = 0$
From this equation, a = 2, b = 1 and c = –21
Hence, product of roots = $\dfrac{c}{a}$

$$= -\frac{21}{2}$$

Let the second root be m.

Since one of the roots is 3, then product of roots which is $-\frac{21}{2}$ is given by:

$3(m) = -\frac{21}{2}$ (Note that the two roots are 3 and m)

$3m = -\frac{21}{2}$

$m = \dfrac{-\frac{21}{2}}{3}$

$= -\frac{21}{2} \times \frac{1}{3}$

$m = -\frac{21}{6}$

$m = -\frac{7}{2}$ (In its lowest term)

Therefore the other root is $-\frac{7}{2}$

Note that we can also factorize the equation $2x^2 + x - 21 = 0$, to obtain the two roots, and hence the other root.

7. If 1 is a root of the equation $5x^2 - kx - 3 = 0$, find the value of K. Hence find the other root.

<u>Solution</u>

$5x^2 - kx - 3 = 0$

In order to find K, substitute 1 for x in the equation. This gives:

$5x^2 - kx - 3 = 0$

$5(1)^2 - k(1) - 3 = 0$

$5(1) - k - 3 = 0$

$5 - k - 3 = 0$

$5 - 3 = k$

$k = 2$

Hence the value of k is 2.

The equation can now be written as $5x^2 - 2x - 3 = 0$

From this equation, a = 5, b = −2 and c = −3

Hence, product of roots $= \dfrac{c}{a}$

$= -\dfrac{3}{5}$

Let the second root be y.

95

Since one of the roots is 1, then product of roots which is $-\dfrac{3}{5}$ is given by:

$$1(y) = -\dfrac{3}{5} \quad \text{(Note that the two roots are 1 and y, hence their product is 1y)}$$

$$y = -\dfrac{3}{5}$$

Therefore the other root is $-\dfrac{3}{5}$

8. 5 and –7 are the roots of the quadratic equation $y^2 - ky + M = 0$. Find the values of k and M.

<u>Solution</u>

$$y^2 - ky + M = 0$$

From this equation, a = 1, b = –k, c = M

Hence sum of roots $= -\dfrac{b}{a}$

$$= -(-\dfrac{k}{1}) \quad \text{(Since b = –k)}$$

$$= k$$

Since the two roots are 5 and –7, and their sum is k, then it follows that:

$$-7 + 5 = k$$
$$-2 = k$$
$$k = -2$$

Also, product of roots $= \dfrac{c}{a}$

$$= \dfrac{M}{1} \quad \text{(Since c = M)}$$

$$= M$$

Since the two roots are 5 and –7, and their product is M, then it follows that:

$$5(-7) = M$$
$$-35 = M$$
$$M = -35$$

Therefore, k = –2 and M = –35

9. If –3 and 8 are the roots of the quadratic equation $2x^2 + Dx - E = 0$. Find the values of the constants D and E.

<u>Solution</u>

$$2x^2 + Dx - E = 0$$

From this equation, a = 2, b = D, c = –E

Hence sum of roots $= -\dfrac{b}{a}$

$$= -\frac{D}{2} \quad \text{(Since } b = D\text{)}$$

Since the two roots are –3 and 8, and their sum is $-\frac{D}{2}$, then it follows that:

$$-3 + 8 = -\frac{D}{2}$$

$$5 = -\frac{D}{2}$$

$$-D = 2 \times 5$$
$$-D = 10$$
$$D = -10 \quad \text{(After dividing both sides by –1)}$$

Also, product of roots $= \frac{c}{a}$

$$= -\frac{E}{2} \quad \text{(Since } c = -E \text{ and } a = 2\text{)}$$

Since the two roots are –3 and 8, and their product is $-\frac{E}{2}$, then it follows that:

$$-3(8) = -\frac{E}{2}$$

$$-24 = -\frac{E}{2}$$

$$-E = -24 \times 2$$
$$-E = -48$$
$$E = 48 \quad \text{(After dividing both sides by –1)}$$

Therefore, D = –10 and E = 48

10. If one root of the equation $y^2 - ky + 18 = 0$ is twice the other, find the two possible values of k. Hence, what are the two possible equations?

Solution
Let one root of the equation be m. Therefore, the other root will be 2m since one root is twice the other.
Hence product of roots = m(2m)
$$= 2m^2$$
The equation is $y^2 - ky + 18 = 0$
From this equation, a = 1, b = –k and c = 18
Product of roots $= \frac{c}{a}$

$$= \frac{18}{1} = 18$$

Hence, $2m^2 = 18$ (Note that product of roots is also $2m^2$)
$$m^2 = \frac{18}{2}$$
$$m^2 = 9$$

$$m = \sqrt{9}$$
$$m = \pm 3$$
$$m = 3 \text{ or } -3$$

When m = 3, the other root, 2m, is 2(3) = 6   (Since one root is twice the other)

When m = −3, then the other root, 2m, is 2(−3) = −6

Hence, the roots are 3 and 6, or −3 and −6.

Recall that the equation is $y^2 - ky + 18 = 0$, and a = 1, b = −k, and c = 18.

Hence, sum of roots $= -\dfrac{b}{a}$

$$= -(-\dfrac{k}{1}) \quad \text{(Since b = −k and a = 1)}$$

$$= k$$

Therefore, when the roots are 3 and 6, then sum of roots = 3 + 6 = 9. But sum of roots is also equal to k.

Hence, k = 9.

When the roots are −3 and −6, then sum of roots = −3 − 6 = −9

Hence, k = −9

Therefore, k = 9 or −9

When k = 9, the equation is: $y^2 - ky + 18 = 0$, which gives:

$$y^2 - 9y + 18 = 0$$

When k = −9, the equation is:

$$y^2 - (-9)y + 18 = 0, \text{ which gives:}$$
$$y^2 + 9y + 18 = 0$$

Completing the Square of Quadratic expressions

Another method of solving quadratic equation is by completing the square. In order to make a quadratic expression a perfect square, simply add the square of half the coefficient of x of the quadratic expression.

Hence, completing the square of a quadratic expression means to make the quadratic expression a perfect square.

Examples

1. What must be added to $x^2 + 10x$ to make it a perfect square?

<u>Solution</u>
$$x^2 + 10x$$

Half the coefficient of x is given by: $\dfrac{10}{2}$ = 5.

Square this value obtained. This gives:

$$5^2 = 25.$$

Hence, 25 must be added to the quadratic expression to make it a perfect square.

If we add 25 to the quadratic expression it gives:

$$x^2 + 10x + 25$$

This means that $x^2 + 10x + 25$ is a perfect square. Its square root is obtained by taking the square root of the first and last terms and adding them together. The square root of x^2 is x

while the square root of 25 is ± 5, but we take $+5$ (take the sign of the coefficient of x in the expression). Therefore the square root of $x^2 + 10x + 25$ is $(x + 5)$. If we square the square root, it gives the perfect square.

Hence, $x^2 + 10x + 25 = (x + 5)^2$

2. What must be added to $x^2 - 18x$ to make it a perfect square?

Solution

$\quad x^2 - 18x$

The coefficient of x is -18. Divide this coefficient by 2. This gives:

$\quad -\dfrac{18}{2} = -9$

Square this value obtained. This gives:

$\quad -9^2 = 81$.

Hence, 81 must be added to the quadratic expression to make it a perfect square.

If we add 81 to the quadratic expression it gives:

$\quad x^2 - 18x + 81$

This means that $x^2 - 18x + 81$ is a perfect square. Its square root is obtained by taking the square roots of the first and last terms and adding them together. The square root of x^2 is x while the square root of 81 is ± 9, but we take -9 (take the sign of the coefficient of x). Therefore the square root of $x^2 - 18x + 81$ is $(x - 9)$. If we square the square root, it gives the perfect square.

Hence, $x^2 - 18x + 81 = (x - 9)^2$

3. Make $y^2 - 5y$ a perfect square.

Solution

$\quad y^2 - 5y$

The coefficient of y is -5. Divide this coefficient by 2. This gives:

$\quad -\dfrac{5}{2}$

Square this value obtained. This gives:

$\quad \left(-\dfrac{5}{2}\right)^2 = \dfrac{25}{4}$

Hence, $\dfrac{25}{4}$ must be added to the quadratic expression to make it a perfect square.

If we add $\dfrac{25}{4}$ to the quadratic expression it gives:

$\quad y^2 - 5y + \dfrac{25}{4}$

This means that $y^2 - 5y + \dfrac{25}{4}$ is a perfect square.

Its square root is obtained by taking the square roots of the first and last terms and adding them together. The square root of y^2 is y while the square root of $\dfrac{25}{4}$ is $\pm\dfrac{5}{2}$, but we take $-\dfrac{5}{2}$

since the coefficient of y in the expression is negative.

Therefore the square root of $y^2 - 5y + \frac{25}{4}$ is $(y - \frac{5}{2})$. If we square the root, it gives the perfect square.

Hence, $y^2 - 5y + \frac{25}{4} = (x - \frac{5}{2})^2$

4. Make $y^2 + \frac{9}{2}y$ a perfect square.

Solution

$$y^2 + \frac{9}{2}y$$

The coefficient of y is $\frac{9}{2}$. Divide this coefficient by 2. This gives:

$$\frac{9}{2} \div 2$$

$$= \frac{9}{2} \times \frac{1}{2}$$

$$= \frac{9}{4}$$

Square this value obtained. This gives:

$$(\frac{9}{4})^2 = \frac{81}{16}$$

Hence, $\frac{81}{16}$ must be added to the quadratic expression to make it a perfect square.

If we add $\frac{81}{16}$ to the quadratic expression it gives:

$$y^2 + \frac{9}{2}y + \frac{81}{16}$$

This means that $y^2 + \frac{9}{2}y + \frac{81}{16}$ is a perfect square.

Therefore the square root of $y^2 + \frac{9}{2}y + \frac{81}{16}$ is $(y + \frac{9}{4})$.

Hence, $y^2 + \frac{9}{2}y + \frac{81}{16} = (y + \frac{9}{4})^2$

5. If $x^2 - 6x + (-3)^2$ is a perfect square, find its square root.

Solution

$$x^2 - 6x + (-3)^2$$

Since this is a perfect square, we find its square root by taking each of the squared term and enclosing them in a bracket. The square root of x^2 is x, while the square root of $(-3)^2$ is -3.
Hence, we put these two square root terms in a bracket. This gives: $(x - 3)$
Therefore the square root of $x^2 - 6x + (-3)^2$ is $(x - 3)$
This means that: $x^2 - 6x + (-3)^2 = (x - 3)^2$

6. If $x^2 - \frac{7}{2}x + (-\frac{7}{4})^2$ is a perfect square, find its square root.

Solution
$$x^2 - \frac{7}{2}x + (-\frac{7}{4})^2$$
Since this is a perfect square, we find its square root by taking each of the squared term and enclosing them in a bracket. The square root of x^2 is x, while the square root of $(-\frac{7}{4})^2$ is $-\frac{7}{4}$. Hence, we put these

two square root terms in a bracket. This gives: $(x - \frac{7}{4})$

Therefore the square root of $x^2 - \frac{7}{2}x + (-\frac{7}{4})^2$ is $(x - \frac{7}{4})$

This means that: $x^2 - \frac{7}{2}x + (-\frac{7}{4})^2 = (x - \frac{7}{4})^2$

7. If $x^2 + 1.5x + (0.75)^2$ is a perfect square, find its square root.

Solution
$$x^2 + 1.5x + (0.75)^2$$
The square root of x^2 is x, while the square root of $(0.75)^2$ is 0.75. Hence, we put these two square root terms in a bracket. This gives: $(x + 0.75)$
Therefore the square root of $x^2 + 1.5x + (0.75)^2$ is $(x + 0.75)$

Solving Quadratic Equation by Completing the Square

Examples
1. Use the method of completing the square to solve the quadratic equation $y^2 + 7y - 30 = 0$

Solution
$$y^2 + 7y - 30 = 0$$
Take the constant term (i.e. -30) to the right hand side of the equation. This gives:
$$y^2 + 7y = 30$$
The next step is to make the left hand side expression a perfect square. In doing this, whatever is added to the left hand side should also be added to the right hand side. Let us do this as follows.
The coefficient of y is 7. Divide this coefficient by 2. This gives:
$$\frac{7}{2}$$
Square the value obtained. This gives:
$$(\frac{7}{2})^2 = \frac{49}{4}$$
Hence, $\frac{49}{4}$ must be added to the left hand expression to make it a perfect square. However, we

are not going to write it as $\frac{49}{4}$. We will leave it as $(\frac{7}{2})^2$. We will now add $(\frac{7}{2})^2$ to the left hand

side, and also add it to the right hand side. This gives:

101

$$y^2 + 7y + (\tfrac{7}{2})^2 = 30 + (\tfrac{7}{2})^2$$

While keeping the left hand side as it is, so that we can easily express it as a square of its square root, we go ahead and simplify the right hand side. This gives:

$$y^2 + 7y + (\tfrac{7}{2})^2 = \frac{30}{1} + \frac{49}{4}$$

$$y^2 + 7y + (\tfrac{7}{2})^2 = \frac{120 + 49}{4} \qquad \text{(Note that 4 x 30 = 120)}$$

$$y^2 + 7y + (\tfrac{7}{2})^2 = \frac{169}{4}$$

Take the square root of the left hand side and square it. This is done by taking the terms in square (i.e. y and $\tfrac{7}{2}$), enclosing them in a bracket and squaring the bracket. This gives:

$$(y + \tfrac{7}{2})^2 = \frac{169}{4}$$

Take the square root of both sides in order to remove the square on the left hand side. This gives:

$$\sqrt{(y + \tfrac{7}{2})^2} = \sqrt{\frac{169}{4}}$$

$$(y + \tfrac{7}{2}) = \pm \frac{13}{2}$$

(Note that $\sqrt{x^2} = x$, hence the removal of square from $(y + \tfrac{7}{2})^2$ to get $(y + \tfrac{7}{2})$. Also, remember to add $\pm$ to the square root of a number).

Let us continue with our solution as follows:

$$(y + \tfrac{7}{2}) = \pm \frac{13}{2}$$

Or, $\quad y + \dfrac{7}{2} = \pm \dfrac{13}{2}$

Take the constant term on the left hand side to the right. This gives:

$$y = -\frac{7}{2} \pm \frac{13}{2}$$

We now use $\pm$ to split the values above into two solutions. When we take only the positive sign of $\pm$, our value above gives:

$$y = -\frac{7}{2} + \frac{13}{2}$$

$$y = \frac{-7 + 13}{2}$$

$$y = \frac{6}{2}$$

$$y = 3$$

When we take only the negative sign of $\pm$, our value above gives:

$$y = -\frac{7}{2} - \frac{13}{2}$$

$$y = \frac{-7 - 13}{2}$$

$$y = -\frac{20}{2}$$

$$y = -10$$

$\therefore \quad y = 3 \text{ or } y = -10$

2. Use the method of completing the square to solve the quadratic equation $x^2 - 11x + 28 = 0$

<u>Solution</u>

$x^2 - 11x + 28 = 0$

Take the constant term (i.e. 28) to the right hand side of the equation. This gives:

$x^2 - 11x = -28$

The coefficient of x is -11. Divide this coefficient by 2 to obtain:

$-\dfrac{11}{2}$

Square $-\dfrac{11}{2}$. This gives:

$\left(-\dfrac{11}{2}\right)^2$

Add $\left(-\dfrac{11}{2}\right)^2$ to both sides of the equation. This gives:

$x^2 - 11x + \left(-\dfrac{11}{2}\right)^2 = -28 + \left(-\dfrac{11}{2}\right)^2$

While keeping the left hand side, simplify the right hand side. This gives:

$x^2 - 11x + \left(-\dfrac{11}{2}\right)^2 = -28 + \dfrac{121}{4}$

$x^2 - 11x + \left(-\dfrac{11}{2}\right)^2 = \dfrac{-112 + 121}{4}$ \qquad (Note that $4 \times -28 = -112$)

$x^2 - 11x + \left(-\dfrac{11}{2}\right)^2 = \dfrac{9}{4}$

Take the square root of the left hand side and square it. This is done by taking the terms in square (i.e. x and $-\dfrac{11}{2}$), enclosing them in a bracket and squaring the bracket. This gives:

$\left(x - \dfrac{11}{2}\right)^2 = \dfrac{9}{4}$

Take the square root of both sides. This gives:

$\sqrt{\left(x - \dfrac{11}{2}\right)^2} = \sqrt{\dfrac{9}{4}}$

$\left(x - \dfrac{11}{2}\right) = \pm\dfrac{3}{2}$

Or, $\quad x - \dfrac{11}{2} = \pm\dfrac{3}{2}$

$$x = \frac{11}{2} \pm \frac{3}{2}$$

When we take only the positive sign of $\pm$, our value above gives:

$$x = \frac{11}{2} + \frac{3}{2}$$

$$x = \frac{11 + 3}{2}$$

$$x = \frac{14}{2}$$

$$x = 7$$

When we take only the negative sign of $\pm$, our value above gives:

$$x = \frac{11}{2} - \frac{3}{2}$$

$$x = \frac{11 - 3}{2}$$

$$x = \frac{8}{2}$$

$$x = 4$$

$\therefore$ $x = 7$ or $x = 4$

3. Solve the quadratic equation $x^2 - 6x - 16 = 0$, by the method of completing the square.

Solution

$$x^2 - 6x - 16 = 0$$

Take the constant term (i.e. -16) to the right hand side of the equation. This gives:

$$x^2 - 6x = 16$$

The coefficient of x is -6. Divide this coefficient by 2 to obtain:

$$-\frac{6}{2} = -3$$

Square the value obtained. This gives:

$$(-3)^2$$

Add $(-3)^2$ to both sides of the equation. This gives:

$$x^2 - 6x + (-3)^2 = 16 + (-3)^2$$
$$x^2 - 6x + (-3)^2 = 16 + 9$$
$$x^2 - 6x + (-3)^2 = 25$$

Take the square root of the left hand side and square it. This gives:

$$(x - 3)^2 = 25$$

Take the square root of both sides. This gives:

$$\sqrt{(x - 3)^2} = \sqrt{25}$$
$$x - 3 = \pm 5$$
$$x = 3 \pm 5$$
$$x = 3 + 5 \quad \text{or} \quad x = 3 - 5$$

$\therefore$ $x = 8$ or $x = -2$

4. Solve the quadratic equation $2x^2 - 5x + 3 = 0$, by the method of completing the square.

Solution

$$2x^2 - 5x + 3 = 0$$

In this case, the coefficient of x^2 is 2. It is not 1 like our first three examples above. Therefore our first step will be to make the coefficient of x^2 to be 1. This is done by dividing each term in the equation by the coefficient of x^2 which is 2. This gives:

$$\frac{2x^2}{2} - \frac{5x}{2} + \frac{3}{2} = \frac{0}{2}$$

$$x^2 - \frac{5}{2}x + \frac{3}{2} = 0$$

Now that we have made the coefficient of x^2 to be 1, we can proceed as usual.

Take the constant term (i.e. $\frac{3}{2}$) to the right hand side of the equation. This gives:

$$x^2 - \frac{5}{2}x = -\frac{3}{2}$$

The coefficient of x is $-\frac{5}{2}$. Divide this coefficient by 2 to obtain:

$$-\frac{5}{2} \div 2 = -\frac{5}{2} \times \frac{1}{2}$$

$$= -\frac{5}{4}$$

Square the value obtained. This gives:

$$(-\frac{5}{4})^2$$

Add $(-\frac{5}{4})^2$ to both sides of the equation. This gives:

$$x^2 - \frac{5}{2}x + (-\frac{5}{4})^2 = -\frac{3}{2} + (-\frac{5}{4})^2$$

$$x^2 - \frac{5}{2}x + (-\frac{5}{4})^2 = -\frac{3}{2} + \frac{25}{16}$$

$$x^2 - \frac{5}{2}x + (-\frac{5}{4})^2 = \frac{-24 + 25}{16}$$

$$x^2 - \frac{5}{4}x + (-\frac{5}{4})^2 = \frac{1}{16}$$

Take the square root of the left hand side and square it. This gives:

$$(x - \frac{5}{4})^2 = \frac{1}{16}$$

Take the square root of both sides. This gives:

$$\sqrt{(x - \frac{5}{4})^2} = \sqrt{\frac{1}{16}}$$

$$x - \frac{5}{4} = \pm\frac{1}{4}$$

$$x = \frac{5}{4} \pm \frac{1}{4}$$

$$x = \frac{5}{4} + \frac{1}{4} \quad \text{or} \quad x = \frac{5}{4} - \frac{1}{4}$$

$$x = \frac{5+1}{4} \quad \text{or} \quad x = \frac{5-1}{4}$$

$$x = \frac{6}{4} \quad \text{or} \quad x = \frac{4}{4}$$

$$x = \frac{3}{2} \quad \text{or} \quad x = 1$$

5   Solve the quadratic equation $3y^2 + 9y + 5 = 0$, by the method of completing the square.

Solution

$$3y^2 + 9y + 5 = 0$$

In this case, the coefficient of y^2 is 3. Therefore our first step will be to make the coefficient of y^2 to be 1. This is done by dividing each term in the equation by 3. This gives:

$$\frac{3y^2}{3} + \frac{9y}{3} + \frac{5}{3} = \frac{0}{3}$$

$$y^2 + 3y + \frac{5}{3} = 0$$

Take the constant term to the right hand side of the equation. This gives:

$$y^2 + 3y = -\frac{5}{3}$$

The coefficient of y is 3. Divide this coefficient by 2 to obtain:

$$\frac{3}{2}$$

Square the value obtained. This gives:

$$\left(\frac{3}{2}\right)^2$$

Add $\left(\frac{3}{2}\right)^2$ to both sides of the equation. This gives:

$$y^2 + 3y + \left(\frac{3}{2}\right)^2 = -\frac{5}{3} + \left(\frac{3}{2}\right)^2$$

$$y^2 + 3y + \left(\frac{3}{2}\right)^2 = -\frac{5}{3} + \frac{9}{4}$$

$$y^2 + 3y + \left(\frac{3}{2}\right)^2 = \frac{-20 + 27}{12}$$

$$y^2 + 3y + \left(\frac{3}{2}\right)^2 = \frac{7}{12}$$

Take the square root of the left hand side and square it. This gives:

$$\left(y + \frac{3}{2}\right)^2 = \frac{7}{12}$$

Take the square root of both sides. This gives:

$$\sqrt{\left(y + \frac{3}{2}\right)^2} = \sqrt{0.5833}$$

$$y + \frac{3}{2} = \pm 0.764$$

$$y + 1.5 = \pm 0.764$$

$$y = -1.5 \pm 0.764$$

$$y = -1.5 + 0.764 \quad \text{or} \quad y = -1.5 - 0.764$$
$$y = -0.736 \quad \text{or} \quad y = -2.264$$

6. Solve the equation $5m^2 + 7m - 3 = 0$

<u>Solution</u>
$$5m^2 + 7m - 3 = 0$$
In this case, the coefficient of m^2 is 5. Therefore our first step will be to make the coefficient of m^2 to be 1. This is done by dividing each term in the equation by 5. This gives:
$$\frac{5m^2}{5} + \frac{7m}{5} - \frac{3}{5} = \frac{0}{5}$$
$$m^2 + \frac{7}{5}m - \frac{3}{5} = 0$$
Take the constant term to the right hand side of the equation. This gives:
$$m^2 + \frac{7}{5}m = \frac{3}{5}$$
The coefficient of m is $\frac{7}{5}$. Divide this coefficient by 2 to obtain:
$$\frac{7}{5} \div 2$$
$$= \frac{7}{5} \times \frac{1}{2}$$
$$= \frac{7}{10}$$
Square the value obtained. This gives:
$$(\frac{7}{10})^2$$
Add $(\frac{7}{10})^2$ to both sides of the equation. This gives:
$$m^2 + \frac{7}{5}m + (\frac{7}{10})^2 = \frac{3}{5} + (\frac{7}{10})^2$$
$$m^2 + \frac{7}{5}m + (\frac{7}{10})^2 = \frac{3}{5} + \frac{49}{100}$$
$$m^2 + \frac{7}{5}m + (\frac{7}{10})^2 = \frac{60+49}{100}$$
$$m^2 + \frac{7}{5}m + (\frac{7}{10})^2 = \frac{109}{100}$$
Take the square root of the left hand side and square it. This gives:
$$(m + \frac{7}{10})^2 = \frac{109}{100}$$
Take the square root of both sides. This gives:
$$\sqrt{(m + \frac{7}{10})^2} = \sqrt{1.09}$$

$$m + \frac{7}{10} = \pm 1.044$$

$$m + 0.7 = \pm 1.044$$

$$m = -0.7 \pm 1.044$$

Separating the sign $\pm$ into + and – gives the two solutions as follows:

$$m = -0.7 + 1.044 \quad \text{or} \quad m = -0.7 - 1.044$$

$$m = 0.344 \quad \text{or} \quad m = -1.744$$

7. Solve the quadratic equation $2x^2 - 8x + 5 = 0$

<u>Solution</u>

$$2x^2 - 8x + 5 = 0$$

The coefficient of x^2 is 2. Therefore our first step will be to make the coefficient of x^2 to be 1. This is done by dividing each term in the equation by 2. This gives:

$$\frac{2x^2}{2} - \frac{8x}{2} + \frac{5}{2} = \frac{0}{2}$$

$$x^2 - 4x + \frac{5}{2} = 0$$

Take the constant term to the right hand side of the equation. This gives:

$$x^2 - 4x = -\frac{5}{2}$$

The coefficient of x is –4. Divide this coefficient by 2 to obtain:

$$-\frac{4}{2} = -2$$

Square the value obtained. This gives:

$$(-2)^2$$

Add $(-2)^2$ to both sides of the equation. This gives:

$$x^2 - 4x + (-2)^2 = -\frac{5}{2} + (-2)^2$$

$$x^2 - 4x + (-2)^2 = -\frac{5}{2} + 4$$

$$x^2 - 4x + (-2)^2 = \frac{-5 + 8}{2}$$

$$x^2 - 4x + (-2)^2 = \frac{3}{2}$$

Take the square root of the left hand side and square it. This gives:

$$(x - 2)^2 = \frac{3}{2}$$

Take the square root of both sides. This gives:

$$\sqrt{(x - 2)^2} = \sqrt{1.5}$$

$$x - 2 = \pm 1.225$$

$$x = 2 \pm 1.225$$

$$x = 2 + 1.225 \quad \text{or} \quad x = 2 - 1.225$$

$$x = 3.225 \quad \text{or} \quad x = 0.775$$

Quadratic Formula

The general expression of a quadratic equation is given by: $ax^2 + bx + c = 0$, where a, b and c are constants. This equation can be solved by completing the square are follows:

$$ax^2 + bx + c = 0$$

The coefficient of x^2 is a. Therefore we make the coefficient of x^2 to be 1. This is done by dividing each term in the equation by a. This gives:

$$\frac{ax^2}{a} + \frac{bx}{a} + \frac{c}{a} = \frac{0}{a}$$

$$x^2 + \frac{b}{a}x + \frac{c}{a} = 0$$

Take the constant term to the right hand side of the equation. This gives:

$$x^2 + \frac{b}{a}x = -\frac{c}{a}$$

The coefficient of x is $\frac{b}{a}$. Divide this coefficient by 2 to obtain:

$$\frac{b}{a} \div 2$$

$$= \frac{b}{a} \times \frac{1}{2}$$

$$= \frac{b}{2a}$$

Square the value obtained. This gives:

$$(\frac{b}{2a})^2$$

Add $(\frac{b}{2a})^2$ to both sides of the equation. This gives:

$$x^2 + \frac{b}{a}x + (\frac{b}{2a})^2 = -\frac{c}{a} + (\frac{b}{2a})^2$$

$$x^2 + \frac{b}{a}x + (\frac{b}{2a})^2 = -\frac{c}{a} + \frac{b^2}{4a^2} \qquad \text{(In } (\frac{b}{2a})^2 \text{, each term in the bracket is squared)}$$

$$x^2 + \frac{b}{a}x + (\frac{b}{2a})^2 = \frac{-4ac + b^2}{4a^2} \qquad \text{(Note that the LCM of } 4a^2 \text{ and a is } 4a^2)$$

$$x^2 + \frac{b}{a}x + (\frac{b}{2a})^2 = \frac{b^2 - 4ac}{4a^2}$$

Take the square root of the left hand side and square it. This gives:

$$(x + \frac{b}{2a})^2 = \frac{b^2 - 4ac}{4a^2}$$

Take the square root of both sides. This gives:

$$\sqrt{(x + \frac{b}{2a})^2} = \frac{b^2 - 4ac}{4a^2}$$

$$x + \frac{b}{2a} = \sqrt{\frac{b^2 - 4ac}{4a^2}} \qquad \text{(Note that } \sqrt{\frac{b}{a}} \text{ can also be expressed as } \frac{\sqrt{b}}{\sqrt{a}})$$

$$x + \frac{b}{2a} = \pm \frac{\sqrt{b^2 - 4ac}}{2a}$$ (Note that the square root of $4a^2$ is $2a$)

$$x = -\frac{b}{2a} \pm \frac{\sqrt{b^2 - 4ac}}{2a}$$

Since their LCM is 2a, we simplify further as follows:

$$x = \frac{-b \pm \sqrt{b^2 - 4ac}}{2a}$$

The above formula is known as quadratic formula. It can be used to solve any quadratic equation.

Examples

1. Solve the following equations by using the quadratic formula:
a. $2x^2 + 3x - 9 = 0$
b. $4x^2 - 8x + 3 = 0$

Solutions
a. $2x^2 + 3x - 9 = 0$
When this equation is compared with $ax^2 + bx + c = 0$, it shows that:
     a = 2, b = 3 and c = −9
We now substitute these values into the quadratic formula as follows:

$$x = \frac{-b \pm \sqrt{b^2 - 4ac}}{2a}$$

$$= \frac{-3 \pm \sqrt{3^2 - [4(2)(-9)]}}{2(2)}$$

$$= \frac{-3 \pm \sqrt{9 - (-72)}}{2 \times 2}$$

$$= \frac{-3 \pm \sqrt{9 + 72}}{4}$$

$$= \frac{-3 \pm \sqrt{81}}{4}$$

$$x = \frac{-3 \pm 9}{4}$$

We now split the sign ± into + and − to obtain the two solutions as follows:

$$x = \frac{-3 + 9}{4} \quad \text{or} \quad x = \frac{-3 - 9}{4}$$

$$x = \frac{6}{4} \quad \text{or} \quad x = \frac{-12}{4}$$

$$x = \frac{3}{2} \quad \text{or} \quad x = -3$$

b. $4x^2 - 8x + 3 = 0$

When this equation is compared with $ax^2 + bx + c = 0$, it shows that:

a = 4, b = –8 and c = 3

We now substitute these values into the quadratic formula as follows:

$$x = \frac{-b \pm \sqrt{b^2 - 4ac}}{2a}$$

$$= \frac{-(-8) \pm \sqrt{(-8)^2 - [4(4)(3)]}}{2(4)} \qquad \text{(Note that –b is –(–8) since b is –8)}$$

$$= \frac{+8 \pm \sqrt{64 - 48}}{8}$$

$$= \frac{8 \pm \sqrt{16}}{8}$$

$$x = \frac{8 \pm 4}{8}$$

We now split the sign $\pm$ into + and – to obtain the two solutions as follows:

$$x = \frac{8 + 4}{8} \quad \text{or} \quad x = \frac{8 - 4}{8}$$

$$x = \frac{12}{8} \quad \text{or} \quad x = \frac{4}{8}$$

$$x = \frac{3}{2} \quad \text{or} \quad x = \frac{1}{2}$$

2. Solve the following equations by using the quadratic formula:

a. $x^2 - 4x - 2 = 0$

b. $4 + x - 5x^2 = 0$

c. $7 - 2x - 3x^2 = 0$

d. $2n^2 - 15n + 18 = 0$

Solutions

a. $x^2 - 4x - 2 = 0$

In this equation, a = 1, b = –4 and c = –2

We now substitute these values into the quadratic formula as follows:

$$x = \frac{-b \pm \sqrt{b^2 - 4ac}}{2a}$$

$$= \frac{-(-4) \pm \sqrt{(-4)^2 - [4(1)(-2)]}}{2(1)}$$

$$= \frac{4 \pm \sqrt{16 - (-8)}}{2}$$

$$= \frac{4 \pm \sqrt{16 + 8}}{2}$$

$$= \frac{4 \pm \sqrt{24}}{2}$$

$$x = \frac{4 \pm 4.90}{2}$$

We now split the sign $\pm$ into $+$ and $-$ to obtain the two solutions as follows:

$$x = \frac{4 + 4.90}{2} \quad \text{or} \quad x = \frac{4 - 4.90}{2}$$

$$x = \frac{8.90}{2} \quad \text{or} \quad x = \frac{-0.90}{2}$$

$$x = 4.45 \quad \text{or} \quad x = -0.45$$

b. $4 + x - 5x^2 = 0$

In this equation, a = –5, b = 1 and c = 4

Note that 'a' is the coefficient of x^2, b is the coefficient of x, while c is the constant term. So, do not be confused by the reversed arrangement of the quadratic equation.

We now substitute the values of a, b and c into the quadratic formula as follows:

$$x = \frac{-b \pm \sqrt{b^2 - 4ac}}{2a}$$

$$= \frac{-1 \pm \sqrt{1^2 - [4(-5)(4)]}}{2(-5)}$$

$$= \frac{-1 \pm \sqrt{1 - (-80)}}{-10}$$

$$= \frac{-1 \pm \sqrt{1 + 80}}{-10}$$

$$= \frac{-1 \pm \sqrt{81}}{-10}$$

$$x = \frac{-1 \pm 9}{-10}$$

We now split the sign $\pm$ into $+$ and $-$ to obtain the two solutions as follows:

$$x = \frac{-1 + 9}{-10} \quad \text{or} \quad x = \frac{-1 - 9}{-10}$$

112

$$x = \frac{8}{-10} \quad \text{or} \quad x = \frac{-10}{-10}$$

$$x = -\frac{4}{5} \quad \text{or} \quad x = 1$$

c. $7 - 2x - 3x^2 = 0$

This arrangement is similar to that of question (b) above. However, let us solve this question by changing this arrangement to the usual arrangement of a quadratic equation. In order to do this, we divide each term in the equation by −1 so that the coefficient of x^2 will become a positive value. This gives:

$$\frac{7}{-1} - \frac{2x}{-1} - \frac{3x^2}{-1} = \frac{0}{-1}$$

$$-7 + 2x + 3x^2 = 0$$

Or, $3x^2 + 2x - 7 = 0$

In this new arrangement, a = 3, b = 2 and c = −7

We now substitute the values of a, b and c into the quadratic formula as follows:

$$x = \frac{-b \pm \sqrt{b^2 - 4ac}}{2a}$$

$$= \frac{-2 \pm \sqrt{2^2 - [4(3)(-7)]}}{2(3)}$$

$$= \frac{-2 \pm \sqrt{4 - (-84)}}{6}$$

$$= \frac{-2 \pm \sqrt{4 + 84}}{6}$$

$$= \frac{-2 \pm \sqrt{88}}{6}$$

$$x = \frac{-2 \pm 9.38}{6}$$

$$x = \frac{-2 + 9.38}{6} \quad \text{or} \quad x = \frac{-2 - 9.38}{6}$$

$$x = \frac{7.38}{6} \quad \text{or} \quad x = \frac{-11.38}{6}$$

$$x = 1.23 \quad \text{or} \quad x = -1.90$$

d. $2n^2 - 15n + 18 = 0$

   a = 2, b = −15 and c = 18

$$n = \frac{-b \pm \sqrt{b^2 - 4ac}}{2a}$$

$$= \frac{-(-15) \pm \sqrt{(-15)^2 - [4(2)(18)}}{2(2)}$$

$$= \frac{15 \pm \sqrt{225 - 144}}{4}$$

$$= \frac{15 \pm \sqrt{81}}{4}$$

$$n = \frac{15 \pm 9}{4}$$

$$n = \frac{15 + 9}{4} \quad \text{or} \quad n = \frac{15 - 9}{4}$$

$$n = \frac{24}{4} \quad \text{or} \quad n = \frac{6}{4}$$

$$n = 6 \quad \text{or} \quad n = \frac{3}{2}$$

Exercise 8

1. Factorize the following quadratic expressions
a. $7a^2 + 9a$
b. $m^2 - 16$
c. $9p^2 - 121$
d. $32b^2 - 200$

2. Factorize the following equations:
a. $3x^2 - 4x - 15$.
b. $x^2 + x - 72$
c. $5x^2 + 22x + 8$
d. $10x^2 - 4x - 6$

3. Determine if the following equations can be factorized or not:
a. $x^2 + 14x + 13 = 0$
b. $3x^2 - 11x + 6 = 0$
c. $5x^2 + 2x - 9 = 0$
d. $2x^2 - 7x - 11 = 0$
e. $8x^2 - 21x + 12 = 0$

4. Solve the following equations by factorization:
a. $3a^2 + 12a = 0$
b. $b^2 - 36 = 0$

114

c. $4y^2 - 49 = 0$

d. $(2x - 5)^2 = 9$

5. Solve the following equations by factorization:

a. $3x^2 - x - 4 = 0$

b. $2y^2 + 2y - 12 = 0$

c. $p^2 - 20p + 36 = 0$

d. $10m^2 = 4 - 3m$

e. $6 - x - 7x^2 = 0$

6. Find the quadratic equation whose roots are:

a. 8 and –5.

b. $\frac{1}{4}$ and -7.

c. –6 and $-\frac{2}{5}$

7. A quadratic equation is given by: $2x^2 - 15x - 8 = 0$. Find:

a. the sum of the roots of the equation

b. the product of the roots of the equation

8. A quadratic equation is given by: $3x^2 + 4x - 11 = 0$. Find:

a. the sum of the roots of the equation

b. the product of the roots of the equation

9. If 2 is a root of the equation $2x^2 + Mx - 18 = 0$, find the value of M. Hence find the other root.

10. If 1 is a root of the equation $3x^2 - kx - 11 = 0$, find the value of K. Hence find the other root.

11. 3 and –9 are the roots of the quadratic equation $b^2 - kb + P = 0$. Find the values of k and P.

12. If –5 and 2 are the roots of the quadratic equation $3x^2 + Ax - B = 0$. Find the values of the constants A and B.

13. If one root of the equation $m^2 + km + 27 = 0$ is thrice the other, find the two possible values of k. Hence, what are the two possible equations?

14. What must be added to $x^2 + 16x$ to make it a perfect square?

15. What must be added to $x^2 - 13x$ to make it a perfect square?

16. Make $a^2 - 6a$ a perfect square.

17. Make $m^2 + \frac{5}{3}m$ a perfect square.

18. If $x^2 - 8x + 16$ is a perfect square, find its square root.

19. If $x^2 + 1.2x + (0.6)^2$ is a perfect square, find its square root.

20. Use the method of completing the square to solve the following quadratic equations:

a. $y^2 + 9y - 22 = 0$

b. $x^2 - 11x + 24 = 0$

c. $2x^2 - 5x - 12 = 0$

d. $3x^2 - 10x + 7 = 0$

e $5y^2 + 11y + 6 = 0$

f. $2m^2 + 3m - 14 = 0$

g. $x^2 - 7x + 8 = 0$

21. Solve the following equations by using the quadratic formula:

a. $x^2 + x - 12 = 0$

b. $7x^2 - 17x + 10 = 0$

c. $2x^2 - 7x - 4 = 0$

d. $9 + 5x - 3x^2 = 0$

e. $5 - 7x - 5x^2 = 0$

f. $4n^2 - 12n + 9 = 0$

CHAPTER 9
WORD PROBLEMS LEADING TO QUADRATIC EQUATIONS

Examples
1. Find two numbers whose difference is 3 and whose product is 54.

<u>Solution</u>
Let the smaller number be x.
Then the larger number will be $x + 3$ (The larger one is greater than the smaller one by 3)
∴ Their product is $x(x + 3) = 54$
   $x^2 + 3x = 54$
   $x^2 + 3x - 54 = 0$
Solving this quadratic equation by factorization gives:
   $(x + 9)(x - 6) = 0$ (Two numbers whose product is −54 and sum is 3 are 9 and −6)
∴ $x = -9$ or $x = 6$ (When each of the bracket above is equated to zero and solved)
When $x = -9$, then the larger number is:
   $x + 3$ (As stated at the beginning)
   $= -9 + 3$ (Since $x = -9$)
   $= -6$
When $x = 6$, then the larger number is:
   $x + 3$
   $= 6 + 3 = 9$
∴   The two numbers are −9 and −6 or 9 and 6.

2. The difference between two numbers is 12. The sum of their squares is 74. Find the numbers.

<u>Solution</u>
Let the smaller number be 'a'
∴   The larger number is a + 12
Their squares are a^2 and $(a + 12)^2$
∴   The sum of their squares is:
   $a^2 + (a + 12)^2 = 74$
   $a^2 + (a + 12)(a + 12) = 74$
   $a^2 + a^2 + 12a + 12a + 144 = 74$
   $2a^2 + 24a + 144 - 74 = 0$
   $2a^2 + 24a + 70 = 0$
Dividing throughout by 2 to simplify it further gives:
   $a^2 + 12a + 35 = 0$
Solving this quadratic equation by factorization gives:
   $(a + 7)(a + 5) = 0$
∴   a = −7 or a = −5
When a = −7, the larger number is:

a + 12

    = −7 + 12

    = 5

When a = −5, the larger number is:

    a + 12

    = −5 + 12

    = 7

∴ The two numbers are −7 and 5 or −5 and 7.

3. A boy is 5 years older than his brother. The product of their ages is 204. Find their ages.

Solution

Let his brother's age be y years.   (Note that his brother's age is smaller since the boy is older)

∴ The boy's age = (y + 5)years

The product of their ages is given by:

    y(y + 5) = 204

∴ y^2 + 5y = 204

    y^2 + 5y − 204 = 0

Solving this quadratic equation by factorization gives:

    (y − 12)(y + 17) = 0   (Note: two numbers whose product is −204 and sum is 5 are −12 and 17)

∴ y = 12 or y = −17

Age cannot be negative, therefore, y = 12

∴ The brother's age is 12 years.

The boy's age is y + 5

            = 12 + 5

            = 17

The boy's age is 17 years.

4. A paper measures 15cm by 10cm. A strip of equal width and having an 'L' shape is cut off from the ends of the paper. If the area of the remaining paper is 84cm^2, find the width of the strip removed.

Solution

Let the width of the strip removed be x cm. Since the L shape affect the length and width, then the new length of the paper is 15 − x, while the new width is 10 − x. This is because a size of x cm has been cut off along a length and along a width of the paper.

∴ The new area (length x width) is given by:

    (15 − x)(10 − x) = 84

    150 − 15x − 10x + x^2 = 84   (Note that −x x −x = +x^2

    x^2 − 15x − 10x + 150 − 84 = 0

    x^2 − 25x + 66 = 0

Factorizing this quadratic equation gives:

$(x - 22)(x - 3) = 0$

$\therefore$ $x = 22$ or 3

The value 22 is not possible since it is greater than the original dimension of the paper.

$\therefore$ $x = 3$.

$\therefore$ The width of the strip removed is 3cm.

5. A rectangular field is 16m long and 8m wide. A path of equal width runs along one side and one end thereby forming an 'L' shape path. If the total area of the field and path is 180m^2, find the width of the path.

Solution

Let the width of the path be y m.

$\therefore$ The new length of the field including the path is (16 + y)m while the new width is (8 + y)m.

$\therefore$ The new area (i.e. total area) is:

$(16 + y)(8 + y) = 180$

$128 + 16y + 8y + y^2 = 180$

$y^2 + 16y + 8y + 128 - 180 = 0$

$y^2 + 24y - 52 = 0$

Factorizing this equation gives:

$(y + 26)(y - 2) = 0$

$\therefore$ $y = -26$ or $y = 2$

$y = -26$ is not possible since length/width cannot be negative

$\therefore$ $y = 2$

$\therefore$ The width of the path is 2m.

6. The ages of two brothers are 7 and 2 years. In how many years time will the product of their ages be 234?

Solution

Let the product of their ages be 234 in x years time.

$\therefore$ In x years time, their ages will be (7 + x) years and (2 + x) years

$\therefore$ Product of their ages then will be

$(7 + x)(2 + x) = 234$

$14 + 7x + 2x + x^2 = 234$

$x^2 + 7x + 2x + 14 - 234 = 0$

$x^2 + 9x - 220 = 0$

Factorizing this equation and solving it gives:

$(x + 20)(x - 11) = 0$

$\therefore$ $x = -20$ or $x = 11$

$x = -20$ is not correct since years cannot be negative.

$\therefore$ $x = 11$years.

$\therefore$ In 11 years time, the product of their ages will be 234.

7. Two sisters are 25 years and 18 years old. How many years ago was the product of their ages 144?

Solution
Let the product of their ages be 144, x years ago.
∴ x years ago, their ages were $(25 - x)$ years and $(18 - x)$years.
∴ The product of their ages then was:
$(25 - x)(18 - x) = 144$
$450 - 25x - 18x + x^2 = 144$
$x^2 - 25x - 18x + 450 - 144 = 0$
$x^2 - 43x + 306 = 0$
Factorizing this quadratic equation gives:
$(x - 34)(x - 9) = 0$
∴ $x = 34$ or $x = 9$
34 is not possible since it is greater than the sisters' ages
∴ $x = 9$
∴ 9 years ago, the product of their ages was 144.

8. Find two consecutive odd numbers whose product is 255.

Solution
Let the smaller number be x. Therefore the larger number is $(x + 2)$
Note that any two consecutive odd numbers (or even numbers) differs by 2. For example, 5 and 7.
∴ Their product is given by:
$x(x + 2) = 255$
$x^2 + 2x - 255 = 0$
Solving the equation by factorizing method gives:
$(x + 17)(x - 15) = 0$
∴ $x = -17$ or $x = 15$
When $x = -17$, the larger number is $(x + 2)$. This gives:
$-17 + 2 = -15$
When $x = 15$, the larger number is $15 + 2 = 17$.
∴ The two consecutive odd numbers are −15 and −17 or 15 and 17.

9. Twice a particular whole number is subtracted from 3 times the square of the number. The answer is 133. What is the number?

Solution
Let the number be x.
∴ Twice the number is $2x$.
The square of the number is x^2.
3 times the square of the number is $3x^2$.

∴   When twice the number is subtracted from 3 times the square of the number to give 133, then this forms an equation as follows:

$$3x^2 - 2x = 133$$
∴ $3x^2 - 2x - 133 = 0$

Apart from factorization method, this equation can also be solved by using quadratic equation formula as follows:

$$3x^2 - 2x - 133 = 0$$

Hence , a = 3, b = −2, and c = −133

∴ $$x = \frac{-b \pm \sqrt{b^2 - 4ac}}{2a}$$

$$= \frac{-(-2) \pm \sqrt{(-2)^2 - [4 \times 3 \times (-133)]}}{2 \times 3}$$

$$= \frac{2 \pm \sqrt{4 - (-1596)}}{6}$$

$$= \frac{2 \pm \sqrt{4 + 1596)}}{6}$$

$$= \frac{2 \pm \sqrt{1600}}{6}$$

$$= \frac{2 \pm 40}{6}$$

$$x = \frac{2 + 40}{6} \quad \text{or} \quad x = \frac{2 - 40}{6}$$

$$x = \frac{42}{6} \quad \text{or} \quad x = \frac{-38}{6}$$

$$x = 7 \text{ or } -6.3$$

∴ $x = 7$ since the number is a whole number

∴   The number is 7.

10   A man is 4 times as old as his daughter. 5 years ago, the product of their ages was 175. Find their present ages.

<u>Solution</u>
Let the daughter's age be y years.
∴ The man's age is 4y years
5 years ago, the daughter's age was (y − 5) years, while the man's age was (4y − 5) years.
∴   The product of their ages then (i.e. 5 years ago) is given by:

$$(4y - 5)(y - 5) = 175$$
$$4y^2 - 20y - 5y + 25 = 175$$
$$4y^2 - 25y + 25 - 175 = 0$$
$$4y^2 - 25y - 150 = 0$$

In order to solve this equation by factorization, multiply the first and last terms on the left hand side. This gives:

$$4y^2 \times (-150) = -600y^2$$

Now look for two numbers in y whose product is $-600y^2$ and sum is $-25y$ (i.e. the middle term in the equation). The two numbers are $-40y$ and $+15y$. Replace $-25y$ in the original equation by these two numbers. This gives:

$$4y^2 - 40y + 15y - 150 = 0$$

Factorizing by grouping gives:

$$4y(y - 10) + 15(y - 10) = 0$$

Take the terms outside each bracket, and then one of the two equal brackets. Note that $(y - 10)$ is common (i.e. the HCF) in the equation above. This gives:

$$(y - 10)(4y + 15) = 0$$

$\therefore$ $y = 10$ or $y = -\dfrac{15}{4}$

$y = 10$ (Ignore $y = -\dfrac{15}{4}$ since age cannot be negative or fraction)

$\therefore$ The daughter's age is 10 years, while the man's age is 40 years (i.e. $4y = 4 \times 10 = 40$)

Exercise 9

1. Find two numbers whose difference is 14 and whose product is 72.

2. The difference between two numbers is 10. The sum of their squares is 58. Find the numbers.

3. A boy is 7 years older than his brother. The product of their ages is 198. Find their ages.

4. A field measures 21cm by 12cm. A path of equal width and having an 'L' shape is at the ends of the field. If the area of the remaining field is 165cm^2, find the width of the path.

5. A rectangular field is 16m long and 8m wide. A path of equal width runs along one side and one end thereby forming an 'L' shape path. If the total area of the field and path is 180m^2, find the width of the path.

6. The ages of two friends are 13 and 9 years. In how many years time will the product of their ages be 480?

7. Two brothers are 28 years and 10 years old. How many years ago was the product of their ages 144?

8. Find two consecutive even numbers whose product is 288.

9. Thrice a particular whole number is added to 2 times the square of the number. The answer is 230. What is the number?

10 A woman is 3 times as old as his daughter. 5 years ago, the product of their ages was 217. Find their present ages.

CHAPTER 10
VARIATION

Direct Variation

Direct variation involves the relationship between two quantities whereby an increase or decrease in one of them leads to an increase or decrease respectively in the other.

The symbol α means 'varies with' or is 'proportional to'.

Examples

1. If x varies directly as y and x =30 when y=12, find:
a. the formula connecting x and y
b. x when y=10
c. y when x =20

Solution

a. $x \, \alpha \, y$ (This means x varies directly as y)

$x = Ky$ (Replacing the proportionality sign with the equals sign introduces a constant K)

So, when $x = 30$ and y = 12, the equation above becomes:

$30 = K \times 12$

$30 = 12K$

$K = \dfrac{30}{12} = \dfrac{5}{2}$

The formula connecting x and y is:

$x = \dfrac{5}{2} y$ (This is obtained by substituting $\dfrac{5}{2}$ for K in the equation above, i.e. $x = Ky$)

b. When y = 10, x is given by:

$x = \dfrac{5}{2} y$

$x = \dfrac{5}{2} \times 10 = \dfrac{50}{2} = 25$

$\therefore \; x = 25$

c. When $x = 20$, y is given by:

$x = \dfrac{5}{2} y$

$20 = \dfrac{5}{2} y$

$5y = 40$

$y = \dfrac{40}{5}$

y = 8

2. If m varies as the square of n and m=27 when n=3, find:
a. the relationship between m and n
b. n when m = 48
c. m when n = $2\frac{1}{3}$

Solution
a. $m \, \alpha \, n^2$ (This means m varies as n^2)
 $m = Kn^2$ (Replacing the proportionality sign with the equals sign introduces a constant K)
So, when m = 27 and n = 3, the equation above becomes:
$27 = K \times 3^2$
$27 = 9K$
$K = \dfrac{27}{9} = 3$
The relationship between m and n is:
$m = 3n^2$ (This is obtained by substituting 3 for K in the equation $m = Kn^2$ above)

b. When m = 48, n is given by:
$m = 3n^2$
$48 = 3 \times n^2$
$48 = 3n^2$
$n^2 = \dfrac{48}{3} = 16$
$n = \sqrt{16}$
$n = 4$

c. When n = $2\frac{1}{3}$, m is given by:
$m = 3n^2$
$m = 3 \times (\dfrac{7}{3})^2$ (Note that $2\frac{1}{3}$ has been converted to $\dfrac{7}{3}$)
$m = 3 \times \dfrac{49}{9}$
Canceling out gives:
$m = \dfrac{49}{3}$

Inverse Variation
In inverse or indirect variation, as one quantity increases the other decreases.

Examples
1. If c varies inversely as d and c=18 when y=4, find:
a. the formula connecting c and d
b. c when d=10
c. d when c=12

Solution

a. $c \alpha \frac{1}{d}$ (This means c varies inversely as d)

$c = \frac{K}{d}$ (Replacing the proportionality sign with the equals sign introduces a constant K)

So, when c = 18 and d = 4, the equation above becomes:

$18 = \frac{K}{4}$

$K = 18 \times 4$

$K = 72$

The formula connecting c and d is:

$c = \frac{72}{d}$ (This is obtained by substituting 72 for K in the equation above, i.e. $c = \frac{K}{d}$)

b. When d = 10, c is given by:

$c = \frac{72}{d}$

$c = \frac{72}{10}$

$c = 7.2$

c. When c = 12, d is given by:

$c = \frac{72}{d}$

$12 = \frac{72}{d}$

$12d = 72$

$d = \frac{72}{12}$

$d = 6$

2. If r varies inversely as the cube root of t and r=6 when t=64, find:
a. the relationship between r and t

b. t when r = 16

c. r when $t = \dfrac{8}{27}$

Solution

a. $r \propto \dfrac{1}{\sqrt[3]{t}}$ (This means r varies inversely as the cube root of t)

$r = \dfrac{K}{\sqrt[3]{t}}$ (Replacing the proportionality sign with the equals sign introduces a constant K)

So, when r = 6 and t = 64, the equation above becomes:

$6 = \dfrac{K}{\sqrt[3]{64}}$

$6 = \dfrac{K}{4}$

K = 6 x 4 = 24

The relationship between r and t is:

$r = \dfrac{24}{\sqrt[3]{t}}$ (This is obtained by substituting 24 for K in the equation $r = \dfrac{K}{\sqrt[3]{t}}$)

b. When r = 16, t is given by:

$r = \dfrac{24}{\sqrt[3]{t}}$

$16 = \dfrac{24}{\sqrt[3]{t}}$

$\sqrt[3]{t} = \dfrac{24}{16}$

$\sqrt[3]{t} = \dfrac{3}{2}$

In order to remove the cube root, take the cube of both sides. This gives:

$(\sqrt[3]{t})^3 = (\dfrac{3}{2})^3$

$\therefore \; t = \dfrac{27}{8}$

c. When $t = \dfrac{8}{27}$, r is given by:

$$r = \dfrac{24}{\sqrt[3]{t}}$$

$$r = \dfrac{24}{\sqrt[3]{8/27}}$$

$$r = \dfrac{24}{2/3}$$

$$r = 24 \times \dfrac{3}{2}$$

After equal division by 2, it gives:

$$r = 12 \times 3$$
$$r = 36$$

Joint Variation

In joint variation, three or more quantities are related directly or inversely or both.

Examples

1. If m varies directly as the square of n and inversely as p, and m=3 when n=2 and p=8, find:
a. the relationship between m, n and p
b. m when n = 3 and p =27
c. p when $m = \dfrac{1}{2}$ and $n = \dfrac{3}{2}$

<u>Solutions</u>

$m \, \alpha \, \dfrac{n^2}{p}$ (This means m varies directly as the square of n and inversely as p)

$$m = \dfrac{Kn^2}{p}$$

So, when m = 3, n = 2, and p = 8, the equation above becomes:

$$3 = \dfrac{K \times 2^2}{8}$$

$$3 = \dfrac{4K}{8}$$

$$4K = 3 \times 8 = 24$$

$$K = \dfrac{24}{4} = 6$$

The relationship between m, n and p is:

$m = \dfrac{6n^2}{p}$ (This is obtained by substituting 6 for K in the equation $m = \dfrac{Kn^2}{p}$)

b. When n = 3 and p = 27, then m is given by:

$$m = \frac{6n^2}{p}$$

$$m = \frac{6 \times 3^2}{27}$$

$$m = \frac{6 \times 9}{27}$$

$$m = \frac{54}{27}$$

$$m = 2$$

c. When $m = \frac{1}{2}$ and $n = \frac{3}{2}$, then p is given by:

$$m = \frac{6n^2}{p}$$

$$\frac{1}{2} = \frac{6 \times \frac{3}{2}}{p}$$

$$\frac{1}{2} = \frac{9}{p}$$

p = 9 x 2
P = 18

2. The weight w of a rod varies jointly as its length L and the square root of its density d. If w =12 when L = 5 and d = 9, find:

a. L in terms of w and d
b. w when L = 8 and d = 25
c. d when L = 20 and w = 4

Solutions

a. $w \, \alpha \, L\sqrt{d}$ (This means w varies jointly as L and the square root of d)

$w = KL\sqrt{d}$

   So, when w = 12, L = 5, and d = 9, the equation above becomes:

$12 = K \times 5 \times \sqrt{9}$

12 = 15K

$K = \frac{12}{15} = \frac{4}{5}$

The formula connecting w, L and d is:

$w = \frac{4}{5}L\sqrt{d}$ (This is obtained by substituting $\frac{4}{5}$ for K in the equation $w = KL\sqrt{d}$)

L can now be expressed in terms of w and d as follows:

$$w = \frac{4}{5} L\sqrt{d}$$

$$5w = 4L\sqrt{d}$$

Dividing both sides of the equation by $4\sqrt{d}$ gives:

$$L = \frac{5w}{4\sqrt{d}}$$

b. When L = 8 and d = 25, then w is given by:

$$w = \frac{4}{5} L\sqrt{d}$$

$$w = \frac{4}{5} \times 8 \times \sqrt{25}$$

$$w = \frac{4}{5} \times 8 \times 5$$

Cancelling out the 5 gives:

$$w = 4 \times 8$$

$$w = 32$$

c. When L = 20 and w = 4, then d is given by:

$$w = \frac{4}{5} L\sqrt{d}$$

$$4 = \frac{4}{5} \times 20 \times \sqrt{d}$$

$$4 \times 5 = 4 \times 20 \times \sqrt{d}$$

$$20 = 80\sqrt{d}$$

$$\sqrt{d} = \frac{20}{80} = \frac{1}{4}$$

Taking the square of both sides gives:

$$(\sqrt{d})^2 = (\frac{1}{4})^2$$

$$d = \frac{1}{16}$$

Partial Variation

The fourth type of variation is called partial variation. In partial variation, one quantity is partly constant and partly varies with the other. Two constants are involved in partial variation.

Examples

1. x is partly constant and partly varies as y. When y=2, x=30, and when y=6, x=50.
a. Find the formula which connects x and y.
b. Find x when y=3

Solutions
a. From the first sentence, we have:

   x = C + Ky   (Let this be equation 1) where C and K are constants.

Substituting y=2 and x=30 in this equation gives:

30 = C + 2K   (Let this be equation 2)

Similarly, when y = 6 and x = 50, we have:

50 = C + 6K   (Let this be equation 3)

Bringing equation 2 and 3 together gives:

$$30 = C + 2K \quad \text{(Equation 2)}$$
$$\underline{50 = C + 6K} \quad \text{(Equation 3)}$$

Equation 3 – Equation 2:   20 =   4K

Divide both sides by 4. This gives:

$$K = \frac{20}{4} = 5$$

K = 5

Substitute 5 for K in equation 2.

30 = C + 2K

30 = C + (2 x 5)

30 = C + 10

30 – 10 = C

C = 20

We now substitute the values of C and K into equation 1 in order to obtain the formula connecting x and y.

The formula connecting x and y is now given by:

x = 20 + 5y

b.   When y = 3, x is obtained by substituting 3 for y in the formula connecting x and y.

x = 20 + 5y

= 20 + (5x3)

= 20 + 15

x = 35

2.   m is partly constant and partly varies as n. When n = 4, m = 5, and when n = 12, m = 14.

a.   Find the formula which connects m and n.

b.   Find m when n = 16

c.   Find n when m = 9

Solutions

a.   From the first sentence, we have:

m = C + Kn   (Let this be equation 1) where C and K are constants.

Substituting n=4 and m=5 in equation 1 gives:

5 = C + 4K   (Let this be equation 2)

Similarly, when n=12 and m=14, we have:

14 = C + 12K   (Let this be equation 3)

Bringing equation 2 and 3 together gives:

$$5 = C + 4K \quad \text{(Equation 2)}$$

$$\underline{14 = C + 12K} \quad \text{(Equation 3)}$$

Equation 3 – Equation 2: $\quad 9 = \quad 8K$

Divide both sides by 8. This gives:

$$K = \frac{9}{8}$$

Substitute $\frac{9}{8}$ for K in equation 2.

$$5 = C + 4K$$

$$5 = C + (4 \times \frac{9}{8})$$

$$5 = C + \frac{9}{2}$$

$$5 - \frac{9}{2} = C$$

$$C = \frac{1}{2}$$

We now substitute the values of C and K into equation 1 in order to obtain the formula connecting m and n.

The formula connecting m and n is given by:

$$m = \frac{1}{2} + \frac{9}{8}n$$

b. When n = 16, m is obtained by substituting 16 for n in the formula connecting m and n.

$$m = \frac{1}{2} + \frac{9}{8}n$$

$$m = \frac{1}{2} + (\frac{9}{8} \times 16)$$

$$= \frac{1}{2} + 18 = \frac{37}{2}$$

$$m = 18\frac{1}{2}$$

c. When m = 9, n is obtained by substituting 9 for m in the formula connecting m and n.

$$m = \frac{1}{2} + \frac{9}{8}n$$

$$9 = \frac{1}{2} + (\frac{9n}{8})$$

$$9 - \frac{1}{2} = \frac{9n}{8}$$

$$\frac{17}{2} = \frac{9n}{8}$$

$$17 \times 8 = 9n \times 2$$

$$136 = 18n$$

$$n = \frac{136}{18} = \frac{68}{9}$$

$$n = 7\frac{5}{9}$$

Exercise 10

1. If x varies directly as y and $x = 10$ when y=8, find:
a. the formula connecting x and y
b. x when y=10
c. y when $x = 16$

2. If h varies as the square root of p and h=5 when p=9, find:
a. the relationship between h and p
b. p when h = 20
c. h when $p = 6\frac{1}{4}$

3. If p varies inversely as q and p = 12 when q = 3, find:
a. the formula connecting p and q
b. q when p = 20
c. p when q = 5

4. If m varies inversely as the cube root of n and m=5 when n = 27, find:
a. the relationship between m and n
b. m when n = 8
c. n when $m = \frac{64}{125}$

5. If a varies directly as the square of b and inversely as c, and when a=4 when b=3 and c=6, find:
a. the formula connecting a, b and c
b. a when b = 5 and c = 10
c. b when $a = \frac{1}{2}$ and c = 8

6. The height h of a box varies jointly as its length L and the square of its width w. If h = 20 when L = 4 and w = 3, find:
a. w in terms of h and L
b. w when h = 12 and L = 4
c. L when h = 8 and w = 5

7. x is partly constant and partly varies as y. When y = 4, $x = 14$, and when y = 5, $x = 17$.
a. Find the relationship between x and y.
b. Find x when y = 8

8. E is partly constant and partly varies as F. When F = 2, E = 25, and when F = 5, E = 55.
a. Find the formula which connects E and F.
b. Find E when F = $2\frac{1}{2}$
c. Find F when E = 40

CHAPTER 11
SIMULTANEOUS LINEAR AND QUADRATIC EQUATIONS

Examples

1. Solve the simultaneous equations: $2x^2 - 5y = -8$ and $3x + y = -1$

<u>Solution</u>

$2x^2 - 5y = -8$ ....Equation (1)

$3x + y = -1$ ....Equation (2)

From equation (2), $y = -1 - 3x$ ....Equation (3)

Substitute $-1-3x$ for y in equation (1)

$2x^2 - 5y = -8$ ....Equation (1)

$2x^2 - 5(-1 - 3x) = -8$

$2x^2 + 5 + 15x = -8$

$2x^2 + 15x + 5 + 8 = 0$

$2x^2 + 15x + 13 = 0$

In order to solve this equation by factorization method, we first multiply the first and last terms. This gives:

$2x^2 \times 13 = 26x^2$

Find two numbers in x whose product is $26x^2$ and whose sum is $15x$ (i.e. the middle term in the original equation). The two numbers in x are $13x$ and $2x$ (their product is $26x^2$ and their sum is $15x$). The two numbers are now used to replace the middle term in the original equation. This gives:

$2x^2 + 13x + 2x + 13 = 0$

Factorize this equation by grouping.

$x(2x + 13) + 1(2x + 13) = 0$

$\therefore$ $(2x + 13)(x + 1) = 0$.   (Note that $(2x + 13)$ is the common factor in the expression above)

Equating each bracket above to zero and solving each of the linear equation formed, gives:

$x = -1$ or $x = -\dfrac{13}{2} = -6\dfrac{1}{2}$

Substitute -1 for x in equation (3)

$y = -1 - 3x$ ....Equation (3)

$= -1 - 3(-1)$

$= -1 + 3$

$y = 2$

Substitute $-\dfrac{13}{2}$ for x in equation (3)

$y = -1 - 3x$ ....Equation (3)

$= -1 - 3(-\dfrac{13}{2})$

$= -1 + \dfrac{39}{2}$

$= \dfrac{-2 + 39}{2}$

$y = \dfrac{37}{2}$

$$y = 18\frac{1}{2}$$

Therefore the solutions are $x = -1$ and $y = 2$ or $x = -6\frac{1}{2}$ and $y = 18\frac{1}{2}$. The solution can be given more neatly in x and y coordinates as follows:

$$(-1, 2), (-6\frac{1}{2}, 18\frac{1}{2})$$

2. Solve the simultaneous equations: $x^2 + 3y^2 = 4$
$$x - 2y = 1$$

<u>Solutions</u>

$x^2 + 3y^2 = 4$ ....Equation (1)

$x - 2y = 1$ ....Equation (2)

From equation (2) $x = 1 + 2y$

Substitute $1 + 2y$ for x in equation (1)

$x^2 + 3y^2 = 4$

$(1 + 2y)^2 + 3y^2 = 4$

$(1 + 2y)(1 + 2y) + 3y^2 = 4$

$1 + 2y + 2y + 4y^2 + 3y^2 = 4$

$7y^2 + 4y + 1 - 4 = 0$

$7y^2 + 4y - 3 = 0$

Solving this quadratic equation by factorizing gives:

$7y^2 + 7y - 3y - 3 = 0$

$7y(y + 1) - 3(y + 1) = 0$

$(y + 1)(7y - 3) = 0$

$\therefore \quad y = -1$ or $y = \dfrac{3}{7}$

Substitute -1 for y in equation (2). This gives:

$x - 2y = 1$ ....Equation (2)

$x - 2(-1) = 1$

$x + 2 = 1$

$x = 1 - 2$

$x = -1$

Similarly, substitute $\dfrac{3}{7}$ for y in equation (2).

$x - 2y = 1$ ....Equation (2)

$x - 2(\dfrac{3}{7}) = 1$

$x - \dfrac{6}{7} = 1$

$x = 1 + \dfrac{6}{7}$

$= \dfrac{7 + 6}{7}$

$x = \dfrac{13}{7} = 1\dfrac{6}{7}$

Therefore the solutions are $x = -1$ and $y = -1$, or $x = 1\frac{6}{7}$ and $y = \frac{3}{7}$

The solutions can also be given as: $(-1, -1)$, $(1\frac{6}{7}, \frac{3}{7})$

3. Solve the equations $3x^2 - xy = 0$, $2y - 5x = 1$

Solution

$\qquad 3x^2 - xy = 0$ ....Equation (1)

$\qquad 2y - 5x = 1$ ....Equation (2)

From equation (1):

$\qquad x(3x - y) = 0$

$\therefore \quad x = 0$ or $3x - y = 0$

$\qquad x = 0 \quad$ or

$\qquad y = 3x$ ....Equation (3)

Substitute 0 for x in equation (2). This gives

$\qquad 2y - 5(0) = 1$ ....Equation (2)

$\qquad 2y = 1$

$\qquad y = \frac{1}{2}$

Similarly, substitute $3x$ for y in equation (2).

$\qquad 2y - 5x = 1$ ....Equation (2)

$\qquad 2(3x) - 5x = 1$

$\qquad 6x - 5x = 1$

$\qquad x = 1$

Substitute 1 for x in equation (3)

$\qquad y = 3x$ ....Equation (3)

$\qquad y = 3(1)$

$\qquad y = 3$

$\therefore \quad x = 0$ and $y = \frac{1}{2} \quad$ or $\quad x = 1$ and $y = 3$

This can also be given by: $(0, \frac{1}{2}) \quad$ or $\quad (1, 3)$

4. Solve simultaneously $3x - 5y = -11$, $xy = -2$

Solution

$\qquad 3x - 5y = -11$ ....Equation (1)

$\qquad xy = -2$ ....Equation (2)

From equation (2), $x = \frac{-2}{y}$ ....Equation (3)

Substitute $\frac{-2}{y}$ for x in equation (1)

$\qquad 3x - 5y = -11$ ....Equation (1)

$\qquad 3(\frac{-2}{y}) - 5y = -11$

$$\frac{-6}{y} - 5y = -11$$

Multiply each term by y in order to clear the fraction. This gives:

$$y\left(\frac{-6}{y}\right) - y(5y) = y(-11)$$
$$-6 - 5y^2 = -11y$$
$$0 = 5y^2 - 11y + 6$$

Or, $5y^2 - 11y + 6 = 0$

Solving this equation by factorization gives:

$$5y^2 - 5y - 6y + 6 = 0$$
$$5y(y-1) - 6(y-1) = 0 \quad \text{(−6 outside the bracket changes the + sign to − sign in the}$$
bracket)
$$(y-1)(5y-6) = 0$$

Equating each bracket to zero and solving each equation gives:

$$y = 1 \text{ or } y = \frac{6}{5} = 1\frac{1}{5}$$

Substitute 1 for y in equation (3)

$$x = \frac{-2}{y} \quad \text{................Equation (3)}$$
$$= \frac{-2}{1}$$
$$= -2$$

Substitute $\frac{6}{5}$ for y in equation (3)

$$x = \frac{-2}{y} \quad \text{................Equation (3)}$$
$$x = \frac{-2}{\frac{6}{5}}$$
$$= -2 \times \frac{5}{6}$$
$$= \frac{-10}{6}$$
$$x = \frac{-5}{3} = -1\frac{2}{3}$$

Therefore, the solutions are $x = -2$ and y = 1, or $x = -1\frac{2}{3}$ and y = $1\frac{1}{5}$. These can be given as: $(-2, 1)$, $(-1\frac{2}{3}, 1\frac{1}{5})$.

5. Solve the equations $9y^2 + 8x = 12$, $2x + 3y = 4$

Solution

$$9y^2 + 8x = 12 \text{ ....Equation (1)}$$
$$2x + 3y = 4 \text{ ....Equation (2)}$$

From equation (2):

$$2x = 4 - 3y$$

$$\therefore \quad x = \frac{4 - 3y}{2} \quad \text{.................Equation (3)}$$

Substitute $\dfrac{4 - 3y}{2}$ for x in equation (1).

$$9y^2 + 8x = 12 \quad \text{.................Equation (1)}$$
$$9y^2 + 8(\frac{4 - 3y}{2}) = 12$$
$$9y^2 + 4(4 - 3y) = 12 \quad \text{(Note that 8 divided by 2 gives 4 which is outside the bracket)}$$
$$9y^2 + 16 - 12y = 12$$
$$9y^2 - 12y + 16 - 12 = 0$$
$$9y^2 - 12y + 4 = 0$$

Solving this equation by factorization gives:

$$9y^2 - 6y - 6y + 4 = 0$$
$$3y(3y - 2) - 2(3y - 2) = 0$$
$$(3y - 2)(3y - 2) = 0$$

Equation each equation to zero and solving each equation gives:

$$y = \frac{2}{3} \text{ in both cases.}$$

Now substitute $\dfrac{2}{3}$ for y in equation (3)

$$x = \frac{4 - 3y}{2} \quad \text{.................Equation (3)}$$

$$= \frac{4 - 3(\frac{2}{3})}{2}$$

$$= \frac{4 - 2}{2} \quad \text{(Note that } 3 \times \frac{2}{3} = 2\text{)}$$

$$x = \frac{2}{2} = 1$$

Therefore, the solutions are $x = 1$ and $y = \dfrac{2}{3}$

6. Solve the equations, $x^2 - y^2 = 20$, $x + y = -2$

Solution

$$x^2 - y^2 = 20 \quad \text{.................Equation (1)}$$
$$x + y = -2 \quad \text{.................Equation (2)}$$

Looking at equation (1) carefully shows that it is a difference of two squares, and that one of the factors of equation (1) is what is given in equation (2).

Recall that a difference of two squares such as $a^2 - b^2$ can be factorized as $a^2 - b^2 = (a + b)(a - b)$. Therefore the equations above can be re-written as follows:

$$(x + y)(x - y) = 20 \quad \text{.................Equation (1)} \quad \text{(Note that } x^2 - y^2 \text{ has been factorized)}$$
$$x + y = -2 \quad \text{.................Equation (2)}$$

Substitute -2 for $x + y$ in equation (1)

$$(x + y)(x - y) = 20 \quad \text{.................Equation (1)}$$
$$-2(x - y) = 20 \quad \text{(Since } (x + y) = -2)$$

Divide both sides by –2 to obtain:

$$x - y = -10 \text{ ....Equation (3)}$$
$$\underline{x + y = -2} \text{ ....Equation (2)}$$

Equation (2) – equation (3)　　$2y = 8$　　(Note that $+y -(-y) = y + y = 2y$ and $-2 -(-10) = -2 + 10 = 8$)

$$y = \frac{8}{2}$$
$$y = 4$$

Substitute 4 for y in equation (2). (Any equation can be used)

$$x + y = -2 \text{ ....Equation (2)}$$
$$x + 4 = -2$$
$$x = -2 - 4$$
$$x = -6$$

Therefore the solution are $x = -6$ and $y = 4$. This can also be given as (–6, 4).

7. Solve the equations, $4x^2 - 9y^2 = -32$, $2x - 3y = -4$

Solution

$$4x^2 - 9y^2 = -32 \text{ ....Equation (1)}$$
$$2x - 3y = -4 \text{ ....Equation (2)}$$

Equation (1) is a difference of two squares. It can be expressed as:

$$(2x)^2 - (3y)^2 = -32$$

Therefore factorizing this difference of two squares gives:

$$(2x - 3y)(2x + 3y) = -32 \text{ ....Equation (1)}$$
$$2x - 3y = -4 \text{ ....Equation (2)}$$

Substitute –4 for $2x - 3y$ in equation (1)

$$(2x - 3y)(2x + 3y) = -32 \text{ ....Equation (1)}$$
$$-4(2x + 3y) = -32 \quad \text{(Since } (2x - 3y) = -4)$$

Divide both sides by –4 to obtain equation (3) as shown below:

$$2x + 3y = 8 \text{ ....Equation (3)}$$
$$\underline{2x - 3y = -4} \text{ ....Equation (2)}$$

Equation (3) – equation (2):　　$6y = 12$　　(Note that $+3y -(-3y) = 6y$ and $8 -(-4) = 8 + 4 = 12$)

$$y = \frac{12}{6}$$
$$y = 2$$

Substitute 2 for y in equation (3). (Note that any equation can be used)

$$2x + 3y = 8 \text{ ....Equation (3)}$$
$$2x + 3(2) = 8$$
$$2x + 6 = 8$$
$$2x = 8 - 6$$
$$2x = 2$$
$$x = \frac{2}{2}$$
$$x = 1$$

Therefore the solution is $x = 1$ and $y = 2$. This can also be represented as (1, 2).

8. A boy is x years old while his father is y years old. The sum of their ages is equal to twice the difference of their ages. The product of their ages is equal to 675. Find their ages.

Solution
The sum of their ages is $x + y$.
The difference of their ages is $y - x$. (The father's age is larger)
Twice the difference of their ages is $2(y - x)$.
The sum of their ages is equal to twice the difference of their ages is given by:

$$x + y = 2(y - x)$$
$$x + y = 2y - 2x$$
$$x + 2x + y - 2y = 0$$
$$3x - y = 0$$

Or $3x = y$ ....Equation (1)
The product of their ages is 675. This is given by:

$$xy = 675$$ ....Equation (2)

Substitute $3x$ for y in equation (2). (Note that $y = 3x$ from equation 1)

$$xy = 675$$ ....Equation (2)
$$x(3x) = 675$$
$$3x^2 = 675$$
$$x^2 = \frac{675}{3}$$
$$x^2 = 225$$
$$x = \sqrt{225}$$
$$x = \pm 15$$
$$x = 15 \quad \text{(Since age cannot be negative)}$$

Substitute 15 for x in equation (1)

$$3x = y$$ ....Equation (1)

Or, $y = 3x$
             $= 3(15)$
             $y = 45$

Therefore the boy is 15 years old (i.e. x), while his father is 45 years old (i.e. y)

9. Twice a number added to another number is 19. The first number added to the square of the second number gives 17. Find the two numbers.
Solution
Let the first number be 'a' and the second number be 'b'.
Twice the first number = 2a
Twice the first number added to the second number is 19. This is given by:

$$2a + b = 19$$ ....Equation (1)

The square of the second number = b^2
The first number added to the square of the second number gives 17. This is represented as:

$$a + b^2 = 17$$ ....Equation (2)

From equation (1), a = $\dfrac{19 - b}{2}$ ....Equation (3)

Substitute $\dfrac{19 - b}{2}$ for 'a' in equation (2)

$\quad$ a + b^2 = 17 ....Equation (2)

$\quad \dfrac{19 - b}{2}$ + b^2 = 17

Multiply each term by 2 (L.C.M) in order to clear the fraction. This gives:

$\quad 2(\dfrac{19 - b}{2})$ + 2(b^2) =2(17)

$\quad$ 19 − b + 2b^2 = 34

$\quad$ 2b^2 − b + 19 − 34 = 0

∴ $\quad$ 2b^2 − b − 15 = 0

Solving this equation by factorization method gives

$\quad$ 2b^2 − 6b + 5b − 15 = 0

$\quad$ 2b(b − 3) + 5(b − 3) = 0

Equating each bracket to zero and solving each equation gives:

$\quad$ (b − 3)(2b + 5) = 0

$\quad$ b = 3 or b = $\dfrac{-5}{2}$ = $-2\dfrac{1}{2}$

From equation (3), when b = 3, we obtain 'a' as follows:

$\quad$ a = $\dfrac{19 - b}{2}$ ....Equation (3)

$\quad$ = $\dfrac{19 - 3}{2}$

$\quad$ = $\dfrac{16}{2}$

$\quad$ a = 8

Similarly, from equation (3), when b = $\dfrac{-5}{2}$, we obtain 'a' as follows:

$\quad$ a = $\dfrac{19 - b}{2}$ ....Equation (3)

$\quad$ = $\dfrac{19 - (\frac{-5}{2})}{2}$

$\quad$ = $\dfrac{19 + \frac{5}{2}}{2}$

$\quad$ = $\dfrac{\frac{38 + 5}{2}}{2}$

$\quad$ = $\dfrac{43}{2}$ ÷ 2

$\quad$ = $\dfrac{43}{2}$ x $\dfrac{1}{2}$

$$= \frac{43}{4}$$

$$a = 10\frac{3}{4}$$

Therefore the solutions are a = 8 and b = 3, or a = $10\frac{3}{4}$ and b = $-2\frac{1}{2}$. These can also be given as: (8, 3) or $(10\frac{3}{4}, -2\frac{1}{2})$.

10. The sum of the squares of the ages of two brothers is 65. If the difference between their ages is 3, find their ages.

<u>Solution</u>
Let the older brother be m years old, and the younger brother be n years old. Therefore the two equations that can be formed from the question are:

$\quad m^2 + n^2 = 65$ ....Equation (1)

$\quad m - n = 3$ ....Equation (2)

From equation (2):

$\quad m = 3 + n$ ....Equation (3)

Substitute 3 + n for m in equation (1). This gives:

$\quad m^2 + n^2 = 65$ ....Equation (1)

$\quad (3 + n)^2 + n^2 = 65$

$\quad (3 + n)(3 + n) + n^2 = 65$

$\quad 9 + 3n + 3n + n^2 + n^2 = 65$

$\quad 2n^2 + 6n + 9 - 65 = 0$

$\quad 2n^2 + 6n - 56 = 0$

Solving this equation by using quadratic equation formula (you can also use factorization method) gives:

$\quad a = 2, b = 6, c = -56$

$$\therefore \quad n = \frac{-b \pm \sqrt{b^2 - 4ac}}{2a}$$

$$= \frac{-6 \pm \sqrt{6^2 - [4 \times 2 \times (-56)]}}{2 \times 2}$$

$$= \frac{-6 \pm \sqrt{36 - (-448)}}{4}$$

$$= \frac{-6 \pm \sqrt{36 + 448}}{4}$$

$$= \frac{-6 \pm \sqrt{484}}{4}$$

$$= \frac{-6 \pm 22}{4}$$

$$n = \frac{-6 + 22}{4} \text{ or } \frac{-6 - 22}{4}$$

$$= \frac{16}{4} \text{ or } \frac{-28}{4}$$

$$n = 4 \text{ or } -7$$

But, age cannot be negative. Therefore:

$$n = 4$$

Substitute 4 for n in equation (3)

$$m = 3 + n \text{ ....Equation (3)}$$

$$= 3 + 4$$

$$m = 7$$

∴   The older brother is 7 years old while the younger brother is 4 years old.

Exercise 11

1. Solve the following simultaneous equations:

a. $x^2 - 2y = -2$
    $5x + y = 13$

b. $x^2 + 2y^2 = 9$
    $2x - y = -4$

c. $5x^2 - 3xy = 26, 3y - 7x = 17$

d. $6x - 5y = -2, 2xy = 1$

e. $5y^2 + 2x = 11, x - 4y = 7$

f. $x^2 - y^2 = 16, x + y = -2$

g. $3x^2 - 2xy + y^2 = 9, 5x - 2y = 8$

2. A boy is x years old while his father is y years old. The sum of their ages is equal to thrice the difference of their ages. The product of their ages is equal to 1800. Find their ages.

3. Four times a number added to twice another number is 34. Twice the first number added to twice the square of the second number gives 108. Find the two numbers.

4. The sum of the squares of the ages of two brothers is 157. If the difference between their ages is 5, find their ages.

CHAPTER 12
LINEAR INEQUALITY

Linear Inequality

If a statement is expressed as $x = 10$, then this is an equation. However, if x is not exactly equal to 10, but can vary in values from 10 and above or from 10 and below, then we say $x < 10$ or $x > 10$. We can also say $x \leq 10$ or $x \geq 10$. These are called inequalities.

The four symbols of inequalities and their meanings are as follows:

$<$ means less than

$>$ means greater than

$\leq$ means less than or equal to

$\geq$ means greater than or equal to

Inequalities can be solved like equations. When solving inequalities, the following points should be noted.

1. When both sides of an inequality are divided by a negative number, the inequality sign should be reversed. This is done in order to keep the inequality true. Consider the example below:

$-3 < 5$ (This is true)

Now let us divide both sides by -3. This gives:

$$\frac{-3}{-3} < \frac{5}{-3}$$

$1 < -\dfrac{5}{3}$ (This is not true because the inequality sign has not been reversed)

Therefore the right thing to do is to reverse the inequality sign when dividing by a negative number. This means that the correct division above should be:

$-3 < 5$

$$\frac{-3}{-3} > \frac{5}{-3}$$

$1 > -\dfrac{5}{3}$ (This is true because the inequality sign has been reversed)

2. When an inequality is cross multiplied, it is advisable to cross multiply from the lower right hand side to the upper left hand side first, and not the other way round first. Let us examine the example below.

$$\frac{2}{7} < \frac{3}{4}$$

The correct way to cross multiply is as follows:

$(2 \times 4) < (7 \times 3)$

$8 < 21$ (This is true)

However, if you cross multiply from the lower left to the upper right first, then it gives:

$(7 \times 3) < (2 \times 4)$

$21 < 8$ (This is not true)

In order to make this kind of cross multiplication true, you have to reverse the sign to give $21 > 8$.

Hence, in order to avoid making mistake, it is better to cross multiply from the lower right hand side to the upper left hand side first. In this case your inequality remains true, and you do not need to reverse the inequality sign.

Examples

1. Solve the inequality $5x - 2 > 19 - 2x$. Represent your result on a number line/line graph.

<u>Solution</u>

$$5x - 2 > 19 - 2x$$

We simply solve this problem just like the way we solve an equation. But we must continue to use the inequality sign.

Let us collect terms in x on one side, and constant terms on the other side.

$$5x - 2 > 19 - 2x$$
$$5x + 2x > 19 + 2 \quad \text{(Note that the sign of any term that crosses the inequality sign must change)}$$
$$7x > 21$$

Divide both sides by 7. This gives:

$$\frac{7x}{7} > \frac{21}{7}$$
$$x > 3$$

In representing inequalities on number lines/line graphs, we simply draw a number line and make sure the arrow starts from the solution (i.e. 3) and points to the right when it is >. It points to the left when the inequality sign is <. The beginning of the arrow line should be a small circle. When the sign is ≤ or ≥, the small circle on the number line where the arrow starts should be shaded to show that the value is included in the possible solutions.

∴ $x > 3$ is represented as shown below:

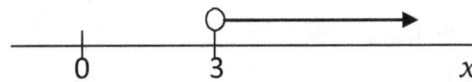

2. Solve the inequality $9x + 5 < 4x - 2(3 - x)$. Represent your result on a number line.

<u>Solution</u>

$$9x + 5 < 4x - 2(3 - x)$$

Let us expand the bracket on the right hand side. This gives:

$$9x + 5 < 4x - 6 + 2x$$

Let us collect terms in x on one side, and constant terms on the other side.

$$9x - 4x - 2x < -6 - 5$$
$$3x < -11$$

Divide both sides by 3. This gives:

$$\frac{3x}{3} < \frac{-11}{3}$$
$$x < -\frac{11}{3}$$

∴ $x < -\dfrac{11}{3}$ is represented on a number line as shown below:

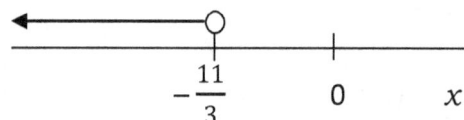

3. Find the range of values of x for which $\dfrac{2x-1}{3} - \dfrac{1}{2}x \leq 8 + x$. Show the result on a line graph.

Solution

$$\dfrac{2x-1}{3} - \dfrac{1}{2}x \leq 8 + x$$

Multiply each term by 6 (i.e. the LCM of 3 and 2 which are the denominators). This is done in order to remove the fractions.

$$6(\dfrac{2x-1}{3}) - 6(\dfrac{x}{2}) \leq 6(8 + x)$$

$2(2x-1) - 3(x) \leq 48 + 6x$ (Note that 2 and 3 outside the brackets are obtained when 6 outside the brackets above divides 3 and 2 (the denominators) respectively).

$4x - 2 - 3x \leq 48 + 6x$

We now collect like terms as follows:

$4x - 3x - 6x \leq 48 + 2$

$-5x \leq 50$

Divide both sides by –5 and reverse the inequality sign. This gives:

$$\dfrac{-5x}{-5} \geq \dfrac{50}{-5}$$ (Take note of the reversal of the sign since we are dividing by a negative number)

$x \geq -10$

We now represent this result on a number line as follows

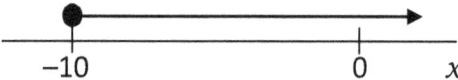

Note that the small circle at the beginning of the arrow is shaded to show that −10 is also part of the solution due to the sign ≥. And remember that greater than represents right hand direction of arrow, while less than represents left hand direction of arrow.

4. Find the range of values of x for which $\dfrac{1}{5} - \dfrac{1-4x}{2} \geq \dfrac{3}{4} + 3x$. Show the result on a line graph.

Solution

$$\dfrac{1}{5} - \dfrac{1-4x}{2} \geq \dfrac{3}{4} + 3x$$

Multiply each term by 20 (i.e. the LCM of 5, 2 and 4 which are the denominators).

$$20(\dfrac{1}{5}) - 20(\dfrac{1-4x}{2}) \geq 20(\dfrac{3}{4}) + 20(3x)$$

$4(1) - 10(1 - 4x) \geq 5(3) + 60x$

Note that 20 has been used to divide the denominators of the fractions.

$$4 - 10 + 40x \geq 15 + 60x$$

We now collect like terms as follows:

$$40x - 60x \geq 15 + 10 - 4$$
$$-20x \geq 21$$

Divide both sides by –20 and reverse the sign. This gives:

$$\frac{-20x}{-20} \leq \frac{21}{-20}$$ (Take note of the reversal of the sign since we are dividing by a negative number)

$$x \leq -\frac{21}{20}$$

We now represent this result on a number line as follows

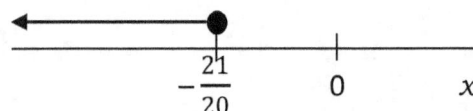

5. Given that m is an integer, find the three greatest values which satisfy the inequality:

$$2(3m + 1) < \frac{1}{2}(2m - 5)$$

Solution

$$2(3m + 1) < \frac{1}{2}(2m - 5)$$

This can also be written as follows:

$$2(3m + 1) < \frac{2m - 5}{2}$$

Cross multiply to obtain:

$$2 \times 2(3m + 1) < 2m - 5$$
$$4(3m + 1) < 2m - 5$$
$$12m + 4 < 2m - 5$$

Collect like terms. This gives:

$$12m - 2m < -5 - 4$$
$$10m < -9$$
$$\therefore \quad m < \frac{-9}{10}$$

An integer is any whole number such as, –1, 7, 0.

Hence, if $m < \frac{-9}{10}$, and m is an integer, then m can be any of the numbers below.

$$m = -1, -2, -3, -4 \ldots\ldots$$

Hence the three greatest values of m are, –1, –2, –3.

Note that the nearest whole number that is just less than $\frac{-9}{10}$ is –1 which is also the greatest whole number for 'm'.

6. Given that x is an integer, find the three lowest values of x which satisfy the following inequality: $2(x - \frac{1}{3}) - \frac{3}{4}(x + 5) > \frac{1}{2}(x - 4)$

Solution

$$2\left(x-\frac{1}{3}\right) - \frac{3}{4}(x+5) > \frac{1}{2}(x-4)$$

Expanding each of the bracket and placing each of the denominator properly gives:

$$2x - \frac{2}{3} - \frac{3x+15}{4} > \frac{x-4}{2}$$

Multiply each term by 12 (i.e. the LCM of the denominators). This gives:

$$12(2x) - 12\left(\frac{2}{3}\right) - 12\left(\frac{3x+15}{4}\right) > 12\left(\frac{x-4}{2}\right)$$

$$24x - 4(2) - 3(3x+15) > 6(x-4) \quad \text{(Note that 12 has been used to divide each denominator)}$$

$$24x - 8 - 9x - 45 > 6x - 24$$

Collect like term to give:

$$24x - 9x - 6x > 8 + 45 - 24$$

$$9x > 29$$

$$x > \frac{29}{9}$$

$$x > 3\frac{2}{9}$$

Since x is an integer (whole number), then x starts from 4 because 4 is the nearest whole number greater than $3\frac{2}{9}$

∴ The three lowest values of x are 4, 5, 6.

7. Given that c is an integer, find the three greatest values which satisfy the inequality:

$$\frac{1}{3}(c-1) > \frac{2}{5}(3c+4)$$

Solution

$$\frac{1}{3}(c-1) > \frac{2}{5}(3c+4)$$

This can also be written as follows:

$$\frac{c-1}{3} > \frac{2(3c+4)}{5}$$

Cross multiply to obtain:

$$5(c-1) > 3 \times 2(3c+4)$$

$$5c - 5 > 6(3c+4)$$

$$5c - 5 > 18c + 24$$

Collect like terms. This gives:

$$5c - 18c > 24 + 5$$

$$-13c > 29$$

∴ $$c < \frac{29}{-13} \qquad$$ (Note that the inequality sign has been reversed due to the division by −13)

Hence, $c < -2\frac{3}{13}$

Hence, if $c < -2\frac{3}{13}$, and c is an integer, then c can be any of the numbers below.

$$c = -3, -4, -5, -6 \ \text{......}$$

Hence the three greatest values of c are, −3, −4, −5.

8. Given that r is an integer, find the three lowest values of r which satisfy the following inequality: $\frac{3}{8} - \frac{1}{2}(3r - 1) < \frac{1}{4}(2 - r)$

Solution

$$\frac{3}{8} - \frac{1}{2}(3r - 1) < \frac{1}{4}(2 - r)$$

Multiply each term by 8 (i.e. the LCM of the denominators). This gives:

$$8(\frac{3}{8}) - \frac{8(3r - 1)}{2} < \frac{8(2 - r)}{4}$$

$3 - 4(3r - 1) < 2(2 - r)$ (Note that 8 has been used to divide each denominator)

$3 - 12r + 4 < 4 - 2r$

Collect like term to give:

$-12r + 2r < 4 - 4 - 3$

$-10r < -3$

$r > \frac{-3}{-10}$ (Take note of the reversal of the inequality sign)

$r > \frac{3}{10}$

Since r is an integer (whole number), then r starts from 1 because 1 is the nearest whole number greater than $\frac{3}{10}$

∴ The three lowest values of r are 1, 2, 3.

9. What is the range of values of y for which $4y - 7 \leq 3y$ and $3y \leq 5y + 8$ are both satisfied? Show your result on a graph.

Solution

We solve each of the inequality to obtain two values of y. This is as shown below.

$4y - 7 \leq 3y$

$4y - 3y \leq 7$

$y \leq 7$

And

$3y \leq 5y + 8$

$3y - 5y \leq 8$

$-2y \leq 8$

$y \geq \frac{8}{-2}$ (Take note of the reversal of the inequality sign due to division by negative number)

∴ $y \geq -4$

Hence, $y \leq 7$ and $y \geq -4$

Combining the two inequalities gives:

$-4 \leq y \leq 7$

When combining the solutions of inequalities, we ensure that the appropriate part of the inequality sign is facing the variable (i.e. y). In the solution, $y \leq 7$, the elbow part of the inequality sign is facing y. This was maintained in the combination above. In the solution y ≥ –4, the open part of the inequality sign is facing y. This was also maintained in the combination above.

Hence, both inequalities are satisfied if: $-4 \leq y \leq 7$

This can be represented on a line graph as shown below.

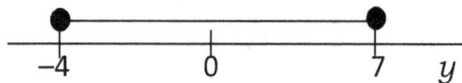

10. Express 3y – 2 < 10 + y < 2 + 5y in the form a < y < b, where a and b are both integers. Show the range of values on a line graph.

<u>Solution</u>

$$3y - 2 < 10 + y < 2 + 5y$$

We can take out two inequalities from the question above and solve. They are:

$3y - 2 < 10 + y$ ……………Inequality (1)

And, $10 + y < 2 + 5y$ ……………Inequality (2)

Let us solve each of them.

$3y - 2 < 10 + y$ ……………Inequality (1)

$3y - y < 10 + 2$

$2y < 12$

$y < \dfrac{12}{2}$

$y < 6$

Now let us solve the second inequality.

$10 + y < 2 + 5y$ ……………Inequality (2)

$y - 5y < 2 - 10$

$-4y < -8$

$y > \dfrac{-8}{-4}$

$y > 2$

Therefore the two solutions are y < 6 and y > 2

These two solutions show that y ranges from 2 to 6.

Hence the solutions can be combined to give:

$$2 < y < 6$$

This can be represented on a line graph as shown below.

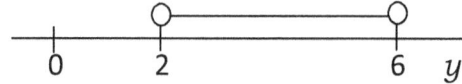

11. Find the range of values of y for which 2y + 1 > 21 or y + 5 ≤ 12 is satisfied. Show your result on a line graph.

Solution
Let us solve the first inequality as follows:

$$2y + 1 > 21$$
$$2y > 21 - 1$$
$$2y > 20$$
$$y > \frac{20}{2}$$
$$y > 10$$

The other inequality is solved as follows:

$$y + 5 \leq 12$$
$$y \leq 12 - 5$$
$$y \leq 7$$

Hence the inequalities are satisfied by either $y > 10$ or $y \leq 7$
Therefore, the line graph is as given below.

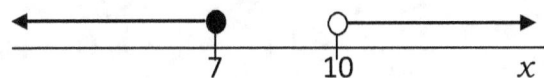

12. Express $3m - 5 < 4m + 1 < 13 + m$ in the form $a < y < b$, where a and b are both integers.

Solution

$$3m - 5 < 4m + 1 < 13 + m$$

Two inequalities that can be obtained from the question above are:

$3m - 5 < 4m + 1$ ....Inequality (1)
And, $4m + 1 < 13 + m$................Inequality (2)
Let us solve each of them.

$3m - 5 < 4m + 1$ ....Inequality (1)
$$3m - 4m < 1 + 5$$
$$-m < 6$$

Divide both sides by −1 and reverse the sign. This gives:

$$\frac{-m}{-1} > \frac{6}{-1}$$
$$m > -6$$

The second inequality is solved as follows:

$4m + 1 < 13 + m$ ....Inequality (2)
$$4m - m < 13 - 1$$
$$3m < 12$$

Hence, $m < \frac{12}{3}$
$$m < 4$$

Therefore the two solutions are $m > -6$ and $m < 4$.
These two solutions show that y ranges from −6 to 4.
Hence the solutions can be combined to give:

$$-6 < m < 4$$

This can be represented on a line graph as shown below

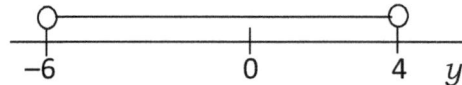

Exercise 12

1. Solve the inequality $9x - 12 > 2 + 2x$. Represent your result on a number line/line graph.

2. Solve the inequality $5x + 8 < 5x + 2(3 - x)$. Represent your result on a number line.

3. Find the range of values of x for which $\dfrac{3x + 1}{4} - \dfrac{3}{5}x \leq 11 - 2x$. Show the result on a line graph.

4. Find the range of values of x for which $\dfrac{3}{4}x - \dfrac{2 - 3x}{2} \geq \dfrac{1}{2} + \dfrac{21}{8}x$. Show the result on a line

5. Given that m is an integer, find the three greatest values which satisfy the inequality:

$$5(7m + 3) < \frac{2}{5}(9m - 2)$$

6. Given that x is an integer, find the three lowest values of x which satisfy the following inequality: $3(2x - \dfrac{3}{4}) - \dfrac{1}{2}(x + 3) > \dfrac{1}{3}(5x - 2)$

7. Given that c is an integer, find the three greatest values which satisfy the inequality:

$$\frac{3}{10}(3c - 4) < \frac{4}{5}(c + 7)$$

8. Given that r is an integer, find the three lowest values of r which satisfy the following inequality:

$$\frac{5}{6} - \frac{3}{2}(4r - 9) < \frac{5}{12}(3 - r)$$

9. What is the range of values of y for which $5y - 4 \leq 3y$ and $9y \leq 15y + 6$ are both satisfied? Show your result on a graph.

10. Express $7y - 4 < 11 + 2y < 5y + 17$ in the form $a < y < b$, where a and b are both integers. Show the range of values on a line graph.

11. Find the range of values of y for which $4y - 1 > -5$ or $5 \geq y + 12$ are both satisfied. Show your result on a line graph.

12. Express $5m - 9 < 7m - 17 < 23 - m$ in the form $a < y < b$, where a and b are both integers.

CHAPTER 13
QUADRATIC INEQUALITY

If we have a factorized quadratic inequality such as $(x + 2)(x - 1) > 0$, then the product $(x + 2)(x - 1)$ has to be positive (> 0). This means that either $(x + 2)$ and $(x - 1)$ are both positive or both negative.

If $(x + 2)(x - 1) < 0$, then the product $(x + 2)(x - 1)$ has to be negative (< 0). This means that only $(x + 2)$ is negative or only $(x - 1)$ is negative.

Therefore it is necessary to test the solutions of a quadratic inequality in order to know the correct solution.

However, there is a general rule that applies to quadratic inequalities (and even other inequalities having two solutions). If the coefficient of x^2 in a quadratic inequality is a positive value, then of the two values of the solutions obtained, the larger value will take the original sign ($<$ or $>$) of the inequality, while the smaller value will take the reverse sign of the inequality. The following examples illustrate this rule.

Examples

1. Solve the inequality: $x^2 - 2x - 8 < 0$

Solution

$$x^2 - 2x - 8 < 0$$

Let us equate this inequality to zero and solve by factorization.

$\therefore \quad x^2 - 2x - 8 = 0$

$\quad (x - 4)(x + 2) = 0$

$\therefore \quad x = 4$ or $x = -2$

In $x^2 - 2x - 8 < 0$, the coefficient of x^2 is positive. Therefore, between the solutions $x = 4$ and $x = -2$, the larger value i.e. 4 will take the sign of the inequality in the question, i.e. $<$. The lower solution i.e. -2 will take $>$ (i.e. the reverse sign).

Therefore, the solutions is $x < 4$ or $x > -2$. These can be combined to give:

$\quad -2 < x < 4$

Hence, $x^2 - 2x - 8 < 0$ is satisfied when $-2 < x < 4$

We can test these solutions to see if our rule and values obtained are correct. When $x < 4$, a possible value is 3 since $3 < 4$. We now substitute $x = 3$ in the original equation as follows:

$\quad x^2 - 2x - 8 < 0$

$\quad (3)^2 - 2(3) - 8 < 0$

$\quad 9 - 6 - 8 < 0$

$\qquad -5 < 0 \qquad$ (This is true)

Hence, $x < 4$ is a correct solution to the inequality.

Similarly, when $x > -2$, a possible value is -1 since $-1 > -2$. We now substitute $x = -1$ in the original inequality as follows:

$\quad x^2 - 2x - 8 < 0$

$\quad (-1)^2 - 2(-1) - 8 < 0$

$\quad 1 + 2 - 8 < 0$

$-5 < 0$ (This is true)

Hence, $x > -2$ is also a correct solution of the inequality.

2. Solve the inequality $2x^2 + 3x - 5 \geq 0$

Solution

$\qquad 2x^2 + 3x - 5 \geq 0$

Equating to zero and solving by factorization gives:

$\qquad 2x^2 + 3x - 5 = 0$

$\qquad 2x^2 + 5x - 2x - 5 = 0$

$\qquad x(2x + 5) - 1(2x + 5) = 0$

$\therefore \qquad (2x + 5)(x - 1) = 0$

This gives: $x = -\dfrac{5}{2}$ or $x = 1$

Of these two values, the larger solution i.e. 1 will take the inequality sign in the question, while $-\dfrac{5}{2}$ will take the reverse sign. Therefore the solution is $x \geq 1$ or $x \leq -\dfrac{5}{2}$

Hence, $2x^2 + 3x - 5 \geq 0$ is satisfied when $x \geq 1$ or $x \leq -\dfrac{5}{2}$

3. Solve the inequality: $3x^2 - 9x + 4 \leq 0$

Solution

$\qquad 3x^2 - 9x + 4 = 0$ (Equating to zero first)

This equation cannot be factorized. So let's solve it by using quadratic formula as follows:

$\qquad 3x^2 - 9x + 4 = 0$

$\qquad a = 3, b = -9, c = 4$

$\therefore \qquad x = \dfrac{-b \pm \sqrt{b^2 - 4ac}}{2a}$

$\qquad = \dfrac{-(-9) \pm \sqrt{(-9)^2 - (4 \times 3 \times 4)}}{2 \times 3}$

$\qquad = \dfrac{9 \pm \sqrt{81 - 48}}{6}$

$\qquad = \dfrac{9 \pm \sqrt{33}}{6}$

$\qquad x = \dfrac{9 + 5.74}{6}$ or $x = \dfrac{9 - 5.74}{6}$

$\qquad x = \dfrac{14.74}{6}$ or $x = \dfrac{3.26}{6}$

$\therefore \qquad x = 2.46 \quad x = 0.54$

Therefore, $x \leq 2.46$ or $x \geq 0.54$ (Note that the larger value takes the original inequality sign)

Hence, $3x^2 - 9x + 4 \leq 0$ is satisfied when $0.54 \leq x \leq 2.46$

4. Solve the inequality $7 + 6x - x^2 > 0$

Solution

$7 + 6x - x^2 > 0$ (Note that the coefficient of x^2 is negative, i.e. −1)

$7 + 6x - x^2 = 0$ (This can be factorized)

$(7 - x)(1 + x) = 0$

$\therefore$ $x = 7$ or $x = -1$

Since the coefficient of x^2 in this inequality is negative, then the smaller value of the solution i.e. −1 will take the sign of the inequality in the question, while the larger value i.e. 7 will take the reverse sign. This is exactly the opposite of what we do when the coefficient of x^2 is positive like in examples 1 to 3.

$\therefore$ $x > -1$ or $x < 7$

Hence, $7 + 6x - x^2 > 0$ is satisfied when $-1 < x < 7$

5. Solve the inequality $5 + 8x - 2x^2 \geq 0$

Solution

$5 + 8x - 2x^2 \geq 0$

Equating to zero and solving by using quadratic formula gives:

$5 + 8x - 2x^2 = 0$

$a = -2, b = 8, c = 5$

$$\therefore \quad x = \frac{-b \pm \sqrt{b^2 - 4ac}}{2a}$$

$$= \frac{-8 \pm \sqrt{8^2 - (4 \times -2 \times 5)}}{2(-2)}$$

$$= \frac{-8 \pm \sqrt{64 + 40}}{-4}$$

$$= \frac{-8 \pm \sqrt{104}}{-4}$$

$$x = \frac{-8 + 10.2}{-4} \quad \text{or} \quad x = \frac{-8 - 10.2}{-4}$$

$$x = -0.55 \quad \text{or} \quad x = 4.55$$

Since the coefficient of x^2 in this inequality is negative, then the smaller value will take the inequality sign in the question.

$\therefore$ $x \geq -0.55$ or $x \leq 4.55$

Hence $5 + 8x - 2x^2 \geq 0$ is satisfied when $-0.55 \leq x \leq 4.55$

Exercise 13

1. Solve the inequality: $x^2 - x - 30 < 0$
2. Solve the inequality $3x^2 + 10x - 8 \geq 0$
3. Solve the inequality: $5x^2 - 9x + 4 \leq 0$
4. Solve the inequality $9 + 9x - 10x^2 > 0$
5. Solve the inequality $11 + 2x - 2x^2 \geq 0$

6. Solve the inequality: $2x^2 - 5x - 18 < 0$
7. Solve the inequality $7x^2 + x - 6 \geq 0$
8. Solve the inequality: $5x^2 + 17x + 14 \leq 0$
9. Solve the inequality $15 + 2x - x^2 > 0$
10. Solve the inequality $22 + 9x - 4x^2 \geq 0$

CHAPTER 14
INTRODUCTORY VECTOR ALGEBRA

Representation of a Vector

A vector is a quantity which has magnitude and direction. A scalar quantity has only magnitude. A vector can be represented by a line such as AB. This line usually has a corresponding x and y values in the Cartesian plane.

A column vector can be written as $\begin{pmatrix} x \\ y \end{pmatrix}$.

Modulus or Magnitude of a Vector

The modulus or magnitude of a vector can be determined by using Pythagoras theorem as follows:

$|AB| = \sqrt{x^2 + y^2}$ (where AB is the vector).

A zero or null vector represented as $\begin{pmatrix} 0 \\ 0 \end{pmatrix}$ has a magnitude of zero.

Two vectors are parallel if they have the same direction. They are equal if they have the same magnitude.

Examples

1. What is the magnitude of the vector AB = $\begin{pmatrix} 5 \\ 12 \end{pmatrix}$

$|AB| = \sqrt{5^2 + 12^2}$

$= \sqrt{25 + 144}$

$= \sqrt{169}$

$\therefore$ $|AB|$ = 13 units

2. If the magnitude of the vector $\begin{pmatrix} x \\ 8 \end{pmatrix}$ is 17, find the value of x.

Solution

Magnitude = $\sqrt{x^2 + 8^2}$

$17 = \sqrt{x^2 + 64}$

Square both sides in order to remove the square root sign. This gives:

$17^2 = (\sqrt{x^2 + 64})^2$

$\therefore$ $289 = x^2 + 64$

$x^2 = 289 - 64$

$x^2 = 225$

Take the square root of both sides. This gives:

$\sqrt{x^2} = \sqrt{225}$

$x = 15$

Vector Algebra

Examples

(1) Given that p = $\begin{pmatrix} -3 \\ 4 \end{pmatrix}$, q = $\begin{pmatrix} -2 \\ -5 \end{pmatrix}$, r = $\begin{pmatrix} 4 \\ 1 \end{pmatrix}$, find:

a. $p - q + r$

b. $2p + q - 3r$

c. $\dfrac{1}{5}(p + q) - \dfrac{1}{2}r$

d. $|3p + q - 2r|$

<u>Solution</u>

a. $p - q + r = \begin{pmatrix} -3 \\ 4 \end{pmatrix} - \begin{pmatrix} -2 \\ -5 \end{pmatrix} + \begin{pmatrix} 4 \\ 1 \end{pmatrix}$

$= \begin{pmatrix} -3 - (-2) + 4 \\ 4 - (-5) + 1 \end{pmatrix}$

$= \begin{pmatrix} -3 + 2 + 4 \\ 4 + 5 + 1 \end{pmatrix}$

$= \begin{pmatrix} 3 \\ 10 \end{pmatrix}$

In solving this problem, simply combine the upper numbers separately and the lower numbers separately. Make sure the signs are well organized. For example, where $4 - (-5)$ should be written, should not be written as $4 - 5$.

`

b. $2p + q - 3r = 2\begin{pmatrix} -3 \\ 4 \end{pmatrix} + \begin{pmatrix} -2 \\ -5 \end{pmatrix} - 3\begin{pmatrix} 4 \\ 1 \end{pmatrix}$

$= \begin{pmatrix} -6 \\ 8 \end{pmatrix} + \begin{pmatrix} -2 \\ -5 \end{pmatrix} - \begin{pmatrix} 12 \\ 3 \end{pmatrix}$

$= \begin{pmatrix} -6 - 2 - 12 \\ 8 - 5 - 3 \end{pmatrix}$

$= \begin{pmatrix} -20 \\ 0 \end{pmatrix}$

Note that a number multiplying a bracket is used to multiply each term in the bracket.

c. $\dfrac{1}{5}(p + q) - \dfrac{1}{2}r = \dfrac{1}{5}\left[\begin{pmatrix} -3 \\ 4 \end{pmatrix} + \begin{pmatrix} -2 \\ -5 \end{pmatrix}\right] - \dfrac{1}{2}\begin{pmatrix} 4 \\ 1 \end{pmatrix}$

$= \dfrac{1}{5}\begin{pmatrix} -3 - 2 \\ 4 - 5 \end{pmatrix} - \begin{pmatrix} 2 \\ \frac{1}{2} \end{pmatrix}$

$= \dfrac{1}{5}\begin{pmatrix} -5 \\ -1 \end{pmatrix} - \begin{pmatrix} 2 \\ \frac{1}{2} \end{pmatrix}$

$= \begin{pmatrix} -1 \\ -\frac{1}{5} \end{pmatrix} - \begin{pmatrix} 2 \\ \frac{1}{2} \end{pmatrix}$

$= \begin{pmatrix} -1 - 2 \\ -\frac{1}{5} - \frac{1}{2} \end{pmatrix}$

$= \begin{pmatrix} -3 \\ -\frac{7}{10} \end{pmatrix}$ (Note that $-\dfrac{1}{5} - \dfrac{1}{2} = -\dfrac{7}{10}$)

d. In order to determine $|3p + q - 2r|$, we have to first determine $3p + q - 2r$

$\therefore \ 3p + q - 2r = 3\begin{pmatrix} -3 \\ 4 \end{pmatrix} + \begin{pmatrix} -2 \\ -5 \end{pmatrix} - 2\begin{pmatrix} 4 \\ 1 \end{pmatrix}$

$= \begin{pmatrix} -9 \\ 12 \end{pmatrix} + \begin{pmatrix} -2 \\ -5 \end{pmatrix} - \begin{pmatrix} 8 \\ 2 \end{pmatrix}$

$= \begin{pmatrix} -9 - 2 - 8 \\ 12 - 5 - 2 \end{pmatrix}$

$$3p + q - 2r = \begin{pmatrix} -19 \\ 5 \end{pmatrix}$$

$$\therefore \quad |3p + q - 2r| = \sqrt{(-19)^2 + 5^2} \quad \text{(Note that we are now calculating the magnitude of } \begin{pmatrix} -19 \\ 5 \end{pmatrix})$$

$$= \sqrt{361 + 25}$$

$$= \sqrt{386}$$

$$= 19.6 \text{ units}$$

2. If $a = \begin{pmatrix} 3 \\ 1 \end{pmatrix}$, $b = \begin{pmatrix} 2 \\ -1 \end{pmatrix}$, $c = \begin{pmatrix} 8 \\ 6 \end{pmatrix}$, find the values of:

a. m and n such that ma + nb = c

b. |d| if d = 2b − a

Solution

(a) ma + nb = c, means that:

$$m\begin{pmatrix} 3 \\ 1 \end{pmatrix} + n\begin{pmatrix} 2 \\ -1 \end{pmatrix} = \begin{pmatrix} 8 \\ 6 \end{pmatrix}$$

$$\begin{pmatrix} 3m \\ m \end{pmatrix} + \begin{pmatrix} 2n \\ -n \end{pmatrix} = \begin{pmatrix} 8 \\ 6 \end{pmatrix}$$

$$\begin{pmatrix} 3m + 2n \\ m - n \end{pmatrix} = \begin{pmatrix} 8 \\ 6 \end{pmatrix}$$

By comparing the upper part of the vectors, and then the lower part, two equations can be obtained as shown below:

$3m + 2n = 8$ ....Equation (1)

$m - n = 6$ ....Equation (2)

Equation (2) x 2 gives:

$2m - 2n = 12$ ....Equation (3)

Bringing equations (1) and (3) together in order to solve them simultaneously gives:

$3m + 2n = 8$ ....Equation (1)

$\underline{2m - 2n = 12}$ ....Equation (3)

Equation (1) + Equation (3): $5m = 20$ (n has been eliminated, since +2n + − 2n = 0

$$m = \frac{20}{5}$$

$m = 4$

Substitute 4 for m in equation (2)

$m - n = 6$ ....Equation (2)

$4 - n = 6$

$4 - 6 = n$

$n = -2$

$\therefore \quad n = -2$ and $m = 4$

b.   d = 2b − a

$$= 2\begin{pmatrix} 2 \\ -1 \end{pmatrix} - \begin{pmatrix} 3 \\ 1 \end{pmatrix}$$

$$= \begin{pmatrix} 4 \\ -2 \end{pmatrix} - \begin{pmatrix} 3 \\ 1 \end{pmatrix}$$

$$= \begin{pmatrix} 4 - 3 \\ -2 - 1 \end{pmatrix}$$

$$d = \begin{pmatrix} 1 \\ -3 \end{pmatrix}$$

$$\therefore \quad |d| = \sqrt{1^2 + (-3)^2}$$
$$= \sqrt{1 + 9}$$
$$|d| = \sqrt{10} \text{ units}$$

Direction of a Vector

If the position vector of a point A is $(x_1 \ y_1)$, and that of point B is $(x_2 \ y_2)$, then the component of the vector AB is given by:

$$AB = (x_2 - x_1, y_2 - y_1)$$

This can be written in column vector as:

$$AB = \begin{pmatrix} x_2 - x_1 \\ y_2 - y_1 \end{pmatrix}$$

If a vector is expressed as CD = $\begin{pmatrix} x \\ y \end{pmatrix}$, then the angle that CD makes with the x-axis (horizontal axis) is given by:

$$\text{Tan } \theta = \frac{y}{x}$$

$$\therefore \quad \theta = \tan^{-1} \frac{y}{x}$$

Examples

1. If u = $\begin{pmatrix} 3 \\ 2 \end{pmatrix}$ and v = $\begin{pmatrix} -2 \\ 5 \end{pmatrix}$

a. Find u − 2v

b. Express u in the form (k, $\theta°$) where k is magnitude of u and θ is the angle u makes with the x-axis.

Solution

a. $\quad$ u − 2v = $\begin{pmatrix} 3 \\ 2 \end{pmatrix} - 2\begin{pmatrix} -2 \\ 5 \end{pmatrix}$

$$= \begin{pmatrix} 3 \\ 2 \end{pmatrix} - \begin{pmatrix} -4 \\ 10 \end{pmatrix}$$

$$= \begin{pmatrix} 3 - -4 \\ 2 - 10 \end{pmatrix}$$

$$= \begin{pmatrix} 3 + 4 \\ 2 - 10 \end{pmatrix}$$

$$= \begin{pmatrix} 7 \\ -8 \end{pmatrix}$$

b. $\quad$ u = $\begin{pmatrix} 3 \\ 2 \end{pmatrix}$

The angle that u makes with the x axis is given by:

$$\text{Tan } \theta = \frac{y}{x}$$

$$\text{Tan } \theta = \frac{2}{3} \qquad \text{(Note that the upper number of the vector is } x \text{ while the lower number is y)}$$

$$\text{Tan } \theta = 0.6667$$

$$\therefore \quad \theta = \tan^{-1} 0.6667$$

$$\theta = 33.7^0 \qquad \text{(From the use of calculator)}$$

Also, $|u| = \sqrt{3^2 + 2^2}$ $\qquad$ [Note that u = $\begin{pmatrix} 3 \\ 2 \end{pmatrix}$]

$$= \sqrt{9 + 4}$$

$= \sqrt{13}$ units

$\therefore$ u in the form (k, θ) is given by: $(\sqrt{13}, 33.7^0)$

2. If $b = \begin{pmatrix} -2 \\ 5 \end{pmatrix}$, find the angle θ that b makes with the horizontal axis. Hence express b in the form (u, θ^0), where u is the magnitude of b.

Solution

$b = \begin{pmatrix} -2 \\ 5 \end{pmatrix}$,

The angle that b makes with the horizontal is given by:

$Tan\ \theta = \dfrac{y}{x}$

$Tan\ \theta = \dfrac{5}{-2}$ (Note that in the vector$\begin{pmatrix} -2 \\ 5 \end{pmatrix}$, -2 represents x, while 5 represents y)

$Tan\ \theta = \dfrac{5}{2}$ (Ignore the negative sign of -2)

$\theta = \tan^{-1} 2.5$

$\therefore \quad \theta = 68.2^0$

Also, $|b| = u = \sqrt{(-2)^2 + 5^2}$ [Since $b = \begin{pmatrix} -2 \\ 5 \end{pmatrix}$]

$= \sqrt{4 + 25}$

$u = \sqrt{29}$

$\therefore$ Expressing b in the form (u, θ^0), gives: $(\sqrt{29}, 68.2^0)$

3. The vertices of a quadrilateral WXYZ are W(2, −1), X(−1, 3), Y(−2, 1), Z(−3, −2). Determine the components and lengths of:
a. WX b. WZ c. −2XY d. YW

Solutions
a. With respect to the origin, each of the points can be expressed as:

$W = \begin{pmatrix} 2 \\ -1 \end{pmatrix}, X = \begin{pmatrix} -1 \\ 3 \end{pmatrix}, Y = \begin{pmatrix} -2 \\ 1 \end{pmatrix}, Z = \begin{pmatrix} -3 \\ -2 \end{pmatrix}$

$\therefore$ WX = X − W   (Take note of the arrangement of the letters, as the second letter comes first)

$= \begin{pmatrix} -1 \\ 3 \end{pmatrix} - \begin{pmatrix} 2 \\ -1 \end{pmatrix}$

$= \begin{pmatrix} -1 - 2 \\ 3 - -1 \end{pmatrix}$

$= \begin{pmatrix} -1 - 2 \\ 3 + 1 \end{pmatrix}$

$\therefore$ WX $= \begin{pmatrix} -3 \\ 4 \end{pmatrix}$

Also, the length of WX is the magnitude of WX. It is given by:

161

$$|WX| = \sqrt{(-3)^2 + 4^2}$$
$$= \sqrt{9 + 16}$$
$$= \sqrt{25}$$
$$= 5$$
$$\therefore \quad |WX| = 5 \text{ units}$$

b. Similarly, WZ = Z − W   (Take note of the arrangement of the letters, as the second letter comes first in the subtraction)

$$WZ = \begin{pmatrix} -3 \\ -2 \end{pmatrix} - \begin{pmatrix} 2 \\ -1 \end{pmatrix}$$

$$= \begin{pmatrix} -3 - 2 \\ -2 - -1 \end{pmatrix}$$

$$= \begin{pmatrix} -3 - 2 \\ -2 + 1 \end{pmatrix}$$

$$\therefore \quad WZ = \begin{pmatrix} -5 \\ -1 \end{pmatrix}$$

Also, the length of WZ is the magnitude of WZ. It is given by:

$$|WZ| = \sqrt{(-5)^2 + (-1)^2}$$
$$= \sqrt{25 + 1}$$
$$|WZ| = \sqrt{26} \text{ units}$$

c.   −2XY = −2(Y − X)

$$= -2\left[\begin{pmatrix} -2 \\ 1 \end{pmatrix} - \begin{pmatrix} -1 \\ 3 \end{pmatrix}\right]$$
$$= -2\left[\begin{pmatrix} -2 - -1 \\ 1 - 3 \end{pmatrix}\right]$$
$$= -2\begin{pmatrix} -2 + 1 \\ -2 \end{pmatrix}$$
$$= -2\begin{pmatrix} -1 \\ -2 \end{pmatrix}$$
$$= \begin{pmatrix} 2 \\ 4 \end{pmatrix} \quad \text{(Take note of the change in sign due to the multiplication by −2)}$$

$$\therefore \quad -2XY = \begin{pmatrix} 2 \\ 4 \end{pmatrix}$$

Also, the length of −2XY is its magnitude. It is given by:

$$\therefore \quad |-2XY| = \sqrt{2^2 + 4^2}$$
$$= \sqrt{4 + 16}$$
$$= \sqrt{20}$$
$$\therefore \quad |-2XY| = \sqrt{20} \text{ units}$$

d.   YW = W − Y

$$YW = \begin{pmatrix} 2 \\ -1 \end{pmatrix} - \begin{pmatrix} -2 \\ 1 \end{pmatrix}$$
$$= \begin{pmatrix} 2 - -2 \\ -1 - 1 \end{pmatrix}$$

$$= \begin{pmatrix} 2+2 \\ -1-1 \end{pmatrix}$$

$$\therefore \quad YW = \begin{pmatrix} 4 \\ -2 \end{pmatrix}$$

Also, $|YW| = \sqrt{4^2 + (-2)^2}$

$$= \sqrt{16 + 4}$$

$$\therefore \quad |YW| = \sqrt{20} \text{ units}$$

4. A(4, 7) is the vertex of a triangle ABC. BA = $\begin{pmatrix} 5 \\ 3 \end{pmatrix}$ and AC = $\begin{pmatrix} 4 \\ -3 \end{pmatrix}$.

a. Find the coordinates of B and C

b. If M is the midpoint of the line BC, find AM.

Solution

a. With respect to the origin, A = $\begin{pmatrix} 4 \\ 7 \end{pmatrix}$

Let the column vector B be, B = $\begin{pmatrix} d \\ e \end{pmatrix}$

$\therefore$ The component, BA = A − B.   Substitute each value into this equation. This gives:

$$BA = A - B$$

$$\begin{pmatrix} 5 \\ 3 \end{pmatrix} = \begin{pmatrix} 4 \\ 7 \end{pmatrix} - \begin{pmatrix} d \\ e \end{pmatrix}$$

$$\begin{pmatrix} 5 \\ 3 \end{pmatrix} = \begin{pmatrix} 4-d \\ 7-e \end{pmatrix}$$

$$\therefore \qquad 5 = 4 - d \quad \text{(By equating the upper parts)}$$

$$d = 4 - 5$$

$$d = -1$$

Similarly, $3 = 7 - e$ (By equating the lower parts)

$$e = 7 - 3$$

$$e = 4$$

Hence, B = $\begin{pmatrix} -1 \\ 4 \end{pmatrix}$

$\therefore$ The coordinates of B are (−1, 4)

Also, let the column vector C be, C = $\begin{pmatrix} f \\ g \end{pmatrix}$

$\therefore$ The component:

$$AC = C - A$$

$$\begin{pmatrix} 4 \\ -3 \end{pmatrix} = \begin{pmatrix} f \\ g \end{pmatrix} - \begin{pmatrix} 4 \\ 7 \end{pmatrix}$$

$$\begin{pmatrix} 4 \\ -3 \end{pmatrix} = \begin{pmatrix} f-4 \\ g-7 \end{pmatrix}$$

$$\therefore \qquad 4 = f - 4 \quad \text{(By equating the upper parts)}$$

$$4 + 4 = f$$

$$f = 8$$

Similarly, $-3 = g - 7$ (By equating the lower parts)

$$-3 + 7 = g$$

$$g = 4$$

Hence, C = $\begin{pmatrix} 8 \\ 4 \end{pmatrix}$

∴ The coordinates of C are (8, 4)

b. The midpoint of BC is M = ½(B + C)

$$M = \frac{1}{2}\left[\begin{pmatrix} -1 \\ 4 \end{pmatrix} + \begin{pmatrix} 8 \\ 4 \end{pmatrix}\right]$$

$$= \frac{1}{2}\left[\begin{pmatrix} -1+8 \\ 4+4 \end{pmatrix}\right]$$

$$= \frac{1}{2}\begin{pmatrix} 7 \\ 8 \end{pmatrix}$$

$$M = \begin{pmatrix} \frac{7}{2} \\ 4 \end{pmatrix}$$

∴ AM = M − A

$$= \begin{pmatrix} \frac{7}{2} \\ 4 \end{pmatrix} - \begin{pmatrix} 4 \\ 7 \end{pmatrix}$$

$$= \begin{pmatrix} \frac{7}{2} - 4 \\ 4 - 7 \end{pmatrix}$$

$$AM = \begin{pmatrix} -\frac{1}{2} \\ -3 \end{pmatrix}$$

Exercise 14

1. What is the magnitude of the vector AB = $\begin{pmatrix} 8 \\ 6 \end{pmatrix}$

2. If the magnitude of the vector $\begin{pmatrix} x \\ 5 \end{pmatrix}$ is 13, find the value of x.

3. Given that p = $\begin{pmatrix} -6 \\ 10 \end{pmatrix}$, q = $\begin{pmatrix} -3 \\ 1 \end{pmatrix}$, r = $\begin{pmatrix} 2 \\ -8 \end{pmatrix}$, find:

a. p + q − 2r

b. p − q + r

c. $\frac{2}{3}$(p − q) − $\frac{1}{4}$r

d. |2p + 5q − 3r|

4. If a = $\begin{pmatrix} 5 \\ 2 \end{pmatrix}$, b = $\begin{pmatrix} 3 \\ -8 \end{pmatrix}$, c = $\begin{pmatrix} -4 \\ -3 \end{pmatrix}$, find the values of:

a. p and q such that pa − qb = 2c

b. |m| if m = 5a − 3b

5. If u = $\begin{pmatrix} 1 \\ 7 \end{pmatrix}$ and v = $\begin{pmatrix} 3 \\ -2 \end{pmatrix}$

a. Find 4u − 5v

b. Express u in the form (K, $\theta°$) where k is magnitude of u and θ is the angle u makes with the x– axis.

6. If $n = \begin{pmatrix} 6 \\ -4 \end{pmatrix}$, find the angle θ that n makes with the horizontal axis. Hence express n in the form (u, θ°), where u is the magnitude of n.

7. The vertices of a quadrilateral ABCD are A(3, –2), B(1, 2), C(–1, 3), D(–3, –4). Determine the components and lengths of:
a. AB b. AD c. –3BC d. BD

8. P(2, 5) is the vertex of a triangle PQR. QP = $\begin{pmatrix} 3 \\ -1 \end{pmatrix}$ and PR = $\begin{pmatrix} 6 \\ 4 \end{pmatrix}$.
a. Find the coordinates of Q and R
b. If M is the midpoint of the line QR, find PM.

9. What is the magnitude of the vector AB = $\begin{pmatrix} 24 \\ 17 \end{pmatrix}$

10. If the magnitude of the vector $\begin{pmatrix} x \\ 14 \end{pmatrix}$ is 20, find the value of x.

11. Given that p = $\begin{pmatrix} -1 \\ 5 \end{pmatrix}$, q = $\begin{pmatrix} -4 \\ 7 \end{pmatrix}$, r = $\begin{pmatrix} 2 \\ 3 \end{pmatrix}$, find:
a. 2p – 3q + r
b. p + q – r
c. $\frac{1}{2}(p + 2q) - \frac{1}{4}r$
d. |5p + 3q – 4r|

12. If a = $\begin{pmatrix} 2 \\ 5 \end{pmatrix}$, b = $\begin{pmatrix} 3 \\ -3 \end{pmatrix}$, c = $\begin{pmatrix} 5 \\ 6 \end{pmatrix}$, find the values of:
a. p and q such that pa + qb = 2c
b. |d| if d = 2a – 3b

13. The vertices of a quadrilateral PQRS are P(3, –2), Q(–5, 1), R(–1, 3), S(–2, –4). Determine the components and lengths of:
a. PR b. RQ c. –5QS d. 2PQ

CHAPTER 15
SIMPLIFICATION OF ALGEBRAIC FRACTIONS

When simplifying single algebraic fractions, factorize the numerator and denominator and then cancel out common terms. However, when two or more algebraic fractions are to be simplified, we treat them like the usual way of dealing with fractions by taking LCM and following the necessary steps.

Examples

1. Simplify $\dfrac{a^2 + ab}{a^2 + ac}$

Solution

Factorize the fraction to give:

$$\frac{a^2 + ab}{a^2 + ac} = \frac{a(a+b)}{a(a+c)}$$

The 'a' will cancel out to give the final answer as follows:

$$\frac{\cancel{a}(a+b)}{\cancel{a}(a+c)} = \frac{a+b}{a+c}$$

Note that we can only cancel out terms linking one another by multiplication. Terms that are linked together by addition sign or subtraction sign cannot be cancelled out. For example, in example 1 above, the fraction $\dfrac{a(a+b)}{a(a+c)}$ can also be written as: $\dfrac{a \times (a+b)}{a \times (a+c)}$. The 'a's linked by multiplication sign can

be cancelled out as shown in our solution above. However, in the final answer given by $\dfrac{a+b}{a+c}$, the a's cannot be cancelled out to give $\dfrac{b}{c}$ since they are linked by addition sign.

2. Simplify $\dfrac{c^2 - 2c - 15}{c^2 - 3c - 10}$

Solution

The numerator and denominator are quadratic expressions. We factorize them as follows:

$$\frac{c^2 - 2c - 15}{c^2 - 3c - 10} = \frac{(c-5)(c+3)}{(c-5)(c+2)}$$

Therefore, $(c-5)$ will cancel out to give our final answer as follows:

$$\frac{\cancel{(c-5)}(c+3)}{\cancel{(c-5)}(c+2)} = \frac{(c+3)}{(c+2)}$$

Refer to my book 'Simplified Basic Algebra' for explanations on factorization of quadratic expression in order to understand how the factorization above was carried out.

3. Simplify $\dfrac{8 - 2a - a^2}{2a^2 - 3a - 2}$

<u>Solution</u>

$$\dfrac{8 - 2a - a^2}{2a^2 - 3a - 2}$$

Let us factorize $8 - 2a - a^2$ as follows:

Multiply the first and last terms to give:

$8(-a^2) = -8a^2$.

Two numbers in 'a' whose product is $-8a^2$ and sum is $-2a$ are $-4a$ and $2a$. Replace $-2a$ in the original expression with these two terms. This gives:

$8 - 4a + 2a - a^2$

We now factorize by grouping as follows:

$8 - 4a + 2a - a^2 = 4(2 - a) + a(2 - a)$

$\qquad\qquad\qquad\qquad = (2 - a)(4 + a)$ (After using $(2 - a)$ as a common term of the expression)

Similarly, we factorize the denominator as follows:

$2a^2 - 3a - 2 = 2a^2 - 4a + a - 2$

$\qquad\qquad\qquad = 2a(a - 2) + 1(a - 2)$

$\qquad\qquad\qquad = (a - 2)(2a + 1)$

$\therefore \quad \dfrac{8 - 2a - a^2}{2a^2 - 3a - 2} = \dfrac{(2-a)(4+a)}{(a - 2)(2a + 1)}$

$\qquad\qquad = \dfrac{-(a - 2)(4+a)}{(a - 2)(2a + 1)}$ [Note that $(2 - a)$ can also be expressed as $-(a -2)$]

Therefore, $(a - 2)$ will cancel out to give:

$\dfrac{-(a - 2)(4+a)}{(a - 2)(2a + 1)} = \dfrac{- \cancel{(a - 2)}(4 + a)}{\cancel{(a - 2)}(2a + 1)}$

$\qquad\qquad = \dfrac{-(4 + a)}{(2a + 1)}$

Note that an expression such as $(a - 2)$ can be converted to $-(2 - a)$. This is done by simply putting a negative sign outside the bracket and changing the signs of the two terms inside the bracket. Similarly, $(5 - x)$ can also be changed to $-(x - 5)$. We only carry out this kind of conversion in order to make it look the same like another term, so that the two terms can be canceled out during the simplification.

4. Simplify: $\dfrac{9a^2 - m^2}{m^2 - 2am - 3a^2}$

<u>Solution</u>

The numerator is a difference of two squares. Recall that a difference of two squares is factorized as follows:

$$a^2 - b^2 = (a + b)(a - b)$$

Similarly, $9a^2 - m^2 = (3a)^2 - m^2$

$$= (3a + m)(3a - m)$$

In order to factorize the denominator, we multiply the first and last terms, since it is a quadratic expression. This gives:

$$m^2(-3a^2) = -3a^2m^2$$

Two numbers in 'am' whose product is $-3a^2m^2$ and sum is $-2am$ are $-3am$ and am.

$$\therefore \quad m^2 - 2am - 3a^2 = m^2 - 3am + am - 3a^2$$

$$= m(m - 3a) + a(m - 3a)$$

$$= (m - 3a)(m + a)$$

$$\therefore \quad \frac{9a^2 - m^2}{m^2 - 2am - 3a^2} = \frac{(3a + m)(3a - m)}{(m - 3a)(m + a)}$$

$$= \frac{-(3a + m)(m - 3a)}{(m - 3a)(m + a)} \qquad \text{(Note that } (3a - m) = -(m - 3a)\text{, as shown)}$$

Hence, $(m - 3a)$ cancels out as follows:

$$\frac{-(3a + m)(m - 3a)}{(m - 3a)(m + a)} = \frac{-(3a + m)\cancel{(m - 3a)}}{\cancel{(m - 3a)}(m + a)}$$

$$= \frac{-(3a + m)}{(m + a)}$$

5. Simplify: $\dfrac{a^2 - am - an + mn}{a^2 - am + an - mn}$

<u>Solution</u>

Factorize both numerator and denominator by grouping terms. This gives:

$$\frac{a^2 - am - an + mn}{a^2 - am + an - mn} = \frac{a(a - m) - n(a - m)}{a(a - m) + n(a - m)}$$

$$= \frac{(a - m)(a - n)}{(a - m)(a + n)}$$

Therefore, $(a - m)$ cancels out as follows:

$$\frac{(a - m)(a - n)}{(a - m)(a + n)} = \frac{\cancel{(a - m)}(a - n)}{\cancel{(a - m)}(a + n)}$$

$$= \frac{(a - n)}{(a + n)}$$

6. Simplify $\dfrac{a+2}{a} - \dfrac{1}{3ab}$

Solution

$$\dfrac{a+2}{a} - \dfrac{1}{3ab}$$

The LCM of a and 3ab is 3ab. Hence divide the LCM by each of the denominator and multiply the value obtained by the respective numerator. This means that $3ab \div a = 3b$. Then multiply 3b by a + 2 (the numerator). Similarly, $3ab \div 3ab = 1$. Then multiply 1 by 1(the numerator). Finally divide these multiplied terms by the LCM. This is as shown below.

$$\dfrac{a+2}{a} - \dfrac{1}{3ab} = \dfrac{3b(a+2) - 1(1)}{3ab}$$

$$= \dfrac{3ab + 6b - 1}{3ab}$$

7. Simplify: $\dfrac{1}{4x-2y} - \dfrac{1}{y-2x}$

Solution

$$\dfrac{1}{4x-2y} - \dfrac{1}{y-2x}$$

A careful look at one of the denominator (i.e. $4x - 2y$), shows that it can be factorized and made to look like the other denominator. This is carried out as follows.

$$\dfrac{1}{4x-2y} - \dfrac{1}{y-2x} = \dfrac{1}{2(2x-y)} - \dfrac{1}{y-2x}$$

$$= \dfrac{1}{2(2x-y)} - \dfrac{1}{-(2x-y)} \qquad \text{[Note that (y – 2x) is also –(2x – y)]}$$

$$= \dfrac{1}{2(2x-y)} + \dfrac{1}{(2x-y)} \qquad \text{(Note that –(–) has become +)}$$

The LCM of 2(2x – y) and (2x – y) is 2(2x – y). Hence we continue our simplification as follows:

$$\dfrac{1}{2(2x-y)} + \dfrac{1}{(2x-y)} = \dfrac{1+2}{2(2x-y)}$$

$$= \dfrac{3}{2(2x-y)}$$

8. Simplify $\dfrac{3}{2ab} + \dfrac{4}{3bc}$

Solution

$$\dfrac{3}{2ab} + \dfrac{4}{3bc}$$

The LCM of 2ab and 3bc is 6abc. Therefore:

$$\frac{3}{2ab} + \frac{4}{3bc} = \frac{3c(3) + 2a(4)}{6abc}$$

$$= \frac{9c + 8a}{6abc}$$

9. Simplify $\dfrac{3x}{x-1} - \dfrac{4}{x+2}$

Solution

$$\frac{3x}{x-1} - \frac{4}{x+2}$$

The LCM of $(x-1)$ and $(x+2)$ is $(x-1)(x+2)$. We now simplify as follows:

$$\frac{3x}{x-1} - \frac{4}{x+2} = \frac{(x+2)3x - 4(x-1)}{(x-1)(x+2)}$$

$$= \frac{3x^2 + 6x - 4x + 4}{(x-1)(x+2)}$$

$$= \frac{3x^2 + 2x + 4}{(x-1)(x+2)}$$

10. Simplify $\dfrac{1}{3xy} + \dfrac{5}{4y^2z} - \dfrac{3y}{2x^3}$

Solution

$$\frac{1}{3xy} + \frac{5}{4y^2z} - \frac{3y}{2x^3}$$

The LCM of $3xy$, $4y^2z$ and $2x^3$ is $12x^3y^2z$. Hence, we use this LCM to simplify the expression as follows:

$$\frac{1}{3xy} + \frac{5}{4y^2z} - \frac{3y}{2x^3} = \frac{1(4x^2yz) + 5(3x^3) - 3y(6y^2z)}{12x^3y^2z}$$

$$= \frac{4x^2yz + 15x^3 - 18y^3z}{12x^3y^2z}$$

11. Simplify $\dfrac{3y}{y^2 - z^2} - \dfrac{3z}{z^2 - y^2}$

Solution

$$\frac{3y}{y^2 - z^2} - \frac{3z}{z^2 - y^2}$$

The denominators are difference of two squares. Let us factorize the denominators and make them look alike.

$$\frac{3y}{y^2-z^2} - \frac{3z}{z^2-y^2} = \frac{3y}{(y-z)(y+z)} - \frac{3z}{(z-y)(z+y)}$$

$$= \frac{3y}{(y-z)(y+z)} - \frac{3z}{-(y-z)(y+z)}$$

Note that $(z-y)$ is also $-(y-z)$. Note also that $(z+y)$ is the same as $(y+z)$ as rearranged above. Hence the $-(-)$ will become $+$ as follows:

$$\frac{3y}{(y-z)(y+z)} + \frac{3z}{(y-z)(y+z)}$$

The LCM of $(y-z)(y+z)$ and $(y-z)(y+z)$ is simply $(y-z)(y+z)$. When terms are the same, just take one of them as the LCM. Hence, we now continue with the simplification as follows:

$$\frac{3y}{(y-z)(y+z)} + \frac{3z}{(y-z)(y+z)} = \frac{3y(1)+3z(1)}{(y-z)(y+z)}$$ (Note that $(y-z)(y+z) \div (y-z)(y+z) = 1$)

$$= \frac{3y+3z}{(y-z)(y+z)}$$

$$= \frac{3(y+z)}{(y-z)(y+z)}$$

$$= \frac{3\,\cancel{(y+z)}}{(y-z)\cancel{(y+z)}}$$

$$= \frac{3}{(y-z)}$$

Exercise 15

1. Simplify $\dfrac{m^2+md}{m^2+mf}$

2. Simplify $\dfrac{x^2-4x-21}{x^2-x-12}$

3. Simplify $\dfrac{6-x-x^2}{3x^2+4x-15}$

4. Simplify $\dfrac{16m^2-n^2}{n^2-3mn-4m^2}$

5. Simplify: $\dfrac{p^2-pq-pr+qr}{p^2-pq+pr-qr}$

6. Simplify $\dfrac{2a + 3}{b} - \dfrac{2}{3b}$

7. Simplify: $\dfrac{3}{10x - 5y} - \dfrac{2}{y - 2x}$

8. Simplify $\dfrac{1}{6mn} + \dfrac{3}{2n}$

9. Simplify $\dfrac{5a}{a - 2} - \dfrac{2}{a + 3}$

10. Simplify $\dfrac{2}{8ab} + \dfrac{1}{3a^2b} - \dfrac{3a}{12b^3}$

11. Simplify $\dfrac{4}{9m^2 - 4n^2} - \dfrac{5n}{4n^2 - 9m^2}$

CHAPTER 16
EQUATIONS AND SUBSTITUTIONS INVOLVING FRACTIONS

Examples

1. Solve the equation $\dfrac{3}{2b-5} - \dfrac{4}{b-3} = 0$

<u>Solution</u>

$$\frac{3}{2b-5} - \frac{4}{b-3} = 0$$

Multiply each term by the LCM, i.e. $(2b-5)(b-3)$. This gives:

$$(2b-5)(b-3)\frac{3}{2b-5} - (2b-5)(b-3)\frac{4}{b-3} = (2b-5)(b-3)0$$

Therefore, we cancel the common denominators as follows:

$$\cancel{(2b-5)}(b-3)\frac{3}{\cancel{2b-5}} - (2b-5)\cancel{(b-3)}\frac{4}{\cancel{b-3}} = 0 \quad \text{(Note that zero multiplies a number to give}$$

zero)

$$(b-3)3 - (2b-5)4 = 0$$

$$3b - 9 - 8b + 20 = 0 \qquad \text{(Take note of the negative sign outside the second bracket)}$$

$$3b - 8b = 9 - 20$$

$$-5b = -11$$

$$b = \frac{-11}{-5}$$

$$b = 2\frac{1}{5} \qquad \text{(Note that the negative sign has cancelled out)}$$

2. Solve the equation $\dfrac{a-4}{7} = \dfrac{2}{3a-1}$

<u>Solution</u>

$$\frac{a-4}{7} = \frac{2}{3a-1}$$

We can multiply each term in the equation by the LCM, i.e. $7(3a-1)$. However, since there is a single fraction on each side of the equation, a simple way to solve this kind of equation is to cross multiply. This gives:

$$(a-4)(3a-1) = 7(2)$$

$$3a^2 - a - 12a + 4 = 14$$

$$3a^2 - 13a + 4 - 14 = 0$$

$$3a^2 - 13a - 10 = 0$$

Solving this quadratic equation by factorization gives:

$$3a^2 - 15a + 2a - 10 = 0$$

Note that $-15a$ and $+2a$ are the two numbers whose product will give $-30a^2$ (i.e. $3a^2 \times (-10) = -30a^2$) and whose sum will give $-13a$. Hence they are used to substitute for $-13a$ in the original quadratic equation. Therefore, we factorize by grouping as follows:

$$3a(a - 5) + 2(a - 5) = 0$$
$$(a - 5)(3a + 2) = 0$$
$$\therefore \quad a = 5 \text{ or } a = -\frac{2}{3}$$

3. Solve the equation $\dfrac{3}{x - 4} = \dfrac{2}{x - 1} - 4$

Solution

$$\frac{3}{x - 4} = \frac{2}{x - 1} - 4$$

Multiply each term by the LCM, i.e. $(x - 4)(x - 1)$. This gives:

$$(x-4)(x-1)\frac{3}{x-4} = (x-4)(x-1)\frac{2}{x-1} - (x-4)(x-1)4$$

We now cancel out the common terms as follows:

$$\cancel{(x-4)}(x-1)\frac{3}{\cancel{x-4}} = (x-4)\cancel{(x-1)}\frac{2}{\cancel{x-1}} - (x-4)(x-1)4$$

$$(x-1)3 = (x-4)2 - 4(x-4)(x-1)$$
$$3x - 3 = 2x - 8 - 4(x^2 - x - 4x + 4)$$
$$3x - 3 = 2x - 8 - 4x^2 + 4x + 16x - 16$$
$$3x - 3 = 22x - 24 - 4x^2$$
$$4x^2 + 3x - 22x - 3 + 24 = 0$$
$$4x^2 - 19x + 21 = 0$$

Solving this quadratic equation by factorization gives:

$$4x^2 - 12x - 7x + 21 = 0$$
$$4x(x - 3) - 7(x - 3) = 0$$
$$(x - 3)(4x - 7) = 0$$
$$x = 3 \text{ or } x = \frac{7}{4}$$
$$x = 3 \text{ or } x = 1\frac{3}{4}$$

Undefined Fractions

A fraction is undefined if the denominator of the fraction is zero.

Examples

1. Find the value of x for which the fraction $\dfrac{8}{15 + 3x}$ is undefined.

174

Solution

$$\frac{8}{15 + 3x}$$

For this fraction to be undefined, the denominator must be equal to zero. Hence:

$15 + 3x = 0$ (Note that only the denominator is equated to zero)

$3x = -15$

$x = \dfrac{-15}{3}$

$x = -5$

$\therefore$ The fraction is undefined when $x = -5$

2. Find the value of x for which the fraction $\dfrac{6x}{2x - 5}$ is not defined.

Solution

For this fraction not to be defined, the denominator must be equal to zero. Hence:

$2x - 5 = 0$

$2x = 5$

$x = \dfrac{5}{2}$

$x = 2\dfrac{1}{2}$

$\therefore$ The fraction is not defined when $x = 2\dfrac{1}{2}$

3. For what value(s) of b is the fraction $\dfrac{2b + 11}{b^2 + b - 20}$

a. not defined

b. equal to zero?

Solution

a. $\dfrac{2b + 11}{b^2 + b - 20}$

For this fraction not to be defined, we equate the denominator to zero as follows:

$b^2 + b - 20 = 0$

Solving this quadratic equation by factorization gives:

$(b + 5)(b - 4) = 0$

Therefore, b = −5 or b = 4

Hence, the expression is not defined when b = −5 or b = 4

b. Recall that a fraction is zero when the numerator of the fraction is zero. Therefore we now equate the numerator to zero as follows:

175

$2b + 11 = 0$

$2b = -11$

$b = \dfrac{-11}{2}$

$b = -5\dfrac{1}{2}$

The fraction is equal to zero when $b = -5\dfrac{1}{2}$

4. For what value(s) of x is the expression $\dfrac{x^2 + 3x + 2}{2x + 3}$

a. undefined

b. zero?

<u>Solution</u>

a. $\dfrac{x^2 + 3x + 2}{2x + 3}$

For this expression to be undefined:

$2x + 3 = 0$

$2x = -3$

$x = \dfrac{-3}{2}$

Therefore, $x = -1\dfrac{1}{2}$

Hence, the expression is undefined when $x = -1\dfrac{1}{2}$

b. This expression is zero if:

$x^2 + 3x + 2 = 0$

Solving this quadratic equation by factorization gives:

$x^2 + 3x + 2 = 0$

$(x + 1)(x + 2) = 0$

$x = -1$ or $x = -2$

$\therefore$ The expression is zero when $x = -1$ or $x = -2$

5. For what values of x is the expression $\dfrac{2x^2 + 3x - 2}{3x^2 - 14x + 8}$

a. zero

b. undefined?

Solution

a. $\dfrac{2x^2 + 3x - 2}{3x^2 - 14x + 8}$

This expression is zero if the numerator is zero. Therefore, we equate it to zero as follows:

$$2x^2 + 3x - 2 = 0$$

Let us solve this quadratic equation by factorization as follows:

$$2x^2 + 3x - 2 = 0$$
$$2x^2 + 4x - x - 2 = 0$$
$$2x(x + 2) - 1(x + 2) = 0$$
$$(x + 2)(2x - 1) = 0$$
$$x = -2 \text{ or } x = \frac{1}{2} \qquad \text{(After equating each bracket to zero and solving for } x\text{)}$$

∴ The expression is zero when $x = -2$ or $x = \dfrac{1}{2}$

b. The expression is undefined if the denominator is zero. Therefore, we equate it to zero as follows:

$$3x^2 - 14x + 8 = 0$$

Let us solve this quadratic equation by factorization as follows:

$$3x^2 - 14x + 8 = 0$$
$$3x^2 - 12x - 2x + 8 = 0$$
$$3x(x - 4) - 2(x - 4) = 0$$
$$(x - 4)(3x - 2) = 0$$
$$x = 4 \text{ or } x = \frac{2}{3} \qquad \text{(After equating each bracket to zero and solving for } x\text{)}$$

∴ The expression is undefined when $x = 4$ or $x = \dfrac{2}{3}$

Substitution in Algebraic Fractions

Examples

1. If $y = \dfrac{a^2 - b}{c - b}$, find the value of y when a = −2, b = −1 and c = 4

Solution

$$y = \frac{a^2 - b}{c - b}$$

We simply substitute the various given values of a, b and c into the equation in order to obtain the value of y. This gives:

$$y = \frac{a^2 - b}{c - b}$$

177

$$= \frac{(-2)^2 - (-1)}{4 - (-1)}$$

$$= \frac{4 + 1}{4 + 1}$$

$$= \frac{5}{5}$$

$$y = 1$$

2. If $x = \dfrac{m^3 + 5n - p}{2mp + m - 3n^2}$, find the value of x when m = −1, n = 3 and p = 6

Solution

$$x = \frac{m^3 + 5n - p}{2mp + m - 3n^2}$$

We simply substitute the various given values of m, n and p into the equation in order to obtain the value of x. This gives:

$$x = \frac{m^3 + 5n - p}{2mp + m - 3n^2}$$

$$= \frac{(-1)^3 + 5(3) - 6}{2(-1)(6) + (-1) - 3(3)^2}$$

$$= \frac{-1 + 15 - 6}{-12 - 1 - 27}$$

$$= \frac{8}{-40}$$

$$x = -\frac{1}{5}$$

3. If p : q = 9 : 5, evaluate $\dfrac{15p - 2q}{5p + 16q}$

Solution

An easy way of solving this problem is to substitute 9 for p and 5 for q in the given expression.

$$\therefore \quad \frac{15p - 2q}{5p + 16q} = \frac{15(9) - 2(5)}{5(9) + 16(5)}$$

$$= \frac{135 - 10}{45 + 80}$$

$$= \frac{125}{125}$$

$$= 1$$

4. If $a : b = 5 : 3$, evaluate $\dfrac{6a + b}{a - \frac{1}{3}b}$

Solution

An easy way of solving this problem is to substitute 5 for a and 3 for b in the given expression.

$$\therefore \quad \frac{6a + b}{a - \frac{1}{3}b} = \frac{6(5) + 3}{5 - \frac{1}{3}(3)}$$

$$= \frac{30 + 3}{5 - 1}$$

$$= \frac{33}{4}$$

$$= 8\frac{1}{4}$$

5. If $\dfrac{m}{n} = \dfrac{3}{4}$, evaluate $\dfrac{7m + n}{2m - \frac{1}{5}n}$

Solution

An easy way of solving this problem is to substitute 3 for m and 4 for q in the given expression.

$$\therefore \quad \frac{7m + n}{2m - \frac{1}{5}n} = \frac{7(3) + 4}{2(3) - \frac{1}{5}(4)}$$

$$= \frac{21 + 4}{6 - \frac{4}{5}}$$

$$= \frac{25}{\frac{30 - 4}{5}}$$

$$= \frac{25}{\frac{26}{5}}$$

$$= \frac{25}{1} \times \frac{5}{26}$$

$$= \frac{125}{26}$$

$$= 4\frac{21}{26}$$

6. If $a = \frac{x+2}{x-1}$, express $\frac{2a+3}{a-1}$ in terms of x

<u>Solution</u>

In order to express $\frac{2a+3}{a-1}$ in terms of x, we simply substitute the expression for 'a' in $\frac{2a+3}{a-1}$.

Since 'a' is expressed in terms x, then $\frac{2a+3}{a-1}$ will become expressed in x if 'a' is substituted in it.

Therefore, let us substitute $\frac{x+2}{x-1}$ for 'a' in $\frac{2a+3}{a-1}$. This gives:

$$\frac{2a+3}{a-1} = \frac{2\left(\frac{x+2}{x-1}\right)+3}{\frac{x+2}{x-1}-1}$$

$$= \frac{\left(\frac{2x+4}{x-1}\right)+3}{\frac{x+2}{x-1}-1}$$

$$= \frac{\left(\frac{2x+4+3(x-1)}{x-1}\right)}{\frac{x+2-1(x-1)}{x-1}}$$

$$= \frac{\left(\frac{2x+4+3x-3}{x-1}\right)}{\frac{x+2-x+1)}{x-1}}$$

$$= \frac{\left(\frac{5x+1}{x-1}\right)}{\frac{3}{x-1}}$$

$$= \frac{5x+1}{x-1} \times \frac{x-1}{3}$$

$$= \frac{5x+1}{\cancel{x-1}} \times \frac{\cancel{x-1}}{3}$$

$$= \frac{5x+1}{3}$$

7. If $m = \dfrac{y+5}{y-3}$, express $\dfrac{5m+2}{m-3}$ in terms of y

Solution

In order to express $\dfrac{5m+2}{m-3}$ in terms of y, we simply substitute the expression for m in $\dfrac{5m+2}{m-3}$.

Since m is expressed in terms y, then $\dfrac{5m+2}{m-3}$ will become expressed in y if m is substituted in it.

Therefore, let us substitute $\dfrac{y+5}{y-3}$ for m in $\dfrac{5m+2}{m-3}$. This gives:

$$\frac{5m+2}{m-3} = \frac{5\left(\dfrac{y+5}{y-3}\right)+2}{\left(\dfrac{y+5}{y-3}\right)-3}$$

$$= \frac{\left(\dfrac{5y+25}{y-3}\right)+2}{\left(\dfrac{y+5}{y-3}\right)-3}$$

$$= \frac{\left(\dfrac{5y+25+2(y-3)}{y-3}\right)}{\dfrac{y+5-3(y-3)}{y-3}}$$

$$= \frac{\left(\dfrac{5y+25+2y-6}{y-3}\right)}{\dfrac{y+5-3y+9)}{y-3}}$$

$$= \frac{\left(\dfrac{7y+19}{y-3}\right)}{\dfrac{14-2y}{y-3}}$$

$$= \frac{7y+19}{y-3} \times \frac{y-3}{14-2y}$$

$$= \frac{7y+19}{\cancel{y-3}} \times \frac{\cancel{y-3}}{14-2y}$$

$$= \frac{7y+19}{14-2y}$$

Exercise 16

1. Solve the equation $\dfrac{2}{m-2} - \dfrac{3}{m-5} = 0$

2. Solve the equation $\dfrac{b-1}{5} = \dfrac{5}{2b-7}$

3. Solve the equation $\dfrac{1}{x-5} = \dfrac{9}{2x-12} - 4$

4. Find the value of x for which the fraction $\dfrac{1}{8+x}$ is undefined.

5. Find the value of x for which the fraction $\dfrac{2x}{3x-7}$ is not defined.

6. For what value(s) of a is the fraction $\dfrac{5a+12}{a^2-a-30}$

   a. zero
   b. undefined?

7. For what value(s) of m is the expression $\dfrac{2m^2+8m+8}{5m+9}$

   a. zero
   b. undefined?

8. For what values of x is the expression $\dfrac{5x^2+13x-6}{2x^2-x-15}$

   a. zero
   b. undefined?

9. If m = $\dfrac{p^2-q}{q-r}$, find the value of m when p = −5, q = 3 and r = -2

10. If $x = \dfrac{2a^3+5b-8c}{7ab+b-3c^2}$, find the value of x when a = −2, b = 5 and c = - 4

11. If m : n = 2 : 7, evaluate $\dfrac{6m+4n}{12m-2n}$

12. If a : b = 9 : 5, evaluate $\dfrac{2a+3b}{3a-\frac{1}{5}b}$

CHAPTER 17
SIMULTANEOUS EQUATIONS INVOLVING FRACTIONS

Examples

1. Solve the simultaneous equation: $\frac{x}{2} - \frac{y}{3} = \frac{1}{6}$

    $\frac{x}{2} - \frac{y}{6} = 5$

Solution

$\frac{x}{2} - \frac{y}{3} = \frac{1}{6}$ ....Equation (1)

$\frac{x}{2} - \frac{y}{6} = 5$ ....Equation (2)

This is a case where the variable is in the numerator. In cases like this, we clear out the fractions by multiplying each term by the LCM of the denominators. The LCM in equation (1) and (2) is 6. Hence we clear out the fractions by multiplying each term in equation (1) and (2) by 6. This gives:

$6(\frac{x}{2}) - 6(\frac{y}{3}) = 6(\frac{1}{6})$ ....Equation (1)

$6(\frac{x}{2}) - 6(\frac{y}{6}) = 6(5)$ ....Equation (2)

These now simplify to give equation 3 and equation 4 respectively, and we solve them simultaneously as shown below:

$3x - 2y = 1$ ....Equation (3)

$\underline{3x - y = 30}$ ....Equation (4)

Equation (4) – equation (3):   y = 29   (Note that –y –(–2y) = –y + 2y = y. Also, 30 – 1 = 29)

In equation (3), substitute 29 for y in order to obtain x. This gives:

$3x - 2y = 1$ ....Equation (3)

$3x - 2(29) = 1$

$3x - 58 = 1$

$3x = 1 + 58$

$3x = 59$

$x = \frac{59}{3}$

Hence, $x = \frac{59}{3}$ and y = 29

2. Solve simultaneously: $\dfrac{1}{x} + \dfrac{1}{y} = 5$

$$\dfrac{1}{y} - \dfrac{1}{x} = 1$$

Solution

METHOD 1

$$\dfrac{1}{x} + \dfrac{1}{y} = 5$$

$$\dfrac{1}{y} - \dfrac{1}{x} = 1$$

In these equations the variables are in the denominators. So, let us take $\dfrac{1}{x} = a$ and $\dfrac{1}{y} = b$. Hence we can rewrite the equations as follows:

    a + b = 5

    b − a = 1

Or,

$$a + b = 5 \ \text{....................Equation (1)}$$

$$\underline{-a + b = 1} \ \text{..................Equation (2)}$$

Equation (1) + equation (2):   2b = 6

$$b = \dfrac{6}{2}$$

    b = 3

Substitute 3 for b in equation (1). This gives:

    a + b = 5 ....Equation (1)

    a + 3 = 5

    a = 5 − 3

    a = 2

Hence a = 2 and b = 3

But recall that $\dfrac{1}{x} = a$ and $\dfrac{1}{y} = b$, and we are actually solving for x and y. Therefore, we now substitute the values of a and b appropriately, in order to obtain x and y as follows:

$$\dfrac{1}{x} = a$$

$$\dfrac{1}{x} = 2 \qquad \text{(Since a = 2)}$$

When we cross multiply it gives:

    $2x = 1$

$$x = \dfrac{1}{2}$$

Similarly, $\dfrac{1}{y} = b$

$\dfrac{1}{y} = 3$ (Since b = 3)

When we cross multiply it gives:

$\qquad 3y = 1$

$\qquad y = \dfrac{1}{3}$

Therefore, $x = \dfrac{1}{2}$ and $y = \dfrac{1}{3}$

METHOD 2

$\qquad \dfrac{1}{x} + \dfrac{1}{y} = 5$

$\qquad \dfrac{1}{y} - \dfrac{1}{x} = 1$

Rearranging the equations gives:

$\qquad \dfrac{1}{x} + \dfrac{1}{y} = 5$ ....Equation (1)

$\qquad -\dfrac{1}{x} + \dfrac{1}{y} = 1$ ....Equation (2)

Since the coefficients of $\dfrac{1}{x}$ are the same in the two equations, but with different signs, then $\dfrac{1}{x}$ can be eliminated by using elimination method as follows:

$$\dfrac{1}{x} + \dfrac{1}{y} = 5 \text{ ....Equation (1)}$$

$$-\dfrac{1}{x} + \dfrac{1}{y} = 1 \text{ ....Equation (2)}$$

Equation (1) + equation (2): $\dfrac{2}{y} = 6$ (Note that $\dfrac{1}{y} + \dfrac{1}{y} = \dfrac{2}{y}$ and $5 + 1 = 6$)

Hence, $y = \dfrac{2}{6}$

$\qquad y = \dfrac{1}{3}$

Substitute $\dfrac{1}{3}$ for y in equation 2 in order to obtain x. This gives:

$\qquad -\dfrac{1}{x} + \dfrac{1}{y} = 1$ ....Equation (2)

$\qquad -\dfrac{1}{x} + \dfrac{1}{\frac{1}{3}} = 1$

$$3 - 1 = \frac{1}{x}$$

$$2 = \frac{1}{x}$$

$$x = \frac{1}{2}$$

Hence, $x = \frac{1}{2}$ and $y = \frac{1}{3}$

3. Solve the simultaneous equations: $\quad \dfrac{3}{c} - \dfrac{4}{d} = \dfrac{1}{3}$

$$\dfrac{2}{c} - \dfrac{5}{d} = 1$$

Solution

$$\dfrac{3}{c} - \dfrac{4}{d} = \dfrac{1}{3} \quad ....Equation\ (1)$$

$$\dfrac{2}{c} - \dfrac{5}{d} = 1 \quad ....Equation\ (2)$$

Let us make the coefficient of $\dfrac{1}{c}$ to be equal in the two equations so that it can be eliminated.

Therefore, multiply equation (1) by 2 and equation (2) by 3. This gives:

$$\dfrac{6}{c} - \dfrac{8}{d} = \dfrac{2}{3} \quad ....Equation\ (3)$$

$$\dfrac{6}{c} - \dfrac{15}{d} = 3 \quad ....Equation\ (4)$$

Equation (3) – Equation (4): $\quad \dfrac{7}{d} = -\dfrac{7}{3}$

Hence, d = $\dfrac{7 \times 3}{-7}$

$$d = -3$$

Substitute –3 for d in equation (2) in order to obtain c. This gives:

$$\dfrac{2}{c} - \dfrac{5}{d} = 1 \quad ....Equation\ (2)$$

$$\dfrac{2}{c} - \dfrac{5}{-3} = 1$$

$$\dfrac{2}{c} + \dfrac{5}{3} = 1$$

$$\dfrac{2}{c} = 1 - \dfrac{5}{3}$$

$$\dfrac{2}{c} = \dfrac{-2}{3}$$

$$c = \frac{2 \times 3}{-2}$$

$$c = -3$$

Therefore, c = –3 and d = –3

4. Solve the simultaneous equations: $\frac{a}{2} - \frac{b}{5} = 1$

$$b - \frac{a}{3} = 8$$

Solution

Rearranging the equations gives:

$$\frac{a}{2} - \frac{b}{5} = 1 \text{ ....Equation (1)}$$

$$-\frac{a}{3} + b = 8 \text{ ....Equation (2)}$$

Let us make the coefficient of a to be equal in the two equations so that it can be eliminated.

Therefore, multiply equation (1) by $\frac{1}{3}$ and equation (2) by $\frac{1}{2}$. This gives:

$$\frac{a}{6} - \frac{b}{15} = \frac{1}{3} \text{ ....Equation (3)}$$

$$-\frac{a}{6} + \frac{b}{2} = 4 \text{ ....Equation (4)}$$

Equation (3) + Equation (4): $\frac{13b}{30} = \frac{13}{3}$

Hence, $b = \frac{30 \times 13}{13 \times 3}$

$$b = 10$$

Substitute 10 for b in equation (1) in order to obtain a. This gives:

$$\frac{a}{2} - \frac{b}{5} = 1 \text{ ....Equation (1)}$$

$$\frac{a}{2} - \frac{10}{5} = 1$$

$$\frac{a}{2} = 1 + 2$$

$$a = 3 \times 2$$

$$a = 6$$

Hence, a = 6 and b = 10

5. Solve simultaneously the equations: $\dfrac{3}{m} + \dfrac{5}{n} = 1$

$$\dfrac{9}{m} - \dfrac{5}{n} = 1$$

Solution

$$\dfrac{3}{m} + \dfrac{5}{n} = 1 \dots\dots\dots\dots\dots\text{Equation(1)}$$

$$\dfrac{9}{m} - \dfrac{5}{n} = 1 \dots\dots\dots\dots\dots\text{Equation(2)}$$

Let us make the coefficient of $\dfrac{1}{m}$ to be equal in the two equations by multiplying equation (1) by 3. This gives:

$$\dfrac{9}{m} + \dfrac{15}{n} = 3 \dots\dots\dots\dots\text{Equation (3)}$$

$$\dfrac{9}{m} - \dfrac{5}{n} = 1 \dots\dots\dots\dots\text{Equation (4)}$$

Equation (3) – Equation (4): $\qquad \dfrac{20}{n} = 2$

Hence, $n = \dfrac{20}{2}$

$\qquad n = 10$

Substitute 10 for n in equation (1) in order to obtain m. This gives:

$$\dfrac{3}{m} + \dfrac{5}{n} = 1 \dots\dots\dots\dots\dots\text{Equation(1)}$$

$$\dfrac{3}{m} + \dfrac{5}{10} = 1$$

$$\dfrac{3}{m} = 1 - \dfrac{1}{2}$$

$$\dfrac{3}{m} = \dfrac{1}{2}$$

$$m = 3 \times 2$$

$$m = 6$$

Hence, m = 6 and n = 10

Exercise 17

1. Solve the simultaneous equation: $\dfrac{x}{3} - \dfrac{y}{4} = -\dfrac{1}{6}$

$$\dfrac{x}{3} - \dfrac{y}{8} = \dfrac{1}{12}$$

2. Solve simultaneously, the equation: $\dfrac{1}{a} + \dfrac{1}{b} = 3$

$$\dfrac{1}{a} - \dfrac{1}{b} = 5$$

3. Solve the simultaneous equations: $\dfrac{2}{c} + \dfrac{3}{d} = 2$

$$\dfrac{3}{c} - \dfrac{1}{d} = 1\dfrac{1}{6}$$

4. Solve the simultaneous equations: $\dfrac{a}{5} - \dfrac{b}{4} = \dfrac{1}{10}$

$$a - \dfrac{b}{2} = 1$$

5. Solve simultaneously the equations: $\dfrac{4}{p} + \dfrac{6}{q} = 4$

$$\dfrac{1}{p} - \dfrac{3}{q} = -8$$

6. Solve the simultaneous equations: $\dfrac{2}{c} - \dfrac{10}{d} = 3$

$$\dfrac{5}{c} - \dfrac{5}{d} = 3\dfrac{1}{2}$$

7. Solve the simultaneous equations: $\dfrac{x}{2} - \dfrac{y}{4} = \dfrac{1}{20}$

$$\dfrac{x}{4} + \dfrac{y}{4} = \dfrac{7}{40}$$

8. Solve simultaneously the equations: $\dfrac{1}{m} + \dfrac{1}{n} = 2$

$$\dfrac{3}{m} - \dfrac{6}{n} = 5$$

CHAPTER 18
ABSOLUTE VALUE EQUATION (MODULUS EQUATION)

The absolute value of a number is a positive (or zero) value of the number. Whether the number was originally positive or negative, its absolute value must be positive. For example, $|7| = 7$ and $|-7| = 7$. This shows that for each absolute value of a number/expression, there are two possible numbers/expressions. Hence, when solving absolute value equations, we split the equation into two possible equations.

Examples

1. Solve $|x| = 5$

Solution

$\quad |x| = 5$

Remove the absolute value bars and split the equation into two cases. The first case should have a positive right hand side, while the second case should have a negative right hand side. This gives:

$\quad x = 5 \quad$ or $\quad x = -5$

2. Solve the following equations:

a. $|3x - 2| = 4$

b. $|2x + 1| = 9$

c. $|7 - 5x| = 2$

Solutions

a. $|3x - 2| = 4$

Simply remove the bars and take a positive right hand value, and then a negative right hand value to obtain two possible equations. Then solve each of the equations separately. This gives:

$3x - 2 = 4$	or	$3x - 2 = -4$
$3x = 4 + 2$		$3x = -4 + 2$
$3x = 6$		$3x = -2$
$x = \dfrac{6}{3}$		$x = -\dfrac{2}{3}$
$x = 2$		

Hence, $x = 2$ or $x = -\dfrac{2}{3}$

Take note of how the two equations were solved separately.

b. $|2x + 1| = 9$

Remove the bars and take a positive right hand value to form one equation. Form a second equation by also removing the bars and taking a negative right hand value. Then solve each of the equations separately. This gives:

$$2x + 1 = 9 \qquad \text{or} \qquad 2x + 1 = -9$$
$$2x = 9 - 1 \qquad\qquad\qquad 2x = -9 - 1$$
$$2x = 8 \qquad\qquad\qquad\quad 2x = -10$$
$$x = \frac{8}{2} \qquad\qquad\qquad\quad x = \frac{-10}{2}$$
$$x = 4 \qquad\qquad\qquad\quad\; x = -5$$

Hence, $x = 4$ or $x = -5$

c. $|7 - 5x| = 2$

$$7 - 5x = 2 \qquad \text{or} \qquad 7 - 5x = -2$$
$$7 - 2 = 5x \qquad\qquad\qquad 7 + 2 = 5x$$
$$5 = 5x \qquad\qquad\qquad\quad 9 = 5x$$
$$x = \frac{5}{5} \qquad\qquad\qquad\quad 5x = 9$$
$$x = 1 \qquad\qquad\qquad\quad x = \frac{9}{5}$$

Hence, $x = 1$ or $x = 1\frac{4}{5}$

3. Solve the following equation:
a. $|4x - 11| = 0$
b. $|2x - 5| = -1$

Solutions

a. $|4x - 11| = 0$

In this case there is only one equation that we can obtain. This is given by:

$$4x - 11 = 0$$
$$4x = 11$$
$$x = \frac{11}{4}$$

Hence, $x = 2\frac{3}{4}$

b. $|2x - 5| = -1$

Recall that an absolute value must be positive. But this equation is telling us that absolute value is negative. This is impossible. An absolute value cannot be negative (i.e. −1 in this case), hence there is no solution to this equation.

4. Solve the following equations:

a. $|x-5| + 8 = 12$

b. $|3x+7| - 14 = -4$

Solution

a. $|x-5| + 8 = 12$

Take the constant term (outside the bars) to the right hand side of the equation. This gives:

$|x-5| = 12 - 8$

$|x-5| = 4$

Let us now remove the bars and split this equation into two possible equations. This gives:

$x-5 = 4$	or	$x-5 = -4$
$x = 4 + 5$		$x = -4 + 5$
$x = 9$		$x = 1$

Hence, $x = 9$ or $x = 1$

b. $|3x+7| - 14 = -4$

Do not think that this equation has no solution since there is a negative value on the right hand side. If we simplify the equation we will get a positive value on the right hand side, with the absolute value bars still in place. Therefore, let us simplify the equation as follows.

Take the constant term to the right hand side of the equation. This gives:

$|3x+7| = -4 + 14$

$|3x+7| = 10$

Now that we have a positive value on the right hand side, let us remove the bars and split this equation into two possible equations. This gives:

$3x+7 = 10$	or	$3x+7 = -10$
$3x = 10 - 7$		$3x = -10 - 7$
$3x = 3$		$3x = -17$
$x = \dfrac{3}{3}$		$x = -\dfrac{17}{3}$
$x = 1$		$x = -5\dfrac{2}{3}$

Hence, $x = 1$ or $x = -5\dfrac{2}{3}$

5. Solve the following absolute value equations:

a. $|x-6| = 2x-3$

b. $|5x+2| = x+4$

c. $|3x-4| = |2x+3|$

Solution

a. $|x-6| = 2x-3$

Let us split the equation into two equations as follows:

$x-6 = 2x-3$ or $x-6 = -(2x-3)$

$x-2x = -3+6$ $x-6 = -2x+3$

     $-x = 3$ $x+2x = 3+6$

      $x = -3$ $3x = 9$

$$x = \frac{9}{3}$$

$$x = 3$$

Hence, $x = -3$ or $x = 3$

Now, in equations such as this, it is necessary for us to test the values obtained to see if they are true solutions. In order to do this, we substitute the values of x into the right side of the equation (the side without the absolute value bars) to see if it will give us a positive value. A negative value means that the value of x obtained is not a solution to the equation. Let us write out the part of the equation without the absolute value bar as follows:

$2x-3$

When $x = -3$, the above expression gives:

$2x-3 = 2(-3) - 3$

         $= -6 - 3 = -9$

Hence $x = -3$ is not a solution since it gives a negative right hand side. This is because an absolute value cannot be negative.

When $x = 3$, the above expression gives:

$2x-3 = 2(3) - 3$

         $= 6 - 3 = 3$

Hence $x = 3$ is a solution since it gives a positive value on the right hand side.

Therefore the overall solution of the equation is $x = 3$

b. $|5x+2| = x+4$

Let us split the equation into two equations as follows:

$5x+2 = x+4$ or $5x+2 = -(x+4)$

$5x-x = 4-2$ $5x+2 = -x-4$

  $4x = 2$ $5x+x = -4-2$

$$x = \frac{2}{4}$$ $6x = -6$

$$x = \frac{1}{2}$$ $$x = \frac{-6}{6}$$

$$x = -1$$

Let us now find out if these two values of x are solutions of the equations. In order to do this, we substitute the values of x into the right side of the equation (the side without the absolute value bars). Let us write out the part of the equation without the absolute value bar as follows:

$$x + 4$$

When $x = \dfrac{1}{2}$, the above expression gives:

$$x + 4 = \dfrac{1}{2} + 4$$

$$= 4\dfrac{1}{2}$$

Hence $x = \dfrac{1}{2}$ is a solution since it gives a positive right hand side.

When $x = -1$, the above expression gives:

$$x + 4 = -1 + 4$$

$$= 3$$

Hence $x = -1$ is a solution since it gives a positive value on the right hand side.

Therefore the overall solutions of the equation are $x = \dfrac{1}{2}$ or $x = -1$

c. $|3x - 4| = |2x + 3|$

Let us split the equation into two equations as follows:

$3x - 4 = 2x + 3$	or	$3x - 4 = -(2x + 3)$
$3x - 2x = 3 + 4$		$3x - 4 = -2x - 3$
$x = 7$		$3x + 2x = -3 + 4$
		$5x = 1$
		$x = \dfrac{1}{5}$

Since this question has the absolute value bars on both side of the equation, then our two values of x are correct. There is no need to verify out solution.

Therefore, $x = 7$ or $x = \dfrac{1}{5}$

6. Solve the following absolute value equations:

a. $||5x + 2| - 5| = 12$

b. $||7 - 2x| - 8| = 5$

Solution

a. $||5x + 2| - 5| = 12$

This is a case where one absolute value expression is nested in another. Working with the outermost bars, let us split the overall equation into two equations. This gives:

$|5x + 2| - 5 = 12$	or	$|5x + 2| - 5 = -12$
$|5x + 2| = 12 + 5$		$|5x + 2| = -12 + 5$
$|5x + 2| = 17$		$|5x + 2| = -7$

Out of these two new equations that we have obtained, one of them has no solution. The equation $|5x+2| = -7$ has no solution since an absolute value cannot be negative. Hence we discard $|5x+2| = -7$. However, the other equation which is $|5x+2| = 17$ can be solved. Hence, let us solve it by splitting it into two equations as follows:

$5x + 2 = 17$ or $5x + 2 = -17$

$5x = 17 - 2$ $5x = -17 - 2$

$5x = 15$ $5x = -19$

$x = \dfrac{15}{5}$ $x = \dfrac{-19}{5}$

$x = 3$ $x = -3\dfrac{4}{5}$

Therefore, $x = 3$ or $x = -3\dfrac{4}{5}$

b. $||7 - 2x| - 8| = 5$

Starting with the outermost bars, we split the overall equation into two equations as follows:

$|7 - 2x| - 8 = 5$ or $|7 - 2x| - 8 = -5$

$|7 - 2x| = 5 + 8$ $|7 - 2x| = -5 + 8$

$|7 - 2x| = 13$ ....Equation (1) $|7 - 2x| = 3$ ....Equation (2)

Now we have to take each of the equations above and split it into two equations. Let us split equation (1) into two equations as follows:

$7 - 2x = 13$ or $7 - 2x = -13$

$-2x = 13 - 7$ $-2x = -13 - 7$

$-2x = 6$ $-2x = -20$

$x = \dfrac{6}{-2}$ $x = \dfrac{-20}{-2}$

$x = -3$ $x = 10$

Hence $x = -3$ or $x = 10$

Let us take equation (2) above and split it into two equations as follows:

$7 - 2x = 3$ or $7 - 2x = -3$

$-2x = 3 - 7$ $-2x = -3 - 7$

$-2x = -4$ $-2x = -10$

$x = \dfrac{-4}{-2}$ $x = \dfrac{-10}{-2}$

$x = 2$ $x = 5$

Hence $x = 2$ or $x = 5$

Therefore the overall solutions to the original equation are: $x = -3, 10, 2$ or 5

7. Solve the following equations:

a. $|2x-3| = |5x+1| + 1$

b. $|x+7| - |3x-1| = -4$

Solution

a. $|2x-3| = |5x+1| + 1$

There are three possible equations that we are going to solve.

(i) The first equation that we are going to solve is obtained by simply dropping the absolute value bars in the equation. This gives:

$$2x - 3 = 5x + 1 + 1$$
$$2x - 5x = 1 + 1 + 3$$
$$-3x = 5$$
$$x = -\frac{5}{3}$$

(ii) The second equation that we are going to solve is obtained by making the two absolute value terms to be negative. This gives:

$$-(2x - 3) = -(5x + 1) + 1$$
$$-2x + 3 = -5x - 1 + 1$$
$$-2x + 5x = -1 + 1 - 3$$
$$3x = -3$$
$$x = \frac{-3}{3}$$
$$x = -1$$

(iii) Lastly, the third equation that we have to solve is obtained by making only the absolute value term on the left hand side to be negative. This gives:

$$-(2x - 3) = 5x + 1 + 1$$
$$-2x + 3 = 5x + 1 + 1$$
$$-2x - 5x = 1 + 1 - 3$$
$$-7x = -1$$
$$x = \frac{-1}{-7}$$
$$x = \frac{1}{7}$$

Hence the three values of x obtained are $-\frac{5}{3}$, -1 and $\frac{1}{7}$.

Now we have to check which of these values are solutions of the equation. We do this by substituting each of the value of x into the equation. The equation is:

$$|2x-3| = |5x+1| + 1$$

When $x = -\frac{5}{3}$, we have:

$$|2(-\tfrac{5}{3}) - 3| = |5(-\tfrac{5}{3}) + 1| + 1$$

$$|-\tfrac{10}{3} - 3| = |-\tfrac{25}{3} + 1| + 1$$

$$|\tfrac{-10 - 9}{3}| = |\tfrac{-25 + 3}{3}| + 1$$

$$|\tfrac{-19}{3}| = |\tfrac{-22}{3}| + 1$$

$$\tfrac{19}{3} = \tfrac{22}{3} + 1 \qquad \text{(Take note of the removal of the negative sign when the bars are removed}$$

$$\tfrac{19}{3} = \tfrac{22+3}{3}$$

$$\tfrac{19}{3} = \tfrac{25}{3}$$

Since both sides of the equation are not equal, then $x = -\dfrac{5}{3}$ is not a solution of the equation.

When $x = -1$, we have:

$$|2x - 3| = |5x + 1| + 1$$

$$|2(-1) - 3| = |5(-1) + 1| + 1$$

$$|-2 - 3| = |-5 + 1| + 1$$

$$|-5| = |-4| + 1$$

$$5 = 4 + 1$$

$$5 = 5$$

Since both sides are equal, then $x = -1$ is a solution of the equation.

When $x = \dfrac{1}{7}$, we have:

$$|2(\tfrac{1}{7}) - 3| = |5(\tfrac{1}{7}) + 1| + 1$$

$$|\tfrac{2}{7} - 3| = |\tfrac{5}{7} + 1| + 1$$

$$|\tfrac{2-21}{7}| = |\tfrac{5+7}{7}| + 1$$

$$|\tfrac{-19}{7}| = |\tfrac{12}{7}| + 1$$

$$\tfrac{19}{7} = \tfrac{12}{7} + 1$$

$$\tfrac{19}{7} = \tfrac{12 + 7}{7}$$

$$\tfrac{19}{7} = \tfrac{19}{7}$$

Since both sides are equal, then $x = \dfrac{1}{7}$ is a solution of the equation.

Therefore the solutions of the equation are $x = -1$ and $x = \dfrac{1}{7}$

b. $|x+7| - |3x-1| = -4$

Let us rearrange the equation in order to have only one of the absolute value expression on one side of the equation. This gives:

$$|x+7| = |3x-1| - 4$$

Let us solve each of the three possible equations from this equation.

(i) The first equation that we are going to solve is obtained by simply dropping the absolute value bars in the equation. This gives:

$x + 7 = 3x - 1 - 4$

$x - 3x = -1 - 4 - 7$

$-2x = -12$

$x = \dfrac{-12}{-2}$

$x = 6$

(ii) The second equation is obtained by making the two absolute value terms to be negative. This gives:

$-(x+7) = -(3x-1) - 4$

$-x - 7 = -3x + 1 - 4$

$-x + 3x = 1 - 4 + 7$

$2x = 4$

$x = \dfrac{4}{2}$

$x = 2$

(iii) The third equation that we have to solve is obtained by making only the absolute value term on the left hand side to be negative. This gives:

$-(x+7) = 3x - 1 - 4$

$-x - 7 = 3x - 1 - 4$

$-x - 3x = -1 - 4 + 7$

$-4x = 2$

$x = \dfrac{2}{-4}$

$x = -\dfrac{1}{2}$

Hence the three values of x obtained are 6, 2 and $-\dfrac{1}{2}$

Now, we have to check which of these values are solutions of the equation. We do this by substituting each of the values of x into the equation. The equation is:

$$|x+7| = |3x-1| - 4$$

When $x = 6$, we have:

$$|6+7| = |3(6) - 1| - 4$$

$$|13| = |18 - 1| - 4$$
$$13 = |17| - 4$$
$$13 = 17 - 4$$
$$13 = 13$$

Since both sides of the equation are equal, then $x = 6$ is a solution of the equation.

When $x = 2$, we have:

$$|2 + 7| = |3(2) - 1| - 4$$
$$|9| = |6 - 1| - 4$$
$$9 = |5| - 4$$
$$9 = 5 - 4$$
$$9 = 1$$

Since both sides of the equation are not equal, then $x = 2$ is not a solution of the equation.

When $x = -\dfrac{1}{2}$, we have:

$$\left|-\frac{1}{2} + 7\right| = \left|3\left(-\frac{1}{2}\right) - 1\right| - 4$$
$$\left|\frac{-1 + 14}{2}\right| = \left|-\frac{3}{2} - 1\right| - 4$$
$$\left|\frac{13}{2}\right| = \left|\frac{-3 - 2}{2}\right| - 4$$
$$\frac{13}{2} = \left|\frac{-5}{2}\right| - 4$$

$$\frac{13}{2} = \frac{5}{2} - 4$$
$$\frac{13}{2} = \frac{5 - 8}{2}$$
$$\frac{13}{2} = -\frac{3}{2}$$

Since both sides of the equation are not equal, then $x = -\dfrac{1}{2}$ is not a solution of the equation.

Therefore the only solution of the equation is $x = 6$

8. Solve the following equations:

a. $-2|4x + 9| = -22$

b. $3|x - 2| + 5 = -2|x - 2| + 10$

Solution

a. $-2|4x + 9| = -22$

Divide both sides by −2. This gives:

$$|4x + 9| = \frac{-22}{-2}$$
$$|4x + 9| = 11$$

We now split this equation into two equations as follows:

$$4x + 9 = 11 \qquad \text{or} \qquad 4x + 9 = -11$$
$$4x = 11 - 9 \qquad\qquad\qquad 4x = -11 - 9$$
$$4x = 2 \qquad\qquad\qquad\qquad 4x = -20$$
$$x = \frac{2}{4} \qquad\qquad\qquad\qquad x = \frac{-20}{4}$$
$$x = \frac{1}{2} \qquad\qquad\qquad\qquad x = -5$$

Therefore, $x = \dfrac{1}{2}$ or $x = -5$

b. $3|x - 2| + 5 = -2|x - 2| + 10$

This equation is similar to $3m + 5 = -2m + 10$, where m is $|x - 2|$. Hence we can solve for m. However, we are not going to use m, we will solve the equation using $|x - 2|$ as a variable. This is done as follows.

$$3|x - 2| + 5 = -2|x - 2| + 10$$

Collect terms in $|x - 2|$ on the left hand side of the equation. This gives:

$3|x - 2| + 2|x - 2| = 10 - 5$ (The left hand side is similar to 3m + 2m which will give 5m)

$5|x - 2| = 5$ (The left side is like 5m which is $5|x - 2|$ if we imagine m = $|x - 2|$)

Divide both sides by 5. This gives:

$$|x - 2| = \frac{5}{5}$$
$$|x - 2| = 1$$

We now split this equation into two equations as follows:

$$x - 2 = 1 \qquad \text{or} \qquad x - 2 = -1$$
$$x = 1 + 2 \qquad\qquad\qquad x = -1 + 2$$
$$x = 3 \qquad\qquad\qquad\qquad x = 1$$

Therefore, $x = 3$ or $x = 1$

9. Solve the following equations:

a. $|2x^2 - 7x + 6| = 0$

b. $|x^2 + 5x - 80| = 4$

Solutions

a. $|2x^2 - 7x + 6| = 0$

Since the right hand side of the equation is zero, we can split this equation to only one equation. This gives:

$$2x^2 - 7x + 6 = 0$$

Solving this quadratic equation by factorization gives:

$$2x^2 - 4x - 3x + 6 = 0$$

$2x(x-2) - 3(x-2) = 0$

$(x-2)(2x-3) = 0$

Hence, $x = 2$ or $\dfrac{3}{2}$.   (After equating each bracket above to zero and solving for x)

b. $|x^2 + 5x - 80| = 4$

We can split this equation into two equations as follows:

   $x^2 + 5x - 80 = 4$ ....Equation (1)

   $x^2 + 5x - 80 = -4$ ....Equation (2)

Let us solve equation (1) as follows:

   $x^2 + 5x - 80 = 4$

   $x^2 + 5x - 80 - 4 = 0$

   $x^2 + 5x - 84 = 0$

Solving this equation by factorization gives:

   $(x \quad)(x \quad) = 0$

We now find two numbers such that their product is −84 and their sum is +5. The two numbers are +12 and −7. The two number are entered into the brackets above to give:

   $(x + 12)(x - 7) = 0$

Hence, $x = -12$ or $x = 7$

Let us now solve equation (2) above as follows:

   $x^2 + 5x - 80 = -4$

   $x^2 + 5x - 80 + 4 = 0$

   $x^2 + 5x - 76 = 0$

Let us use quadratic formula to solve this equation. From the equation, a = 1, b = 5 and c = −76

$x = \dfrac{-b \pm \sqrt{b^2 - 4ac}}{2a}$

$= \dfrac{-5 \pm \sqrt{5^2 - [4(1)(-76)]}}{2(1)}$

$= \dfrac{-5 \pm \sqrt{25 - (-304)}}{2}$

$= \dfrac{-5 \pm \sqrt{25 + 304}}{2}$

$= \dfrac{-5 \pm \sqrt{329}}{2}$

$= \dfrac{-5 \pm 18.14}{2}$

$$x = \frac{-5 + 18.14}{2} \quad \text{or} \quad x = \frac{-5 - 18.14}{2}$$

$$= \frac{13.14}{2} \quad \text{or} \quad \frac{-23.14}{2}$$

$$x = 6.57 \quad \text{or} \quad x = -11.57$$

Therefore, the four solutions of the equation are $x = -12, 7, 6.57$ and -11.57.

Exercise 18

1. Solve $|x| = 9$

2. Solve the following equations:
   a. $|5x + 7| = 17$
   b. $|3x + 1| = 8$
   c. $|15 - 4x| = 18$

3. Solve the following equation:
   a. $|7x + 13| = 8$
   b. $|5x - 27| = -14$

4. Solve the following equations:
   a. $|4x - 3| + 6 = 17$
   b. $|x + 9| - 11 = -7$

5. Solve the following absolute value equations:
   a. $|3x - 6| = x - 2$
   b. $|8x + 5| = 7x + 40$
   c. $|2x - 5| = |10x - 1|$

6. Solve the following absolute value equations:
   a. $||2x + 7| - 5| = 6$
   b. $||9 - 4x| - 8| = 1$

7. Solve the following equations:
   a. $|3x - 1| = |2x + 5| + 6$
   b. $|5x + 2| - |11x - 3| = -1$

8. Solve the following equations:

a. $-5|4x + 9| = 25$

b. $2|2x - 3| + 1 = -|2x - 3| + 22$

9. Solve the following equations:

a. $|3x^2 - 7x - 6| = 0$

b. $|x^2 + 14x - 68| = 4$

10. Solve the following equations:

a. $|5x^2 - 13x + 6| = 0$

b. $|2x^2 + 9x - 12| = 3$

CHAPTER 19
INEQUALITIES INVOLVING ABSOLUTE VALUES, QUOTIENT AND SQUARE FUNCTIONS

The modulus or absolute value of a number is the size of the number without its sign. It is denoted by a vertical line enclosing a number. For example $|5| = 5$ and $|-5| = 5$. A negative number enclosed in the vertical lines is regarded as positive.

If an inequality is given by : $|x| < 2$, it means that x is in the range: $-2 < x < 2$, i.e. $x < 2$ and $x > -2$

Examples

1. Solve the inequality: $|5x - 2| < 13$

<u>Solution</u>

$|5x - 2| < 13$

Note that this also means $|-(5x - 2)| < 13$.

The range of the term in modulus is given by:

$-13 < 5x - 2 < 13$

We now take them separately and solve. Taking the first part (left hand part) gives:

$-13 < 5x - 2$

$-13 + 2 < 5x$

$-11 < 5x$

$-\dfrac{11}{5} < x$

Or $x > -\dfrac{11}{5}$ (Make sure the open side of the inequality sign still faces x when rearranging it).

Let us now take the other part of the inequality range and solve. This gives:

$5x - 2 < 13$

$5x < 13 + 2$

$5x < 15$

$x < \dfrac{15}{5}$

$x < 3$

We now combine the two results to obtain the range of the solution as follows:

$-\dfrac{11}{5} < x < 3$

Note that the opened end of the inequality sign is facing x in the solution $x > -\dfrac{11}{5}$, while the closed end (elbow end) of the inequality sign is facing x in the solution $x < 3$. Therefore, when combining the solution (i.e. $-\dfrac{11}{5} < x < 3$) you have to ensure that the right part of the inequality sign is facing x.

2. Solve the inequality $|3x + 7| \geq 10$

<u>Solution</u>

$$|3x + 7| \geq 10$$

This means: $(3x + 7) \geq 10$. This directly gives: $3x + 7 \geq 10$

It also means $-(3x + 7) \geq 10$. When we divide both sides by -1, it gives

$$\dfrac{-(3x + 7)}{-1} \geq \dfrac{10}{-1}$$

$\therefore$ $3x + 7 \leq -10$

Take note of the reversal of the inequality sign since both sides of the inequality were divided by a negative number.

Hence, the two inequalities obtained from $|3x + 7| \geq 10$ are:

$$3x + 7 \geq 10 \text{ and } 3x + 7 \leq -10$$

We now take each inequality and solve. This gives:

$$3x + 7 \geq 10$$
$$3x \geq 10 - 7$$
$$3x \geq 3$$
$\therefore$ $x \geq \dfrac{3}{3}$
$$x \geq 1$$

Solving the second inequality obtained gives:

$$3x + 7 \leq -10$$
$$3x \leq -10 - 7$$
$$3x \leq -17$$
$$x \leq -\dfrac{17}{3}$$
$$x \leq -5\dfrac{2}{3}$$

$\therefore$ $|3x + 7| \geq 10$ is satisfied when $x \geq 1$ and $x \leq -5\dfrac{2}{3}$

Note that these two results cannot be combined together since the solutions do not range from one to the other. This is because $x \geq 1$ means x can be 1, 2, 3 ..., while $x \leq -5\dfrac{2}{3}$ means that x can be $-5\dfrac{2}{3}$, -6, -7.... Hence, the two set of values do not meet and so cannot be combined together. A solution such as $x \leq 1$ and $x \geq -3$ can be combined together to give $-3 \leq x \leq 1$. This is because the values meet as follows: $-3, -2, -1, 0, 1$. Therefore, take note of this point before combining any pair of solutions.

3. Solve: $|4x - 9| \leq 3$

Solution

$\quad |4x - 9| \leq 3$

The two inequalities that can be obtained from this are:

$\quad 4x - 9 \leq 3$

And $\quad 4x - 9 \geq -3$ (This is obtained by simply reversing the inequality sign and changing the positive number on the right to be negative)

We now solve each of the inequality as follows:

$\quad 4x - 9 \leq 3$

$\quad 4x \leq 3 + 9$

$\quad 4x \leq 12$

$\quad x \leq \dfrac{12}{4}$

$\quad x \leq 3$

And: $\quad 4x - 9 \geq -3$

$\quad 4x \geq -3 + 9$

$\quad 4x \geq \dfrac{6}{4}$

$\quad x \geq \dfrac{3}{2}$

Wee can now combine the solutions to obtain: $\dfrac{3}{2} \leq x \leq 3$

$\therefore$ The inequality $|4x - 9| \leq 3$ is satisfied within the range $\dfrac{3}{2} \leq x \leq 3$

4. Solve the inequality $|3m - 5| > 4$

Solution

$\quad |3m - 5| > 4$

The two inequalities that can be obtained from this are:

$\quad 3m - 5 > 4$

And $\quad 3m - 5 < -4$ (Simply reverse the inequality sign and assign a negative sign to the number on the right hand side)

We now solve each of the inequality as follows:

$\quad 3m - 5 > 4$

$\quad 3m > 4 + 5$

$\quad 3m > 9$

$\quad m > \dfrac{9}{3}$

$\quad m > 3$

And $\quad 3m - 5 < -4$

$\quad 3m < -4 + 5$

$$3m < 1$$

$$m < \frac{1}{3}$$

∴ The inequality $|3m - 5| > 4$ is satisfied when $m > 3$ and $m < \frac{1}{3}$ (They cannot be combined)

Inequalities Involving Quotients

If $\frac{x}{y} > 0$, then $\frac{x}{y}$ has to be a positive value.

For $\frac{x}{y}$ to be positive either both x and y must be positive or both x and y must be negative.

If $\frac{x}{y} < 0$, then $\frac{x}{y}$ has to be a negative value. For $\frac{x}{y}$ to be negative, only x or only y must be negative.

However, for inequalities such as $\frac{x}{y} \geq 0$ or $\frac{x}{y} \leq 0$, the same rules above applies to the numerator (i.e. x), but $y \geq 0$ or $y \leq 0$ cannot be used for the denominator so that it will not render the fraction undefined. Hence we use $y > 0$ or $y < 0$ for the denominator. This is because a fraction such as $\frac{2}{0}$ is undefined. So, in $\frac{x}{y} \geq 0$ or $\frac{x}{y} \leq 0$, $y \geq 0$ or $y \leq 0$ is not possible since it will make the fraction undefined.

Examples

1. Solve the inequality $\frac{x-2}{3x+9} > 0$

<u>Solution</u>

$$\frac{x-2}{3x+9} > 0$$

For the inequality above to be greater than zero, (i.e. > 0) it means that it has to be positive. For it to be positive, either the numerator and denominator are both positive (i.e. > 0) or they are both negative (i.e. < 0). This means that:

a. When both numerator and denominator are positive, then:

$$x - 2 > 0$$

and $3x + 9 > 0$

Solving each of these gives:

$$x - 2 > 0$$

∴ $x > 2$

and $3x + 9 > 0$

$$3x > -9$$

$$x > -\frac{9}{3}$$

$$x > -3$$

Therefore, we now have two possible solutions. But we have to test each of the solution to see if they are true. This is done by substituting a possible value in each solution into the original inequality.

Let us test the solution $x > 2$. A possible value here is 3 (since $3 > 2$). Now substitute 3 for x in the original inequality. This gives:

$$\frac{x-2}{3x+9} > 0$$

$$\frac{3-2}{3(3)+9} > 0$$

$$\frac{1}{9+9} > 0$$

$$\frac{1}{18} > 0 \quad \text{(This is true)}$$

Hence, the solution $x > 2$ is correct.

Let us test the solution $x > -3$. A possible value here is -2 (Since $-2 > -3$). Now substitute -2 for x in the original inequality. This gives:

$$\frac{x-2}{3x+9} > 0$$

$$\frac{-2-2}{3(-2)+9} > 0$$

$$\frac{-4}{-6+9} > 0$$

$$\frac{-4}{3} > 0$$

$$-\frac{4}{3} > 0 \quad \text{(This is not true since 0 is greater)}$$

Hence, the solution $x > -3$ is not correct.

b. Now we are through with when numerator and denominator are positive. When numerator and denominator are negative, then:

$$x - 2 < 0$$

And $3x + 9 < 0$

Solving each of them gives:

$$x - 2 < 0$$

$$x < 2$$

And $3x + 9 < 0$

$$3x < -9$$

$$x < -\frac{9}{3}$$

$$x < -3$$

Hence, $x < 2$ or $x < -3$

Let us now test each of the solution to see if they are true.

When $x < 2$, a possible value here is 1 (since $1 < 2$). Now substitute 1 for x in the inequality. This gives:

$$\frac{x-2}{3x+9} > 0$$

$$\frac{1-2}{3(1)+9} > 0$$

$$\frac{-1}{3+9} > 0$$

$$-\frac{1}{12} > 0 \quad \text{(This is not true since 0 is greater than } -\frac{1}{12}\text{)}$$

Hence, the solution $x < 2$ is not correct.

Similarly, when $x < -3$, a possible value here is -4 (since $-4 < -3$). We now substitute -4 for x in the inequality. This gives:

$$\frac{x-2}{3x+9} > 0$$

$$\frac{-4-2}{3(-4)+9} > 0$$

$$\frac{-6}{-12+9} > 0$$

$$\frac{-6}{-3} > 0$$

$$2 > 0 \quad \text{(This is true)}$$

Hence, the solution $x < -3$ is correct.

Hence, in all our four solutions, the two that are correct are $x > 2$ and $x < -3$

Therefore, $\frac{x-2}{3x+9} > 0$ is satisfied when $x > 2$ and $x < -3$

2. Solve the inequality $\frac{2m+6}{m-4} < 0$

<u>Solution</u>

$$\frac{2m+6}{m-4} < 0$$

Example (1) above appears complex because I took my time to thoroughly explain it. In order to avoid making this example as long as example (1), I will use a more direct procedure.

Here, let us solve each of the numerators and denominators using the equality sign as follows:

$$2m + 6 = 0$$

$$2m = -6$$

$$m = -\frac{6}{2}$$

$$m = -3$$

Applying inequality to the solution above implies that:

$$m < -3 \text{ or } m > -3 \quad \text{(Use the two inequality signs)}$$

Let us test each of them to find the correct solution.

When m < –3, a possible value here is – 3.5 (since –3.5 < –3. Testing with a value that just begins the solution puts us on a safer side). Now, substitute –3.5 into the inequality. This gives:

$$\frac{2m + 6}{m - 4} < 0$$

$$\frac{2(-3.5) + 6}{-3.5 - 4} < 0$$

$$\frac{-7 + 6}{-3.5 - 4} < 0$$

$$\frac{-1}{-7.5} < 0$$

$$\frac{1}{7.5} < 0 \qquad \text{(This is not true)}$$

When m > –3, a possible value here is –2.5. Substituting this value into the inequalities gives:

$$\frac{2m + 6}{m - 4} < 0$$

$$\frac{2(-2.5) + 6}{-2.5 - 4} < 0$$

$$\frac{-5 + 6}{-2.5 - 4} < 0$$

$$\frac{1}{-6.5} < 0 \qquad \text{(This is true)}$$

Hence m > –3 is correct.

Let us now carry out the same procedure with the denominator.

$$m - 4 = 0$$

∴ $\qquad m = 4$

Hence, m < 4 or m > 4 $\qquad$ (When the two inequality signs are applied)

When m < 4, let us substitute 3.5 into the inequality as follows:

$$\frac{2m + 6}{m - 4} < 0$$

$$\frac{2(3.5) + 6}{3.5 - 4} < 0$$

$$\frac{7 + 6}{3.5 - 4} < 0$$

$$\frac{13}{-0.5} < 0$$

$$-26 < 0 \qquad \text{(This is true)}$$

Hence, m < 4 is correct

Since this is correct, it means that the other solution of m > 4 is wrong since we already have our two possible solutions. However, let us test the other value.

When m > 4, let us substitute 4.5 into the inequality as follows:

$$\frac{2m + 6}{m - 4} < 0$$

$$\frac{2(4.5) + 6}{4.5 - 4} < 0$$

$$\frac{9+6}{0.5} < 0$$

$$\frac{15}{0.5} < 0$$

$$30 < 0 \qquad \text{(This is not true)}$$

Therefore, our two possible solutions are m > –3 or m < 4. They can be combined to give:

$$-3 < x < 4$$

Therefore, $\frac{2m+6}{m-4} < 0$, is satisfied when $-3 < x < 4$

3. Solve the inequality $\frac{2x-3}{x+2} \leq 1$

Solution

$$\frac{2x-3}{x+2} \leq 1$$

Let us first make the right hand side to be zero.

$$\therefore \quad \frac{2x-3}{x+2} - 1 \leq 0$$

Simplifying with $x + 2$ as the LCM gives:

$$\frac{2x-3-1(x+2)}{x+2} \leq 0$$

$$\frac{2x-3-x-2}{x+2} \leq 0$$

$$\frac{2x-x-3-2}{x+2} \leq 0$$

$$\frac{x-5}{x+2} \leq 0$$

At this point, we can now take the numerator and solve for x as follows:

$$x - 5 = 0$$

$$x = 5$$

Applying inequality to this solution gives:

$$x \leq 5 \text{ or } x \geq 5$$

If we test the inequality when $x = 4.5$ (for $x \leq 5$) and $x = 5.5$ (for $x \geq 5$) respectively, we will find out that the true solution is $x \leq 5$

Taking the denominator gives:

$$x + 2 = 0$$

$$x = -2$$

Applying inequality to this solution gives:

$$x < -2 \text{ or } x > -2$$

Note that we cannot use $\leq$ or $\geq$ for the denominator since the denominator cannot be zero. If this is done, it will make the denominator undefined since fractions such as $\frac{5}{0}$ is undefined.

Now, back to $x < -2$ or $x > -2$. If we test the inequality with $x = -2.5$ (for $x < -2$) and $x = -1.5$ (for $x > -2$), respectively, we will find out that the true solution is $x > -2$.

∴   The solutions are $x \leq 5$ or $x > -2$

Hence, $\dfrac{2x-3}{x+2} \leq 1$ is satisfied when $-2 < x \leq 5$

4. Solve the inequality $\dfrac{2-b}{b+3} \geq 4$

Solution

$$\dfrac{2-b}{b+3} \geq 4$$

Let us first make the right hand side to be zero.

$$\dfrac{2-b}{b+3} \geq 4$$

$$\dfrac{2-b}{b+3} - 4 \geq 0$$

$$\dfrac{2-b-4(b+3)}{b+3} \geq 0$$

$$\dfrac{2-b-4b-12}{b+3} \geq 0$$

$$\dfrac{-5b-10}{b+3} \geq 0$$

Taking the numerator gives:

$$-5b - 10 = 0$$

$$-5b = 10$$

$$b = \dfrac{10}{-5}$$

$$b = -2$$

∴ $b \geq -2$ or $b \leq -2$

Testing possible values (-1.5 and -2.5) in each of these two solutions shows that $b \leq -2$ is the correct solution.

Taking the denominator gives:

$$b + 3 = 0$$

$$b = -3$$

∴ $b < -3$ or $b > -3$ (Note that $\geq$ or $\leq$ cannot be used for the denominator)

Testing values (-3.5 and -2.5) from each of the solution above shows that $b > -3$ is the correct solution.

Hence the solution are $b \leq -2$ or $b > -3$

Therefore, $\dfrac{2-b}{b+3} \geq 4$ is satisfied when $-3 < b \leq -2$

Inequalities Involving Square Functions

If $x^2 < c$, then $x < \sqrt{c}$ or $x > -\sqrt{c}$

Similarly, if $x^2 > c$, then $x > \sqrt{c}$ or $x < -\sqrt{c}$

These rules are applied in inequalities involving square functions.

Examples

1. Solve the inequality $x^2 > 16$

<u>Solution</u>

$$x^2 > 16$$

$\therefore \quad x > \sqrt{16}$

$\quad\quad x > 4$

Or $x < -\sqrt{16}$ (Reverse the inequality sign and introduce a negative sign)

$\quad\quad x < -4$

$\therefore \quad x > 4$ or $x < -4$

Therefore $x^2 > 16$ is satisfied when $x > 4$ or $x < -4$

2. Solve the inequality: $y^2 < 25$

<u>Solution</u>

$$y^2 < 25$$

$\therefore \quad y < \sqrt{25}$

$\quad\quad y < 5$

Or $\quad y > -\sqrt{25}$ (Reverse the inequality sign and introduce a negative sign)

$\quad\quad y > -5$

Therefore $y^2 < 25$ is satisfied when $-5 < y < 5$

3. Solve: $(2a - 3)^2 \geq 49$

<u>Solution</u>

$$(2a - 3)^2 \geq 49$$

$\quad\quad 2a - 3 \geq \sqrt{49}$

$\quad\quad 2a - 3 \geq 7$

$\quad\quad 2a \geq 7 + 3$

$\quad\quad 2a \geq 10$

$\quad\quad a \geq \dfrac{10}{2}$

$\quad\quad a \geq 5$

Or $\quad (2a - 3) \leq -\sqrt{49}$ (Reverse the inequality sign and introduce a negative sign)

$\quad\quad 2a - 3 \leq -7$

$$2a \leq -7 + 3$$
$$2a \leq -4$$
$$a \leq -\frac{4}{2}$$
$$a \leq -2$$

Therefore $(2a - 3)^2 \geq 49$ is satisfied when $x \geq 5$ or $x \leq -2$

4. Solve: $7 - 4y^2 \leq -29$

Solution

$$7 - 4y^2 \leq -29$$
$$-4y^2 \leq -29 - 7$$
$$-4y^2 \leq -36$$
$$y^2 \geq \frac{-36}{-4}$$ (Take note of the reversal of the inequality sign due to division by a negative

number)

$$y^2 \geq 9$$

Hence, $y \geq \sqrt{9}$
$$y \geq 3$$
Or $y \leq -\sqrt{9}$
$$y \leq -3$$

$\therefore$ $7 - 4y^2 \leq -29$ is satisfied when $y \geq 3$ or $y \leq -3$

5. Solve $5m^2 + 2 \geq 12$

Solution

$$5m^2 + 2 \geq 12$$
$$5m^2 \geq 12 - 2$$
$$5m^2 \geq 10$$
$$m^2 \geq \frac{10}{5}$$
$$m^2 \geq 2$$

$\therefore$ $m \geq \sqrt{2}$

Or $m \leq -\sqrt{2}$

Therefore $5m^2 + 2 \geq 12$ is satisfied when $m \geq \sqrt{2}$ or $m \leq -\sqrt{2}$

Exercise 19

Solve the following inequalities:

1. $|5x + 3| \geq 7$

2. $|3x - 2| < 17$

3. $|x - 8| \leq 11$

4. $|7m + 5| > 2$

5. $\dfrac{x - 5}{2x + 6} > 0$

6. $\dfrac{3m + 9}{m - 2} < 0$

7. $\dfrac{2x - 7}{x - 2} \leq 1$

8. $\dfrac{9 - x}{2x + 4} \geq 5$

9. $2x^2 > 18$

10. $y^2 < 36$

11. $(3a - 5)^2 \geq 125$

12. $5 - 2y^2 \leq -13$

13. $9m^2 - 1 \geq 8$

14. $\dfrac{19 - 2x}{2x + 1} \geq 3$

15. $7 - 5y^2 \leq -73$

CHAPTER 20
INDICIAL EQUATIONS

An indicial equation is an equation in which the unknown variable is a power (index) in the equation. Usually the knowledge of indices and logarithms is applied in solving this type of equation.

Examples

1. Solve: $3^{3x+4} = 27^{2x+5}$

Solution

$$3^{3x+4} = 27^{2x+5}$$

In order to solve indicial equation, express both sides of the equation in the same base, and then equate the powers. The left hand side of the equation above has a base of 3. It is clear that the right hand side can also be expressed as a base of 3 because $27 = 3^3$. Hence, the equation now simplifies as follows:

$$3^{3x+4} = (3^3)^{2x-5}$$
$$3^{3x+4} = 3^{3(2x-5)} \quad \text{(From indices, } (a^x)^y = a^{xy})$$
$$3^{3x+4} = 3^{6x-15}$$

Since the bases on both sides of the equation are equal, it implies that the powers are also equal. Hence we ignore the bases and equate their powers as follows:

$$3x + 4 = 6x - 15$$
$$4 + 15 = 6x - 3x$$
$$19 = 3x$$
$$\therefore \quad x = \frac{19}{3}$$
$$x = 6\frac{1}{3}$$

2. Solve the equation: $\dfrac{8^{x+3}}{4^{3x-1}} = 32^{2x+7}$

Solution

$$\frac{8^{x+3}}{4^{3x-1}} = 32^{2x+7}$$

We can express each of the bases in the equation as a base of two. This gives:

$$\frac{(2^3)^{x+3}}{(2^2)^{3x-1}} = (2^5)^{2x+7} \quad \text{(Note that } 8 = 2^3, 4 = 2^2 \text{ and } 32 = 2^5)$$

$$\frac{2^{3(x+3)}}{2^{2(3x-1)}} = 2^{5(2x+7)} \quad \text{(Since } (a^x)^y = a^{xy})$$

$$\frac{2^{3x+9}}{2^{6x-2}} = 2^{10x+35}$$

216

$\therefore$ $2^{3x+9-(6x-2)} = 2^{10x+35}$ (From indices, $\frac{a^x}{a^y} = a^{x-y}$, i.e. subtraction of powers)

$2^{3x+9-6x+2} = 2^{10x+35}$

$2^{-3x+11} = 2^{10x+35}$

Since the bases are equal, it also means that the powers are equal. Hence, we equate the powers as follows:

$-3x + 11 = 10x + 35$

$-3x - 10x = 35 - 11$

$-13x = 24$

$\therefore$ $x = -\frac{24}{13}$

$x = -1\frac{11}{13}$

3. Solve: $\frac{125^{x-5}}{625^{2x}} = \frac{5^{3x+1}}{256^{6x-1}}$

Solution

$\frac{125^{x-5}}{625^{2x}} = \frac{5^{3x+1}}{256^{6x-1}}$

Let us express each base above as a base of 5 as follows:

$\frac{(5^3)^{x-5}}{(5^4)^{2x}} = \frac{5^{3x+1}}{(5^2)^{6x-1}}$

$\frac{5^{3(x-5)}}{5^{4(2x)}} = \frac{5^{3x+1}}{5^{2(6x-1)}}$

$\frac{5^{3x-15)}}{5^{8x}} = \frac{5^{3x+1}}{5^{12x-2}}$

Cross multiply to obtain:

$5^{3x-15} \times 5^{12x-2} = 5^{3x+1} \times 5^{8x}$

$5^{3x-15+12x-2} = 5^{3x+1+8x}$ (From indices, $a^x \times a^y = a^{x+y}$ i.e. addition of powers)

$5^{15x-17} = 5^{11x+1}$

We now equate the powers since the bases are equal.

$\therefore$ $15x - 17 = 11x + 1$

$15x - 11x = 1 + 17$

$4x = 18$

$x = \frac{18}{4}$

$= \frac{9}{2}$

$\therefore$ $x = 4\frac{1}{2}$

4. Solve the equation $3^{2x} = 12$

Solution

$$3^{2x} = 12$$

A careful look at this equation shows that the two sides of the equation cannot be expressed in the same base. Hence, we take the logarithm of both sides of the equation as follows:

$$3^{2x} = 12$$

$$\text{Log } 3^{2x} = \log 12$$

∴ $2x\log 3 = \log 12$ (Note that from logarithm, $\log x^y = y\log x$)

$2x \,(0.4771) = 1.0792$ (From calculator, $\log 3 = 0.4771$ and $\log 12 = 1.0792$)

$0.9542x = 1.0792$

∴ $x = \dfrac{1.0792}{0.9542}$

$x = 1.13$

5. Solve the equation $5^{4x-1} = 18^{2x+3}$

Solution

$$5^{4x-1} = 18^{2x+3}$$

It is clear that 5 and 18 cannot be expressed in the same base in whole number. Hence, we take the logarithm of both sides of the equation as follows:

$$5^{4x-1} = 18^{2x+3}$$

$$\text{Log} 5^{4x-1} = \text{Log} 18^{2x+3}$$

$(4x - 1)\log 5 = (2x + 3)\log 18$

$(4x - 1)(0.6990) = (2x + 3)(1.2553)$ (Note that $\log 5 = 0.6990$ and $\log 18 = 1.2553$)

$2.796x - 0.6990 = 2.5106x + 3.7659$

$2.796x - 2.5106x = 3.7659 + 0.669$

$0.2854x = 4.4649$

∴ $x = \dfrac{4.4649}{0.2854}$

$x = 15.64$

6. Solve $3.2^x \times 7.5^{3x+2} = 4^{2x+5}$

Solution

$$3.2^x \times 7.53^{3x+2} = 4^{2x+5}$$

Taking the logarithm of both sides gives:

$\text{Log}(3.2^x \times 7.53^{3x+2}) = \text{Log} 4^{2x+5}$

$\text{Log} 3.2^x + \text{Log} 7.53^{3x+2} = \text{Log} 4^{2x+5}$ (Note that from logarithm, $\text{Log}(AB) = \text{Log}A + \text{Log}B$)

$x\text{Log} 3.2 + (3x + 2)\text{Log} 7.53 = (2x + 5)\text{Log} 4$

$x(0.5051) + (3x + 2)(0.8573) = (2x + 5)(0.6021)$

$$0.5051x + 2.5719x + 1.7146 = 1.2042x + 3.0105$$
$$0.5051x + 2.5719x - 1.2042x = 3.0105 - 1.7146$$
$$1.8728x = 1.2959$$
$$x = \frac{1.2959}{1.8728}$$
$$x = 0.692$$

7. Solve the equation $2^{x^2} = 4^{3x-4}$

Solution
$$2^{x^2} = 4^{3x-4}$$

Expressing both sides in the same base (i.e. 2) gives:
$$2^{x^2} = (2^2)^{3x-4}$$
$$2^{x^2} = 2^{2(3x-4)}$$
$$2^{x^2} = 2^{6x-8}$$

Since the bases are equal, we now equate the powers as follows:
$$x^2 = 6x - 8$$
$$x^2 - 6x + 8 = 0$$

Solving this equation by factorization gives:
$$(x-4)(x-2) = 0$$

This gives: $x = 4$, or $x = 2$

8. Solve the equation $9^{x^2} = 3^{5x-3}$

Solution
$$9^{x^2} = 3^{5x-3}$$

Expressing both sides in the same base of 3 gives:
$$(3^2)^{x^2} = 3^{5x-3}$$
$$3^{2x^2} = 3^{5x-3}$$

Equating the powers gives:
$$2x^2 = 5x - 3$$
$$2x^2 - 5x + 3 = 0$$

Solving this equation by factorization gives:
$$2x^2 - 2x - 3x + 3 = 0$$
$$2x(x-1) - 3(x-1) = 0$$
$$(x-1)(2x-3) = 0$$
$$\therefore \quad x = 1 \text{ or } x = \frac{3}{2}$$

9. Solve the equation: $4^x - 3 \times 2^x + 2 = 0$

Solution

This equation is slightly different from others since it contains the addition/subtraction of three terms. The terms having x as their powers can be expressed in the same base as follows.

$(2^2)^x - 3(2^x) + 2 = 0$

$(2^x)^2 - 3(2^x) + 2 = 0$ (Note that $(2^2)^x = (2^x)^2$, since $2 \times x$ is the same as $x \times 2$)

Since $(2^x)^2$ means (2^x) raise to the power 2, it means that this equation can be expressed as a quadratic equation.

Let $2^x = y$ ....Equation (1)

Substitute y for 2^x in the simplified equation above as follows

$(2^x)^2 - 3(2^x) + 2 = 0$

$y^2 - 3y + 2 = 0$ (Since $2^x = y$)

Solving this equation by factorization gives:

$(y - 1)(y - 2) = 0$

Hence, y = 1 or 2

When y = 1, we substitute 1 for y in equation (1) in order to find x. This gives:

$2^x = y$ ....Equation (1)

$2^x = 1$

$2^x = 2^0$ (Note that $2^0 = 1$)

Equating the powers gives:

$x = 0$

When y = 2, we substitute 2 for y in equation (1) in order to find x. This gives:

$2^x = y$ ....Equation (1)

$2^x = 2$

$2^x = 2^1$ (Note that $2^1 = 2$)

Equating the powers gives:

$x = 1$

Hence, $x = 0$ or $x = 1$.

10. Solve the equation $2^{2x} - 2^{1+x} - 8 = 0$

Solution

$2^{2x} - 2^{1+x} - 8 = 0$

In this equation, 2^{1+x} can be expressed as $2^1 \times 2^x$, since from indices $a^x \times a^y = a^{x+y}$. This also means that $a^{x+y} = a^x \times a^y$. Hence substituting $2^1 \times 2^x$ for 2^{1+x} gives:

$2^{2x} - 2^{1+x} - 8 = 0$

$2^{2x} - 2^1 \times 2^x - 8 = 0$

$(2^x)^2 - 2(2^x) - 8 = 0$

Now, let $2^x = b$ ....Equation (1)

Substitute b for 2^x in the equation above. This gives:

$$(2^x)^2 - 2(2^x) - 8 = 0$$
$$b^2 - 2b - 8 = 0 \qquad \text{(Since } 2^x = b\text{)}$$

Solving this equation by factorization gives:

$$(b - 4)(b + 2) = 0$$
$$\therefore \quad b = 4 \text{ or } b = -2$$

When b = 4, we substitute 4 for b in equation (1). This gives:

$$2^x = b \ldots\ldots\ldots\ldots\ldots\text{Equation (1)}$$
$$2^x = 4$$
$$\therefore \quad 2^x = 2^2$$

Equating powers gives:

$$x = 2$$

Note that the other solution of b = −2 has been discarded since substituting −2 for b in equation (1) cannot be solved, i.e.:

$$2^x = -2 \text{ has no solution}$$

11. Solve: $3^{1+2x} + 2 \times 3^x - 1 = 0$

<u>Solution</u>

$$3^{1+2x} + 2 \times 3^x - 1 = 0$$
$$3^1 \times 3^{2x} + 2 \times 3^x - 1 = 0 \qquad \text{(Note that } 3^1 \times 3^{2x} = 3^{1+2x} \text{ from indices)}$$
$$3(3^x)^2 + 2(3^x) - 1 = 0$$

Let $3^x = y$ ……………Equation (1)

Substitute y for 3^x in the equation above. This gives:

$$3(3^x)^2 + 2(3^x) - 1 = 0$$
$$3y^2 + 2y - 1 = 0 \qquad \text{(Since } 3^x = y\text{)}$$

Solving this equation by factorization gives:

$$3y^2 + 3y - y - 1 = 0$$
$$3y(y + 1) - 1(y + 1) = 0$$
$$(y + 1)(3y - 1) = 0$$
$$\therefore \quad y = -1 \text{ or } y = \frac{1}{3}$$

When $y = \frac{1}{3}$, we substitute $\frac{1}{3}$ for y in equation (1). This gives:

$$3^x = y \ldots\ldots\ldots\ldots\ldots\text{Equation (1)}$$
$$3^x = \frac{1}{3}$$
$$\therefore \quad 3^x = 3^{-1} \qquad \text{(Note that } 3^{-1} = \frac{1}{3}\text{)}$$

Equating powers gives:

$$x = -1$$

Note that the other solution of y = –1 has been discarded since we cannot use it to solve for x.

12. Solve the simultaneous equation: $9^{2x-y} = 3$ and $16^{x+y} = 8$

<u>Solution</u>

Let us simplify each of the equation above one after the other.

$$9^{2x-y} = 3$$

Expressing both sides of the equation in the same base gives:

$$(3^2)^{2x-y} = 3^1$$
$$3^{2(2x-y)} = 3^1$$
$$3^{4x-2y} = 3^1$$

Equating the powers gives:

$$4x - 2y = 1 \text{ ....Equation (1)}$$

Similarly, $16^{x+y} = 8$ can be expressed in base 2 as follows:

$$(2^4)^{x+y} = 2^3$$
$$2^{4(x+y)} = 2^3$$
$$2^{4x+4y} = 3^3$$

Equating the powers gives:

$$4x + 4y = 3 \text{ ....Equation (2)}$$

Bring equation 1 and 2 together in order to solve them simultaneously. This gives:

$$4x - 2y = 1 \text{ ....Equation (1)}$$
$$\underline{4x + 4y = 3} \text{ ....Equation (2)}$$

Equation (2) – (1): $6y = 2$ (Note that 4y – (–2y) = 6y, and 3 – 1 = 2)

$$y = \frac{2}{6}$$
$$y = \frac{1}{3}$$

Substitute $\frac{1}{3}$ for y in equation (1).

$$4x - 2y = 1 \text{ ....Equation (1)}$$
$$4x - 2\left(\frac{1}{3}\right) = 1$$
$$4x - \frac{2}{3} = 1$$
$$4x = 1 + \frac{2}{3}$$
$$4x = 1\frac{2}{3}$$
$$4x = \frac{5}{3}$$
$$x = \frac{\frac{5}{3}}{4}$$
$$= \frac{5}{3} \div 4$$

$$= \frac{5}{3} \times \frac{1}{4} \qquad \text{(Note that 4 also means } \frac{4}{1}\text{)}$$

$$x = \frac{5}{12}$$

$$\therefore \quad x = \frac{5}{12} \text{ and } y = \frac{1}{3}$$

13. Solve simultaneously, the equations: $4^{x-2y} = 64$ and $25^{4x-3y} = 625$

Solution

$$4^{x-2y} = 64$$

Expressing both sides of the equation in the same base gives:

$$4^{x-2y} = 4^3$$

Equating the powers gives:

$$x - 2y = 3 \text{ ....Equation (1)}$$

Similarly, $25^{4x-3y} = 625$ can be expressed in base 25 as follows:

$$25^{4x-3y} = 25^2$$

Equating the powers gives:

$$4x + 3y = 2 \text{ ....Equation (2)}$$

From equation (1), $x = 3 + 2y$ ....Equation (3)

Substitute $3 + 2y$ for x in equation (2). This gives:

$$4x + 3y = 2 \text{ ....Equation (2)}$$

$$4(3 + 2y) + 3y = 2$$

$$12 + 8y + 3y = 2$$

$$11y = 2 - 12$$

$$11y = -10$$

$$y = \frac{-10}{11}$$

$$y = -\frac{10}{11}$$

Substitute $-\frac{10}{11}$ for y in equation (3) (Note that any of the equations can be used)

$$x = 3 + 2y \text{ ....Equation (3)}$$

$$= 3 + 2(-\frac{10}{11})$$

$$= 3 - \frac{20}{11}$$

$$x = \frac{33 - 20}{11}$$

$$x = \frac{13}{11}$$

$$x = 1\frac{2}{11}$$

$$\therefore \quad x = 1\frac{2}{11} \text{ and } y = -\frac{10}{11}$$

Exercise 20

1. Solve: $2^{5x+1} = 32^{2x-3}$

2. Solve the equation: $\dfrac{4^{2x+5}}{16^{2x-7}} = 64^{3x+2}$

3. Solve: $\dfrac{6^{3x-5}}{216^x} = \dfrac{216^{1-3x}}{36^{2x-3}}$

4. Solve the equation $5^{4x} = 15$

5. Solve the equation $3^{2x-11} = 20^{x+7}$

6. Solve $4.5^{2x} \times 10.2^{2-5x} = 8^{9x-4}$

7. Solve the equation $5^{x^2} = 2^{5x+2}$

8. Solve the equation $25^{x^2} = 125^{3x-1}$

9. Solve the equation: $4^x - 3 \times 2^x - 40 = 0$

10. Solve the equation $3^{2x} - 3^{x-1} - 78 = 0$

11. Solve: $5^{1+2x} + 2 \times 3^x - 7 = 0$

12. Solve the simultaneous equation: $9^{x-y} = 81$ and $25^{2x+y} = 25$

13. Solve simultaneously, the equations: $2^{5x-3y} = 32$ and $27^{2x-y} = 9^x$

14. Solve the simultaneous equation: $4^{x-2y} = 64$ and $25^{2x+5y} = 625$

15. Solve the equation $64^{x^2-2} = 16^{x+1}$

CHAPTER 21
ROOTS OF QUADRATIC EQUATIONS (USE OF ALPHA AND BETA)

A quadratic equation can have three cases of roots (solution of a quadratic equation). Consider a quadratic equation given by:

$$ax^2 + bx + c = 0$$

1. The roots are real and different if $b^2 - 4ac > 0$
2. The roots are real and equal if $b^2 - 4ac = 0$ or $b^2 = 4ac$
3. The roots are complex if $b^2 - 4ac < 0$

If each term of the quadratic equation above is divided by 'a', it gives:

$$x^2 + \frac{b}{a}x + \frac{c}{a} = 0 \text{ .... (1)}$$

If α and β are the roots of the equation, then the equation can be written as:

$$x^2 - (\alpha + \beta)x + \alpha\beta = 0 \text{ .... (2)}$$

Or

$$x^2 - (\text{sum of roots})x + (\text{product of roots}) = 0 \text{ ....(3)}$$

Comparing equations (1) to (3) above shows that:

Sum of roots $= \alpha + \beta = -\dfrac{b}{a}$

And product of roots $= \alpha\beta = \dfrac{c}{a}$

The roots of quadratic equations can be expressed as functions of α and β.

In order to apply α and β in quadratic equations, it is important to know the following identities.

1. $\alpha^2 + \beta^2 = (\alpha + \beta)^2 - 2\alpha\beta$
2. $(\alpha - \beta)^2 = (\alpha + \beta)^2 - 4\alpha\beta$
3. $\alpha - \beta = \sqrt{(\alpha + \beta)^2 - 4\alpha\beta}$
4. $\alpha^2 - \beta^2 = (\alpha + \beta)(\alpha - \beta)$
$$= (\alpha + \beta)\sqrt{(\alpha + \beta)^2 - 4\alpha\beta}$$
5. $\alpha^3 - \beta^3 = (\alpha - \beta)^3 + 3\alpha\beta(\alpha - \beta)$
6. $\alpha^3 + \beta^3 = (\alpha + \beta)^3 - 3\alpha\beta(\alpha + \beta)$

Examples
1. If α and β are the roots of the quadratic equation $2x^2 - 7x + 3 = 0$, find:

a. $\alpha + \beta$

b. $\alpha\beta$

c. $\alpha^2 + \beta^2$

d. $\dfrac{\alpha}{\beta} + \dfrac{\beta}{\alpha}$

e. $\dfrac{1}{\alpha} + \dfrac{1}{\beta}$

f. $\dfrac{1}{\alpha^2} + \dfrac{1}{\beta^2}$

g. $\alpha^3 + \beta^3$

<u>Solution</u>

a. $2x^2 - 7x + 3 = 0$

Recall the form: $ax^2 + bx + c = 0$

Comparing the two equations above shows that:

  $a = 2, b = -7, c = 3$

But $\alpha + \beta = -\dfrac{b}{a} = -\left(\dfrac{-7}{2}\right)$　　　(Take note of the use of negative sign)

$\therefore\ \ \alpha + \beta = \dfrac{7}{2}$

And $\alpha\beta = \dfrac{c}{a} = \dfrac{3}{2}$

b. $\alpha\beta = \dfrac{3}{2}$ as shown in (a) above

c. $\alpha^2 + \beta^2$

Recall that $\alpha^2 + \beta^2 = (\alpha + \beta)^2 - 2\alpha\beta$, as given in the identities above.
Hence, $\alpha^2 + \beta^2 = (\alpha + \beta)^2 - 2\alpha\beta$

$$= \left(\dfrac{7}{2}\right)^2 - 2\left(\dfrac{3}{2}\right) \quad \text{(By substituting } \dfrac{7}{2} \text{ for } (\alpha + \beta) \text{ and } \dfrac{3}{2} \text{ for } \alpha\beta)$$

$$= \dfrac{49}{4} - 3$$

$$= \dfrac{49 - 12}{4}$$

$$= \dfrac{37}{4}$$

d. $\dfrac{\alpha}{\beta} + \dfrac{\beta}{\alpha} = \dfrac{\alpha^2 + \beta^2}{\alpha\beta}$　　　(Note that the LCM of α and β is $\alpha\beta$)

$$= \dfrac{(\alpha + \beta)^2 - 2\alpha\beta}{\alpha\beta} \quad \text{(Note that } \alpha^2 + \beta^2 = (\alpha + \beta)^2 - 2\alpha\beta)$$

$$= \dfrac{\dfrac{37}{4}}{\dfrac{3}{2}} \quad \text{(Note that from (c) above } \alpha^2 + \beta^2 = \dfrac{37}{4})$$

$$= \dfrac{37}{4} \times \dfrac{2}{3}$$

$$= \frac{37}{2} \times \frac{1}{3} \qquad \text{(After division by 2)}$$

$$= \frac{37}{6}$$

e. $\dfrac{1}{\alpha} + \dfrac{1}{\beta} = \dfrac{\beta + \alpha}{\alpha \beta}$

$$= \frac{\alpha + \beta}{\alpha \beta}$$

$$= \frac{\frac{7}{2}}{\frac{3}{2}}$$

$$= \frac{7}{2} \times \frac{2}{3}$$

$$= \frac{7}{3} \qquad \text{(After 2 cancels out)}$$

f. $\dfrac{1}{\alpha^2} + \dfrac{1}{\beta^2} = \dfrac{\beta^2 + \alpha^2}{\alpha^2 \beta^2}$

$$= \frac{\alpha^2 + \beta^2}{(\alpha \beta)^2}$$

$$= \frac{\frac{37}{4}}{(\frac{3}{2})^2}$$

$$= \frac{\frac{37}{4}}{\frac{9}{4}}$$

$$= \frac{37}{4} \times \frac{4}{9}$$

$$= \frac{37}{9} \qquad \text{(Since 4 cancels out)}$$

g. $\alpha^3 + \beta^3 = (\alpha + \beta)^3 - 3\alpha\beta(\alpha + \beta)$

$$= (\frac{7}{2})^3 - 3(\frac{3}{2})(\frac{7}{2})$$

$$= \frac{343}{8} - \frac{9}{2}(\frac{7}{2})$$

$$= \frac{343}{8} - \frac{63}{4}$$

$$= \frac{343 - 126}{8}$$

$$= \frac{217}{8}$$

2. If the roots of the quadratic equation, $3x^2 + 5x - 9 = 0$, are α and β, find the equation whose roots are α^2 and β^2

Solution

$$3x^2 + 5x - 9 = 0$$

From this equation: a = 3, b = 5 and c = –9

But, $\alpha + \beta = -\dfrac{b}{a}$

$$= -\dfrac{5}{3} \qquad \text{(Since a = 3 and b = 5)}$$

Also, $\alpha\beta = \dfrac{c}{a}$

$$= \dfrac{-9}{3} \qquad \text{(Since a = 3 and c = – 9)}$$

$$= -3$$

Now, roots of new equation are α^2 and β^2

∴ Sum of roots of new equation = $\alpha^2 + \beta^2$

$$= (\alpha + \beta)^2 - 2\alpha\beta \quad \text{(From the identities given above)}$$

$$= \left(-\dfrac{5}{3}\right)^2 - 2(-3)$$

$$= \dfrac{25}{9} + 6$$

$$= \dfrac{25 + 54}{9}$$

Sum of roots $(\alpha + \beta) = \dfrac{79}{9}$

And product of roots of new equation = $\alpha^2 \times \beta^2$

$$= \alpha^2\beta^2$$

$$= (\alpha\beta)^2$$

$$= (-3)^2$$

Product of roots $(\alpha\beta) = 9$

Recall that a quadratic equation is represented as:

$$x^2 - (\text{sum of roots})x + (\text{product of roots}) = 0$$

Or, $x^2 - (\alpha + \beta)x + (\alpha\beta) = 0$

When the values obtained for $\alpha + \beta$ and $\alpha\beta$ are substituted into the equation above, it gives:

$$x^2 - \dfrac{79}{9}x + 9 = 0$$

When each term is multiplied by 9 in order to clear out the fraction, it gives:

$$9x^2 - 79x + 81 = 0$$

∴ The equation whose roots are α^2 and β^2 is $9x^2 - 79x + 81 = 0$

3. If the roots of the quadratic equation $5x^2 + x + 4 = 0$, are α and β, find the equation whose roots are $\dfrac{\alpha}{\beta}$ and $\dfrac{\beta}{\alpha}$

Solution

$$5x^2 + x + 4 = 0$$

From this equation: $a = 5$, $b = 1$ and $c = 4$

But, $\alpha + \beta = -\dfrac{b}{a}$

$$= -\dfrac{1}{5} \qquad \text{(Since } a = 5 \text{ and } b = 1)$$

Also, $\alpha\beta = \dfrac{c}{a}$

$$= \dfrac{4}{5} \qquad \text{(Since } a = 5 \text{ and } c = 4)$$

Now, roots of new equation are $\dfrac{\alpha}{\beta}$ and $\dfrac{\beta}{\alpha}$

∴   Sum of roots of new equation $= \dfrac{\alpha}{\beta} + \dfrac{\beta}{\alpha}$

$$= \dfrac{\alpha^2 + \beta^2}{\alpha\beta} \qquad \text{(From the identities given above)}$$

$$= \dfrac{(\alpha + \beta)^2 - 2\alpha\beta}{\alpha\beta}$$

$$= \dfrac{\left(-\frac{1}{5}\right)^2 - 2\left(\frac{4}{5}\right)}{\frac{4}{5}}$$

$$= \dfrac{\frac{1}{25} - \frac{8}{5}}{\frac{4}{5}}$$

$$= \dfrac{\frac{1-40}{25}}{\frac{4}{5}}$$

$$= \dfrac{-\frac{39}{25}}{\frac{4}{5}}$$

$$= -\dfrac{39}{25} \times \dfrac{5}{4}$$

$$= -\dfrac{39}{5} \times \dfrac{1}{4} \qquad \text{(After division by 5)}$$

$$= -\dfrac{39}{20}$$

Sum of roots $(\alpha + \beta) = -\dfrac{39}{20}$

And product of roots of new equation $= \dfrac{\alpha}{\beta} \times \dfrac{\beta}{\alpha}$

$$= \dfrac{\alpha\beta}{\alpha\beta}$$

$$= 1$$

Product of roots $(\alpha\beta) = 1$

Recall that a quadratic equation is represented as:

x^2 – (sum of roots)x + (product of roots) = 0

Or, $x^2 - (\alpha + \beta)x + (\alpha\beta) = 0$

When the values obtained for $\alpha + \beta$ and $\alpha\beta$ are substituted into the equation above, it gives:

$$x^2 - (-\frac{39}{20})x + 1 = 0$$

$$x^2 + \frac{39}{20}x + 1 = 0$$

When each term is multiplied by 20 in order to clear out the fraction, it gives:

$$20x^2 + 39x + 20 = 0$$

∴ The equation whose roots are $\frac{\alpha}{\beta}$ and $\frac{\beta}{\alpha}$ is $20x^2 + 39x + 20 = 0$

4. Given that the quadratic equation $2x^2 - 5x - 3 = 0$, has roots α and β, find the equation whose roots are $\frac{1}{\alpha}$ and $\frac{1}{\beta}$

Solution

$$2x^2 - 5x - 3 = 0$$

From this equation: a = 2, b = –5 and c = –3

But, $\alpha + \beta = -\frac{b}{a}$

$= -(\frac{-5}{2})$ (Since a = 2 and b = –5)

$= \frac{5}{2}$

Also, $\alpha\beta = \frac{c}{a}$

$= \frac{-3}{2}$ (Since a = 2 and c = –3)

Now, roots of new equation are $\frac{1}{\alpha}$ and $\frac{1}{\beta}$

∴ Sum of roots of new equation $= \frac{1}{\alpha} + \frac{1}{\beta}$

$= \frac{\beta + \alpha}{\alpha\beta}$

$= \frac{\alpha + \beta}{\alpha\beta}$

$= \dfrac{\frac{5}{2}}{-\frac{3}{2}}$

$= \frac{5}{2} \times (-\frac{2}{3})$

$$= -\frac{5}{3}$$

Sum of roots of new equation $(\alpha + \beta) = -\frac{5}{3}$

And product of roots of new equation $= \frac{1}{\alpha} \times \frac{1}{\beta}$

$$= \frac{1}{\alpha\beta}$$

$$= \frac{1}{-\frac{3}{2}}$$

$$= \frac{1}{1} \times (-\frac{2}{3})$$

Product of roots of new equation $(\alpha\beta) = -\frac{2}{3}$

Recall that a quadratic equation is represented as:

$x^2 - (\text{sum of roots})x + (\text{product of roots}) = 0$

Or, $x^2 - (\alpha + \beta)x + (\alpha\beta) = 0$

When the values obtained for $\alpha + \beta$ and $\alpha\beta$ are substituted into the equation above, it gives:

$$x^2 - (-\frac{5}{3})x + (-\frac{2}{3}) = 0$$

$$x^2 + \frac{5}{3}x - \frac{2}{3} = 0$$

When each term is multiplied by 3 in order to clear out the fraction, it gives:

$$3x^2 + 5x - 2 = 0$$

$\therefore$ The equation whose roots are $\frac{1}{\alpha}$ and $\frac{1}{\beta}$ is $3x^2 + 5x - 2 = 0$

5. If α and β are the roots of the quadratic equation $x^2 - 7x + 9 = 0$, find the quadratic equation whose roots are $\frac{1}{\alpha^2}$ and $\frac{1}{\beta^2}$

Solution

$$x^2 - 7x + 9 = 0$$

From this equation: $a = 1$, $b = -7$ and $c = 9$

But, $\alpha + \beta = -\frac{b}{a}$

$$= -(\frac{-7}{1}) \qquad (\text{Since } a = 1 \text{ and } b = -7)$$

$$= 7$$

Also, $\alpha\beta = \frac{c}{a}$

$$= \frac{9}{1} \qquad (\text{Since } a = 1 \text{ and } c = 9)$$

$$= 9$$

Now, roots of new equation are $\dfrac{1}{\alpha^2}$ and $\dfrac{1}{\beta^2}$

∴   Sum of roots of new equation $= \dfrac{1}{\alpha^2} + \dfrac{1}{\beta^2}$

$$= \dfrac{\alpha^2 + \beta^2}{\alpha^2 \beta^2}$$

$$= \dfrac{(\alpha + \beta)^2 - 2\alpha\beta}{(\alpha\beta)^2}$$

$$= \dfrac{7^2 - 2(9)}{9^2}$$

$$= \dfrac{49 - 18}{81}$$

Sum of roots $(\alpha + \beta) = \dfrac{31}{81}$

And product of roots of new equation $= \dfrac{1}{\alpha^2} \times \dfrac{1}{\beta^2}$

$$= \dfrac{1}{\alpha^2 \beta^2}$$

$$= \dfrac{1}{(\alpha\beta)^2}$$

$$= \dfrac{1}{9^2}$$

Product of roots $(\alpha\beta) = \dfrac{1}{81}$

Recall that a quadratic equation is represented as:

$x^2 -$ (sum of roots)$x +$ (product of roots) $= 0$

Or, $x^2 - (\alpha + \beta)x + (\alpha\beta) = 0$

When the values obtained for $\alpha + \beta$ and $\alpha\beta$ are substituted into the equation above, it gives:

$$x^2 - (\dfrac{31}{81})x + \dfrac{1}{81} = 0$$

When each term is multiplied by 81 in order to clear out the fractions, it gives:

$$81x^2 - 31x + 1 = 0$$

∴   The equation whose roots are $\dfrac{1}{\alpha^2} + \dfrac{1}{\beta^2}$ is $81x^2 - 31x + 1 = 0$

6. If α and β are the roots of the quadratic equation $2x^2 + 3x + 5 = 0$, find the quadratic equation whose roots are α^3 and β^3

Solution

$$2x^2 + 3x + 5 = 0$$

From this equation: a = 2, b = 3 and c = 5

But, $\alpha + \beta = -\dfrac{b}{a}$

$= -\dfrac{3}{2}$

Also, $\alpha\beta = \dfrac{c}{a}$

$= \dfrac{5}{2}$

Now, roots of new equation are α^3 and β^3

∴   Sum of roots of new equation $= \alpha^3 + \beta^3$

$= (\alpha + \beta)^3 - 3\alpha\beta(\alpha + \beta)$

$= (-\dfrac{3}{2})^3 - 3(\dfrac{5}{2})(-\dfrac{3}{2})$

$= -\dfrac{27}{8} + \dfrac{45}{4}$

$= \dfrac{-27 + 90}{8}$

$= \dfrac{63}{8}$

Sum of roots $(\alpha + \beta) = \dfrac{63}{8}$

And product of roots of new equation $= \alpha^3 \times \beta^3$

$= (\alpha\beta)^3$

$= (\dfrac{5}{2})^3$

$= \dfrac{125}{8}$

Product of roots of new equation $(\alpha\beta) = \dfrac{125}{8}$

Recall that a quadratic equation is represented as:

$x^2 - (\text{sum of roots})x + (\text{product of roots}) = 0$

Or, $x^2 - (\alpha + \beta)x + (\alpha\beta) = 0$

When the values obtained for $\alpha + \beta$ and $\alpha\beta$ are substituted into the equation above, it gives:

$x^2 - (\dfrac{63}{8})x + \dfrac{125}{8} = 0$

When each term is multiplied by 8 in order to clear out the fractions, it gives:

$8x^2 - 63x + 125 = 0$

∴   The equation whose roots are α^3 and β^3 is $8x^2 - 63x + 125 = 0$

7. If α and β are the roots of the equation $4x^2 - 12x + 7 = 0$, find the values:

a. $\alpha - \beta$

b. $\alpha^2 - \beta^2$

<u>Solution</u>

$$4x^2 - 12x + 7 = 0$$

From this equation: a = 4, b = –12 and c = 7

But, $\alpha + \beta = -\dfrac{b}{a}$

$$= -(\dfrac{-12}{4})$$

$$= 3$$

Also, $\alpha\beta = \dfrac{c}{a}$

$$= \dfrac{7}{4}$$

Recall that: $(\alpha - \beta)^2 = (\alpha + \beta)^2 - 4\alpha\beta$

Taking the square root of both sides gives $\alpha - \beta$ as follows:

$$\alpha - \beta = \sqrt{(\alpha + \beta)^2 - 4\alpha\beta}$$

$$= \sqrt{3^2 - 4(\dfrac{7}{4})}$$

$$= \sqrt{9 - 7}$$

$$= \sqrt{2}$$

$\therefore \;\; \alpha - \beta = \sqrt{2} \;$ or $\; 1.41$

b. Recall that: . $\alpha^2 - \beta^2 = (\alpha + \beta)(\alpha - \beta)$ (This is a difference of two squares)

$$= (3)(\sqrt{2}) \quad \text{(Since } \alpha + \beta = 3 \text{ and } \alpha - \beta = \sqrt{2} \text{ from (a) above)}$$

$$= 3\sqrt{2}$$

$\therefore \;\; \alpha^2 - \beta^2 = 3\sqrt{2} \;$ or $\; 4.24$

8. If α and β are the roots of the equation $3x^2 + 11x - 8 = 0$, find the value of $(\alpha - \beta)^2$

<u>Solution</u>

$$3x^2 + 11x - 8 = 0$$

From this equation: a = 3, b = 11 and c = –8

But, $\alpha + \beta = -\dfrac{b}{a}$

$$= -\dfrac{11}{3}$$

Also, $\alpha\beta = \dfrac{c}{a}$

$$= -\dfrac{8}{3}$$

Recall that: $(\alpha - \beta)^2 = (\alpha + \beta)^2 - 4\alpha\beta$

$$= (-\frac{11}{3})^2 - 4(-\frac{8}{3})$$

$$= \frac{121}{9} + \frac{32}{3}$$

$$= \frac{121 + 96}{9}$$

$$= \frac{217}{9}$$

$$\therefore \quad (\alpha - \beta)^2 = \frac{217}{9}$$

9. If the roots of the equation $2x^2 - 3x - 1 = 0$ are α and β, find the value of $(\alpha - \beta)^3$

Solution

$$2x^2 - 3x - 1 = 0$$

From this equation: a = 2, b = –3 and c = –1

But, $\alpha + \beta = -\frac{b}{a}$

$$= -(\frac{-3}{2})$$

$$= \frac{3}{2}$$

Also, $\alpha\beta = \frac{c}{a}$

$$= -\frac{1}{2}$$

Let us first find the value of $\alpha - \beta$ in order to determine $(\alpha - \beta)^3$

Recall that: $\alpha - \beta = \sqrt{(\alpha + \beta)^2 - 4\alpha\beta}$

$$= \sqrt{(\frac{3}{2})^2 - 4(-\frac{1}{2})}$$

$$= \sqrt{\frac{9}{4} + 2}$$

$$= \sqrt{\frac{9 + 8}{4}}$$

$$= \sqrt{\frac{17}{4}}$$

$$\alpha - \beta = 2.0616$$

$$\therefore \quad (\alpha - \beta)^3 = (2.0616)^3$$

$$= 8.76$$

10. If α and β are the roots of the quadratic equation $3x^2 - 5x - 1 = 0$, form an equation whose roots are:

a. $(\alpha^2 + \frac{1}{\beta})$ and $(\beta^2 + \frac{1}{\alpha})$

b. $(\alpha - \frac{1}{\beta})$ and $(\beta - \frac{1}{\alpha})$

Solution

$$3x^2 - 5x - 1 = 0$$

From this equation: a = 3, b = –5 and c = –1

Hence, $\alpha + \beta = -\frac{b}{a}$

$$= -(\frac{-5}{3})$$

$$= \frac{5}{3}$$

Also, $\alpha\beta = \frac{c}{a}$

$$= -\frac{1}{3}$$

Now, roots of new equation are $(\alpha^2 + \frac{1}{\beta})$ and $(\beta^2 + \frac{1}{\alpha})$

∴ Sum of roots of new equation $= \alpha^2 + \frac{1}{\beta} + \beta^2 + \frac{1}{\alpha}$

$$= \alpha^2 + \beta^2 + \frac{1}{\beta} + \frac{1}{\alpha} \qquad \text{(By rearrangement)}$$

$$= (\alpha + \beta)^2 - 2\alpha\beta + \frac{\alpha + \beta}{\alpha\beta} \qquad \text{(Note that } \alpha^2 + \beta^2 = (\alpha + \beta)^2 - 2\alpha\beta)$$

$$= (\frac{5}{3})^2 - 2(-\frac{1}{3}) + \frac{\frac{5}{3}}{-\frac{1}{3}}$$

$$= \frac{25}{9} + \frac{2}{3} - (\frac{5}{3} \times \frac{3}{1})$$

$$= \frac{25}{9} + \frac{2}{3} - 5$$

$$= \frac{25 + 6 - 45}{9}$$

$$= -\frac{14}{9}$$

Sum of roots $(\alpha + \beta) = -\frac{14}{9}$

And product of roots of new equation $= (\alpha^2 + \frac{1}{\beta})(\beta^2 + \frac{1}{\alpha})$

$$= \alpha^2\beta^2 + \frac{\alpha^2}{\alpha} + \frac{\beta^2}{\beta} + \frac{1}{\alpha\beta}$$

$$= (\alpha\beta)^2 + \alpha + \beta + \frac{1}{\alpha\beta}$$

$$= \left(-\frac{1}{3}\right)^2 + \frac{5}{3} + \frac{1}{-\frac{1}{3}}$$

$$= \frac{1}{9} + \frac{5}{3} - 3$$

$$= \frac{1 + 15 - 27}{9}$$

$$= -\frac{11}{9}$$

Product of roots of new equation $(\alpha\beta) = -\frac{11}{9}$

Recall that a quadratic equation is represented as:

$x^2 - $ (sum of roots)$x +$ (product of roots) $= 0$

Or, $x^2 - (\alpha + \beta)x + (\alpha\beta) = 0$

When the values obtained for $\alpha + \beta$ and $\alpha\beta$ are substituted into the equation above, it gives:

$$x^2 - \left(-\frac{14}{9}\right)x - \frac{11}{9} = 0$$

When each term is multiplied by 9 in order to clear out the fractions, it gives:

$$9x^2 + 14x - 11 = 0$$

$\therefore$ The equation whose roots are $(\alpha^2 + \frac{1}{\beta})$ and $(\beta^2 + \frac{1}{\alpha})$ is $9x^2 + 14x - 11 = 0$

b. Since roots of new equation are $(\alpha - \frac{1}{\beta})$ and $(\beta - \frac{1}{\alpha})$,

$\therefore$ Sum of roots of new equation $= \alpha - \frac{1}{\beta} + \beta - \frac{1}{\alpha}$

$$= \alpha + \beta - \left(\frac{1}{\beta} + \frac{1}{\alpha}\right) \quad \text{(By rearrangement)}$$

$$= \alpha + \beta - \left(\frac{\alpha + \beta}{\alpha\beta}\right)$$

$$= \frac{5}{3} - \frac{\frac{5}{3}}{-\frac{1}{3}}$$

$$= \frac{5}{3} + \left(\frac{5}{3} \times \frac{3}{1}\right)$$

$$= \frac{5}{3} + 5$$

$$= \frac{5 + 15}{3}$$

Sum of roots $(\alpha + \beta) = \frac{20}{3}$

And product of roots of new equation $= (\alpha - \frac{1}{\beta})(\beta - \frac{1}{\alpha})$

$$= \alpha\beta - \frac{\alpha}{\alpha} - \frac{\beta}{\beta} + \frac{1}{\alpha\beta}$$

$$= \alpha\beta - 1 - 1 + \frac{1}{\alpha\beta}$$

$$= \alpha\beta - 2 + \frac{1}{\alpha\beta}$$

$$= -\frac{1}{3} - 2 + \frac{1}{-\frac{1}{3}}$$

$$= -\frac{1}{3} - 2 - 3$$

$$= -\frac{1}{3} - 5$$

$$= \frac{-1 - 15}{3}$$

$$= -\frac{16}{3}$$

Product of roots of new equation $(\alpha\beta) = -\frac{16}{3}$

Recall that a quadratic equation is represented as:

$$x^2 - (\text{sum of roots})x + (\text{product of roots}) = 0$$

Or, $x^2 - (\alpha + \beta)x + (\alpha\beta) = 0$

When the values obtained for $\alpha + \beta$ and $\alpha\beta$ are substituted into the equation above, it gives:

$$x^2 - (\frac{20}{3})x - \frac{16}{3} = 0$$

When each term is multiplied by 3 in order to clear out the fractions, it gives:

$$3x^2 - 20x - 16 = 0$$

∴ The equation whose roots are $(\alpha - \frac{1}{\beta})$ and $(\beta - \frac{1}{\alpha})$ is $3x^2 - 20x - 16 = 0$

Maximum and Minimum Values of a Quadratic Function

The maximum or minimum value of a quadratic function is given by:

$$\frac{4ac - b^2}{4a}$$

The equation of the line of symmetry of a quadratic curve is given by: $x = -\frac{b}{2a}$

This line is called the axis of symmetry of a quadratic curve.

For a quadratic equation to have equal roots, the following condition must be met:

$b^2 = 4ac$ (From $b^2 - 4ac = 0$)

A quadratic equation which has equal roots is also a perfect square. Therefore, if a quadratic equation is a perfect square, then it follows that:

$b^2 = 4ac$

Examples

1. If $(2x + p)(3x - q) = 6x^2 - 7x + 2$, where p and q are constant, find the possible values of p.

Solution

$$(2x + p)(3x - q) = 6x^2 - 7x + 2$$

Let us expand the left hand side. This gives

$6x^2 - 2qx + 3px - pq = 6x^2 - 7x + 2$

$6x^2 - (2q - 3p)x - pq = 6x^2 - 7x + 2$ [Note: $-2qx + 3px$ has been factorized to give $-(2q - 3p)x$]

Comparing the terms in x on both sides of the equation shows that:

$-(2p - 3p) = -7$ (The terms in x)

∴ $2p - 3p = 7$ (After dividing both sides by -1)

∴ $2p - 3p = 7$ ....Equation (1)

Similarly, comparing the constant terms on both sides of the equation shows that:

$-pq = 2$

∴ $pq = -2$ (After dividing both sides by -1)

$pq = -2$ ....Equation(2)

From equation (2):

$$q = -\frac{2}{p}$$

Substitute $-\frac{2}{p}$ for q in equation (1)

$2q - 3p = 7$ ....Equation(1)

$2(-\frac{2}{p}) - 3p = 7$

$-\frac{4}{p} - 3p = 7$

Multiply each term by p in order to clear the fraction.

$p(-\frac{4}{p}) - p(3p) = p(7)$

$-4 - 3p^2 = 7p$

$0 = 3p^2 + 7p + 4$

∴ $3p^2 + 7p + 4 = 0$

Solving this equation by factorization method gives:

$3p^2 + 3p + 4p + 4 = 0$

$3p(p + 1) + 4(p + 1) = 0$

∴ $(p + 1)(3p + 4) = 0$

Equating each bracket to zero and solving each equation gives $p = -1$ or $p = -\frac{4}{3}$

∴   The possible values of p are -1 and $-\frac{4}{3}$

2. If $2x^2 - (m - 4)x - 4(m + 2) = 0$ has equal roots, find the possible values of m

<u>Solution</u>

$$2x^2 - (m - 4)x - 4(m + 2) = 0$$

From the equation, the values of a, b and c are given by a = 2, b = $-(m - 4)$, c = $-4(m + 2)$

Or, a = 2, b = $-m + 4$, c = $-4m - 8$ (When we expand the brackets above)

For a quadratic equation to have equal roots, the following condition must be met:

$$b^2 = 4ac \quad \text{(From } b^2 - 4ac = 0\text{)}$$

Substituting the values above gives:

$$b^2 = 4ac$$

$(-m + 4)^2 = 4 \times 2 \times (-4m - 8)$ (By substituting the values of a, b and c)

$(4 - m)^2 = 8(-4m - 8)$ (Note that $-m + 4 = 4 - m$)

$(4 - m)(4 - m) = -32m - 64$

$16 - 4m - 4m + m^2 = -32m - 64$

$m^2 - 8m + 32m + 16 + 64 = 0$

$m^2 + 24m + 80 = 0$

∴ $(m + 20)(m + 4) = 0$ (After factorization)

∴   m = -20 or m = -4

Hence, the possible values of m are -20 and -4.

3. Find the value of the constant k for which the expression $4x^2 + 20x + (12 + k)$ is a perfect square.

<u>Solution</u>

$4x^2 + 20x + (12 + k)$

From this equation, a, b and c are:

a = 4, b = 20, c = 12 + k

For a quadratic equation to be a perfect square, the condition below must be met.

$$b^2 = 4ac$$

$20^2 = 4 \times 4 \times (12 + k)$

$400 = 16(12 + k)$

$400 = 192 + 16K$

$400 - 192 = 16K$

∴ $k = \dfrac{208}{16}$

   k = 13

4. Find the maximum value of $5 - 11x - 2x^2$

Solution

Recall that maximum value is given by:

$$\frac{4ac - b^2}{4a}$$

From the equation $5 - 11x - 2x^2$:

$$a = -2, b = -11, c = 5$$

$\therefore$ Maximum value $= \dfrac{4ac - b^2}{4a}$

$$= \frac{4(-2 \times 5) - (-11)^2}{4(-2)}$$

$$= \frac{-40 - 121}{-8}$$

$$= \frac{-161}{-8}$$

$$= \frac{161}{8}$$

5. Given the function, $y = 3x^2 + 7x - 10$, find:

a. the minimum value of y

b. the value of x for which y is minimum

Solution

a. $y = 3x^2 + 7x - 10$

From the equation:

$$a = 3, b = 7, c = -10$$

The minimum value of y is given by:

$$y = \frac{4ac - b^2}{4a}$$

$$= \frac{4(3)(-10) - (7)^2}{4(3)}$$

$$= \frac{-120 - 49}{12}$$

$$= \frac{-169}{12}$$

$$= -\frac{169}{12}$$

b. The value of x for which y is minimum is given by:

$$x = \frac{-b}{2a}$$

$$= \frac{-7}{2(3)}$$

241

$$x = \frac{-7}{6}$$ (Note that this is also the line of symmetry of the quadratic curve)

6. Given the function, $y = 12 - 9x - 2x^2$, find:
a. the maximum value of y
b. the value of x for which y is maximum

Solution

a. $y = 12 - 9x - 2x^2$

From the equation:

$a = -2, b = -9, c = 12$

The maximum value of y is given by:

$$y = \frac{4ac - b^2}{4a}$$
$$= \frac{4(-2)(12) - (-9)^2}{4(-2)}$$
$$= \frac{-96 - 81}{-8}$$
$$= \frac{-177}{-8}$$
$$= \frac{177}{8}$$

b. The value of x for which y is maximum is given by:

$$x = \frac{-b}{2a}$$
$$= \frac{-(-9)}{2(-2)}$$
$$x = \frac{9}{-4}$$ (Note that this is also the line of symmetry of the quadratic curve)
$$x = -\frac{9}{4}$$

7. One root of the quadratic equation $x^2 - (4 + p)x + 12 = 0$ is three times the other. Find:
a. the roots of the equation
b. the possible values of p

Solution

a. $x^2 - (4 + p)x + 12 = 0$

Let one of the roots be α. Therefore, the other root is 3α.

From the equation:

Product of roots = 12   (From $\frac{c}{a}$ in the equation. Note that a = 1)

$\therefore$ $\alpha(3\alpha) = 12$

242

$3\alpha^2 = 12$

$\alpha^2 = 4$ (After dividing both sides by 3)

∴ $\alpha = \sqrt{4}$

$\alpha = \pm 2$

∴ $\alpha = 2$ or -2

When $\alpha = 2$, the other root is $3\alpha = 3 \times 2 = 6$

When $\alpha = -2$, the other root is $3\alpha = 3(-2) = -6$

Therefore, the roots are either 2 and 6, or −6 and −2

b. From the equation: $x^2 - (4 + p)x + 12 = 0$:

$a = 1, b = -(4 + p), c = 12$

Sum of the roots = $-b/a$

$\alpha + 3\alpha = -[-(4 + p)]/1$ (Note that α and 3α are the two roots)

$4\alpha = 4 + p$

$4\alpha - 4 = p$

∴   When $\alpha = 2$, p is given by

$4\alpha - 4 = p$

$4(2) - 4 = p$

$8 - 4 = p$

∴ $p = 4$

Also, when $\alpha = -2$, p is given by:

$4\alpha - 4 = p$

$4(-2) - 4 = p$

$-8 - 4 = p$

$p = -12$

∴   The possible values of p are 4 and −12.

Exercise 21

1. If α and β are the roots of the quadratic equation $3x^2 - 8x + 5 = 0$, find:

a. $\alpha + \beta$

b. $\alpha\beta$

c. $\alpha^2 + \beta^2$

d. $\dfrac{\alpha}{\beta} + \dfrac{\beta}{\alpha}$

e. $\dfrac{1}{\alpha} + \dfrac{1}{\beta}$

f. $\dfrac{1}{\alpha^2} + \dfrac{1}{\beta^2}$

g. $\alpha^3 + \beta^3$

2. If the roots of the quadratic equation, $2x^2 + 12x - 11 = 0$, are α and β, find the equation whose roots are α^2 and β^2

3. If the roots of the quadratic equation $4x^2 + 3x - 10 = 0$, are α and β, find the equation whose roots are $\dfrac{\alpha}{\beta}$ and $\dfrac{\beta}{\alpha}$

4. Given that the quadratic equation $x^2 - 15x - 2 = 0$, has roots α and β, find the equation whose roots are $\dfrac{1}{\alpha}$ and $\dfrac{1}{\beta}$

5. If α and β are the roots of the quadratic equation $3x^2 - 15x + 7 = 0$, find the quadratic equation whose roots are $\dfrac{1}{\alpha^2}$ and $\dfrac{1}{\beta^2}$

6. If α and β are the roots of the quadratic equation $2x^2 + 5x - 6 = 0$, find the quadratic equation whose roots are α^3 and β^3

7. If α and β are the roots of the equation $5x^2 - 8x + 2 = 0$, find the values of:

a. $\alpha - \beta$

b. $\alpha^2 - \beta^2$

8. If α and β are the roots of the equation $10x^2 + 15x - 8 = 0$, find the value of $(\alpha - \beta)^2$

9. If the roots of the equation $4x^2 - 2x - 3 = 0$ are α and β, find the value of $(\alpha - \beta)^3$

10. If α and β are the roots of the quadratic equation $x^2 - 2x - 3 = 0$, form an equation whose roots are:

a. $(\alpha^2 + \dfrac{1}{\beta})$ and $(\beta^2 + \dfrac{1}{\alpha})$

b. $(\alpha - \dfrac{1}{\beta})$ and $(\beta - \dfrac{1}{\alpha})$

11. If $(3x + m)(x - n) = 3x^2 - 11x + 6$, where m and n are constants, find the possible values of m and n.

12. If $2x^2 - (p + 3)x - 2(p - 2) = 0$ has equal roots, find the possible values of p.

13. Find the value of the constant k for which the expression $3x^2 + 4x - (8 + k)$ is a perfect square.

14. Find the maximum value of $7 - 14x - 5x^2$

15. Given the function, $y = 2x^2 + 11x - 12$, find:

a. the minimum value of y

b. the value of x for which y is minimum

16. Given the function, $y = 18 - 15x - 4x^2$, find:

a. the maximum value of y

b. the value of x for which y is maximum

17. One root of the quadratic equation $2x^2 - (2 - k)x + 6 = 0$ is three times the other. Find:

a. the roots of the equation

b. the possible values of k

18. If α and β are the roots of the equation $2x^2 - 6x + 3 = 0$, find the values of $\alpha^2 - \beta^2$

19. Find the value of the constant k for which the expression $5x^2 + 10x - (3 + k)$ is a perfect square.

20. One root of the quadratic equation $4x^2 - (9 - k)x + 10 = 0$ is twice the other. Find the roots of the equation.

CHAPTER 22
FUNCTIONS

A function, f, whose output is y, and input or variable is x, is a rule which describes how a value of x is used to obtain a value of y. A function is usually represented in the form of an equation as follows:

$$y = f(x)$$

The rule of a function must imply that for each input of x, there must be only one output of y. For example, the equation:

$$y = 5x - 1$$

shows that for every value of x, there is only one value of y. Hence this rule is a function. However, an equation such as:

$$y = \sqrt{x} \quad \text{or} \quad y = x^{\frac{1}{2}}$$

is a rule that can give two possible values of y. If $x = 4$, then $y = 4^{\frac{1}{2}}$. Which gives:

$$y = \sqrt{4}$$
$$y = \pm 2 \text{ which means +2 or } -2.$$

This shows that for a positive value of x, there are two possible values of y. Hence, $y = x^{\frac{1}{2}}$ is not a function. A function should give only one value of the output y.

Examples

Determine if each of the following is a function or not.

1. $y = x^{\frac{1}{4}} - 2$
2. $y = 3x^2 - 5\sqrt[3]{x}$
3. $y = \sqrt{20 - x^2}$
4. $y = \dfrac{1}{2x - 1}$

<u>Solutions</u>

1. $y = x^{\frac{1}{4}} - 2$

In order to test if this is a function, we can put a value of x and see if we will get only one value of y. Let us take $x = 16$ and put it in the equation. This gives:

$$y = x^{\frac{1}{4}} - 2$$
$$y = 16^{\frac{1}{4}} - 2$$
$$= \sqrt[4]{16} - 2$$
$$= \pm 2 - 2$$

$$= +2 - 2 \text{ or } -2 - 2$$

Hence, $y = 0$ or $y = -4$

Therefore, $y = x^{\frac{1}{4}} - 2$ is not a function since a value of x produces two values of y.

2. $y = 3x^2 - 5\sqrt[3]{x}$

Let us put $x = -8$ into the equation. Note that any convenient value of can be used.

$$\begin{aligned}
y &= 3x^2 - 5\sqrt[3]{x} \\
&= 3(-8)^2 - 5\sqrt[3]{-8} \\
&= 3(64) - 5(-2) \\
&= 192 + 10 \\
&= 202
\end{aligned}$$

Hence, $y = 3x^2 - 5\sqrt[3]{x}$ is a function since only one value of y is produced.

3. $y = \sqrt{20 - x^2}$

Let us put $x = 3$ into the equation. Be careful to put in the square root sign, only values of x that will give a positive value. This is because we cannot evaluate values such as $\sqrt{-10}$, but only positive values such as $\sqrt{10}$

$$\begin{aligned}
\text{Hence, } y &= \sqrt{20 - x^2} \\
&= \sqrt{20 - 3^2} \quad \text{(Since } x = 3 \text{ as stated above)} \\
&= \sqrt{20 - 9} \\
&= \sqrt{11} \\
y &= \pm\sqrt{11}
\end{aligned}$$

Therefore, $y = +\sqrt{11}$ or $-\sqrt{11}$

Hence, $y = \sqrt{20 - x^2}$ is not a function since there are two possible values is y.

4. $y = \dfrac{1}{2x - 1}$

Let us put $x = -2$ into the equation as follows:

$$\begin{aligned}
y &= \frac{1}{2x - 1} \\
&= \frac{1}{2(-2) - 1} \\
&= \frac{1}{-4 - 1} \\
&= \frac{1}{-5} \\
y &= -0.2
\end{aligned}$$

Hence, $y = \dfrac{1}{2x - 1}$ is a function

Domain and Range of a Function

Domain are all the values of x that a function can take and process.

Range is the set of values y obtained from each domain of a function. Range is also called co–domain.

For example, if y = $\sqrt{4 - x^2}$, where x and y are real numbers, and x is a whole number, then the domain is the set of values –2, –1, 0, 1, 2. These are the values of x that will not give negative values in the square root sign. The range (i.e. the corresponding values of y from the domain values of x) is the set of values 0, $\sqrt{3}$, 2, $\sqrt{3}$, 0, or simply 0, $\sqrt{3}$, 2, when avoiding repetition.

A function can be defined by a given domain of values.

For example if a function is defined as follows:

$$y = 2x, \quad -5 < x < 1$$

then from the defined values, the domain of the function is:

$$-5 < x < 1$$

When each of the extreme values (i.e. –5 and 1) is substituted into the function, then the range of the function will be obtained as:

$$y = 2(-5) = -10 \qquad \text{(When } x = -5)$$

and $y = 2(1) = 2 \qquad$ (When $x = 1$)

This gives the range: $\quad -10 < x < 2$

If a function is given by:

$$y = \frac{1}{(x - 2)(x + 5)}$$

then the domain can be any value of x except $x = 2$ and $x = -5$. These two values of x will give a denominator of zero which makes the function undefined. Let us see the result of these two values of x. When $x = 2$, then:

$$y = \frac{1}{(x - 2)(x + 5)}$$

$$= \frac{1}{(2 - 2)(2 + 5)}$$

$$= \frac{1}{(0)(7)} = \frac{1}{0} \qquad \text{(This is undefined)}$$

When $x = -5$, then:

$$y = \frac{1}{(x - 2)(x + 5)}$$

$$= \frac{1}{(-5 - 2)(-5 + 5)}$$

$$= \frac{1}{(-7)(0)} = \frac{1}{0} \qquad \text{(This is undefined)}$$

Hence, $x = 2$ and $x = -5$ are two values that should not be in the domain. Hence, the range of the function should not be calculated when $x = 2$ and $x = -5$.

Arithmetic Operations of Function

Examples

1. If $f(x) = 2x^2 + 1$, find:

a. $f(-2)$

b. $f(5)$

c. $f(0.3)$

d. $f(x - 1)$

e. $f(2x + 3)$

<u>Solutions</u>

a. $f(x) = 2x^2 + 1$

In order to obtain f(−2), we simply substitute −2 for x in the given function.

Hence, $f(x) = 2x^2 + 1$

$$f(-2) = 2(-2)^2 + 1$$
$$= 2(4) + 1 = 8 + 1$$
$$= 9$$

Therefore, $f(-2) = 9$

b. $f(x) = 2x^2 + 1$

$$f(5) = 2(5)^2 + 1$$
$$= 2(25) + 1 = 50 + 1$$
$$= 51$$

Therefore, $f(5) = 51$

c. $f(x) = 2x^2 + 1$

$$f(0.3) = 2(0.3)^2 + 1$$
$$= 2(0.09) + 1$$
$$= 0.18 + 1$$
$$= 1.18$$

Therefore, $f(0.3) = 1.18$

d. $f(x) = 2x^2 + 1$

$$f(x - 1) = 2(x - 1)^2 + 1$$
$$= 2(x - 1)(x - 1) + 1$$
$$= 2(x^2 - x - x + 1) + 1$$
$$= 2(x^2 - 2x + 1) + 1$$
$$= 2x^2 - 4x + 2 + 1$$
$$= 2x^2 - 4x + 3$$

Therefore, $f(x - 1) = 2x^2 - 4x + 3$

e. $f(x) = 2x^2 + 1$

$f(2x + 3) = 2(2x + 3)^2 + 1$

$= 2(2x + 3)(2x + 3) + 1$

$= 2(4x^2 + 6x + 6x + 9) + 1$

$= 2(4x^2 + 12x + 9) + 1$

$= 8x^2 + 24x + 18 + 1$

$= 8x^2 + 24x + 19$

Therefore, $f(2x + 3) = 8x^2 + 24x + 19$

2. A function f is defined by $f(x) = 5x - 8$

a. Find $f(\frac{1}{2})$

b. If $f(3m - 1) = 14$, find the value of m^2

c. Find $5f(-2)$

Solution

a. $f(x) = 5x - 8$

$f(\frac{1}{2}) = 5(\frac{1}{2}) - 8$

$= \frac{5}{2} - 8 = \frac{5 - 16}{2}$

$= \frac{-11}{2}$

Therefore, $f(\frac{1}{2}) = -5\frac{1}{2}$

b. $f(x) = 5x - 8$

$f(3m - 1) = 5(3m - 1) - 8$

$14 = 15m - 5 - 8$ (Note that $f(3m - 1) = 14$)

$14 = 15m - 13$

$14 + 13 = 15m$

$27 = 15m$

$m = \frac{27}{15}$

$m = \frac{9}{5}$

Therefore, $m^2 = (\frac{9}{5})^2$

$m^2 = \frac{81}{25}$

c. $f(x) = 5x - 8$

$f(-2) = 5(-2) - 8$

$$= -10 - 8$$
$$f(-2) = -18$$

Therefore, 5f(−2) is obtained by simply multiplying f(−2) by 5.

Therefore, $5f(-2) = 5(-18)$
$$= -90$$

3. If $f(x) = 2x + 5$ and $g(x) = x - 2$, find:
a. $f(x) + g(x)$
b. $g(x) - 2f(x)$
c. $f(x) \times g(x)$
d. $f(2) \div g(5)$
e. $h(x) = f(x - 1) - g(-3)$
f. $h(-1)$

Solutions
a. $f(x) + g(x)$

This is the addition of the two functions. It is given as follows:
$$f(x) + g(x) = 2x + 5 + x - 2$$
$$= 3x + 3$$

b. $g(x) - 2f(x)$

In this case, 2f(x) has to be evaluated first. The overall solution is given by:
$$g(x) - 2f(x) = x - 2 - [2(2x + 5)]$$
$$= x - 2 - (4x + 10)$$
$$= x - 2 - 4x - 10$$
$$= -3x - 12$$

c. $f(x) \times g(x) = (2x + 5)(x - 2)$
$$= 2x^2 - 4x + 5x - 10$$
$$= 2x^2 + x - 10$$

d. $f(2) \div g(5)$

Let us determine f(2) and g(5) separately as follows:
$$f(x) = 2x + 5$$
$$f(2) = 2(2) + 5$$
$$= 4 + 5$$
$$= 9$$
$$g(x) = x - 2$$

$$g(5) = 5 - 2$$
$$= 3$$
$$\therefore \quad f(2) \div g(5) = \frac{9}{3}$$
$$= 3$$

e. $h(x) = f(x - 1) - g(-3)$

Let us find $f(x - 1)$ as follows:

$$f(x) = 2x + 5$$

Hence, $f(x - 1) = 2(x - 1) + 5$

$$= 2x - 2 + 5$$
$$f(x - 1) = 2x + 3$$

Also, let us find $g(-3)$ as follows:

$$g(x) = x - 2$$
$$g(-3) = -3 - 2$$
$$g(-3) = -5$$
$$\therefore \quad h(x) = f(x - 1) - g(-3)$$
$$= 2x + 3 - (-5) \qquad \text{(Since } f(x - 1) = 2x + 3 \text{ and } g(-3) = -5\text{)}$$
$$= 2x + 3 + 5$$
$$h(x) = 2x + 8$$

f. $h(x) = 2x + 8$

$$h(-1) = 2(-1) + 8$$
$$= -2 + 8$$
$$h(-1) = 6$$

Composing Functions

Chains of functions can be obtained when two or more functions combine together. For example if $f(x) = 2x$ and $g(x) = x - 1$, then a third function such as $h(x) = g[f(x)]$ can be obtained by the combination of $f(x)$ and $g(x)$. It is written as $h(x) = g \circ f(x)$, and is read as h of x equals g of f of x. h is said to be a function of function. Note that any letter can be used to represent a function.

Examples

1. If $a(x) = 2x - 1$ and $b(x) = -3x$, find

a. $f(x) = a[b(x)]$

b. $g(x) = b[a(x)]$

Solutions

a. $f(x) = a[b(x)]$

252

This can also be written as: f(x) = a o b(x)

In order to find f(x), we simply substitute b(x) (i.e. $-3x$) for x in a(x). This means to put $x = -3x$ in a(x). This gives:

$\quad$ f(x) = a[b(x)]

$\qquad$ = a($-3x$) $\qquad$ (Since b(x) = $-3x$)

But, a(x) = $2x - 1$

$\therefore$ a($-3x$) = $2(-3x) - 1$

$\qquad$ = $-6x - 1$

Hence, f(x) = $-6x - 1$ $\qquad$ [Since f(x) = a($-3x$)]

b. g(x) = b[a(x)]

In order to find g(x), we simply substitute a(x) (i.e. $2x - 1$) for x in b(x). This means we put $x = 2x - 1$ in b(x). This gives:

$\quad$ g(x) = b[a(x)]

$\qquad$ = b($2x - 1$) $\qquad$ (Since a(x) = $2x - 1$)

But, b(x) = $-3x$

$\therefore$ b($2x - 1$) = $-3(2x - 1)$

$\qquad$ = $-6x + 3$

Hence, g(x) = $-6x + 3$ $\qquad$ [Since g(x) = b($2x - 1$)]

2. If a(x) = x^2, b(x) = $5x$ and c(x) = $x - 1$, find:

a. f(x) = b[a[c(x)]]

b. g(x) = c[b[c(x)]]

c. h(x) = a[b[b(x)]]

Solution

a. f(x) = b[a[c(x)]]

This can also be written as b o a o c(x).

In order to find f(x), we start from the innermost bracket. Hence, let us first find a[c(x)]. This simply means to substitute c(x) in a(x)

c(x) = $x - 1$ and a(x) = x^2.

$\therefore$ a[c(x)] = a($x - 1$) $\qquad$ (Since c(x) = $x - 1$)

But, $\quad$ a(x) = x^2

Hence, a($x - 1$) = $(x - 1)^2$

$\qquad\qquad$ = $(x - 1)(x - 1)$

$\qquad\qquad$ = $x^2 - x - x + 1$

$\quad$ a($x - 1$) = $x^2 - 2x + 1$

$\therefore$ a[c(x)] = $x^2 - 2x + 1$ $\quad$ (Since a[c(x)] = a($x - 1$)

253

Let us now determine the final output i.e. b[a[c(x)]]. This means to substitute a[c(x)] in b(x)

$a[c(x)] = x^2 - 2x + 1$ and $b(x) = 5x$

Hence, $b[a[c(x)]] = b(x^2 - 2x + 1)$ (Since $a[c(x)] = x^2 - 2x + 1$)

But, $b(x) = 5x$

Therefore, $b(x^2 - 2x + 1) = 5(x^2 - 2x + 1)$

$$= 5x^2 - 10x + 5$$

Hence, $b[a[c(x)]] = 5x^2 - 10x + 5$ [Since $b[a[c(x)]] = b(x^2 - 2x + 1)$]

Therefore, $f(x) = 5x^2 - 10x + 5$ (Note that $f(x) = b[a[c(x)]]$)

b. $g(x) = c[b[c(x)]]$

Let us first find b[c(x)]. This simply means to substitute c(x) in b(x)

$c(x) = x - 1$ and $b(x) = 5x$

∴ $b[c(x)] = b(x - 1)$ (Since $c(x) = x - 1$)

But, $b(x) = 5x$

Hence, $b(x - 1) = 5(x - 1)$

$b(x - 1) = 5x - 5$

∴ $b[c(x)] = 5x - 5$ (Since $b[c(x)] = b(x - 1)$)

Let us now determine the final output i.e. c[b[c(x)]]. This means to substitute b[c(x)] in c(x)

Hence, $b[c(x)] = 5x - 5$ and $c(x) = x - 1$

Hence, $c[b[c(x)]] = c(5x - 5)$ [Since $b[c(x)] = 5x - 5)$]

But, $c(x) = x - 1$

Therefore, $c(5x - 5) = (5x - 5) - 1$

$c(5x - 5) = 5x - 6$

Hence, $c[b[c(x)]] = 5x - 6$ [Since $c[b[c(x)]] = c(5x - 5)$]

Therefore, $g(x) = 5x - 6$ (Note that $g(x) = c[b[c(x)]]$)

c. $h(x) = a[b[b(x)]]$

Note that a[b[b(x)]] is interpreted as a of b of b of x or a o b o b(x).

Let us first find b[b(x)]. This simply means to substitute b(x) in b(x)

$b(x) = 5x$

∴ $b[b(x)] = b(5x)$

But, $b(x) = 5x$

Hence, $b(5x) = 5(5x)$

$b(5x) = 25x$

∴ $b[b(x)] = 25x$

Let us now find a[b[b(x)]]. This means to substitute b[b(x)] in a(x)

Recall that: $b[b(x)] = 25x$ and $a(x) = x^2$

Hence, $a[b[b(x)]] = a(25x)$ [Since $b[b(x)] = 25x$]

But, $a(x) = x^2$

Therefore, $a(25x) = (25x)^2$

$$a(25x) = 625x^2$$

Hence, $a[b[b(x)]] = 625x^2$ [Since $a[b[b(x)]] = a(25x)$]

Therefore, $h(x) = 625x^2$ (Since $h(x) = a[b[b(x)]]$)

Continuous and Discontinuous Functions

If a graph of a function is drawn without having to take ones hand off the graph paper, then the function is a continuous function. The graph of a continuous function has no sudden jump or break. Sine and cosine functions are continuous functions.

Some graphs of some functions make jumps at a point or some points in the interval. Such functions are called discontinuous functions. $y = \tan x$ is a discontinuous function.

Even Functions

A function $f(x)$ is said to be even if $f(x) = f(-x)$ for all values of x. The graphs of even functions always have their line of symmetry as the y–axis. This means that the vertical line $x = 0$ divide the graph into two equal parts.

$f(x) = x^2$ and $f(x) = \cos x$ are examples of even functions.

Odd Functions

A function $f(x)$ is said to be odd if $f(-x) = -f(x)$ for all values of x. Graphs of odd functions usually pass through the origin as a point of symmetry. This means that the origin, (0, 0) divides the graph into two equal parts. $F(x) = x^3$ and $f(x) = \sin x$ are examples of odd functions.

Note that some functions are neither even nor odd.

Examples

1. Classify the following into even and odd functions:

a. $f(x) = \tan x$

b. $f(x) = x^4$

Solution

a. $f(x) = \tan x$

Let us take a value of x such as $60°$ and use it to test the function as follows:

$\quad F(60°) = \tan 60$

$\quad F(60°) = 1.732$

Now, find $f(-60°)$ to see if the value obtained will be the same as that of $f(60°)$.

$\quad f(-60°) = \tan -60$

$\quad f(-60°) = -1.732$

Comparing the two results [i.e. f(60°) and f(−60°)] shows that f(60°) = 1.732, while f(−60°) = − 1.732. Hence the function is an odd function since:

$$F(-x) = -f(x), \text{ i.e. } f(-60°) = -f(60°)$$

Note that once the two values obtained are the same size, but different signs, then the function is an odd function.

b. $f(x) = x^4$

Let us take a value of x such as 2.

Hence, $f(2) = 2^4$

$f(2) = 16$

Now use the negative value of x i.e. −2. This gives:

$f(-2) = -2^4$

$F(-2) = 16$

Therefore, $f(2) = f(-2) = 16$

Since the two values obtained are equal and of the same sign, then the function is an even function.

2. Classify the following into even and odd functions:

a. $3x^5$

b. $\cos^3 x$

c. $4x - 1$

Solutions

a. $3x^5$

Let us use $x = 1$

Hence, $3x^5 = 3(1)^5$

$\qquad = 3$

When $x = -1$, we have:

$3x^5 = 3(-1)^5$

$\qquad = 3 \times (-1) = -3 \quad$ (Note that $(-1)^5 = -1$)

Since the values are the same but have opposite signs, then the function is an odd function.

b. $\cos^3 x$

Let $x = 60°$

$\cos^3 x = (\cos x)^3$

$\qquad = (\cos 60)^3$

$\qquad = (0.5)^3$

$\qquad = 0.125$

When $x = -60°$, we have:

$$\text{Cos}^3 x = (\cos x)^3$$
$$= [\cos(-60)]^3$$
$$= (0.5)^3$$
$$= 0.125$$

The two values obtained are equal. Therefore, the function is an even function.

c. $4x - 1$

Let $x = 5$

$$4x - 1 = 4(5) - 1$$
$$= 20 - 1$$
$$= 19$$

When x = –5, we have

$$4x - 1 = 4(-5) - 1$$
$$= -20 - 1$$
$$= -21$$

The two values obtained are not equal. They are entirely different in values. Hence, the function is neither even nor odd.

Inverse of a Function

The inverse of a function f(x), is another function denoted by $f^{-1}(x)$ which reverses the function f(x). For an inverse $f^{-1}(x)$ that exists, the following is true:

$$f[\,f^{-1}(x)] = f^{-1}[f(x)] = x$$

The graph of $f^{-1}(x)$ will be a reflection of f(x) in the line y = x.

In order to find the inverse of a function, interchange the positions of x and y, and then make y the subject of the formula.

Examples

1. Find the inverse of the following functions:
a. $f(x) = 5x - 1$
b. $f(x) = 2x^2 + 3$
c. $f(x) = 5x^3$

Solution

a. $f(x) = 5x - 1$

Express the function as y = f(x). This gives:

$$y = 5x - 1$$

Interchange the positions of x and y. This gives:

$$x = 5y - 1$$

Now make y the subject of the formula.

$$x = 5y - 1$$
$$x + 1 = 5y$$
$$y = \frac{x + 1}{5} \quad \text{(After dividing both sides by 5)}$$

Therefore, the inverse of $f(x)$ is:

$$f^{-1}(x) = \frac{x + 1}{5} \quad \text{[Simply by replacing y with } f^{-1}(x)]$$

b. $f(x) = 2x^2 + 3$

As solved in (a) above, only three steps are involved. They are:

1. equate the function to y
2. interchange the positions of x and y
3. make y the subject of the formula to obtain the inverse of the function.

Note that step 1 is not needed if the function is already equated to y.

Let us now continue as follows:

$$f(x) = 2x^2 + 3$$
$$y = 2x^2 + 3$$
$$x = 2y^2 + 3$$
$$x - 3 = 2y^2$$
$$y^2 = \frac{x - 3}{2}$$
$$y = \sqrt{\frac{x - 3}{2}} \quad \text{(This is the required inverse)}$$

Therefore, the inverse of $f(x)$ is:

$$f^{-1}(x) = \sqrt{\frac{x - 3}{2}}$$

c. $f(x) = 5x^3$

$$y = 5x^3$$
$$x = 5y^3$$
$$\frac{x}{5} = y^3$$
$$y = \sqrt[3]{\frac{x}{5}} \quad \text{(This is the required inverse)}$$

Therefore, the inverse of $f(x)$ is:

$$f^{-1}(x) = \sqrt[3]{\frac{x}{5}}$$

2. Find the inverse of the following functions:

a. $f(x) = \left(\frac{2x - 1}{5}\right)^2$

b. $f(x) = (\dfrac{x+1}{3x-4})^{\frac{2}{3}}$

Solutions

a. $f(x) = (\dfrac{2x-1}{5})^2$

$y = (\dfrac{2x-1}{5})^2$

$x = (\dfrac{2y-1}{5})^2$

Taking the square root of both sides gives:

$\sqrt{x} = \sqrt{(\dfrac{2y-1}{5})^2}$

$\sqrt{x} = \dfrac{2y-1}{5}$

Cross multiply to obtain:

$2y - 1 = 5\sqrt{x}$

$2y = 5\sqrt{x} + 1$

$y = \dfrac{5\sqrt{x}+1}{2}$

Therefore, the inverse of $f(x)$ is:

$f^{-1}(x) = \dfrac{5\sqrt{x}+1}{2}$

b. $f(x) = (\dfrac{x+1}{3x-4})^{\frac{2}{3}}$

$y = (\dfrac{x+1}{3x-4})^{\frac{2}{3}}$

$x = (\dfrac{y+1}{3y-4})^{\frac{2}{3}}$

Raise both sides of the equation to a power of $\dfrac{3}{2}$ i.e. the inverse of $\dfrac{2}{3}$ in order to make the power of $(\dfrac{y+1}{3y-4})^{\frac{2}{3}}$ to become 1. This gives:

$x^{\frac{3}{2}} = ((\dfrac{y+1}{3y-4})^{\frac{2}{3}})^{\frac{3}{2}}$

$x^{\frac{3}{2}} = (\dfrac{y+1}{3y-4})^1$ (Note that the powers are multiplied to give 1, i.e. $\dfrac{3}{2} \times \dfrac{2}{3} = 1$)

$(\sqrt{x})^3 = \dfrac{y+1}{3y-4}$ (Note that from indices, $a^{\frac{3}{2}} = (\sqrt{a})^3 = \sqrt{a^3}$

$\sqrt{x^3} = \dfrac{y+1}{3y-4}$

Cross multiply to obtain:

$$\sqrt{x^3}(3y - 4) = y + 1$$
$$3y\sqrt{x^3} - 4\sqrt{x^3} = y + 1$$

Collect terms in y to obtain:
$$3y\sqrt{x^3} - y = 1 + 4\sqrt{x^3}$$

Factorizing the left hand side gives:
$$y(3\sqrt{x^3} - 1) = 1 + 4\sqrt{x^3}$$

Divide both sides by $3\sqrt{x^3} - 1$ to obtain y as follows:

$$y = \frac{1 + 4\sqrt{x^3}}{3\sqrt{x^3} - 1} \qquad \text{(This is the required inverse)}$$

Therefore, the inverse of $f(x)$ is:

$$f^{-1}(x) = \frac{1 + 4\sqrt{x^3}}{3\sqrt{x^3} - 1}$$

3. If $f(x) = \dfrac{3x + 1}{x + 5}$, find:

a. $f^{-1}(x)$

b. $f^{-1}(-2)$

Solution

a. $f(x) = \dfrac{3x + 1}{x + 5}$

$$y = \frac{3x + 1}{x + 5}$$

$$x = \frac{3y + 1}{y + 5}$$

Cross multiply to obtain:
$$3y + 1 = x(y + 5)$$
$$3y + 1 = xy + 5x$$

Collect terms in y on the left hand side of the equation.
$$3y - xy = 5x - 1$$

Factorize the left hand side to obtain:
$$y(3 - x) = 5x - 1$$

Dividing both sides of the equation by $3 - x$ gives y as follows:

$$y = \frac{5x - 1}{3 - x}$$

Or, $y = \dfrac{-(1 - 5x)}{-(x - 3)}$ (After factorizing by taking −1 as a factor)

$$y = \frac{1 - 5x}{x - 3}$$ (After the negative signs cancel out)

Hence, $y = \dfrac{5x - 1}{3 - x}$ or $y = \dfrac{1 - 5x}{x - 3}$

Therefore, the inverse of $f(x)$ is:

$$f^{-1}(x) = \frac{5x - 1}{3 - x} \quad \text{or} \quad f^{-1}(x) = \frac{1 - 5x}{x - 3} \qquad \text{(Note that any two of the inverses is correct)}$$

b. $f^{-1}(x) = \dfrac{5x - 1}{3 - x}$

$$f^{-1}(-2) = \frac{5(-2) - 1}{3 - (-2)}$$

$$= \frac{-10 - 1}{3 + 2}$$

$$= \frac{-11}{5}$$

$$\therefore \quad f^{-1}(-2) = -2\frac{1}{5}$$

Further Worked Examples on Functions

1. If $f(x - 1) = x^2 - 4x + 3$, find:
a. $f(x)$
b. $f(3)$

Solution

a. $f(x - 1) = x^2 - 4x + 3$

Let $y = x - 1$

Make x the subject of the formula. This gives:

$$x = y + 1$$

We now substitute $y + 1$ for x in the function above. This gives:

$$F(y) = f(x - 1) = x^2 - 4x + 3$$
$$= (y + 1)^2 - 4(y + 1) + 3$$
$$= (y + 1)(y + 1) - 4(y + 1) + 3$$
$$= y^2 + 2y + 1 - 4y - 4 + 3$$
$$F(y) = y^2 - 2y$$

Now, put $y = x$ into $f(y)$ in order to obtain $f(x)$ as follows:

$$F(x) = x^2 - 2x$$

b. $F(x) = x^2 - 2x$

$$f(3) = (3)^2 - 2(3)$$
$$= 9 - 6$$
$$= 3$$

2. If $f(2x + 1) = 5x - 3$, find $f(x + 3)$

Solution

$$f(2x + 1) = 5x - 3$$

Let $y = 2x + 1$

Make x the subject of the formula. This gives:

$$x = \frac{y - 1}{2}$$

We now substitute $\frac{y-1}{2}$ for x in the function above. This gives:

$$F(y) = f(2x + 1) = 5x - 3$$

$$= 5\left(\frac{y-1}{2}\right) - 3$$

$$= \frac{5y - 5}{2} - 3$$

$$= \frac{5y - 5 - 6}{2}$$

$$F(y) = \frac{5y - 11}{2}$$

This can now be written as a function of x by substituting x for y

$$F(x) = \frac{5x - 11}{2}$$

Substitute $x + 3$ for x in order to obtain $f(x + 3)$ as follows:

$$F(x) = \frac{5x - 11}{2}$$

$$F(x + 3) = \frac{5(x + 3) - 11}{2}$$

$$= \frac{5x + 15 - 11}{2}$$

$$F(x + 3) = \frac{5x + 4}{2}$$

3. Given that $f(x + 3) = x^2 - 7$, find $f(-1)$

Solution

$$f(x + 3) = x^2 - 7$$

We are going to use a more direct method which is different from that used in examples 1 and 2 above.

$$f(x + 3) = x^2 - 7$$

In order to find $f(-1)$, we simply equate the two terms in bracket and make x the subject of the formula. Hence, we have:

$$f(x + 3) = f(-1)$$

$$(x + 3) = -1$$

$$x = -1 - 3$$

$$x = -4$$

We now substitute -4 for x in the original function in order to obtain $f(-1)$. This gives:

$F(-1) = f(-4)$ in the function of $f(x + 3)$

Hence, $f(x + 3) = x^2 - 7$

$F(-4) = (-4)^2 - 7$

$= 16 - 7 = 9$

Therefore, $f(-1) = 9$ [Since $f(-1) = f(-4)$ in the function of $f(x + 3)$]

4. If $f(x + 2) = 2x - 5$, find:

a. $f(x)$

b. $f^{-1}(x)$

c. $f(-3)$

d. $f^{-1}(-3)$

e. $f(x - 3)$

Solution

a. $f(x + 2) = 2x - 5$

In order to find $f(x)$, equate the two terms in each of the brackets

     $f(x + 2) = f(x)$

Hence, $x + 2 = x$

      $x = x - 2$

Be sure to make the x in the first bracket i.e. from the original function, to be the subject of the formula.

     Hence $f(x) = f(x - 2)$ in the function of $f(x + 2)$

Now substitute $x - 2$ for x in the given function.

    $F(x + 2) = 2x - 5$

    $F(x) = f(x - 2) = 2(x - 2) - 5$ ($x - 2$ has been substituted for x in the function)

      $= 2x - 4 - 5$

    $F(x) = 2x - 9$

b. $F(x) = 2x - 9$

Let us find $f^{-1}(x)$ i.e. the inverse of $f(x)$. This is done as follows:

     $y = 2x - 9$

     $x = 2y - 9$

     $x + 9 = 2y$

     $y = \dfrac{x + 9}{2}$

Therefore, $f^{-1}(x) = \dfrac{x + 9}{2}$

c. $F(x) = 2x - 9$

$$F(-3) = 2(-3) - 9$$
$$= -6 - 9$$
$$F(-3) = -15$$

d. $f^{-1}(x) = \dfrac{x + 9}{2}$

$f^{-1}(-3) = \dfrac{-3 + 9}{2}$

$= \dfrac{6}{2}$

$f^{-1}(-3) = 3$

e. $F(x) = 2x - 9$

$F(x - 3) = 2(x - 3) - 9$

$= 2x - 6 - 9$

$F(x - 3) = 2x - 15$

5. If $f(3x - 1) = 6x + 5$, find:

a. $f(2x + 5)$

b. $f(x^2)$

<u>Solution</u>

a. $f(3x - 1) = 6x + 5$

$f(2x + 5)$ is obtained as follows:

$3x - 1 = 2x + 5$ (By equating only terms in the brackets)

$3x = 2x + 5 + 1$

$3x = 2x + 6$

$x = \dfrac{2x + 6}{3}$

Hence, $f(2x + 5) = f(\dfrac{2x + 6}{3})$ in the function of $f(3x - 1)$

Therefore substitute $\dfrac{2x + 6}{3}$ for x in the original function above. This gives:

$f(3x - 1) = 6x + 5$

$f(2x + 5) = f(\dfrac{2x + 6}{3}) = 6(\dfrac{2x + 6}{3}) + 5$

$= 2(2x + 6) + 5$ (Note that $6 \div 3 = 2$)

$= 4x + 12 + 5$

$F(2x + 5) = 4x + 17$

b. $f(3x - 1) = 6x + 5$

264

$F(x^2)$ will be obtained from $f(3x-1)$ as follows

$$3x - 1 = x^2 \quad \text{(By equating only terms in the brackets)}$$
$$3x = x^2 + 1$$
$$x = \frac{x^2 + 1}{3}$$

Hence, $f(x^2) = f(\frac{x^2 + 1}{3})$ in the function of $f(3x - 1)$

Therefore substitute $\frac{x^2 + 1}{3}$ for x in the original function above. This gives:

$$f(3x - 1) = 6x + 5$$
$$f(x^2) = f(\frac{x^2 + 1}{3}) = 6(\frac{x^2 + 1}{3}) + 5$$
$$= 2(x^2 + 1) + 5$$
$$= 2x^2 + 2 + 5$$
$$F(x^2) = 2x^2 + 7$$

Exercise 22

1. Determine if each of the following is a function or not.

a. $y = x^{\frac{1}{2}} + 5$

b. $y = 2x^3 - \sqrt[4]{x}$

c. $y = \sqrt{5} - x^5$

d. $y = \frac{3}{5x - 2}$

2. If $f(x) = 3x^3 - 2$, find:

a. $f(-1)$

b. $f(2)$

c. $f(0.5)$

d. $f(3x - 2)$

e. $f(x + 3)$

3. A function f is defined by $f(x) = 2x - 3$

a. Find $f(\frac{1}{5})$

b. If $f(2p + 3) = 10$, find the value of p

c. Find $9f(-\frac{1}{5})$

4. If $f(x) = x - 3$ and $g(x) = 2x + 1$, find:

a. $f(x) + g(x)$

b. $g(x) - 5f(x)$

c. $f(x) \times g(x)$

d. $f(-1) \div g(2)$

e. $h(x) = f(x-2) - g(5)$

f. $h(2)$

5. If $a(x) = x - 2$ and $b(x) = 2x$, find

a. $f(x) = b[a(x)]$

b. $g(x) = a[b(x)]$

6. If $a(x) = 2x^2$, $b(x) = 3x$ and $c(x) = 2x + 1$, find:

a. $f(x) = a[b[c(x)]]$

b. $g(x) = b[b[a(x)]]$

c. $h(x) = c[a[b(x)]]$

7. Classify the following into even and odd functions:

a. $f(x) = \cos x$

b. $f(x) = 2x^3$

8. Classify the following into even and odd functions:

a. $4x^2$

b. $\cos^2 3x$

c. $9x - 4$

d. $\sin^2 x$

9. Find the inverse of the following functions:

a. $f(x) = 5x - 1$

b. $f(x) = 9x^3 + 2$

c. $f(x) = 2x^5$

10. Find the inverse of the following functions:

a. $f(x) = (\dfrac{x-1}{2})^3$

b. $f(x) = (\dfrac{2x+1}{5x-2})^{1/2}$

11. If $f(x) = \dfrac{5x-1}{2x+1}$, find:

a. $f^{-1}(x)$

b. $f^{-1}(-1)$

12. If $f(x-2) = 2x^2 - 5x - 7$, find:

a. $f(x)$

b. $f(-2)$

13. If $f(5x-1) = 2x + 7$, find $f(x-2)$

14. Given that $f(x+4) = x^2 - 2$, find $f(-3)$

15. If $f(2x+1) = 4x - 3$, find:

a. $f(x)$

b. $f^{-1}(x)$

c. $f(-2)$

d. $f^{-1}(-1)$

e. $f(2x-1)$

16. If $f(x-3) = 7x + 2$, find:

a. $f(x-5)$

b. $f(2x^2)$

CHAPTER 23
POLYNOMIALS

A polynomial is an expression which is a sum of terms containing a variable or variables whose power starts from one and above.

The highest power of a variable in a polynomial is called the degree of the polynomial.

Addition and Subtraction of Polynomials

When adding or subtracting polynomials, the like terms are added or subtracted as the case may be. Note that like terms are terms whose variables have the same power.

Examples

1. If $A = 2x^3 + 7x^2 - 5$, $B = 5x^3 - 11$, and $C = x^3 + 2x^2 + 5x + 3$, find:

a. $A + B$

b. $C - B$

c. $2B + A$

d. $A - 2C + 3B$

e. $A - B - C$

Solutions

a. $A + B = (2x^3 + 7x^2 - 5) + (5x^3 - 11)$

$\qquad = 2x^3 + 5x^3 + 7x^2 - 5 - 11$ (Take note of how like terms are brought together)

$\qquad = 7x^3 + 7x^2 - 16$

Take note of the arrangement of terms in ascending order of powers of the variables.

b. $C - B = (x^3 + 2x^2 + 5x + 3) - (5x^3 - 11)$

$\qquad = x^3 + 2x^2 + 5x + 3 - 5x^3 + 11$

$\qquad = x^3 - 5x^3 + 2x^2 + 5x + 3 + 11$

$\qquad = -4x^3 + 2x^2 + 5x + 14$ (Note that x^3 also means $1x^3$)

c. $2B + A = 2(5x^3 - 11) + (2x^3 + 7x^2 - 5)$

$\qquad = 10x^3 - 22 + 2x^3 + 7x^2 - 5$

$\qquad = 10x^3 + 2x^3 + 7x^2 - 22 - 5$

$\qquad = 12x^3 + 7x^2 - 27$

d. $A - 2C + 3B = 2x^3 + 7x^2 - 5 - 2(x^3 + 2x^2 + 5x + 3) + 3(5x^3 - 11)$

$\qquad = 2x^3 + 7x^2 - 5 - 2x^3 - 4x^2 - 10x - 6 + 15x^3 - 33$

Take note of how a negative sign outside a bracket changes all the signs in the bracket.

$\qquad = 2x^3 - 2x^3 + 15x^3 + 7x^2 - 4x^2 - 10x - 5 - 6 - 33$

$\qquad = 15x^3 + 3x^2 - 10x - 44$

e. $A - B - C = (2x^3 + 7x^2 - 5) - (5x^3 - 11) - (x^3 + 2x^2 + 5x + 3)$

$= 2x^3 + 7x^2 - 5 - 5x^3 + 11 - x^3 - 2x^2 - 5x - 3$

$= 2x^3 - 5x^3 - x^3 + 7x^2 - 2x^2 - 5x - 5 + 11 - 3$

$= -4x^3 + 5x^2 - 5x + 3$

2. If $f(x) = 2x^3 - x^2 + 6x - 5$, find:

a. $f(-2)$

b. $f(0)$

Solutions

a. $f(x) = 2x^3 - x^2 + 6x - 5$

In order to find f(−2), simply substitute −2 for x in the given function. This gives:

$F(-2) = 2(-2)^3 - (-2)^2 + 6(-2) - 5$

$F(-2) = 2(-8) - (4) - 12 - 5$

$= -16 - 4 - 12 - 5$

$= -37$

b. $f(x) = 2x^3 - x^2 + 6x - 5$

$f(0) = 2(0)^3 - (0)^2 + 6(0) - 5$

$= 0 - 0 + 0 - 5$

$= -5$

Multiplication of Polynomials

When multiplying polynomials, remember to add the powers of the variable according to the multiplication law of indices.

Examples

1. If $A = 2x^3 - 5x^2 + 3x$ and $B = 3x^3 + x^2 - 7x - 5$, find AB.

Solution

METHOD 1

In order to carry out this multiplication, use each term in A to multiply all the terms in B. The use of bracket in doing this is necessary. Also, remember to carry the sign of each term in A.

$\therefore$ $AB = (2x^3 - 5x^2 + 3x)(3x^3 + x^2 - 7x - 5)$

$= 2x^3(3x^3 + x^2 - 7x - 5) - 5x^2(3x^3 + x^2 - 7x - 5) + 3x(3x^3 + x^2 - 7x - 5)$

$= 6x^6 + 2x^5 - 14x^4 - 10x^3 - 15x^5 - 5x^4 + 35x^3 + 25x^2 + 9x^4 + 3x^3 - 21x^2 - 15x$

Take note of the addition of the powers of x. We now bring like terms together, i.e. terms having the same powers of x. After that we add/subtract the like terms. This gives:

$AB = 6x^6 + 2x^5 - 15x^5 - 14x^4 - 5x^4 + 9x^4 - 10x^3 + 35x^3 + 3x^3 + 25x^2 - 21x^2 - 15x$

$$AB = 6x^6 - 13x^5 - 10x^4 + 28x^3 + 4x^2 - 15x$$

METHOD 2

In order to carry out this method, we arrange the terms in columns with the polynomial having the higher number of terms above the one having lower number of terms. B has higher number of terms, so we place B above A. Note that AB = BA. Also, arrange like terms above each other when carrying out the initial arrangement and during the multiplication of terms. Note that we use each term in the lower column to multiply all the terms in the higher column by starting from the right hand side of the arrangement. This means that we work from the right hand side to the left hand side. It is similar to the way we carry out multiplication of numbers.

Let us now multiply A and B as follows:

$$
\begin{array}{r}
3x^3 + x^2 - 7x - 5 \\
2x^3 - 5x^2 + 3x \\
\hline
+ 9x^4 + 3x^3 - 21x^2 - 15x \\
- 15x^5 - 5x^4 + 35x^3 + 25x^2 \\
+ 6x^6 + 2x^5 - 14x^4 - 10x^3 \\
\hline
6x^6 - 13x^5 - 10x^4 + 28x^3 + 4x^2 - 15x
\end{array}
$$

2. If $M = x^3 - 8x^2 + 2x + 1$ and $N = 5x^3 + 2x^2 - 4x - 7$, find MN.

<u>Solution</u>

METHOD 1

$MN = (x^3 - 8x^2 + 2x + 1)(5x^3 + 2x^2 - 4x - 7)$

$\quad = x^3(5x^3 + 2x^2 - 4x - 7) - 8x^2(5x^3 + 2x^2 - 4x - 7) + 2x(5x^3 + 2x^2 - 4x - 7) + 1(5x^3 + 2x^2 - 4x - 7)$

$\quad = 5x^6 + 2x^5 - 4x^4 - 7x^3 - 40x^5 - 16x^4 + 32x^3 + 56x^2 + 10x^4 + 4x^3 - 8x^2 - 14x + 5x^3 + 2x^2 - 4x - 7$

$\quad = 5x^6 + 2x^5 - 40x^5 - 4x^4 - 16x^4 + 10x^4 - 7x^3 + 32x^3 + 4x^3 + 5x^3 + 56x^2 - 8x^2 + 2x^2 - 14x - 4x - 7$

$MN = 5x^6 - 38x^5 - 10x^4 + 34x^3 + 50x^2 - 18x - 7$

METHOD 2

Ensure you multiply the signs of any two terms multiplied together. The working is as shown below.

$$
\begin{array}{r}
x^3 - 8x^2 + 2x + 1 \\
5x^3 + 2x^2 - 4x - 7 \\
\hline
- 7x^3 + 56x^2 - 14x - 7 \\
- 4x^4 + 32x^3 - 8x^2 - 4x \\
2x^5 - 16x^4 + 4x^3 + 2x^2 \\
+ 5x^6 - 40x^5 + 10x^4 + 5x^3 \\
\hline
5x^6 - 38x^5 - 10x^4 + 34x^3 + 50x^2 - 18x - 7
\end{array}
$$

3. Given that $f(x) = 2x^3 - 3$ and $g(x) = 3x^3 - 2x^2 + x - 5$, find $f(x) \cdot g(x)$

Solution

METHOD 1

$$\begin{aligned}
f(x) \times g(x) &= (2x^3 - 3)(3x^3 - 2x^2 + x - 5) \\
&= 2x^3(3x^3 - 2x^2 + x - 5) - 3(3x^3 - 2x^2 + x - 5) \\
&= 6x^6 - 4x^5 + 2x^4 - 10x^3 - 9x^3 + 6x^2 - 3x + 15 \\
&= 6x^6 - 4x^5 + 2x^4 - 19x^3 + 6x^2 - 3x + 15
\end{aligned}$$

METHOD 2

Since $g(x)$ has more terms, we will place $g(x)$ above $f(x)$. Also arrange like terms above each other as shown below

$$
\begin{array}{r}
3x^3 - 2x^2 + x - 5 \\
\underline{2x^3 \qquad\qquad - 3} \\
-9x^3 + 6x^2 - 3x + 15 \\
\underline{+\ 6x^6 - 4x^5 + 2x^4 - 10x^3 \qquad\qquad\qquad} \\
6x^6 - 4x^5 + 2x^4 - 19x^3 + 6x^2 - 3x + 15
\end{array}
$$

Division of Polynomials

If we divide $2x^3 - 2x^2 + x - 5$ by $x - 5$, then $2x^3 - 2x^2 + x - 5$ is called dividend, while $x - 5$ is called the divisor. The result obtained after the division is called the quotient, while what is left at the end of the division is called the remainder.

Examples

1. Divide $2x^2 + 3x - 5$ by $x - 1$

Solution

The layout is as shown below. Note that 'like terms' are arranged in columns above each other when carrying out the division.

STEP 1: Divide the first term of the dividend by the first term of the divisor. This gives: $\dfrac{2x^2}{x} = 2x$.

Then write the $2x$ at the top of the division sign as shown below.

$$
\begin{array}{r}
2x \\
x - 1 \overline{) 2x^2 + 3x - 5}
\end{array}
$$

STEP 2: Use the $2x$ obtained above to multiply $x - 1$ and write your answer under the dividend. Note that $2x(x - 1)$ will give $2x^2 - 2x$. This is now written below the corresponding like terms of the dividend as shown below.

$$
\begin{array}{r}
2x \\
x - 1 \overline{) 2x^2 + 3x - 5} \\
2x^2 - 2x
\end{array}
$$

STEP 3: Subtract the like terms as arranged above. This means: $2x^2 - 2x^2 = 0x^2$, while $3x - (-2x)$ $= 3x + 2x = 5x$. We now write $5x$ under its corresponding column and ignore the zero under $2x^2$ since there is no need of writing $0x^2$. This is as shown below.

$$\begin{array}{r} 2x \\ x-1{\overline{)2x^2 + 3x - 5}} \\ \underline{2x^2 - 2x} \\ 5x \end{array}$$

STEP 4: Bring down the next term of the dividend which is –5. This is as shown below.

$$\begin{array}{r} 2x \\ x-1{\overline{)2x^2 + 3x - 5}} \\ \underline{2x^2 - 2x} \\ 5x - 5 \end{array}$$

STEP 5: After bringing down –5, we now have a new dividend of $5x - 5$. Use this new dividend to repeat step 1 to step 4 above. This means:

Step 1: $\dfrac{5x}{x}$ = 5. Write this as +5 at the top of the division sign.

Step 2: Use the 5 above to multiply $x - 1$. This gives $5x - 5$. Write this below the corresponding like terms of our new dividend which is also $5x - 5$.

Step 3: Subtract the like terms arranged above. This will give zero since $5x - 5$ subtracted from $5x - 5$ gives zero.

Step 4: When your subtraction in step 3 gives zero, and there is no more term to bring down from the original dividend, then you have arrived at your answer. These steps give the final solution as shown below:

$$\begin{array}{r} 2x + 5 \\ x-1{\overline{)2x^2 + 3x - 5}} \\ \underline{2x^2 - 2x} \\ 5x - 5 \\ \underline{5x - 5} \\ - \ - \ - \end{array}$$

$\therefore \quad (2x^2 + 3x - 5) \div (x - 1) = 2x + 5$

Check your answer by multiplying $(x - 1)$ by $(2x + 5)$. It will give $2x^2 + 3x - 5$, which is the dividend.

Notice that the steps from 1 to 4 above can be formulated into an acronym written as: DMSBd. D means divide, M means multiply, S means subtract, while Bd means bring down. This acronym can always remind you of the next step to take.

2. Divide $5x^3 + x^2 - 8x - 4$ by $x + 1$

272

<u>Solution</u>

The question can also be written as:

$$\frac{5x^3 + x^2 - 8x - 4}{x + 1}$$

The workings are explained below.

STEP 1: Divide the first term of the dividend by the first term of the divisor. This gives: $\frac{5x^3}{x} = 5x^2$.

Then write the $5x^2$ at the top of the division sign as shown below.

$$x + 1 \overline{)5x^3 + x^2 - 8x - 4} \quad \text{with } 5x^2 \text{ on top}$$

STEP 2: Use the $5x^2$ obtained above to multiply $x + 1$ and write your answer under the dividend. Note that $5x^2(x + 1)$ will give $5x^3 + 5x^2$. This is now written below the corresponding like terms of the dividend as shown below.

$$
\begin{array}{r}
5x^2 \\
x + 1 \overline{)5x^3 + x^2 - 8x - 4} \\
5x^3 + 5x^2
\end{array}
$$

STEP 3: Subtract the like terms as arranged above. This means: $5x^3 - 5x^3 = 0x^3$, while $x^2 - (+5x^2)$ $= x^2 - 5x^2 = -4x^2$. We now write $-4x^2$ under its corresponding column and ignore the zero under $5x^3$ since there is no need of writing $0x^3$. This is as shown below.

$$
\begin{array}{r}
5x^2 \\
x + 1 \overline{)5x^3 + x^2 - 8x - 4} \\
\underline{5x^3 + 5x^2} \\
-4x^2
\end{array}
$$

STEP 4: Bring down the next term of the dividend which is $-8x$. This is as shown below.

$$
\begin{array}{r}
5x^2 \\
x + 1 \overline{)5x^3 + x^2 - 8x - 4} \\
\underline{5x^3 + 5x^2} \\
-4x^2 - 8x
\end{array}
$$

STEP 5: After bringing down $-8x$, we now have a new dividend of $-4x^2 - 8x$ as shown above. Use this new dividend to repeat step 1 to step 4 above. This means:

Step 1: $\frac{-4x^2}{x} = -4x$. Write this at the top of the division sign as shown below.

Step 2: Use the $-4x$ above to multiply $x + 1$. This gives $-4x^2 - 4x$. Write this below the corresponding like terms of our new dividend.

Step 3: Subtract the like terms arranged from step 2 above. This will give $-4x$.

Step 4: Bring down the next term of the original dividend which is -4.

Step 5: After bringing down -4, we now have a new dividend of $-4x - 4$. Use this new dividend to repeat step 1 to step 4 above.

Now let us use the acronym DMSBd to complete the remaining part of the division as follows.

D: $-4x \div x = -4$ (This is written on the division sign)

M: $-4(x + 1) = -4x - 4$ (Write this under $-4x - 4$ which is our present dividend)

S: $-4x - 4 - (-4x - 4)$ will give zero.

Bd: There is nothing more to bring down. Hence our division is complete and we now have our final answer.

All the division processes explained above are as shown below.

$$
\begin{array}{r}
5x^2 - 4x - 4 \\
x + 1)\overline{5x^3 + x^2 - 8x - 4} \\
\underline{5x^3 + 5x^2} \\
-4x^2 - 8x \\
\underline{-4x^2 - 4x} \\
-4x - 4 \\
\underline{-4x - 4} \\
-\ \ -\ \ -
\end{array}
$$

$\therefore$ $(5x^3 + x^2 - 8x - 4) \div (x + 1) = 5x^2 - 4x - 4$

Check your answer by multiplying $(x + 1)$ by $(5x^2 - 4x - 4)$. It will give $5x^3 + x^2 - 8x - 4$, which is the dividend.

Points to note as a reminder:

D: During division, only the first term of the dividend is used to divide the first term of the divisor.

M: During multiplication, the answer obtained during division is used to multiply all the terms in the divisor.

S: During subtraction, only the like terms are subtracted. The sign of each term must be carried along with it. Be careful of the negative sign of terms, and the subtraction sign used to carry out the operation.

Bd: Bring down the next term along with its sign. When there is nothing else to bring down, then the division process has ended and the terms on the division sign become the answer.

3. Evaluate $\dfrac{x^3 - 7x - 6}{x - 3}$

<u>Solution</u>

A close look at the dividend shows that there is no term in x^2. Therefore, in order to avoid mistake in our division, it is advisable to include $0x^2$ at the right position in the dividend. Hence the question can be re–written as:

$$\frac{x^3 - 0x^2 - 7x - 6}{x - 3}$$

We now set out our division as shown below.

$$\frac{x^2 + 3x + 2}{x - 3\overline{)x^3 + 0x^2 - 7x - 6}}$$

$$\underline{x^3 - 3x^2}$$
$$3x^2 - 7x$$
$$\underline{3x^2 - 9x}$$
$$2x - 6$$
$$\underline{2x - 6}$$
$$-\ -\ -$$

WORKING

D: $\dfrac{x^3}{x} = x^2$ (This is written on the division sign)

M: $x^2(x - 3) = x^3 - 3x^2$ (Write this under $x^3 + 0x^2$)

S: $0x^2 - (-3x^2) = 0x^2 + 3x^2 = 3x^2$.

Bd: Bring down $-7x$ to obtain $3x^2 - 7x$ as the new dividend.

We now repeat the process using $3x^2 - 7x$

D: $\dfrac{3x^2}{x} = 3x$ (This is written as $+3x$ on the division sign)

M: $3x(x - 3) = 3x^2 - 9x$ (Write this under $3x^2 - 7x$)

S: $3x^2 - 7x - (3x^2 - 9x) = 2x$.

Bd: Bring down -6 to meet $2x$. This gives $2x - 6$ as the new dividend.

Finally, we repeat the process by using $2x - 6$ as dividend.

D: $\dfrac{2x}{x} = 2$ (This is written as $+2$ on the division sign)

M: $2(x - 3) = 2x - 6$ (Write this under $2x - 6$)

S: Their subtraction gives zero

Bd: Nothing more to bring down. Hence we have our answer as $x^2 + 3x + 2$.

Therefore, $\dfrac{x^3 - 7x - 6}{x - 3} = x^2 + 3x + 2$

4. Divide $2x^3 - 11x^2y + 3xy^2 + y^3$ by $2x - y$

<u>Solution</u>

$$\frac{x^2 - 5xy - y^2}{2x - y\overline{)2x^3 - 11x^2y + 3xy^2 + y^3}}$$

$$\underline{2x^3\ -\ x^2y}$$
$$-10x^2y + 3xy^2$$
$$\underline{-10x^2y + 5xy^2}$$
$$-2xy^2 + y^3$$
$$\underline{-2xy^2 + y^3}$$
$$-\ -\ -$$

5. Simplify: $\dfrac{x^3 - y^3}{x - y}$

<u>Solution</u>

A careful look at the dividend shows that some terms are not present. That is, they are zero. The missing terms are x^2y and xy^2. One method of knowing the missing term is to raise the divisor to the highest power (degree) of the dividend. This means that if $(x - y)^3$ is evaluated, you will see the missing terms.

Hence we represent the missing terms by $0x^2y$ and $0xy^2$. We now carry out the division as shown below.

$$
\begin{array}{r}
x^2 + xy + y^2 \\
x - y \overline{)\,x^3 + 0x^2y + 0xy^2 - y^3} \\
\underline{x^3 - x^2y} \quad\quad\quad\quad\quad \\
x^2y + 0xy^2 \quad\quad \\
\underline{x^2y - xy^2} \quad\quad \\
xy^2 - y^3 \\
\underline{xy^2 - y^3} \\
- \quad - \quad -
\end{array}
$$

Hence, $\dfrac{x^3 - y^3}{x - y} = x^2 + xy + y^2$

6. Find the quotient and the remainder when $4x^3 - 6x^2 + 8x - 5$ is divided by $2x + 1$

<u>Solution</u>

This is a case where we will have a remainder. In the course of our division, whenever we carry out a subtraction, and there is nothing else to bring down, whatever is left becomes the remainder. Let us now carry out the division as follows.

$$
\begin{array}{r}
2x^2 - 4x + 6 \\
2x + 1 \overline{)\,4x^3 - 6x^2 + 8x - 5} \\
\underline{4x^3 + 2x^2} \quad\quad\quad\quad\quad \\
-8x^2 + 8x \quad\quad \\
\underline{-8x^2 - 4x} \quad\quad \\
12x - 5 \\
\underline{12x + 6} \\
-11
\end{array}
$$

There is nothing else left to bring down. This ends the division.

$\therefore$ $(4x^3 - 6x^2 + 8x - 5) \div (2x + 1) = 2x^2 - 4x + 6$, remainder -11.

Hence, the quotient is $2x^2 - 4x + 6$, while the remainder is -11.

This division can be written as: $\dfrac{4x^3 - 6x^2 + 8x - 5}{2x + 1} = 2x^2 - 4x + 6 - \dfrac{11}{2x + 1}$

7. Find the quotient and the remainder when $3x^3 - 7x^2 - x + 9$ is divided by $x - 5$

<u>Solution</u>

Let us carry out the division as follows.

$$
\begin{array}{r}
3x^2 + 8x + 39 \\
x - 5{\overline{\smash{\big)}\,3x^3 - 7x^2 - x + 9}} \\
\underline{3x^3 - 15x^2} \\
8x^2 - x \\
\underline{8x^2 - 40x} \\
39x + 9 \\
\underline{39x - 195} \\
204
\end{array}
$$

There is nothing else left to bring down. This ends the division.

∴ The quotient is $3x^2 + 8x + 39$, while the remainder is 204.

This division can be written as: $\dfrac{3x^3 - 7x^2 - x + 9}{x - 5} = 3x^2 + 8x + 39 - \dfrac{204}{x - 5}$

Zeros of Polynomials

A zero of a function f(x) is the root of the equation f(x) = 0.

Examples

Find the zeros of the following polynomial functions:

1. f(x) = $x^2 - 5x + 6$
2. f(x) = $2x^2 + 5x - 3$

<u>Solution</u>

1. In order to find the zeros of the function, we equate the function to zero and solve for x as follows:

$$x^2 - 5x + 6 = 0$$
$$(x - 2)(x - 3) = 0$$

∴ $x = 2$ or $x = 3$

Hence, the zeros of the function are 2 and 3.

2. $2x^2 + 5x - 3 = 0$

Solving this equation by factorization gives:

$$2x^2 - x + 6x - 3 = 0$$

$$x(2x - 1) + 3(2x - 1) = 0$$

$$(2x - 1)(x + 3)$$

$$\therefore \quad x = \frac{1}{2} \text{ or } x = -3$$

Hence, the zeros of the function are $\frac{1}{2}$ and -3.

The Factor Theorem

Let us solve the quadratic equation below by factorization.

$$x^2 - 8x - 20 = 0$$

$$(x - 10)(x + 2) = 0$$

$$\therefore \quad x = 10 \text{ or } x = -2$$

This shows that the factor $(x - 10)$ gives a root of 10, while the factor $(x + 2)$ gives a root of -2. This is what the factor theorem means.

Therefore the factor theorem state that:

If $x = a$ is a root of the equation f(x) = 0, then $(x - a)$ is a factor of f(x). This also means that f(a) = 0 (i.e. substituting 'a' for x in the function gives zero).

The Remainder Theorem

The remainder theorem states that if f(x) is divided by $(x - a)$, the remainder is equal to f(a).

Generally, we can say that, if f(x) is divided by a$x - b$, the remainder is equal to f$(\frac{b}{a})$.

Example

1. Factorize $x^3 - 2x^2 - x + 2$ and use it to solve the cubic equation $x^3 - 2x^2 - x + 2 = 0$

<u>Solution</u>

Let us represent the expression as a function of x.

$$F(x) = x^3 - 2x^2 - x + 2$$

We have to employ the method of trial and error to obtain one factor of f(x). Try numbers such as 1, -1, 2, -2, 3 etc.

$$F(x) = x^3 - 2x^2 - x + 2$$

If $x = 1$, then f$(1) = (1)^3 - 2(1)^2 - (1) + 2$

$$= 1 - 2 - 1 + 2$$

$$= 0$$

Since f(1) = 0, then $(x - 1)$ is a factor of f(x) (factor theorem). Note that f(1) = 0 means that $x = 1$, which can be rearranged to give $x - 1 = 0$ (by taking 1 to the left hand side of the equation). Hence we can see that $x - 1$ is a factor of f(x).

Let us now use the polynomial to divide $(x - 1)$ in order to obtain the other factors. This is shown below.

$$
\begin{array}{r}
x^2 - x - 2 \\
x - 1 \overline{)x^3 - 2x^2 - x + 2} \\
\underline{x^3 - x^2} \\
-x^2 - x \\
\underline{-x^2 + x} \\
-2x + 2 \\
\underline{-2x + 2} \\
- \quad - \quad -
\end{array}
$$

Hence, $\dfrac{x^3 - 2x^2 - x + 2}{x - 1} = x^2 - x - 2$

Or, $x^3 - 2x^2 - x + 2 = (x - 1)(x^2 - x - 2)$

The quadratic part $x^2 - x - 2$ can be factorized to give:

$x^2 - x - 2 = (x - 2)(x + 1)$

$\therefore$ $x^3 - 2x^2 - x + 2 = (x - 1)(x - 2)(x + 1)$

Let us now use the factorized expression above to solve the given equation as follows:

$x^3 - 2x^2 - x + 2 = 0$

$(x - 1)(x - 2)(x + 1) = 0$

Hence, $x = 1$ or 2 or -1 (When each bracket is equated to zero and solved)

2. Solve the equation $2x^3 + 13x^2 + 13x - 10 = 0$

<u>Solution</u>

Let $f(x) = 2x^3 + 13x^2 + 13x - 10$

Let us put numbers such as -1, 1, -2, 2, -3 etc, into the function in order to obtain one of the factors.

Hence, if $x = 1$, $f(1) = 2(1)^3 + 13(1)^2 + 13(1) - 10$

$= 2 + 13 + 13 - 10$

$= 18$

Therefore, $(x - 1)$ is not a factor

If $x = -1$, $f(-1) = 2(-1)^3 + 13(-1)^2 + 13(-1) - 10$

$= -2 + 13 - 13 - 10$

$= -12$

Therefore, $(x + 1)$ is not a factor. Note that from $x = -1$, we have $x + 1$ by taking -1 to the left hand side.

If $x = 2$, $f(2) = 2(2)^3 + 13(2)^2 + 13(2) - 10$

$= 16 + 52 + 26 - 10$

$$= 84$$

Therefore, $(x - 2)$ is not a factor.

If $x = -2$, $f(-2) = 2(-2)^3 + 13(-2)^2 + 13(-2) - 10$

$$= -16 + 52 - 26 - 10$$

$$= 0$$

Therefore, $(x + 2)$ is a factor of $f(x)$

We now divide $f(x)$ by $(x + 2)$ in order to get the other factors. This gives:

$$
\require{enclose}
\begin{array}{r}
2x^2 + 9x - 5 \\[2pt]
x + 2 \enclose{longdiv}{2x^3 + 13x^2 + 13x - 10} \\[2pt]
\underline{2x^3 + 4x^2} \\[2pt]
9x^2 + 13x \\[2pt]
\underline{9x^2 + 18x} \\[2pt]
-5x - 10 \\[2pt]
\underline{-5x - 10} \\[2pt]
-\quad -\quad -
\end{array}
$$

Hence, $\dfrac{2x^3 + 13x^2 + 13x - 10}{x + 2} = 2x^2 + 9x - 5$

Or, $2x^3 + 13x^2 + 13x - 10 = (x + 2)(2x^2 + 9x - 5)$

The quadratic part $2x^2 + 9x - 5$ can be factorized to give:

$$2x^2 + 9x - 5 = 2x^2 + 10x - x - 5$$

$$= 2x(x + 5) - 1(x + 5)$$

$$= (x + 5)(2x - 1)$$

$\therefore \quad 2x^3 + 13x^2 + 13x - 10 = (x + 2)(x + 5)(2x - 1)$

Let us now use the factorized expression above to solve the given equation as follows:

$$2x^3 + 13x^2 + 13x - 10 = 0$$

$$(x + 2)(x + 5)(2x - 1) = 0$$

Hence, $x = -2$ or -5 or $\dfrac{1}{2}$ (When each bracket is equated to zero and solved)

3. Find the remainder when $x^3 - 2x^2 + 5x + 9$ is divided by $x + 2$.

Solution

Let $f(x) = x^3 - 2x^2 + 5x + 9$

According to the remainder theorem, when this polynomial is divided by $x + 2$, the remainder is obtained from $f(-2)$. Note that -2 is obtained when we set $x + 2 = 0$ and solve it to get $x = -2$.

$\therefore \quad f(x) = x^3 - 2x^2 + 5x + 9$

$\quad f(-2) = (-2)^3 - 2(-2)^2 + 5(-2) + 9$

$$= -8 - 8 - 10 + 9$$

$$= -17$$

Hence the remainder is −17

4. Find the remainder when $2x^2 + 8x - 3$ is divided by $3x - 1$

Solution

Let $f(x) = 2x^2 + 8x - 3$

According to the remainder theorem, when this polynomial is divided by $3x - 1$, the remainder is obtained from $f(\frac{1}{3})$. Note that $\frac{1}{3}$ is obtained when we set $3x - 1 = 0$ and solve it to get $x = \frac{1}{3}$.

$\therefore \quad f(x) = 2x^2 + 8x - 3$

$$f(\tfrac{1}{3}) = 2(\tfrac{1}{3})^2 + 8(\tfrac{1}{3}) - 3$$

$$= 2(\tfrac{1}{9}) + \frac{8}{3} - 3$$

$$= \frac{2}{9} + \frac{8}{3} - 3$$

$$= \frac{2 + 24 - 27}{9}$$

$$= -\frac{1}{9}$$

Hence the remainder is $-\frac{1}{9}$

5. If $(x + 1)$ and $(3x + 2)$ are factors of $3x^3 + 2x^2 - 3x - 2$, find the third factor.

Solution

Since $(x + 1)$ and $(3x + 2)$ are factors of $3x^3 + 2x^2 - 3x - 2$, then the third factor can be obtained by dividing $3x^3 + 2x^2 - 3x - 2$ by $(x + 1)(3x + 2)$. This is the simple logic:

$(x + 1)(3x + 2)(\quad) = 3x^3 + 2x^2 - 3x - 2$ (Where ( ) is the third factor)

Hence, $(\quad) = \dfrac{3x^3 + 2x^2 - 3x - 2}{(x + 1)(3x + 2)}$ (When both sides of the equation are divide by $(x + 1)(3x + 2)$

$(\quad) = \dfrac{3x^3 + 2x^2 - 3x - 2}{3x^2 + 5x + 2}$ (After expanding the denominator)

We now carry out the division as shown below

$$
\begin{array}{r}
x - 1 \\
3x^2 + 5x + 2\overline{)\,3x^3 + 2x^2 - 3x - 2} \\
\underline{3x^3 + 5x^2 + 2x} \\
-3x^2 - 5x - 2 \\
\underline{-3x^2 - 5x - 2} \\
-\quad -\quad -\quad -
\end{array}
$$

Hence the third factor is $x - 1$

6. If $(3x - 1)$ is a factor of the polynomial $f(x) = 6x^2 + kx - 1$, where k is a constant, find the zeros of $f(x)$.

Solution

Since $3x - 1$ is a factor, then we obtain x as follows:

$$3x - 1 = 0$$
$$3x = 1$$
$$x = \frac{1}{3}$$

Hence, $f(\frac{1}{3}) = 0$

$$f(x) = 6x^2 + kx - 1$$
$$f(\frac{1}{3}) = 6(\frac{1}{3})^2 + k(\frac{1}{3}) - 1$$
$$= 6(\frac{1}{9}) + \frac{k}{3} - 1$$
$$= \frac{6}{9} + \frac{k}{3} - 1$$
$$= \frac{2}{3} + \frac{k}{3} - 1$$
$$= \frac{2 + k - 3}{3}$$
$$= \frac{k - 1}{3}$$

But $f(\frac{1}{3}) = 0$

Therefore, $\frac{k - 1}{3} = 0$

$$k - 1 = 3(0)$$
$$k - 1 = 0$$
$$k = 1$$

Hence the polynomial is $f(x) = 6x^2 + x - 1$ (Since k = 1)

In order to obtain the zeros of the polynomial, we solve the polynomial equation by factorization as follows:

$$6x^2 + x - 1 = 0$$
$$6x^2 + 3x - 2x - 1 = 0$$
$$3x(2x + 1) - 1(2x + 1) = 0$$
$$(2x + 1)(3x - 1) = 0$$

Therefore, $x = -\frac{1}{2}$ or $x = \frac{1}{3}$ (i.e. the roots of the equation)

Hence the zeros of $f(x)$ are $-\frac{1}{2}$ and $x = \frac{1}{3}$

7. Given that $x + 2$ is a factor of the polynomial $2x^3 - x^2 - 7x + 6$, find the other two factors.

Solution

Let us first divide $2x^2 - x^2 - 7x + 6$ by $x + 2$. This is as shown below.

282

$$2x^2 - 5x + 3$$
$$x + 2 \overline{)2x^3 - x^2 - 7x + 6}$$
$$\underline{2x^3 + 4x^2}$$
$$-5x^2 - 7x$$
$$\underline{-5x^2 - 10x}$$
$$3x + 6$$
$$\underline{3x + 6}$$
$$- \quad - \quad -$$

Hence the quadratic factor of the polynomial is $2x^2 - 5x + 3$. Let us factorize this quadratic expression in order to find the other two linear factors as follows:

$$2x^2 - 5x + 3 = 2x^2 - 3x - 2x + 3$$
$$= x(2x - 3) - 1(2x - 3)$$
$$= (2x - 3)(x - 1)$$

Therefore the other two factors are $(2x - 3)$ and $(x - 1)$

8. The remainder when the polynomial $f(x) = ax^3 + bx^2 + x - 5$ is divided by $x + 2$ is -39, and when it is divided by $x - 1$ the remainder is -3. Determine the values of a and b.

Solution

$$f(x) = ax^3 + bx^2 + x - 5$$

When $f(x)$ is divided by $x + 2$, then $x = -2$ (When we set $x + 2 = 0$, then $x = -2$)

Hence, $f(-2) = a(-2)^3 + b(-2)^2 + (-2) - 5$

$-39 = a(-8) + b(4) - 2 - 5$ (Note that -39 is the remainder)

$-39 = -8a + 4b - 7$

$8a - 4b = 39 - 7$

$8a - 4b = 32$

$2a - b = 8$ (After dividing each term by 8)

$2a - b = 8$ ....Equation (1)

Similarly, when $f(x)$ is divided by $x - 1$, then $x = 1$ (When we set $x - 1 = 0$, then $x = 1$)

Hence, $f(1) = a(1)^3 + b(1)^2 + (1) - 5$

$-3 = a(1) + b(1) + 1 - 5$ (Note that -3 is the remainder)

$-3 = a + b - 4$

$4 - 3 = a + b$

$a + b = 1$ ....Equation (2)

From equation (2), $a = 1 - b$ ....Equation (3)

Substitute $1 - b$ for a in equation (1) as follows.

$2a - b = 8$ ....Equation (1)

$2(1 - b) - b = 8$

$2 - 2b - b = 8$

$2 - 3b = 8$

$-3b = 8 - 2$

$-3b = 6$

$b = \dfrac{6}{-3}$

$b = -2$

Substitute -2 for b in equation (3) in order to find a.

$a = 1 - b$ ....Equation (3)

$= 1 - (-2)$

$= 1 + 2$

$a = 3$

Therefore, $a = 3$ and $b = -2$

9. The polynomial $x^3 + 2x^2 + mx + n$ is divisible by $x + 1$. It leaves a remainder of 12 when it is divided by $x - 2$.

a. Find m and n

b. Factorize the polynomial completely

c. Find the zeros of the polynomial

Solution

a. $x^3 + 2x^2 + mx + n$

Since the polynomial is divisible by $x + 1$, it means that $x + 1$ is a factor of the polynomial. Hence, $x = -1$ is a root of the polynomial equation. Hence substituting $x = -1$, gives zero as follows.

$x^3 + 2x^2 + mx + n$

$(-1)^3 + 2(-1)^2 + m(-1) + n = 0$

$-1 + 2 - m + n = 0$

$-m + n = -1$ ....Equation (1)

Also, since the polynomial leaves a remainder of 12 when divided by $(x - 2)$, it means that if we put $x = 2$ in the polynomial equation, we will obtain 12. This is done as follows.

$x^3 + 2x^2 + mx + n$

$(2)^3 + 2(2)^2 + m(2) + n = 0$

$8 + 8 + 2m + n = 12$

$2m + n = 12 - 8 - 8$

$2m + n = -4$ ....Equation (2)

From equation (1), $n = m - 1$ ....Equation (3)

Substitute $m - 1$ for n in equation (2)

$2m + n = -4$ ....Equation (2)

$2m + (m - 1) = -4$

$2m + m - 1 = -4$

$3m = -4 + 1$

284

$3m = -3$

$m = \dfrac{-3}{3}$

$m = -1$

Substitute −1 for m in equation (3)

$n = m - 1$ ....Equation (3)

$= -1 - 1$

$n = -2$

Hence, m = −1 and n = −2

b. The polynomial is $x^3 + 2x^2 + mx + n$. When we substitute the values of m and n the polynomial becomes:

$x^3 + 2x^2 - x - 2$ (Since m = −1 and n = −2)

From the question, $(x + 1)$ is a factor of the polynomial. Hence, let us divide the polynomial by $(x + 1)$ in order to get the other factors. The working is as shown below.

$$\begin{array}{r} x^2 + x - 2 \\ x + 1\overline{)x^3 + 2x^2 - x - 2} \\ \underline{x^3 + x^2} \\ x^2 - x \\ \underline{x^2 + x} \\ -2x - 2 \\ \underline{-2x - 2} \\ -\ -\ - \end{array}$$

Hence the quadratic factor is $x^2 + x - 2$. We now factorize it as follows:

$x^2 + x - 2 = (x + 2)(x - 1)$

Let us now write the factorized polynomial as follows:

$x^3 + 2x^2 - x - 2 = (x + 1)(x + 2)(x - 1)$

c. Set the factorized polynomial to be equal to zero. This gives:

$x^3 + 2x^2 - x - 2 = 0$

$(x + 1)(x + 2)(x - 1) = 0$

Hence, x = −1, −2 or 1   (When each of the bracket above is equated to zero and solved)

Therefore the zeros of the polynomial are −1, −2 and 1.

10. If $2x^2 - 7x - 4$ is a factor of the polynomial $f(x) = 2x^4 - 5x^3 - 15x^2 + px + q$, where p and q are constants:

a. find the values of p and q

b. factorize $f(x)$ completely.

<u>Solution</u>

a. Let us factorize $2x^2 - 7x - 4$ as follows:

$2x^2 - 7x - 4 = 2x^2 - 8x + x - 4$

$\qquad = 2x(x - 4) + 1(x - 4)$

$\qquad = (x - 4)(2x + 1)$

Hence $2x^2 - 7x - 4 = (x - 4)(2x + 1)$

Therefore $(x - 4)$ and $(2x + 1)$ are factors of f(x)

From these two factors, $x = 4$ and $x = -\dfrac{1}{2}$

Hence, f$(4) = 0$ and f$(-\dfrac{1}{2}) = 0$

$f(x) = 2x^4 - 5x^3 - 15x^2 + px + q$

$f(4) = 2(4)^4 - 5(4)^3 - 15(4)^2 + p(4) + q$

Since f$(4) = 0$, then it follows that:

$2(4)^4 - 5(4)^3 - 15(4)^2 + p(4) + q = 0$

$2(256) - 5(64) - 15(16) + 4p + q = 0$

$512 - 320 - 240 + 4p + q = 0$

$-48 + 4p + q = 0$

$4p + q = 48$ ....Equation (1)

Also, $f(x) = 2x^4 - 5x^3 - 15x^2 + px + q$

$f(-\dfrac{1}{2}) = 2(-\dfrac{1}{2})^4 - 5(-\dfrac{1}{2})^3 - 15(-\dfrac{1}{2})^2 + p(-\dfrac{1}{2}) + q$

Since f$(-\dfrac{1}{2}) = 0$, then it follows that:

$2(-\dfrac{1}{2})^4 - 5(-\dfrac{1}{2})^3 - 15(-\dfrac{1}{2})^2 + p(-\dfrac{1}{2}) + q = 0$

$2(\dfrac{1}{16}) - 5(-\dfrac{1}{8}) - 15(\dfrac{1}{4}) - \dfrac{p}{2} + q = 0$

$\dfrac{1}{8} + \dfrac{5}{8} - \dfrac{15}{4} - \dfrac{p}{2} + q = 0$

$\dfrac{1 + 5 - 30 - 4p + 8q}{8} = 0$

$-24 - 4p + 8q = 8(0)$

$-24 = 4p - 8q$

$4p - 8q = -24$

$p - 2q = -6$ 　　　(After dividing each term by 4)

$p - 2q = -6$ ....Equation (2)

From equation (2), $p = 2q - 6$ ....Equation (3)

Substitute $2q - 6$ for p in equation 1.

$4p + q = 48$ ....Equation (1)

$4(2q - 6) + q = 48$

$8q - 24 + q = 48$

$9q = 48 + 24$

$9q = 72$

$q = \dfrac{72}{9}$

$q = 8$

Substitute 8 for q in equation (3)

$\quad p = 2q - 6 \ldots\ldots\ldots\ldots$Equation (3)

$\quad\quad = 2(8) - 6$

$\quad\quad = 16 - 6$

$\quad p = 10$

Hence, p = 10 and q = 8

b. Substitute 10 for p and 8 for q in order to obtain the polynomial as follows:

$\quad f(x) = 2x^4 - 5x^3 - 15x^2 + 10x + 8$

Since $2x^2 - 7x - 4$ is a factor of the polynomial, let us divide f(x) by $2x^2 - 7x - 4$ as shown below.

$$
\begin{array}{r}
x^2 + x - 2 \\
2x^2 - 7x - 4\overline{)2x^4 - 5x^3 - 15x^2 - 10x + 8} \\
\underline{2x^4 - 7x^3 - 4x^2} \\
2x^3 - 11x^2 + 10x \\
\underline{2x^3 - 7x^2 - 4x} \\
-4x^2 + 14x + 8 \\
\underline{-4x^2 + 14x + 8} \\
- \ - \ - \ - \ - \ -
\end{array}
$$

Hence the other quadratic factor is $x^2 + x - 2$

We now factorize this as follows:

$\quad x^2 + x - 2 = (x + 2)(x - 1)$

Recall that: $2x^2 - 7x - 4 = (x - 4)(2x + 1)$ (From our first step in (a) above)

Therefore the polynomial is completely factorized by using all our linear factors as follows:

$\quad 2x^4 - 5x^3 - 15x^2 + 10x + 8 = (x + 2)(x - 1)(x - 4)(2x + 1)$

Exercise 23

1. If $A = 3x^3 + 5x^2 - 1$, $B = 2x^3 + 7$, and $C = 2x^3 - x^2 + 2x - 4$, find:

a. $A + B$

b. $2C - 2B$

c. $3C - A$

d. $2B + A - 2C$

e. $A + 2B + C$

2. If $f(x) = x^3 - 4x^2 + 3x - 10$, find:

a. $f(-3)$

b. $f(0)$

3. If $A = 5x^3 - 2x^2 + x$ and $B = 2x^3 + 3x^2 - 5x - 2$, find AB.

4. If $M = 2x^3 - 3x^2 + 6x + 5$ and $N = 3x^3 - x^2 + 2x - 4$, find MN.

5. Given that $f(x) = x^3 - 5$ and $g(x) = 2x^3 - 3x^2 + 6x - 9$, find $f(x) \cdot g(x)$

6. Divide $3x^2 + 2x - 8$ by $3x - 4$

7. Divide $2x^3 + x^2 - 7x + 24$ by $x + 3$

8. Evaluate $\dfrac{6x^3 + 3x^2 - 10x - 5}{2x + 1}$

9. Divide $4x^3 - 12x^2y + 11xy^2 + 15y^3$ by $2x - 5y$

10. Simplify: $\dfrac{5x^3 + 8x^2y - 5xy^2 - 2y^3}{x + 2y}$

11. Find the quotient and the remainder when $2x^3 - 7x^2 + 12x - 9$ is divided by $x - 5$

12. Find the quotient and the remainder when $5x^3 + 2x^2 - 11x + 1$ is divided by $x + 2$

13. Find the zeros of the following polynomial functions:

a. $f(x) = 2x^2 - 19x + 30$

b. $f(x) = 4x^2 + 4x - 3$

14. Factorize $x^3 - 4x^2 + x + 6$ and use it to solve the cubic equation $x^3 - 4x^2 + x + 6 = 0$

15. Solve the equation $2x^4 + x^3 - 20x^2 - 13x + 30 = 0$

16. Find the remainder when $2x^3 - 9x^2 + 6x + 3$ is divided by $x - 4$.

17. Find the remainder when $5x^2 - 2x - 7$ is divided by $2x + 3$

18. If $(x + 1)$ and $(x - 2)$ are factors of $4x^4 + 4x^3 - 13x^2 - 19x - 6$, find the other two factors.

19. If $(2x - 1)$ is a factor of the polynomial $f(x) = 12x^3 + kx$, where k is a constant, find the zeros of $f(x)$.

20. Given that $2x + 1$ is a factor of the polynomial $6x^3 + 3x^2 - 10x - 5$, find the quadratic factor.

21. The remainder when the polynomial $f(x) = ax^3 + bx^2 + 5x - 9$ is divided by $x - 1$ is -10, and when it is divided by $x - 3$ the remainder is -12. Determine the values of a and b.

22. The polynomial $x^3 - 2x^2 + mx + n$ is divisible by $x - 1$. It leaves a remainder of -8 when divided by $x - 2$.

a. Find m and n

b. Find the quadratic factor

c. Find the zeros of the polynomial

23. If $2x^2 - x - 6$ is a factor of the polynomial $f(x) = 2x^4 + x^3 - 20x^2 + mx + n$, where m and n are constants. Find the values of m and n

CHAPTER 24
PARTIAL FRACTION

Consider the addition of the algebraic fractions below.

$$\frac{3}{x-2} + \frac{2}{x+5} = \frac{3(x+5) + 2(x-2)}{(x-2)(x+5)}$$ [Note that the LCM is $(x-2)(x+5)$]

$$= \frac{3x + 15 + 2x - 4}{(x-2)(x+5)}$$

$$= \frac{5x + 11}{(x-2)(x+5)}$$

Hence, $\frac{3}{x-2}$ and $\frac{2}{x+5}$ added up to give $\frac{5x+11}{(x-2)(x+5)}$

We say that $\frac{3}{x-2}$ and $\frac{2}{x+5}$ are the partial fractions of $\frac{5x+11}{(x-2)(x+5)}$. The result obtained i.e.

$\frac{5x+11}{(x-2)(x+5)}$ is called the compound fraction. Since we know how to move from partial

fractions to compound fraction, it is also important that we know how to move from compound fraction to partial fractions. This is what we want to deal with here.

Resolving Algebraic Fractions into Partial Fractions

The steps below are applied in resolving algebraic fraction into partial fractions:

1. The polynomial must be factorized or already be expressed in factors.

2. The highest power of the variable in the numerator must be at least one less than that of the denominator.

3. When the highest power of the variable in the numerator is equal to, or higher than the highest power of the variable in the denominator, then the numerator should be divided by the denominator before resolving the remainder (expressed as fraction) into partial fractions.

Types of Partial Fraction

1. **LINEAR FACTORS:** Algebraic fractions whose denominators contain linear factors are resolved into partial fractions as shown below.

$$\frac{f(x)}{(x+a)(x+b)} = \frac{A}{x+a} + \frac{B}{x+b}$$

2. **REPEATED LINEAR FACTORS:** Fractions with repeated linear denominators are resolved as follows:

$$\frac{f(x)}{(x+a)^2} = \frac{A}{x+a} + \frac{B}{(x+a)^2}$$

And:

$$\frac{f(x)}{(x+a)^3} = \frac{A}{x+a} + \frac{B}{(x+a)^2} + \frac{C}{(x+a)^3}$$

3. **QUADRATIC FACTORS:** When a denominator of an algebraic fraction contains a quadratic expression that cannot be factorized into linear factors, then it is resolved into partial fractions as follows:

$$\frac{f(x)}{(ax^2 + bx + c)(x + d)} = \frac{Ax + B}{(ax^2 + bx + c)} + \frac{C}{x + d}$$

Note that in all three cases above, the highest power of the variable in $f(x)$ must be less than the highest power of the variable in the denominator. Otherwise we must divide first before we proceed.

Examples

1. Express $\dfrac{x + 21}{(x + 3)(x + 1)}$ in partial fractions

Solution

METHOD 1

Let $\dfrac{x + 21}{(x + 3)(x + 1)} = \dfrac{A}{x + 3} + \dfrac{B}{x + 1}$

The LCM or LCD (lowest common denominator) of the right hand side is $(x + 3)(x + 1)$. Hence we carry out the addition of the right hand side as follows:

$$\frac{x + 21}{(x + 3)(x + 1)} = \frac{A(x + 1) + B(x + 3)}{(x + 3)(x + 1)}$$

Hence, since the denominators are the same on both sides of the equation, then the numerators should also be the same or identical. Hence for the numerators, we write:

$x + 21 \equiv A(x + 1) + B(x + 3)$

This is an identity. It means that both sides are equal for all values of x, hence the use of the equivalent symbol. In order to find A and B, we take a value of x that will make the term in any of the bracket to be zero. In order to find the number that will make A to be zero, we simply equate the bracket attached to A to be equal to zero and solve for x. For example, $x + 1 = 0$, which will give $x = -1$ when we solve for x. Hence if we put $x = -1$, A will be eliminated so that we can obtain B. This gives:

$x + 21 \equiv A(x + 1) + B(x + 3)$

$-1 + 21 = A(-1 + 1) + B(-1 + 3)$

$20 = A(0) + 2B$

$20 = 2B$ (Since A x 0 = 0)

$B = \dfrac{20}{2}$

$B = 10$

If we put $x = -3$ (from $x + 3 = 0$, then $x = -3$), B will be eliminated, so that we can obtain A. This gives:

$x + 21 \equiv A(x + 1) + B(x + 3)$

$-3 + 21 = A(-3 + 1) + B(-3 + 3)$

$$18 = A(-2) + B(0)$$

$$18 = -2A \quad \text{(Since B x 0 = 0)}$$

$$A = \frac{18}{-2}$$

$$A = -9$$

Hence, A = −9 and B = 10

Therefore, $\dfrac{x + 21}{(x + 3)(x + 1)} = \dfrac{-9}{x + 3} + \dfrac{10}{x + 1}$

Or, $\dfrac{x + 21}{(x + 3)(x + 1)} = \dfrac{10}{x + 1} - \dfrac{9}{x + 3}$ (When we rearrange the answer above)

METHOD 2

In this method, we follow the same procedure as in method 1 until we get to the point written below:

$$x + 21 \equiv A(x + 1) + B(x + 3)$$

Now we expand the brackets above, then simplify and collect like terms as carried out below.

$$x + 21 \equiv Ax + A + Bx + 3B$$

$$x + 21 \equiv Ax + Bx + A + 3B \quad \text{(By collecting like terms together)}$$

$$x + 21 \equiv (A + B)x + A + 3B \quad \text{(After factorizing terms having like variables)}$$

By comparing coefficients of like terms on both sides of the identity, we have:

$$A + B = 1$$

On the left hand side of the identity above, we have $1x$, while on the right hand side we have $(A + B)x$. Hence the coefficients of x are 1 and A + B. Therefore, A + B = 1.

Similarly, A + 3B = 21

The constant term on the left hand side is 21, while the constant terms on the right hand side are A and 3B. Hence we equate them to obtain A + 3B = 21.

Therefore our two equations above are:

$$A + B = 1 \ \text{................Equation (1)}$$

$$\underline{A + 3B = 21} \ \text{.................Equation (2)}$$

Equation (2) − equation (1):   2B = 20

$$B = \frac{20}{2}$$

$$B = 10$$

Substitute 10 for B in equation (1).

$$A + B = 1 \ \text{................Equation (1)}$$

$$A + 10 = 1$$

$$A = 1 - 10$$

$$A = -9$$

Therefore, A = −9 and B = 10

Hence, $\dfrac{x+21}{(x+3)(x+1)} = \dfrac{-9}{x+3} + \dfrac{10}{x+1}$

Or, $\dfrac{x+21}{(x+3)(x+1)} = \dfrac{10}{x+1} - \dfrac{9}{x+3}$ (As obtained in method 1 above)

It is obvious that method 1 is more direct or shorter than method 2, since method 2 leads to a simultaneous equation. Hence, in most of our examples we will be using method 1. However in some cases where only method 1 cannot be used, we will apply the two methods or only method 2, as the case may be.

2. Resolve $\dfrac{4x-16}{x^2-2x-3}$ into partial fractions.

Solution

The denominator can be factorized into linear factors as follows:

$x^2 - 2x - 3 = (x-3)(x+1)$

Hence the fraction is now written as follows:

$\dfrac{4x-16}{x^2-2x-3} = \dfrac{4x-16}{(x-3)(x+1)}$

We now resolve the factorized one into partial fractions as follows:

Let $\dfrac{4x-16}{(x-3)(x+1)} = \dfrac{A}{x-3} + \dfrac{B}{x+1}$

$\dfrac{4x-16}{(x-3)(x+1)} = \dfrac{A(x+1) + B(x-3)}{(x-3)(x+1)}$

Since the denominators are the same, then the numerators are identical and can be brought out to give:

$4x - 16 \equiv A(x+1) + B(x-3)$

Now substitute $x = -1$ into the identity in order to obtain B as follows:

$4x - 16 \equiv A(x+1) + B(x-3)$

$4(-1) - 16 = A(-1+1) + B(-1-3)$

$-4 - 16 = A(0) - 4B$

$-20 = -4B$ (Since A x 0 = 0)

$B = \dfrac{-20}{-4}$

$B = 5$

Put $x = 3$ in order to obtain A as follows:

$4x - 16 \equiv A(x+1) + B(x-3)$

$4(3) - 16 = A(3+1) + B(3-3)$

$12 - 16 = A(4) + B(0)$

$-4 = 4A$ (Since B x 0 = 0)

$A = \dfrac{-4}{4}$

$A = -1$

Hence, A = −1 and B = 5

Therefore, $\dfrac{4x - 16}{x^2 - 2x - 3} = -\dfrac{1}{x - 3} + \dfrac{5}{x + 1}$

$\qquad\qquad = \dfrac{5}{x + 1} - \dfrac{1}{x - 3}$ (When we rearrange the answer above)

3. Express $\dfrac{15}{x^2 - 9}$ in partial fractions.

Solution

The denominator is a difference of two squares. Recall that a difference of two squares such as $a^2 - b^2$ can be factorized as follows:

$\qquad a^2 - b^2 = (a + b)(a - b)$

Similarly, the denominator above can be expressed as a difference of two squares and factorized as follows:

$\qquad x^2 - b^2 = x^2 - 3^2 = (x + 3)(x - 3)$ (Note that 9 has been expressed as 3^2)

Hence, $\dfrac{15}{(x + 3)(x - 3)} = \dfrac{A}{x + 3} + \dfrac{B}{x - 3}$

$\qquad \dfrac{15}{(x + 3)(x - 3)} = \dfrac{A(x - 3) + B(x + 3)}{(x + 3)(x - 3)}$

$\qquad 15 \equiv A(x - 3) + B(x + 3)$

Substitute $x = 3$ into the identity to obtain B as follows:

$\qquad 15 = A(3 - 3) + B(3 + 3)$ (Note that the right hand side remains 15 since there is no x there)

$\qquad 15 = 0 + 6B$

$\qquad 15 = 6B$

$\qquad B = \dfrac{15}{6}$

$\qquad B = \dfrac{5}{2}$ (In its lowest term)

Put $x = -3$ in order to eliminate B and obtain A as follows:

$\qquad 15 \equiv A(x - 3) + B(x + 3)$

$\qquad 15 = A(-3 - 3) + B(-3 + 3)$

$\qquad 15 = -6A + 0$

$\qquad 15 = -6A$

$\qquad A = \dfrac{15}{-6}$

$\qquad A = -\dfrac{5}{2}$

Hence, $A = -\dfrac{5}{2}$ and $B = \dfrac{5}{2}$

Therefore, $\dfrac{15}{x^2 - 9} = \dfrac{-\frac{5}{2}}{x + 3} + \dfrac{\frac{5}{2}}{x - 3}$

$$= -\frac{5}{2(x+3)} + \frac{5}{2(x-3)}$$

$$= \frac{5}{2(x-3)} - \frac{5}{2(x+3)}$$

4. Resolve $\dfrac{2x-9}{x(x-3)}$ into partial fractions.

Solution

$$\frac{2x-9}{x(x-3)} = \frac{A}{x} + \frac{B}{x-3}$$

$$\frac{2x-9}{x(x-3)} = \frac{A(x-3)+B(x)}{x(x-3)}$$

Equating the numerators gives:

$$2x-9 \equiv A(x-3) + B(x)$$

Substitute $x = 3$ into the identity to obtain B as follows:

$$2(3) - 9 = A(3-3) + B(3)$$

$$6 - 9 = 0 + 3B$$

$$-3 = 3B$$

$$B = \frac{-3}{3}$$

$$B = -1$$

Put $x = 0$ in order to eliminate B and obtain A as follows:

$$2x - 9 \equiv A(x-3) + B(x)$$

$$2(0) - 9 = A(0-3) + B(0)$$

$$0 - 9 = -3A + 0$$

$$-9 = -3A$$

$$A = \frac{-9}{-3}$$

$$A = 3$$

Hence, A = 3 and B = −1

Therefore, $\dfrac{2x-9}{x(x-3)} = \dfrac{3}{x} - \dfrac{1}{x-3}$

5. Resolve $\dfrac{2x+12}{(x+5)^2}$ into partial fractions.

Solution

The denominator is a repeated linear factor. Hence we resolve into partial factions as follows:

$$\frac{2x+12}{(x+5)^2} = \frac{A}{x+5} + \frac{B}{(x+5)^2}$$

$$\frac{2x+12}{(x+5)^2} = \frac{A(x+5) + B(1)}{(x+5)^2} \qquad [\text{Note that } (x+5)^2 = (x+5)(x+5)]$$

Equating the numerators gives:

$$2x + 12 \equiv A(x+5) + B$$

Substitute –5 for x in order to solve for B

$2(-5) + 12 = A(-5 + 5) + B$

$-10 + 12 = A(0) + B$

$2 = B$

$B = 2$

Since we already know the value of B, we can substitute any value for x into the identity above and obtain the value of A. Note that an identity is true for all values of the variable. Hence let us substitute 1 for x (any value of x can be used) into the identity while also substituting 2 for B as we have already obtained above. This will give us A as follows:

$2x + 12 \equiv A(x + 5) + B$

$2(1) + 12 = A(1 + 5) + 2$

$2 + 12 = 6A + 2$

$14 - 2 = 6A$

$12 = 6A$

$A = \dfrac{12}{6}$

$A = 2$

Hence, A = 2 and B = 2

Therefore, $\dfrac{2x + 12}{(x + 5)^2} = \dfrac{2}{x + 5} + \dfrac{2}{(x + 5)^2}$

6. Express $\dfrac{3x - 5}{x^2(x-2)}$ in partial fractions.

<u>Solution</u>

x^2 in the denominator is like a repeated linear factor. With that in mind, we express in partial fraction as follows:

$\dfrac{3x - 5}{x^2(x - 2)} = \dfrac{A}{x} + \dfrac{B}{x^2} + \dfrac{C}{x - 2}$

$\dfrac{3x - 5}{x^2(x - 2)} = \dfrac{A[x(x - 2)] + B(x - 2) + C(x^2)}{x^2(x - 2)}$

Equating the numerators gives:

$3x - 5 \equiv A[x(x - 2)] + B(x - 2) + Cx^2$

Observing the right hand side of the identity shows that if we substitute $x = 0$, we will obtain the value of B as follows:

$3x - 5 \equiv A[x(x - 2)] + B(x - 2) + Cx^2$

$3(0) - 5 = A[0(0 - 2)] + B(0 - 2) + C(0^2)$

$0 - 5 = A[0(-2)] + B(-2) + 0$

$-5 = 0 - 2B + 0$

$-5 = -2B$

$$B = \frac{-5}{-2}$$

$$B = \frac{5}{2}$$

Similarly, substituting 2 for x in the identity above will give us the value of C as follows:

$$3x - 5 \equiv A[x(x-2)] + B(x-2) + Cx^2$$

$$3(2) - 5 = A[2(2-2)] + B(2-2) + C(2^2)$$

$$6 - 5 = A[2(0)] + B(0) + 4C$$

$$1 = 0 + 0 + 4C$$

$$1 = 4C$$

$$C = \frac{1}{4}$$

Now, since we already know the values of B and C, we can substitute any value of x (apart from 0 and 2 used above) into the identity in order to obtain the last letter which is A. Note that this can be done when finding the last letter in any identity. Hence, let us put $x = 1$ into the identity as follows:

$$3x - 5 \equiv A[x(x-2)] + B(x-2) + Cx^2$$

$$3(1) - 5 = A[1(1-2)] + \frac{5}{2}(1-2) + \frac{1}{4}(1^2)$$

$$3 - 5 = A[1(-1)] + \frac{5}{2}(-1) + \frac{1}{4}(1)$$

$$-2 = -A - \frac{5}{2} + \frac{1}{4}$$

$$A = 2 - \frac{5}{2} + \frac{1}{4}$$

$$A = \frac{8 - 10 + 1}{4}$$

$$A = -\frac{1}{4}$$

Hence, $A = -\frac{1}{4}$, $B = \frac{5}{2}$ and $C = \frac{1}{4}$

Therefore, $\dfrac{3x - 5}{x^2(x-2)} = \dfrac{-\frac{1}{4}}{x} + \dfrac{\frac{5}{2}}{x^2} + \dfrac{\frac{1}{4}}{x-2}$

$$= -\frac{1}{4x} + \frac{5}{2x^2} + \frac{1}{4(x-2)}$$

This can also be rearranged to avoid starting with a negative value as follows:

$$\frac{3x - 5}{x^2(x-2)} = \frac{5}{2x^2} - \frac{1}{4x} + \frac{1}{4(x-2)}$$

7. Resolve $\dfrac{x+6}{(x+2)(x-3)^2}$ into partial fractions.

Solutions

The denominator contains repeated factors. Hence we resolve as follows:

$$\frac{x+6}{(x+2)(x-3)^2} = \frac{A}{x+2} + \frac{B}{x-3} + \frac{C}{(x-3)^2}$$

$$\frac{x+6}{(x+2)(x-3)^2} = \frac{A(x-3)^2 + B(x+2)(x-3) + C(x+2)}{(x+2)(x-3)^2}$$

Equating the numerators gives:

$$x + 6 \equiv A(x-3)^2 + B(x+2)(x-3) + C(x+2)$$

Substitute 3 for x in order to obtain C as follows:

$$3 + 6 = A(3-3)^2 + B(3+2)(3-3) + C(3+2)$$

$$9 = A(0) + B(5)(0) + C(5)$$

$$9 = 0 + 0 + 5C$$

$$9 = 5C$$

$$C = \frac{9}{5}$$

Substitute −2 for x in the identity above in order to find A. This gives:

$$x + 6 \equiv A(x-3)^2 + B(x+2)(x-3) + C(x+2)$$

$$-2 + 6 = A(-2-3)^2 + B(-2+2)(-2-3) + C(-2+2)$$

$$4 = A(-5)^2 + B(0)(-5) + C(0)$$

$$4 = A(25) + 0 + 0$$

$$4 = 25A$$

$$A = \frac{4}{25}$$

Since we now know the values of A and C, we can substitute any value for x in the identity above in order to get B. However, I want us to use the method of comparing coefficient to find B. In order to use this method, we have to expand the brackets on the right hand side of the identity. Let us expand and then compare coefficients as follows:

$$x + 6 \equiv A(x-3)^2 + B(x+2)(x-3) + C(x+2)$$

$$x + 6 \equiv A(x-3)(x-3) + B(x+2)(x-3) + C(x+2)$$

$$x + 6 \equiv A(x^2-6x+9) + B(x^2-x-6) + Cx + 2C$$

$$x + 6 \equiv Ax^2 - 6Ax + 9A + Bx^2 - Bx - 6B + Cx + 2C$$

$$x + 6 \equiv (A+B)x^2 - (6A+B-C)x + (9A-6B+2C)$$

By comparing the coefficient of x^2 on both sides of the identity shows that:

$$A + B = 0$$

Since there is no term in x^2 on the left hand side of the identity, it means that the coefficient of x^2 on the left hand side is zero. The coefficient of x^2 on the right hand side is A + B. Hence equating the two coefficients gives A + B = 0, as written above.

We already know the value of A from our answer above, hence we calculate B as follows:

$$A + B = 0$$

$$\frac{4}{25} + B = 0$$

$$B = -\frac{4}{25}$$

Hence, $A = \frac{4}{25}$, $B = -\frac{4}{25}$ and $C = \frac{9}{5}$

Therefore, $\dfrac{x+6}{(x+2)(x-3)^2} = \dfrac{4}{25(x+2)} - \dfrac{4}{25(x-3)} + \dfrac{9}{5(x-3)^2}$

Or, $\dfrac{x+6}{(x+2)(x-3)^2} = \dfrac{1}{5}\left[\dfrac{4}{5(x+2)} - \dfrac{4}{5(x-3)} + \dfrac{9}{(x-3)^2}\right]$

8. Resolve $\dfrac{2x^2 - 3x - 4}{x(x^2 - 2)}$ into partial fractions.

Solution

Note that the denominator contains a quadratic factor which does not factorize. This means that when the quadratic factor is resolved to partial fraction, its numerator will contain x and two constant terms. Hence we resolve into partial fractions as follows:

$$\frac{2x^2 - 3x - 4}{x(x^2 - 2)} = \frac{A}{x} + \frac{Bx + C}{x^2 - 2}$$

$$\frac{2x^2 - 3x - 4}{x(x^2 - 2)} = \frac{A(x^2 - 2) + (Bx+C)(x)}{x(x^2 - 2)}$$

Equating the numerators gives:

$$2x^2 - 3x - 4 \equiv A(x^2 - 2) + (Bx + C)(x)$$

Substitute 0 for x in the identity above. This gives:

$$2(0)^2 - 3(0) - 4 = A((0)^2 - 2) + [B(0) + C](0)$$

$$0 - 0 - 4 = A(0 - 2) + 0$$

$$-4 = -2A$$

$$A = \frac{-4}{-2}$$

$$A = 2$$

In the identity above, we can no longer substitute any value of x in order to obtain B or C. Hence we have to use the method of comparing coefficients to find B and C. In order to use this method, we expand the bracket in the identity above. This gives:

$$2x^2 - 3x - 4 \equiv A(x^2 - 2) + (Bx + C)(x)$$
$$\equiv Ax^2 - 2A + Bx^2 + Cx$$
$$\equiv Ax^2 + Bx^2 + Cx - 2A$$

$$2x^2 - 3x - 4 \equiv (A + B)x^2 + Cx - 2A \quad \text{[Note that } Ax^2 + Bx^2 \text{ has been factorized to give } (A + B)x^2\text{]}$$

Comparing the coefficients of x^2 on both sides of the identity shows that $A + B = 2$. This is because on the left hand side we have $2x^2$ and on the right hand side we have $(A + B)x^2$. Hence $2x^2 = (A + B)x^2$, which shows that $A + B = 2$.

Let us solve this equation as follows:

$$A + B = 2$$

$$2 + B = 2 \quad \text{(Since } A = 2 \text{ as obtained above)}$$

B = 2 − 2

B = 0

Similarly, comparing the coefficients of x in the identity above, shows that:

C = −3   (Since −3x = Cx)

Hence, A = 2, B = 0 and C = −3

Therefore, $\dfrac{2x^2 - 3x - 4}{x(x^2 - 2)} = \dfrac{2}{x} + \dfrac{0x - 3}{x^2 - 2}$

This gives: $\dfrac{2x^2 - 3x - 4}{x(x^2 - 2)} = \dfrac{2}{x} - \dfrac{3}{x^2 - 2}$

9. Resolve $\dfrac{2x^2 - 1}{(x^2 - 1)(x - 2)}$ into partial fractions

Solution

$$\dfrac{2x^2 - 1}{(x^2 - 1)(x - 2)} = \dfrac{Ax + B}{(x^2 - 1)} + \dfrac{C}{(x - 2)}$$

$$\dfrac{2x^2 - 1}{(x^2 - 1)(x - 2)} = \dfrac{(Ax + B)(x - 2) + C(x^2 - 1)}{(x^2 - 1)(x - 2)}$$

Equating the numerators gives:

$2x^2 - 1 \equiv (Ax + B)(x - 2) + C(x^2 - 1)$

Substitute 2 for x in order to get C. This gives:

$2(2)^2 - 1 = (A(2) + B)(2 - 2) + C(2^2 - 1)$

$2(4) - 1 = (2A + B)(0) + C(4 - 1)$

$8 - 1 = 0 + 3C$

$7 = 3C$

$C = \dfrac{7}{3}$

Now let us expand the bracket in the identity above.

$2x^2 - 1 \equiv (Ax + B)(x - 2) + C(x^2 - 1)$

$\equiv Ax^2 - 2Ax + Bx - 2B + Cx^2 - C$

$\equiv Ax^2 + Cx^2 - 2Ax + Bx - 2B - C$

$2x^2 - 1 \equiv (A + C)x^2 + (-2A + B)x - (2B + C)$ (After factorization)

Comparing coefficient of terms on both sides of the identity shows that:

A + C = 2 ....Equation (1)

−2A + B = 0 ....Equation (2)

Note that there is no term in x on the left hand side, hence the coefficient of x on the left hand side is zero. This is why equation (2) is equated to zero as shown above.

−(2B + C) = −1   (This is obtained by comparing the coefficients of the constant terms)

Or, 2B + C = 1 ....Equation (3)   (After dividing both sides by −1)

Hence from equation (1), we have:

A + C = 2 ....Equation (1)

$A + \dfrac{7}{3} = 2$ (Since $C = \dfrac{7}{3}$)

$A = 2 - \dfrac{7}{3}$

$\quad = \dfrac{6 - 7}{3}$

$A = -\dfrac{1}{3}$

From equation (2) we have:

−2A + B = 0 ....Equation (2)

$-2\left(-\dfrac{1}{3}\right) + B = 0$

$\dfrac{2}{3} + B = 0$

$B = -\dfrac{2}{3}$

Hence, $A = -\dfrac{1}{3}$, $B = -\dfrac{2}{3}$ and $C = \dfrac{7}{3}$

Therefore, $\dfrac{2x^2 - 1}{(x^2 - 1)(x - 2)} = \dfrac{-\dfrac{1}{3}x - \dfrac{2}{3}}{(x^2 - 1)} + \dfrac{\dfrac{7}{3}}{(x - 2)}$

$\qquad\qquad\qquad\qquad = \dfrac{-\dfrac{1}{3}(x + 2)}{(x^2 - 1)} + \dfrac{7}{3(x - 2)}$

$\dfrac{2x^2 - 1}{(x^2 - 1)(x - 2)} = \dfrac{7}{3(x - 2)} - \dfrac{(x + 2)}{3(x^2 - 1)}$

Note that the method of comparing coefficients is often used when we have a quadratic factor in the denominator.

10. Resolve $\dfrac{6x^2 - 24x - 3}{(2x - 1)(x^2 + 5x + 4)}$ into partial fractions

<u>Solution</u>

A careful look at the quadratic part of the denominator shows that it can be factorized into linear factors as follows:

$x^2 + 5x + 4 = (x + 4)(x + 1)$

Hence the fraction in the question above can be rewritten as:

$\dfrac{6x^2 - 24x - 3}{(2x - 1)(x + 4)(x + 1)}$

We now resolve into partial fractions as follows:

$\dfrac{6x^2 - 24x - 3}{(2x - 1)(x + 4)(x + 1)} = \dfrac{A}{2x - 1} + \dfrac{B}{x + 4} + \dfrac{C}{x + 1}$

$\dfrac{6x^2 - 24x - 3}{(2x - 1)(x + 4)(x + 1)} = \dfrac{A(x + 4(x + 1) + B(2x - 1)(x + 1) + C(2x - 1)(x + 4)}{(2x - 1)(x + 4)(x + 1)}$

300

Equating the numerators gives:

$$6x^2 - 24x - 3 \equiv A(x + 4)(x + 1) + B(2x - 1)(x + 1) + C(2x - 1)(x + 4)$$

Substitute $x = -4$ in order to obtain B. This gives:

$$6(-4)^2 - 24(-4) - 3 = A(-4 + 4)(-4 + 1) + B(2(-4) - 1)(-4 + 1) + C(2(-4) - 1)(-4 + 4)$$

$$6(16) + 96 - 3 = A(0)(-3) + B(-8 - 1)(-3) + C(-8 - 1)(0)$$

$$96 + 96 - 3 = 0 + 27B + 0$$

$$189 = 27B$$

$$B = \frac{189}{27}$$

$$B = 7$$

Substitute -1 for x in the identity above. This will give C as follows:

$$6x^2 - 24x - 3 \equiv A(x + 4)(x + 1) + B(2x - 1)(x + 1) + C(2x - 1)(x + 4)$$

$$6(-1)^2 - 24(-1) - 3 = A(-1 + 4)(-1 + 1) + B(2(-1) - 1)(-1 + 1) + C(2(-1) - 1)(-1 + 4)$$

$$6 + 24 - 3 = A(3)(0) + B(-2 - 1)(0) + C(-2 - 1)(3)$$

$$27 = 0 + 0 + C(-3)(3)$$

$$27 = -9C$$

$$C = \frac{27}{-9}$$

$$C = -3$$

Since we already know B and C we can substitute any value for x in the identity and obtain A. Let us substitute 1 for x in order to obtain A. This gives:

$$6x^2 - 24x - 3 \equiv A(x + 4)(x + 1) + B(2x - 1)(x + 1) + C(2x - 1)(x + 4)$$

$$6(1)^2 - 24(1) - 3 = A(1 + 4)(1 + 1) + B(2(1) - 1)(1 + 1) + C(2(1) - 1)(1 + 4)$$

$$6 - 24 - 3 = A(5)(2) + B(2 - 1)(2) + C(2 - 1)(5)$$

$$-21 = 10A + 2B + 5C$$

$$-21 = 10A + 2(7) + 5(-3) \qquad \text{(Note that B = 7 and C = -3)}$$

$$-21 = 10A + 14 - 15$$

$$-21 = 10A - 1$$

$$-21 + 1 = 10A$$

$$-20 = 10A$$

$$A = \frac{-20}{10}$$

$$A = -2$$

Hence A = -2, B = 7 and C = -3

Therefore, $\dfrac{6x^2 - 24x - 3}{(2x - 1)(x^2 + 5x + 4)} = -\dfrac{2}{2x - 1} + \dfrac{7}{x + 4} - \dfrac{3}{x + 1}$

Or, $\dfrac{6x^2 - 24x - 3}{(2x - 1)(x^2 + 5x + 4)} = \dfrac{7}{x + 4} - \dfrac{2}{2x - 1} - \dfrac{3}{x + 1}$

11. Express $\dfrac{6x^2 + 19x - 11}{(x + 1)(x^2 + 5x - 2)}$ in partial fractions

Solution

The quadratic factor in the denominator cannot be factorized into linear factors. Therefore we resolve into partial fractions as follows:

$$\frac{6x^2 + 19x - 11}{(x + 1)(x^2 + 5x - 2)} = \frac{A}{x + 1} + \frac{Bx + C}{(x^2 + 5x - 2)}$$

$$\frac{6x^2 + 19x - 11}{(x + 1)(x^2 + 5x - 2)} = \frac{A(x^2 + 5x - 2) + (Bx + C)(x + 1)}{(x+1)(x^2 + 5x - 2)}$$

Equating the numerators gives:

$$6x^2 + 19x - 11 \equiv A(x^2 + 5x - 2) + (Bx + C)(x + 1)$$

Substitute −1 for x in order to obtain A as follows:

$$6(-1)^2 + 19(-1) - 11 = A((-1)^2 + 5(-1) - 2) + (B(-1) + C)(-1 + 1)$$

$$6 - 19 - 11 = A(1 - 5 - 2) + (-B + C)(0)$$

$$-24 = -6A + 0$$

$$6A = 24$$

$$A = \frac{24}{6}$$

$$A = 4$$

Since we have a quadratic factor in this question, we will have to apply the method of comparing coefficients. Hence, let us expand the brackets in the identity above. This gives:

$$6x^2 + 19x - 11 \equiv A(x^2 + 5x - 2) + (Bx + C)(x + 1)$$

$$6x^2 + 19x - 11 \equiv Ax^2 + 5Ax - 2A + Bx^2 + Bx + Cx + C$$

$$\equiv Ax^2 + Bx^2 + 5Ax + Bx + Cx - 2A + C$$

$$6x^2 + 19x - 11 \equiv (A + B)x^2 + (5A + B + C)x - 2A + C$$

Comparing the coefficients of like terms on both sides of the equation shows that:

A + B = 6 .... Equation (1)

5A + B + C = 19 ....Equation (2)

We have already calculated the value of A to be 4 as done above. Hence from equation (1) we have:

A + B = 6 ....Equation (1)

4 + B = 6   (Since A = 4)

B = 6 − 4

B = 2

From equation (2) we have:

5A + B + C = 19 ....Equation (2)

5(4) + 2 + C = 19

20 + 2 + C = 19

C = 19 − 22

C = –3

Hence, A = 4, B = 2 and C = –3

$$\frac{6x^2 + 19x - 11}{(x+1)(x^2 + 5x - 2)} = \frac{A}{x+1} + \frac{Bx + C}{(x^2 + 5x - 2)}$$

Therefore, $$\frac{6x^2 + 19x - 11}{(x+1)(x^2 + 5x - 2)} = \frac{4}{x+1} + \frac{2x - 3}{(x^2 + 5x - 2)}$$

12. Resolve $\dfrac{2x^2 - 5x + 1}{(x-2)^3}$ into partial fractions

Solution

This is a case of repeated linear factor. Hence we resolve into partial fractions as follows:

$$\frac{2x^2 - 5x + 1}{(x-2)^3} = \frac{A}{(x-2)} + \frac{B}{(x-2)^2} + \frac{C}{(x-2)^3}$$

$$\frac{2x^2 - 5x + 1}{(x-2)^3} = \frac{A(x-2)^2 + B(x-2) + C}{(x-2)^3}$$

Equating the numerators gives:

$$2x^2 - 5x + 1 \equiv A(x-2)^2 + B(x-2) + C$$

Substitute 2 for x in order to eliminate A and B and obtain the value of C. This gives:

$$2(2)^2 - 5(2) + 1 = A(2-2)^2 + B(2-2) + C$$

$$2(4) - 10 + 1 = A(0) + B(0) + C$$

$$8 - 10 + 1 = 0 + 0 + C$$

$$-1 = C$$

$$C = -1$$

In order to find A and B, we need to compare coefficients. Let us expand the brackets in the identity above as follows:

$$2x^2 - 5x + 1 \equiv A(x-2)^2 + B(x-2) + C$$
$$\equiv A(x-2)(x-2) + B(x-2) + C$$
$$\equiv A(x^2 - 4x + 4) + Bx - 2B + C$$
$$\equiv Ax^2 - 4Ax + 4A + Bx - 2B + C$$
$$2x^2 - 5x + 1 \equiv Ax^2 + (-4A + B)x + 4A - 2B + C$$

Comparing the coefficients of like terms on both sides of the equation shows that:

A = 2   (Since $2x^2 \equiv Ax^2$)

Similarly, –4A + B = –5   (From $-5x \equiv (-4A + B)x$)

–4(2) + B = –5   (Note that A = 2 as obtained above)

–8 + B = –5

B = –5 + 8

B = 3

Hence, A = 2, B = 3 and C = –1

Therefore, $\dfrac{2x^2 - 5x + 1}{(x-2)^3} = \dfrac{2}{(x-2)} + \dfrac{3}{(x-2)^2} - \dfrac{1}{(x-2)^3}$

13. Resolve $\dfrac{5x^3 + 3x^2 + 2x - 1}{x^2(x^2 + 1)}$ into partial fractions

Solution

Let $\dfrac{5x^3 + 3x^2 + 2x - 1}{x^2(x^2 + 1)} = \dfrac{A}{x} + \dfrac{B}{x^2} + \dfrac{Cx + D}{x^2 + 1}$

$\dfrac{5x^3 + 3x^2 + 2x - 1}{x^2(x^2 + 1)} = \dfrac{A[x(x^2 + 1)] + B(x^2 + 1) + (Cx + D)x^2}{x^2(x^2 + 1)}$

Equating the numerators gives:

$$5x^3 + 3x^2 + 2x - 1 \equiv A[x(x^2 + 1)] + B(x^2 + 1) + (Cx + D)(x^2)$$
$$\equiv A(x^3 + x) + B(x^2 + 1) + Cx^3 + Dx^2$$
$$\equiv Ax^3 + Ax + Bx^2 + B + Cx^3 + Dx^2$$
$$\equiv Ax^3 + Cx^3 + Bx^2 + Dx^2 + Ax + B$$
$$5x^3 + 3x^2 + 2x - 1 \equiv (A + C)x^3 + (B + D)x^2 + Ax + B$$

Comparing the coefficients of terms on both sides of the identity, gives:

$A = 2$ (From coefficients of x)

$B = -1$ (The constant term on both sides of the identity)

$A + C = 5$ (From coefficients of x^3)

$2 + C = 5$ (Since $A = 2$)

$C = 5 - 2$

$C = 3$

Similarly, $B + D = 3$ (From coefficients of x^2)

$-1 + D = 3$

$D = 3 + 1$

$D = 4$

Hence, $A = 2$, $B = -1$, $C = 3$ and $D = 4$

Therefore, $\dfrac{5x^3 + 3x^2 + 2x - 1}{x^2(x^2 + 1)} = \dfrac{2}{x} - \dfrac{1}{x^2} + \dfrac{3x + 4}{x^2 + 1}$

14 Resolve $\dfrac{2x^2 + 19x + 47}{x^2 + 9x + 20}$ into partial fractions.

Solutions

From the question the highest power of x in the numerator and denominator are equal (i.e. x^2 in both cases). Hence we have to divide the numerator by the denominator in order to reduce the power of x in the numerator. Let us divide it as follows:

2

304

$$x^2 + 9x + 20\overline{)2x^2 + 19x + 47}$$
$$\underline{2x^2 + 18x + 40}$$
$$x + 7$$

We cannot divide further, hence:

$$\frac{2x^2 + 19x + 47}{x^2 + 9x + 20} = 2 + \frac{x + 7}{x^2 + 9x + 20}$$

Note that 2 is the quotient and $x + 7$ is the remainder in the division carried out above.

Hence we now resolve $\dfrac{x + 7}{x^2 + 9x + 20}$ into partial fractions.

The denominator factorizes into linear factors as follows:

$$x^2 + 9x + 20 = (x + 4)(x + 5)$$

Hence the fraction above can be written as:

$$\frac{x + 7}{(x+4)(x+5)}$$

Therefore, $\dfrac{x + 7}{(x+4)(x+5)} = \dfrac{A}{(x+4)} + \dfrac{B}{(x+5)}$

$$\frac{x + 7}{(x+4)(x+5)} = \frac{A(x+5) + B(x+4)}{(x+4)(x+5)}$$

Equating the numerators gives:

$$x + 7 \equiv A(x + 5) + B(x + 4)$$

Substitute $x = -5$ in order to get B. This gives:

$$-5 + 7 = A(-5 + 5) + B(-5 + 4)$$
$$2 = A(0) + B(-1)$$
$$2 = 0 - B$$
$$B = -2$$

Substitute $x = -4$ in order to find A. This gives:

$$x + 7 \equiv A(x + 5) + B(x + 4)$$
$$-4 + 7 = A(-4 + 5) + B(-4 + 4)$$
$$3 = A(1) + B(0)$$
$$3 = A + 0$$
$$A = 3$$

Hence A = 3 and B = −2

Therefore, $\dfrac{2x^2 + 19x + 47}{x^2 + 9x + 20} = 2 + \dfrac{x + 7}{x^2 + 9x + 20}$ (Note: 2 was obtained from the division above)

$$= 2 + \frac{3}{(x+4)} - \frac{2}{(x+5)}$$

15. Resolve $\dfrac{x^3 + 2x^2 - x - 11}{x^2 - x - 2}$ into partial fractions

<u>Solution</u>

From the question, the highest power of x in the numerator is higher than the highest power of x in the denominator (i.e. x^3 is greater than x^2). In such a case, we have to divide first before we proceed. Let us carry out the division as follows:

$$
\begin{array}{r}
x + 3 \\
x^2 - x - 2\overline{)\, x^3 + 2x^2 - x - 11} \\
\underline{x^3 - x^2 - 2x} \\
3x^2 + x - 11 \\
\underline{3x^2 - 3x - 6} \\
4x - 5
\end{array}
$$

Hence, using the remainder which is $4x - 5$, the fraction in our question can be written as:

$$\frac{x^3 + 2x^2 - x - 11}{x^2 - x - 2} = x + 3 + \frac{4x - 5}{x^2 - x - 2}$$

We can now resolve $\dfrac{4x - 5}{x^2 - x - 2}$ into partial fractions. The denominator can be factorized to give:

$x^2 - x - 2 = (x - 2)(x + 1)$. Hence, the fraction to resolve can be written as $\dfrac{4x - 5}{(x - 2)(x + 1)}$

Therefore we resolve as follows:

$$\frac{4x - 5}{(x - 2)(x + 1)} = \frac{A}{x - 2} + \frac{B}{x + 1}$$

$$\frac{4x - 5}{(x - 2)(x + 1)} = \frac{A(x + 1) + B(x - 2)}{(x - 2)(x + 1)}$$

Equating the numerators gives:

$4x - 5 \equiv A(x + 1) + B(x - 2)$

Substitute $x = -1$ in order to get B. This gives:

$4(-1) - 5 = A(-1 + 1) + B(-1 - 2)$

$-4 - 5 = A(0) + B(-3)$

$-9 = 0 - 3B$

$3B = 9$

$B = \dfrac{9}{3}$

$B = 3$

Substitute $x = 2$ in order to find A. This gives:

$4x - 5 \equiv A(x + 1) + B(x - 2)$

$4(2) - 5 = A(2 + 1) + B(2 - 2)$

$8 - 5 = A(3) + B(0)$

$3 = 3A + 0$

$A = \dfrac{3}{3}$

$A = 1$

Hence A = 1 and B = 3

Therefore, $\dfrac{x^3 + 2x^2 - x - 11}{x^2 - x - 2} = x + 3 + \dfrac{1}{x - 2} + \dfrac{3}{x + 1}$

16. If $\dfrac{x^2 + x - 1}{(x + 1)(x - 1)} = A + \dfrac{B}{x + 1} + \dfrac{C}{x - 1}$ where A, B and C are constants, find A + 2B − C

Solution

If we expand the denominators, it gives:

$$(x + 1)(x - 1) = x^2 - x + x - 1$$
$$= x^2 - 1$$

This shows that the numerator and the denominator are of the same degree (i.e. both have the same highest power of x). Hence we have to divide first. This is as shown below.

$$
\begin{array}{r}
1 \\
x^2 - 1 \overline{)\, x^2 + x - 1\,} \\
\underline{x^2 + 0x - 1} \\
x
\end{array}
$$ (Note that $0x$ is added since there is no term in x)

Therefore with x as the remainder, we now write the fraction as follows:

$\dfrac{x^2 + x - 1}{(x + 1)(x - 1)} = 1 + \dfrac{x}{(x + 1)(x - 1)}$ (This shows that A = 1)

Let us now resolve $\dfrac{x}{(x + 1)(x - 1)}$ into partial fractions as follows.

$$\dfrac{x}{(x + 1)(x - 1)} = \dfrac{B}{x + 1} + \dfrac{C}{x - 1}$$

$$\dfrac{x}{(x + 1)(x - 1)} = \dfrac{B(x - 1) + C(x + 1)}{(x + 1)(x - 1)}$$

Equating the numerators gives:

$$x \equiv B(x - 1) + C(x + 1)$$

If we put $x = 1$, we obtain C as follows:

$$1 = B(1 - 1) + C(1 + 1)$$
$$1 = B(0) + C(2)$$
$$1 = 0 + 2C$$
$$1 = 2C$$
$$C = \dfrac{1}{2}$$

If we put $x = -1$, we obtain B as follows:

$$-1 = B(-1 - 1) + C(-1 + 1)$$
$$-1 = B(-2) + C(0)$$
$$-1 = -2B + 0$$
$$2B = 1$$
$$B = \dfrac{1}{2}$$

Hence, $B = \dfrac{1}{2}$ and $C = \dfrac{1}{2}$. And recall that A = 1

Therefore, $A + 2B - C = 1 + 2\left(\dfrac{1}{2}\right) - \dfrac{1}{2}$

$$= 1 + 1 - \dfrac{1}{2}$$

$$= 2 - \dfrac{1}{2}$$

$$= 1\dfrac{1}{2}$$

17. If $\dfrac{3x^2 - 7}{x^3 + 2x^2 - 8x} \equiv \dfrac{7}{8x} + \dfrac{P}{x + 4} + \dfrac{Q}{x - 2}$, find P + Q

<u>Solution</u>

Looking at the denominators on both sides of the identity shows that the factors of $x^3 + 2x^2 - 8x$ are x, $(x + 4)$ and $(x - 2)$. This means that:

$$x^3 + 2x^2 - 8x = x\,(x + 4)(x - 2)$$

This shows that $\dfrac{7}{8x}$ should be written as $\dfrac{\frac{7}{8}}{x}$. If we use $\dfrac{7}{8x}$, then the denominators on both sides of the identity will not be identical. For the two denominators to be identical, we must write the question above as follows:

$$\dfrac{3x^2 - 7}{x^3 + 2x^2 - 8x} \equiv \dfrac{\frac{7}{8}}{x} + \dfrac{P}{x + 4} + \dfrac{Q}{x - 2}$$

We now continue our working as follows:

$$\dfrac{3x^2 - 7}{x(x + 4)(x - 2)} \equiv \dfrac{\frac{7}{8}(x + 4)(x - 2) + P(x)(x - 2) + Q(x)(x + 4)}{x(x + 4)(x - 2)}$$

Equating the numerators gives:

$$3x^2 - 7 \equiv \tfrac{7}{8}(x + 4)(x - 2) + P(x)(x - 2) + Q(x)(x + 4)$$

Substitute −4 for x to obtain P as follows

$$3(-4)^2 - 7 \equiv \tfrac{7}{8}(-4 + 4)(-4 - 2) + P(-4)(-4 - 2) + Q(-4)(-4 + 4)$$

$$3(16) - 7 = \tfrac{7}{8}(0)(-6) + P(-4)(-6) + Q(-4)(0)$$

$$48 - 7 = 0 + 24P + 0$$

$$41 = 24P$$

$$P = \dfrac{41}{24}$$

Let us put $x = 2$ in order to get Q.

$$3x^2 - 7 \equiv \tfrac{7}{8}(x + 4)(x - 2) + P(x)(x - 2) + Q(x)(x + 4)$$

$$3(2)^2 - 7 \equiv \tfrac{7}{8}(2 + 4)(2 - 2) + P(2)(2 - 2) + Q(2)(2 + 4)$$

$$3(4) - 7 = \frac{7}{8}(6)(0) + P(2)(0) + Q(2)(6)$$

$$12 - 7 = 0 + 0 + 12Q$$

$$5 = 12Q$$

$$Q = \frac{5}{12}$$

Therefore, $P = \frac{41}{24}$, and $Q = \frac{5}{12}$

Hence, $P + Q = \frac{41}{24} + \frac{5}{12}$

$$= \frac{41 + 10}{24}$$

$$= \frac{51}{24}$$

$$\therefore \quad P + Q = \frac{17}{8} \qquad \text{(In its lowest term)}$$

18. Resolve $\dfrac{1 - 4x + 7x^2 - x^3 - x^4}{(x+3)(x^2+2)}$ into partial fractions.

Solutions

If we expand the denominator, it will give us a degree (highest power of x) of 3 (i.e. x^3). However the numerator contains x^4. Hence the fraction is an improper fraction. Therefore we have to divide the fraction before we proceed.

Let us expand the denominator as follows:

$$(x+3)(x^2+2) = x^3 + 2x + 3x^2 + 6$$

$$= x^3 + 3x^2 + 2x + 6$$

We also have to rearrange the numerator so that the powers of x will be in descending order. This is necessary for easier division of the polynomial. Hence the fraction can be written as follows:

$$\frac{-x^4 - x^3 + 7x^2 - 4x + 1}{x^3 + 3x^2 + 2x + 6}$$

Let us now carry out the division as follows:

$$
\begin{array}{r}
-x + 2 \\
x^3 + 3x^2 + 2x + 6 \overline{)\; -x^4 - x^3 + 7x^2 - 4x + 1} \\
\underline{-x^4 - 3x^3 - 2x^2 - 6x} \\
2x^3 + 9x^2 + 2x + 1 \\
\underline{2x^3 + 6x^2 + 4x + 12} \\
3x^2 - 2x - 11
\end{array}
$$

Note that we can no longer continue the division when there is no more term to bring down and the highest power of x in the remainder is smaller than that of the divisor. Hence, the fraction has been broken down to give:

$$\frac{1 - 4x + 7x^2 - x^3 - x^4}{(x+3)(x^2+2)} = 2 - x + \frac{3x^2 - 2x - 11}{(x+3)(x^2+2)}$$ (Note that $-x + 2$ has been written as $2 - x$)

Let us now resolve $\dfrac{3x^2 - 2x - 11}{(x+3)(x^2+2)}$ into partial fractions as follows:

$$\frac{3x^2 - 2x - 11}{(x+3)(x^2+2)} = \frac{A}{x+3} + \frac{Bx + C}{(x^2+2)}$$

$$\frac{3x^2 - 2x - 11}{(x+3)(x^2+2)} = \frac{A(x^2+2) + (Bx + C)(x+3)}{(x+3)(x^2+2)}$$

Equating the numerators gives:

$$3x^2 - 2x - 11 \equiv A(x^2 + 2) + (Bx + C)(x + 3)$$

Substitute -3 for x in order to find A. This gives:

$$3(-3)^2 - 2(-3) - 11 = A((-3)^2 + 2) + (B(-3) + C)(-3 + 3)$$

$$3(9) + 6 - 11 = A(9 + 2) + (-3B + C)(0)$$

$$27 + 6 - 11 = 11A + 0$$

$$22 = 11A$$

$$A = \frac{22}{11}$$

$$A = 2$$

Let us expand the brackets in the identity above.

$$3x^2 - 2x - 11 \equiv A(x^2 + 2) + (Bx + C)(x + 3)$$

$$\equiv Ax^2 + 2A + Bx^2 + 3Bx + Cx + 3C$$

$$\equiv Ax^2 + Bx^2 + 3Bx + Cx + 2A + 3C$$

$$3x^2 - 2x - 11 \equiv (A + B)x^2 + (3B + C)x + 2A + 3C$$

Comparing coefficients of like terms on both sides of the identity shows that:

A + B = 3 ....Equation (1)

3B + C = −2 ....Equation (2)

Since we have obtained the value of A as 2, let us substitute this in equation (1) as follows:

A + B = 3 ....Equation (1)

2 + B = 3   (Note that A = 2 as obtained above)

B = 3 − 2

B = 1

From equation (2) we have that:

3B + C = −2 ....Equation (2)

3(1) + C = −2

3 + C = −2

C = −2 − 3

C = −5

Hence, A = 2, B = 1 and C = −5

Therefore, $$\frac{1 - 4x + 7x^2 - x^3 - x^4}{(x+3)(x^2+2)} = 2 - x + \frac{2}{x+3} + \frac{x-5}{(x^2+2)}$$

310

Exercise 24

1. Express $\dfrac{5x-1}{(x+3)(x-5)}$ in partial fractions

2. Resolve $\dfrac{2x+11}{2x^2-3x-9}$ into partial fractions.

3. Express $\dfrac{8}{x^2-16}$ in partial fractions.

4. Resolve $\dfrac{3x-10}{2x(x-5)}$ into partial fractions.

5. Resolve $\dfrac{x+7}{(x-2)^2}$ into partial fractions.

6. Express $\dfrac{6+4x-5x^2}{x^2(2x+3)}$ in partial fractions.

7. Resolve $\dfrac{2x^2-5x+11}{(x+1)(x-2)^2}$ into partial fractions.

8. Resolve $\dfrac{5x^2-x+2}{x(x^2+3)}$ into partial fractions.

9. Resolve $\dfrac{3x^2+2x+8}{(x^2+2)(x+1)}$ into partial fractions

10. Resolve $\dfrac{23x-31}{(x-2)(x^2+2x-3)}$ into partial fractions

11. Express $\dfrac{5x^2+2x-9}{(x+3)(x^2-3x-3)}$ in partial fractions

12. Resolve $\dfrac{x^2+x-7}{(x-1)^3}$ into partial fractions

13. Resolve $\dfrac{2x^3+3x^2-x-5}{x^2(x^2-1)}$ into partial fractions

14. Resolve $\dfrac{x^2-3x-1}{x^2-2x-3}$ into partial fractions.

15. Resolve $\dfrac{x^3+8x^2+13x-29}{x^2+3x-4}$ into partial fractions

16. If $\dfrac{5x^2+13x-14}{(x-1)(x+3)} = A + \dfrac{B}{x-1} + \dfrac{C}{x+3}$ where A, B and C are constants, find $2A + 5B - 7C$

17. If $\dfrac{2x^2+5}{2x^3+14x^2+20x} \equiv \dfrac{1}{2x} + \dfrac{P}{x+5} + \dfrac{Q}{x+2}$, find the values of P and Q

18. Resolve $\dfrac{4-2x+5x^2+2x^3-x^4}{(x-1)(x^2+1)}$ into partial fractions.

CHAPTER 25
RADICAL EQUATIONS

Equations which consist of square root signs are called radical equations. In solving such equations, we keep squaring both sides of the equation until the square root sign is finally removed. Then we solve the resulting equation after the removal of the root sign. However, it is important that we recall the following rules when simplifying surds:

1. $\sqrt{a} \times \sqrt{b} = \sqrt{ab}$

2. $\dfrac{\sqrt{a}}{\sqrt{b}} = \sqrt{\dfrac{a}{b}}$

3. $\sqrt{a} \times \sqrt{a} = a$

Examples

1. Solve the equation $\sqrt{x-2} = 3$

Solution

$\sqrt{x-2} = 3$

Square both sides in order to remove the root sign. This gives:

$(\sqrt{x-2})^2 = 3^2$

$x - 2 = 9$

Note that the expression $(\sqrt{x-2})^2$ gives $x - 2$, i.e. simply take the term in the root sign.

$x = 9 + 2$

$x = 11$

2. Solve $\sqrt{x+14} = \sqrt{6-x}$

Solution

$\sqrt{x+14} = \sqrt{6-x}$

Squaring both sides gives:

$(\sqrt{x+14})^2 = (\sqrt{6-x})^2$

$x + 14 = 6 - x$

$x + x = 6 - 14$

$2x = -8$

$x = \dfrac{-8}{2}$

$x = -4$

3. Solve the equation $\sqrt{x} + \sqrt{2x-2} = 7$

<u>Solution</u>

$$\sqrt{x} + \sqrt{2x-2} = 7$$

Square both sides of the equation in order to remove the root sign. This gives:

$$(\sqrt{x} + \sqrt{2x-2})^2 = 7^2$$

$$(\sqrt{x} + \sqrt{2x-2})(\sqrt{x} + \sqrt{2x-2}) = 49$$

In order to expand the left hand side, simply use each term in the first bracket to multiply the second bracket. This is as shown below.

$$(\sqrt{x})(\sqrt{x} + \sqrt{2x-2}) + \sqrt{2x-2}(\sqrt{x} + \sqrt{2x-2}) = 49$$

$$x + \sqrt{x}\sqrt{2x-2} + \sqrt{x}\sqrt{2x-2} + 2x - 2 = 49$$

$$x + 2x - 2 + \sqrt{x(2x-2)} + \sqrt{x(2x-2)} = 49$$

$$3x - 2 + 2\sqrt{x(2x-2)} = 49 \qquad [\text{Note that } \sqrt{x(2x-2)} + \sqrt{x(2x-2)} = 2\sqrt{x(2x-2)}]$$

$$3x + 2\sqrt{2x^2 - 2x} = 49 + 2$$

$$3x + 2\sqrt{2x^2 - 2x} = 51$$

$$2\sqrt{2x^2 - 2x} = 51 - 3x$$

Now that we have a single root sign, we have to square both sides of the equation to remove the root sign again. This gives:

$$(2\sqrt{2x^2 - 2x})^2 = (51 - 3x)^2$$

$$(2)^2(\sqrt{2x^2 - 2x})^2 = (51 - 3x)(51 - 3x)$$

$$4(2x^2 - 2x) = 2601 - 153x - 153x + 9x^2$$

$$8x^2 - 8x = 2601 - 306x + 9x^2$$

$$0 = 9x^2 - 8x^2 + 8x - 306x + 2601$$

$$0 = x^2 - 298x + 2601$$

Or, $x^2 - 298x + 2601$

Solving this equation by factorization gives:

$$(x - 9)(x - 289) = 0$$

$$\therefore \quad x = 9 \text{ or } x = 289$$

After solving a radical equation it is necessary to check if the values obtained are actually the true solutions of the equation. Hence, let us substitute each of the values of x obtained above into the equation. The equation is:

$$\sqrt{x} + \sqrt{2x-2} = 7$$

When $x = 9$, the left hand side of the equation gives:

$$\sqrt{9} + \sqrt{2(9) - 2} = 3 + \sqrt{18 - 2}$$

$$= 3 + \sqrt{16}$$

$$= 3 + 4$$

$$= 7$$

Since 7 is also the value on the right hand side of the equation, then $x = 9$ is a solution of the equation.

Recall that the equation is $\sqrt{x} + \sqrt{2x - 2} = 7$

When $x = 289$, the left hand side of the equation gives:

$$\sqrt{289} + \sqrt{2(289) - 2} = 17 + \sqrt{578 - 2}$$
$$= 17 + \sqrt{576}$$
$$= 17 + 24$$
$$= 41$$

Since 41 is not what we have on the right hand side of the equation, then $x = 289$ is not a solution of the equation.

Therefore, the solution of the equation is $x = 9$

Note that 289 which is not a solution of the equation is called an extraneous root of the equation.

4. Solve the equation $\sqrt{3x + 13} - x = 1$

<u>Solution</u>

$$\sqrt{3x + 13} - x = 1$$

Collect terms containing the root sign (radical) on one side of the equation, and move other terms to the other side of the equation. This means we have to isolate the radical term. This gives:

$$\sqrt{3x + 13} = 1 + x$$

We can now square both sides of the equation as follows:

$$(\sqrt{3x + 13})^2 = (1 + x)^2$$
$$3x + 13 = (1 + x)(1 + x)$$
$$3x + 13 = 1 + x + x + x^2$$
$$3x + 13 = x^2 + 2x + 1$$
$$0 = x^2 + 2x - 3x + 1 - 13$$
$$0 = x^2 - x - 12$$
$$x^2 - x - 12 = 0$$

Solving this equation by factorization gives:

$$(x - 4)(x + 3) = 0$$

Hence, $x = 4$ or $x = -3$

Let us check the values of x obtained in order to know if they are solutions of the equation. The equation is:

$$\sqrt{3x + 13} - x = 1$$

When $x = 4$, the left hand side of the equation gives:

$$\sqrt{3x + 13} - x = \sqrt{3(4) + 13} - 4$$

$$= \sqrt{12 + 13} - 4$$
$$= \sqrt{25} - 4$$
$$= 5 - 4 = 1$$

Since 1 is also the value on the right hand side of the equation, the $x = 4$ is a solution of the equation.

When $x = -3$, the left hand side of the equation gives:

$$\sqrt{3x + 13} - x = \sqrt{3(-3) + 13} - (-3)$$
$$= \sqrt{-9 + 13} + 3$$
$$= \sqrt{4} + 3$$
$$= 2 + 3$$
$$= 5$$

Since 5 is not what we have on the right hand side of the equation, then $x = -3$ is not a solution of the equation.

Therefore, the solution of the equation is $x = 4$

5. Solve for x if $\sqrt{x} + 2\sqrt{x + 8} = 7$

Solution

$$\sqrt{x} + 2\sqrt{x + 8} = 7$$

Squaring both sides of the equation gives:

$$(\sqrt{x} + 2\sqrt{x + 8})^2 = 7^2$$
$$(\sqrt{x} + 2\sqrt{x + 8})(\sqrt{x} + 2\sqrt{x + 8}) = 49$$
$$x + 2\sqrt{x(x + 8)} + 2\sqrt{x(x + 8)} + 4(x + 8) = 49$$
$$x + 4\sqrt{x(x + 8)} + 4x + 32 = 49$$
$$4\sqrt{x^2 + 8x} = 49 - 32 - 4x - x$$
$$4\sqrt{x^2 + 8x} = 17 - 5x$$

Squaring both sides again gives:

$$(4\sqrt{x^2 + 8x})^2 = (17 - 5x)^2$$
$$4^2(\sqrt{x^2 + 8x})^2 = (17 - 5x)(17 - 5x)$$
$$16(x^2 + 8x) = 289 - 85x - 85x + 25x^2$$
$$16x^2 + 128x = 289 - 170x + 25x^2$$
$$0 = 25x^2 - 16x^2 - 170x - 128x + 289$$
$$0 = 9x^2 - 298x + 289$$
$$9x^2 - 298x + 289 = 0$$

Let us solve this equation by factorization (Note that quadratic formula can also be used) as follows:

$$9x^2 - 9x - 289x + 289 = 0$$

$9x(x-1) - 289(x-1) = 0$

$(x-1)(9x-289) = 0$

Hence, $x = 1$ or $x = \dfrac{289}{9}$

$\dfrac{289}{9}$ cannot be the solution to the equation since it is a fraction and it is too large (the right hand side is just 7).

Therefore the solution to the equation is $x = 1$

6. Solve for x if $\sqrt{x+7} + 3x = 9$

Solution

$\sqrt{x+7} + 3x = 9$

Collect term containing the radical on the left hand side of the equation. This gives:

$\sqrt{x+7} = 9 - 3x$

Square both sides of the equation

$(\sqrt{x+7})^2 = (9-3x)^2$

$x + 7 = (9-3x)(9-3x)$

$x + 7 = 81 - 27x - 27x + 9x^2$

$0 = 81 - 54x + 9x^2 - x - 7$

$0 = 9x^2 - 55x + 74$

$9x^2 - 55x + 74 = 0$

Let us solve this equation by using quadratic formula. From the equation:

$a = 9$, $b = -55$ and $c = 74$

$x = \dfrac{-b \pm \sqrt{b^2 - 4ac}}{2a}$

$= \dfrac{-(-55) \pm \sqrt{(-55)^2 - (4 \times 9 \times 74)}}{2 \times 9}$

$= \dfrac{55 \pm \sqrt{3025 - 2664}}{18}$

$= \dfrac{55 \pm \sqrt{361}}{18}$

$= \dfrac{55 \pm 19}{18}$

$x = \dfrac{55 + 19}{18}$ or $x = \dfrac{55 - 19}{18}$

$x = \dfrac{74}{18}$ or $x = \dfrac{36}{18}$

$x = \dfrac{37}{9}$ or $x = 2$

$\dfrac{37}{9}$ cannot be the solution of the equation if substituted into the equation.

Therefore the solution to the equation is $x = 2$

7. Find the value of x in the equation $\sqrt{x+1} + \sqrt{x+6} = 5$

Solution

$$\sqrt{x+1} + \sqrt{x+6} = 5$$

Squaring both sides of the equation gives:

$$(\sqrt{x+1} + \sqrt{x+6})^2 = 5^2$$

$$(\sqrt{x+1} + \sqrt{x+6})(\sqrt{x+1} + \sqrt{x+6}) = 25$$

$$(\sqrt{x+1} + \sqrt{x+6})(\sqrt{x+1} + \sqrt{x+6}) = 25$$

$$x+1+\sqrt{(x+1)(x+6)} + \sqrt{(x+1)(x+6)} + x+6 = 25$$

$$x+1+2\sqrt{(x+1)(x+6)} + x + 6 = 25$$

$$2x+7+2\sqrt{(x+1)(x+6)} = 25$$

Isolating the radical term on the left hand side gives:

$$2\sqrt{(x+1)(x+6)} = 25 - 2x - 7$$

$$2\sqrt{x^2+7x+6} = 18 - 2x$$

Divide both sides by 2. This gives:

$$\sqrt{x^2+7x+6} = 9 - x$$

Squaring both sides of the equation gives:

$$(\sqrt{x^2+7x+6})^2 = (9-x)^2$$

$$x^2+7x+6 = (9-x)(9-x)$$

$$x^2+7x+6 = 81 - 18x + x^2$$

$$x^2 - x^2 + 7x + 18x = 81 - 6$$

$$25x = 75$$

$$x = \frac{75}{25}$$

$$x = 3$$

8. Find the value of x if $\sqrt{2x^2+7} = 3$

Solution

$$\sqrt{2x^2+7} = 3$$

Squaring both sides of the equation gives:

$$(\sqrt{2x^2+7})^2 = 3^2$$

$$2x^2 + 7 = 9$$

$$2x^2 = 9 - 7$$

$$2x^2 = 2$$

$$x^2 = 1$$

$$x = \sqrt{1}$$

$$x = \pm 1$$

$$x = 1 \quad \text{or} \quad x = -1$$

Let us check if the two values of x are solutions of the equation. The equation is:

$$\sqrt{2x^2 + 7} = 3$$

When $x = 1$, the left hand side of the equation gives:

$$\sqrt{2x^2 + 7} = \sqrt{2(1)^2 + 7}$$
$$= \sqrt{2 + 7}$$
$$= \sqrt{9}$$
$$= 3$$

Hence $x = 1$ is a solution of the equation.

When $x = -1$, the left hand side of the equation gives:

$$\sqrt{2x^2 + 7} = \sqrt{2(-1)^2 + 7}$$
$$= \sqrt{2(1) + 7}$$
$$= \sqrt{2 + 7}$$
$$= \sqrt{9} = 3$$

Hence $x = -1$ is a solution of the equation.

Therefore both $x = 1$, and $x = -1$ are solutions of the equation.

9. Solve the equation $\sqrt{5x^2 - 2x} = 4$

Solution

$$\sqrt{5x^2 - 2x} = 4$$

Squaring both sides of the equation gives:

$$(\sqrt{5x^2 - 2x})^2 = 4^2$$
$$5x^2 - 2x = 16$$
$$5x^2 - 2x - 16 = 0$$

Solving this equation by factorization gives:

$$5x^2 - 10x + 8x - 16 = 0$$
$$5x(x - 2) + 8(x - 2) = 0$$
$$(x - 2)(5x + 8) = 0$$
$$x = 2 \quad \text{or} \quad x = -\frac{8}{5}$$

If we substitute $x = 2$ and $x = -\frac{8}{5}$ into the equation, we will find out that the two values of x are solutions of the equation.

Therefore, both $x = 2$ and $x = -\frac{8}{5}$ are solutions of the equation.

10. Solve $\sqrt{x^2 - 9} - x = -1$

Solution

Take $-x$ to the other side of the equation in order to isolate the radical term. This gives:

$$\sqrt{x^2 - 9} = x - 1$$

Squaring both sides of the equation gives:

$$(\sqrt{x^2 - 9})^2 = (x - 1)^2$$
$$x^2 - 9 = x^2 - 2x + 1$$
$$x^2 - x^2 + 2x = 1 + 9$$
$$2x = 10$$
$$x = \frac{10}{2}$$
$$x = 5$$

Exercise 25

1. Solve the equation $\sqrt{2x - 3} = 7$

2. Solve $\sqrt{2x - 9} = \sqrt{x - 2}$

3. Solve the equation $\sqrt{2x} - \sqrt{4x - 2} = -1$

4. Solve the equation $\sqrt{2x + 16} - x = -4$

5. Solve for x if $\sqrt{4x} + 5\sqrt{x + 3} = 12$

6. Solve for x if $\sqrt{2x - 6} + 3x = 17$

7. Find the value of x in the equation $\sqrt{x - 1} + \sqrt{x + 4} = 5$

8. Find the value of x if $\sqrt{3x^2 + 13} = 11$

9. Solve the equation $\sqrt{2x^2 - 4x} = 4$

10. Solve $\sqrt{5x^2 + 4} - 2x = 1$

CHAPTER 26
LIMIT OF A FUNCTION

The concept of limits is very important in differential calculus. A function has a limiting value when its variable approaches a certain value. For example, if the limiting value of f(x) as x approaches 4 is 16, it is written as:

$$\lim_{x \to 4} f(x) = 16$$

In evaluating limit, some problems are simply solved by just putting in the values of the variables, while some problems are solved by applying certain rules.

Some important limits

1. $\lim_{x \to 0} \dfrac{\sin x}{x} = 1$ and $\lim_{x \to 0} \dfrac{\tan x}{x} = 1$ (Or $\lim_{x \to 0} \dfrac{\sin ax}{ax} = 1$ and $\lim_{x \to 0} \dfrac{\tan ax}{ax} = 1$)

2. $\lim_{x \to 0} \dfrac{1 - \cos x}{x} = 0$ and $\lim_{x \to 0} \dfrac{1 - \cos ax}{x} = 0$

3. $\lim_{x \to \infty} \left(1 + \dfrac{1}{x}\right)^x = e$

4. $\lim_{x \to 0} \dfrac{a^x - 1}{a^x} = 1$

5. $\lim_{x \to 0} \dfrac{e^x - 1}{x} = 1$

6. $\lim_{x \to 0} \dfrac{1 + x}{x} = 1$

7. $\lim_{x \to a} \dfrac{x^n - a^n}{x - a} = na^{n-1}$

8. $\lim_{x \to 0} \dfrac{a^x - 1}{x} = \ln a$

9. $\lim_{x \to \infty} \dfrac{\ln x}{x} = 0$

10. $\lim_{x \to \infty} x^{\frac{1}{x}} = 0$

11. $\lim_{x \to 0} (1 + x)^{\frac{1}{x}} = e$

12. $\lim_{x \to 0} (1 + \sin x)^{\frac{1}{x}} = e$

13. $\lim_{x \to a} c = c$

Examples

1. Evaluate $\lim_{x \to 0} 3x^3 - 5x^2 + 2x + 6$

Solution

$$\lim_{x \to 0} 3x^3 - 5x^2 + 2x + 6$$

Simply substitute zero for x in the given expression. This gives:

$$= 3(0^3) - 5(0^2) + 2(0) + 6$$
$$= 0 - 0 + 0 + 6$$
$$= 6$$

2. Evaluate $\lim_{x \to 0} \dfrac{x^2 + 3x - 11}{4x^2 - 5x - 5}$

Solution

$$\lim_{x \to 0} \frac{x^2 + 3x - 11}{4x^2 - 5x - 5}$$

Substituting 0 for x into the expression gives:

$$= \frac{0^2 + 3(0) - 11}{4(0)^2 - 5(0) - 5}$$
$$= \frac{0 + 0 - 11}{0 - 0 - 5}$$
$$= \frac{-11}{-5}$$
$$= \frac{11}{5}$$
$$= 2\frac{1}{5}$$

3. Evaluate $\lim_{x \to 4} \dfrac{x^2 - 2x - 8}{x - 4}$

Solution

A close look at the expression shows that if 4 is substituted for x, it will give $\dfrac{0}{0}$. This has no value as it is indeterminate. Therefore, in order to solve a limit such as this, we have to factorize the denominator and then simplify the expression. This is done as follows:

$$\lim_{x \to 4} \frac{x^2 - 2x - 8}{x - 4} = \lim_{x \to 4} \frac{(x+2)(x-4)}{x - 4}$$

Cancelling out $(x - 4)$ gives:

$$\lim_{x \to 4} (x + 2)$$

We now substitute 4 for x to obtain our final answer as follows:

$$\lim_{x \to 4} (x + 2) = 4 + 2$$
$$= 6$$

Therefore, $\lim_{x \to 4} \dfrac{x^2 + x - 8}{x - 4} = 6$

321

4. Evaluate $\lim\limits_{x \to 2} \dfrac{x^2 - 4}{x - 2}$

Solution

This is similar to example 3 above since substituting 2 for x in the expression will $\dfrac{0}{0}$ which has no value. Hence we factorize the numerator and simplify as follows:

$\lim\limits_{x \to 2} \dfrac{x^2 - 4}{x - 2} = \lim\limits_{x \to 2} \dfrac{(x-2)(x+2)}{x - 2}$ [Note that $x^2 - 4 = x^2 - 2^2$ and recall that $a^2 - b^2 = (a + b)(a - b)$]

Cancelling out $(x - 2)$ gives:

$\lim\limits_{x \to 2} (x + 2)$

We now substitute 2 for x to obtain our final answer as follows:

$\lim\limits_{x \to 2} (x + 2) = 2 + 2$

$= 4$

Therefore, $\lim\limits_{x \to 2} \dfrac{x^2 - 4}{x - 2} = 4$

5. Evaluate $\lim\limits_{x \to 0} (2x - 5)(x + 2)(3x - 1)$

Solution

$\lim\limits_{x \to 0} (2x - 5)(x + 2)(3x - 1)$

Substituting 0 for x in the expression gives:

$\lim\limits_{x \to 0} (2x - 5)(x + 2)(3x - 1) = [2(0) - 5][(0) + 2][3(0) - 1]$

$= (-5)(2)(-1)$

$= 10$

6. Evaluate $\lim\limits_{x \to \infty} \dfrac{x^3 + 2x^2 + 5x + 3}{x^3 + x^2 + 3x + 7}$

Solution

In this case of x approaching infinity, we evaluate it by first dividing each term in the numerator and denominator by the variable having the highest exponent (power). Hence we divide each term by x^3. This is done as follows:

$\lim\limits_{x \to \infty} \dfrac{x^3 + 2x^2 + 5x + 3}{x^3 + x^2 + 3x + 7} = \lim\limits_{x \to \infty} \dfrac{\frac{x^3}{x^3} + \frac{2x^2}{x^3} + \frac{5x}{x^3} + \frac{3}{x^3}}{\frac{x^3}{x^3} + \frac{x^2}{x^3} + \frac{3x}{x^3} + \frac{7}{x^3}}$

$= \lim\limits_{x \to \infty} \dfrac{1 + \frac{2}{x} + \frac{5}{x^2} + \frac{3}{x^3}}{1 + \frac{1}{x} + \frac{3}{x^2} + \frac{7}{x^3}}$

$= \dfrac{1 + \frac{2}{\infty} + \frac{5}{\infty^2} + \frac{3}{\infty^3}}{1 + \frac{1}{\infty} + \frac{3}{\infty^2} + \frac{7}{\infty^3}}$

$$= \frac{1+0+0+0}{1+0+0+0}$$ (Note that a number divided by ∞ gives 0)

$$= \frac{1}{1}$$

$$= 1$$

7. Evaluate $\displaystyle\lim_{x\to\infty} \frac{2x^4 - 3x^2 + 8x - 1}{x^4 - 5x + 3}$

<u>Solution</u>

The variable in its highest exponent (power) is x^4. Hence we divide each term by x^4 as follows:

$$\lim_{x\to\infty} \frac{2x^4 - 3x^2 + 8x - 1}{x^4 - 5x + 3} = \lim_{x\to\infty} \frac{\frac{2x^4}{x^4} - \frac{3x^2}{x^4} + \frac{8x}{x^4} - \frac{1}{x^4}}{\frac{x^4}{x^4} - \frac{5x}{x^4} + \frac{3}{x^4}}$$

$$= \lim_{x\to\infty} \frac{2 - \frac{3}{x^2} + \frac{8}{x^3} - \frac{1}{x^4}}{1 - \frac{5}{x^3} + \frac{3}{x^4}}$$

$$= \frac{2 - \frac{3}{\infty} + \frac{8}{\infty} - \frac{1}{\infty}}{1 - \frac{5}{\infty} + \frac{3}{\infty}}$$

$$= \frac{2 - 0 + 0 - 0}{1 - 0 + 0}$$ (Note that a number divided by ∞ gives 0)

$$= \frac{2}{1}$$

$$= 2$$

8. Evaluate $\displaystyle\lim_{x\to 4} \frac{2x^2 - 7x - 4}{x^2 - 3x - 4}$

<u>Solution</u>

If 4 is substituted for x in the function above, it will give $\frac{0}{0}$ which is an indeterminate value.

Hence we factorize the expression and simplify as follows:

$$\lim_{x\to 4} \frac{2x^2 - 7x - 4}{x^2 - 3x - 4} = \lim_{x\to 4} \frac{(2x+1)(x-4)}{(x+1)(x-4)}$$

Cancelling out $(x - 4)$ gives:

$$\lim_{x\to 4} \frac{2x + 1}{x + 1}$$

We now substitute 4 for x to obtain our final answer as follows:

$$\lim_{x\to 4} \frac{2x + 1}{x + 1} = \frac{2(4) + 1}{(4) + 1}$$

$$= \frac{8+1}{4+1}$$

$$= \frac{9}{5}$$

$$= 1\frac{4}{5}$$

Therefore, $\lim_{x \to 4} \frac{2x^2 - 7x - 4}{x^2 - 3x - 4} = 1\frac{4}{5}$

9. Evaluate $\lim_{x \to 3} \frac{x^2 - 9}{x^2 + 6x + 9}$

Solution

Substituting 3 for x in the function can give us the final answer as follows:

$$\lim_{x \to 3} \frac{x^2 - 9}{x^2 + 6x + 9} = \frac{3^2 - 9}{3^2 + 6(3) + 9}$$

$$= \frac{9 - 9}{9 + 18 + 9}$$

$$= \frac{0}{36}$$

$$= 0$$

Hence, $\lim_{x \to 3} \frac{x^2 - 9}{x^2 + 6x + 9} = 0$

Note that even if this expression had been factorized and simplified, the final answer would still give zero.

10. Determine the limiting value of $\frac{x + 2}{x^2 + 6x + 8}$ as x tends to -2.

Solution

The question can also be written as $\lim_{x \to -2} \frac{x + 2}{x^2 + 6x + 8}$

If -2 is substituted for x in the function above, it will give $\frac{0}{0}$. Hence we factorize the expression and simplify as follows:

$$\lim_{x \to -2} \frac{x + 2}{x^2 + 6x + 8} = \lim_{x \to -2} \frac{x + 2}{(x + 2)(x + 4)}$$

Cancelling out $(x + 2)$ gives:

$$\lim_{x \to -2} \frac{1}{x + 4}$$

We now substitute -2 for x as follows:

$$\lim_{x \to -2} \frac{1}{x + 4} = \frac{1}{-2 + 4}$$

$$= \frac{1}{2}$$

324

11. Find the limit of $\dfrac{x^3 - 125}{x - 5}$ as $x \to 5$

Solution

The question can also be written as $\displaystyle\lim_{x \to 5} \dfrac{x^3 - 125}{x - 5}$

We factorize the expression and simplify as follows:

$$\lim_{x \to 5} \dfrac{x^3 - 125}{x - 5} = \lim_{x \to 5} \dfrac{x^3 - 5^3}{x - 5}$$

$$= \lim_{x \to 5} \dfrac{(x - 5)(x^2 + 5x + 5^2)}{x - 5} \quad \text{[Recall the identity } a^3 - b^3 = (a - b)(a^2 + ab + b^2)\text{]}$$

Cancelling out $(x - 5)$ gives:

$$\lim_{x \to 5} (x^2 + 5x + 5^2)$$

We now substitute 5 for x as follows:

$$\lim_{x \to 5} (x^2 + 5x + 5^2) = (5^2 + 5(5) + 25)$$

$$= 25 + 25 + 25$$

$$= 75$$

12. Find the limit of $\dfrac{x^3 + 64}{x + 4}$ as $x \to -4$

Solution

The question can also be written as $\displaystyle\lim_{x \to -4} \dfrac{x^3 + 4^3}{x + 4}$

We factorize the expression and simplify as follows:

$$\lim_{x \to -4} \dfrac{x^3 + 64}{x + 4} = \lim_{x \to -4} \dfrac{x^3 + 4^3}{x + 4}$$

$$= \lim_{x \to -4} \dfrac{(x + 4)(x^2 - 4x + 4^2)}{x + 4} \quad \text{[Recall the identity } a^3 + b^3 = (a + b)(a^2 - ab + b^2)\text{]}$$

Cancelling out $(x + 4)$ gives:

$$\lim_{x \to -4} (x^2 - 4x + 4^2)$$

We now substitute -4 for x as follows:

$$\lim_{x \to -4} (x^2 - 4x + 4^2) = (-4^2 - 4(-4) + 16)$$

$$= 16 + 16 + 16$$

$$= 48$$

13. Evaluate $\displaystyle\lim_{x \to 9} \dfrac{3 - \sqrt{x}}{9 - x}$

Solution

If 9 is substituted for x in the expression above, it will give $\dfrac{0}{0}$ which is not the desired answer.

Hence a way to go around this problem is to multiply the top and bottom by the conjugate of the surd at the top. The conjugate of $3 - \sqrt{x}$ is $3 + \sqrt{x}$ (only a difference in their middle signs).

Hence we multiply top and bottom by $3 + \sqrt{x}$ and simplify as follows:

$$\lim_{x \to 9} \frac{3 - \sqrt{x}}{9 - x} = \lim_{x \to 9} \frac{(3 - \sqrt{x})(3 + \sqrt{x})}{(9 - x)(3 + \sqrt{x})}$$

$$= \lim_{x \to 9} \frac{3^2 - (\sqrt{x})^2}{(9 - x)(3 + \sqrt{x})} \qquad \text{(Note that the top was simplified by using } (a - b)(a + b) = a^2 - b^2)$$

$$= \lim_{x \to 9} \frac{9 - x}{(9 - x)(3 + \sqrt{x})}$$

$$= \lim_{x \to 9} \frac{1}{(3 + \sqrt{x})} \qquad \text{(Since } 9 - x \text{ cancels out)}$$

We now substitute 9 for x as follows:

$$\lim_{x \to 9} \frac{1}{(3 + \sqrt{x})} = \frac{1}{(3 + \sqrt{9})}$$

$$= \frac{1}{3 + 3}$$

$$= \frac{1}{6}$$

14. Evaluate $\lim_{x \to 1} \frac{x^2 - 1}{x^5 - 1}$

<u>Solution</u>

If 2 is substituted for x in the expression above, it will give $\frac{0}{0}$ which is not a good answer. Hence a way to go around this problem is to divide the top and bottom by $x - 1$ (i.e. the difference between the variable and 1). This is done as follows:

$$\lim_{x \to 1} \frac{x^2 - 1}{x^5 - 1} = \lim_{x \to 1} \frac{\dfrac{x^2 - 1}{x - 1}}{\dfrac{x^5 - 1}{x - 1}} \qquad \text{(Note that this has not changed the original fraction)}$$

Note that $\lim_{x \to a} \frac{x^n - a^n}{x - a} = na^{n-1}$. With this rule we now simplify the expression above as follows:

$$\lim_{x \to 1} \frac{\dfrac{x^2 - 1}{x - 1}}{\dfrac{x^5 - 1}{x - 1}} = \frac{2 \times 1^{(2 - 1)}}{5 \times 1^{(5 - 1)}}$$

$$= \frac{2 \times 1^1}{5 \times 1^5}$$

$$= \frac{2}{5}$$

Continuity of a function

A function is said to be continuous if the three conditions below are satisfied.

1. f(a) exists

2. $\displaystyle\lim_{x \to a} f(x)$ exists

3. $\displaystyle\lim_{x \to a} f(x) = f(a)$ (i.e. if the values of conditions 1 and 2 above are equal)

Examples

1. Determine if the function $f(x) = 2x^2 - 8x + 5$ is continuous at the point $x = -1$

Solution

When $x = -1$, the f(−1) is obtained as follows:

$$f(x) = 2x^2 - 8x + 5$$
$$f(-1) = 2(-1)^2 - 8(-1) + 5$$
$$= 2 + 8 + 5$$
$$= 15$$

Since f(−1) = 15, it means that f(−1) exists.

The next step is to find the value of $\displaystyle\lim_{x \to -1} f(x)$ as follows:

$$\lim_{x \to -1} 2x^2 - 8x + 5 = 2(-1)^2 - 8(-1) + 5$$
$$= 2 + 8 + 5$$
$$= 15$$

Hence f(−1) = $\displaystyle\lim_{x \to -1} f(x) = 15$

Therefore, the function is continuous.

2. Determine if the function $f(x) = \dfrac{2x + 1}{x^2 - 3x + 7}$ is continuous at $x = 2$

Solution

When $x = 2$, the f(2) is obtained as follows:

$$f(x) = \frac{2x + 1}{x^2 - 3x + 7}$$
$$f(2) = \frac{2(2) + 1}{2^2 - 3(2) + 7}$$
$$= \frac{4 + 1}{4 - 6 + 7}$$
$$= \frac{5}{5}$$
$$= 1$$

Since f(2) = 1, it means that f(2) exists.

The next step is to find the value of $\displaystyle\lim_{x \to 2} f(x)$ as follows:

$$\lim_{x \to 2} \frac{2x + 1}{x^2 - 3x + 7} = \frac{2(2) + 1}{2^2 - 3(2) + 7}$$
$$= \frac{4 + 1}{4 - 6 + 7}$$
$$= \frac{5}{5}$$

$$= 1$$

Hence $f(2) = \underset{x \to 2}{lim} f(x) = 1$

Therefore, the function is continuous.

3. Determine if the function $f(x) = \dfrac{2x^2 - 8}{x + 4}$ is continuous at $x = -4$.

When $x = -4$, the f(−4) is obtained as follows:

$$f(x) = \frac{2x^2 - 8}{x + 4}$$

$$f(-4) = \frac{2(-4^2) - 8}{-4 + 4}$$

$$= \frac{2(16) - 8}{0}$$

$$= \frac{24}{0} \quad \text{(Undefined)}$$

Hence f(−4) is undefined, and does not exist. This shows that the function is not continuous. It is discontinuous as $x = -4$.

4. Determine if the function $f(x) = \dfrac{x^2 - 9}{x - 3}$ is continuous at $x = 3$.

When $x = 3$, the f(3) is obtained as follows:

$$f(x) = \frac{x^2 - 9}{x - 3}$$

$$f(3) = \frac{3^2 - 9}{3 - 3}$$

$$= \frac{0}{0} \quad \text{(This has no value)}$$

Hence f(3) does not exist.

Let us determine the value of $\underset{x \to 3}{lim} f(x)$ as follows:

$$\underset{x \to 3}{lim} \frac{x^2 - 9}{x - 3} = \underset{x \to 3}{lim} \frac{(x - 3)(x + 3)}{x - 3}$$

$$= \underset{x \to 3}{lim} (x + 3) \quad \text{(Since } x - 3 \text{ cancels out)}$$

$$= 3 + 3 \quad \text{(When 3 is substituted for } x\text{)}$$

$$= 6$$

Hence, f(3) has no value while $\underset{x \to 3}{lim} f(x) = 6$

This shows that $f(3) \neq \underset{x \to 3}{lim} f(x)$

Therefore the function discontinuous.

5. Test for the continuity of the function: $f(x) = \begin{cases} -5x + 7 & \text{if } x > 1 \\ x^2 + 2 & \text{if } x \leq 1 \end{cases}$

Solution

$$f(x) = \begin{cases} -5x + 7 & \text{if } x > 1 \\ x^2 + 2 & \text{if } x \leq 1 \end{cases}$$

When $x = 1$, we use the expression $x^2 + 2$ to evaluate f(1). Note that $x \leq 1$ means $x < 1$ or $x = 1$.

Hence, $f(x) = x^2 + 2$ (for $x = 1$)

$\qquad f(1) = 1^2 + 2$

$\qquad\qquad = 3$

Since f(1) = 3, it means that f(1) exist.

Let us now determine the limiting value of f(x).

If x approaches 1 from the left, which is $x \leq 1$ (left hand elbow) we write it as $\lim\limits_{x \to 1^-} f(x)$, and we pick the expression given by:

$\qquad f(x) = x^2 + 2$

Hence, $\lim\limits_{x \to 1} f(x) = 1^2 + 3$

$\qquad\qquad = 3$

If x approaches 1 from the right, which is $x > 1$ (right hand elbow), we write it as $\lim\limits_{x \to 1^+} f(x)$, and we pick the expression given by:

$\qquad f(x) = -5x + 7$

Hence, $\lim\limits_{x \to 1} f(x) = -5(1) + 7$

$\qquad\qquad = 2$

Hence, $\lim\limits_{x \to 1^-} f(x) = 3$ while $\lim\limits_{x \to 1^+} f(x) = 2$. They are not equal. This shows that $\lim\limits_{x \to 1} f(x)$ does not exist. Therefore, the function is not continuous.

6. Determine the continuity of the function: $f(x) = \begin{cases} x^2 & \text{if } x \geq -2 \\ x+6 & \text{if } x < -2 \end{cases}$

Solution

$\qquad f(x) = \begin{cases} x^2 & \text{if } x \geq -2 \\ x+6 & \text{if } x < -2 \end{cases}$

A direct way of determining the continuity of functions such as these is to find the limits of the two expressions in the function as x tends to the values given in the question.

Therefore, when x tends to -2 in the function $f(x) = x^2$, we have:

$\qquad f(x) = x^2$

Hence, $\lim\limits_{x \to -2} f(x) = (-2)^2$

$\qquad\qquad = 4$

Also, when x tends to -2 in the function $f(x) = x + 6$, we have:

$\qquad f(x) = x + 6$

Hence, $\lim\limits_{x \to -2} f(x) = -2 + 6$

$\qquad\qquad = 4$

Hence, $\lim\limits_{x \to -2} x^2 = \lim\limits_{x \to -2} x + 6 = 4$. Their limits give the same value. This shows that $\lim\limits_{x \to -2} f(x)$ exist. Therefore, the function is continuous.

7. Determine if the function below is continuous or not.

$$f(x) = \begin{cases} \dfrac{x^2 - 16}{x + 4} & \text{if } x \geq -4 \\ \dfrac{x + 4}{2x - 3} & \text{if } x < -4 \end{cases}$$

Solution

As x tends to -4 in the function $f(x) = \dfrac{x^2 - 16}{x + 4}$ we simplify as follows:

$$f(x) = \frac{(x + 4)(x - 4)}{x + 4}$$

Hence, $\lim\limits_{x \to -4} f(x) = (x - 4)$ (Note that $(x + 4)$ cancels out)

$$= -4 - 4$$
$$= -8$$

Also, when x tends to -4 in the function $f(x) = \dfrac{x + 4}{2x - 3}$, we have:

$$\lim\limits_{x \to -4} f(x) = \frac{-4 + 4}{2(-4) - 3}$$

$$= \frac{-4 + 4}{-8 - 3}$$

$$= \frac{0}{-11}$$

$$= 0$$

Their limits give different values. This shows that $\lim\limits_{x \to -4} f(x)$ does not exist. Therefore, the function is not continuous.

8. If $f(x) = \begin{cases} \dfrac{x^2 - 4}{x - 2} \\ 4 \ \ \text{if } x = 2 \end{cases}$ test for the continuity of the function at $x = 2$.

Solution

$$f(x) = \begin{cases} \dfrac{x^2 - 4}{x - 2} \\ 4 \ \ \text{if } x = 2 \end{cases}$$

Let us find the limits of the two expressions in the function as x tends 2.

From the first expression, as x tends to 2 we have:

$$f(x) = \frac{x^2 - 4}{x - 2}$$

Hence, $\lim\limits_{x \to 2} f(x) = \dfrac{(x - 2)(x + 2)}{x - 2}$ (When the numerator is factorized)

$$\lim\limits_{x \to 2} f(x) = (x + 2) \quad (x - 2 \text{ has cancelled out})$$

$$\lim\limits_{x \to 2} f(x) = 2 + 2$$

$$= 4$$

Also, when x tends to 2 in the function $f(x) = 4$, we have:

$$f(x) = 4$$

Hence, $\lim_{x \to 2} 4 = 4$ (Since $\lim_{x \to a} c = c$)

The question also tells us that when $x = 2$, $f(x) = 4$

Hence, $\lim_{x \to 2} 4 = \lim_{x \to 2} \dfrac{x^2 - 4}{x - 2} = 4$. Their limits give the same value of 4. This shows that $f(x)$ exist,

and that $f(2) = \lim_{x \to 2} f(x)$. Therefore, the function is continuous.

Examples on Limits of Trigonometric Functions

1. Evaluate $\lim_{x \to 0} \dfrac{\sin 2x}{5x}$

Solution

$\lim_{x \to 0} \dfrac{\sin 2x}{5x}$

Substituting zero directly into the expression above will give $\dfrac{0}{0}$. Hence we have to apply the

appropriate rule of limit. Let us first make an adjustment to the expression as follows.

Multiply the expression by $\dfrac{2x}{5x}$ and change the denominator to $2x$ just like the numerator, as

follows:

$\lim_{x \to 0} \dfrac{\sin 2x}{5x} = \lim_{x \to 0} \left(\dfrac{\sin 2x}{2x} \times \dfrac{2x}{5x} \right)$

Note that when $2x$ cancels out, the original expression remains the same.

Recall that $\lim_{x \to 0} \dfrac{\sin ax}{ax} = 1$

Hence, $\lim_{x \to 0} \dfrac{\sin 2x}{2x} = 1$

Substituting 1 for $\lim_{x \to 0} \dfrac{\sin 2x}{2x}$ in the simplification given above to gives:

$\lim_{x \to 0} \left(\dfrac{\sin 2x}{2x} \times \dfrac{2x}{5x} \right) = \lim_{x \to 0} \dfrac{\sin 2x}{2x} \times \lim_{x \to 0} \dfrac{2x}{5x}$

$= 1 \times \lim_{x \to 0} \dfrac{2}{5}$ (x cancels out)

$= 1 \times \dfrac{2}{5}$ (Since $\lim_{x \to a} c = c$)

$= \dfrac{2}{5}$

2. 1. Evaluate $\lim_{x \to 0} \dfrac{\sin 5x}{7x}$

Solution

$\lim_{x \to 0} \dfrac{\sin 5x}{7x}$

Multiply the expression by $\dfrac{5x}{7x}$ and change the denominator to $5x$ just like the numerator. This

gives:

$$\lim_{x \to 0} \frac{\sin 5x}{7x} = \lim_{x \to 0} \left(\frac{\sin 5x}{5x} \times \frac{5x}{7x} \right)$$

Hence, $\lim_{x \to 0} \dfrac{\sin 5x}{5x} = 1$

Substituting 1 for $\lim_{x \to 0} \dfrac{\sin 5x}{5x}$ in the simplification given above to gives:

$$\lim_{x \to 0} \left(\frac{\sin 5x}{5x} \times \frac{5x}{7x} \right) = \lim_{x \to 0} \frac{\sin 5x}{5x} \times \lim_{x \to 0} \frac{5x}{7x}$$

$$= 1 \times \lim_{x \to 0} \frac{5}{7} \quad (x \text{ cancels out})$$

$$= 1 \times \frac{5}{7} \quad (\text{Since } \lim_{x \to a} c = c)$$

$$= \frac{5}{7}$$

The two examples above show that $\lim_{x \to 0} \dfrac{\sin a\, x}{x} = a$ or $\lim_{x \to 0} \dfrac{\sin a\, x}{bx} = \dfrac{a}{b}$. This rule also applies to tangent as $\lim_{x \to 0} \dfrac{\tan a\, x}{x} = a$ or $\lim_{x \to 0} \dfrac{\tan a\, x}{bx} = \dfrac{a}{b}$.

3. Find the value of $\lim_{x \to \frac{\pi}{2}} \dfrac{\cos 2x}{\sin 3x}$

Solution

$$\lim_{x \to \frac{\pi}{2}} \frac{\cos 2x}{\sin 3x}$$

We can simply substitute in the value of x and obtain our answer.

$$\lim_{x \to \frac{\pi}{2}} \frac{\cos 2x}{\sin 3x} = \frac{\cos 2(\frac{\pi}{2})}{\sin 3(\frac{\pi}{2})}$$

$$= \frac{\cos \pi}{\sin \frac{3\pi}{2}}$$

$$= \frac{\cos 180}{\sin 270} \quad (\text{Note that } \pi \text{ radians} = 180°)$$

$$= \frac{-1}{-1}$$

$$= 1$$

This problem can also be solved by using the angles directly in radians, but I prefer to work in degrees. Note that $\cos \pi$ in radians = –1, and $\sin \dfrac{3\pi}{2}$ in radians = –1 as obtained above.

4. Evaluate $\lim_{x \to 0} \dfrac{\cos x}{\sin x - 3}$

Solution

$$\lim_{x \to 0} \frac{\cos x}{\sin x - 3} = \frac{\cos 0}{\sin 0 - 3}$$

$$= \frac{1}{0-3}$$

$$= -\frac{1}{3}$$

5. Evaluate $\underset{x\to 0}{lim} \dfrac{\sin 3x}{\sin 2x}$

Solution

Multiply by $\dfrac{x}{\sin 2x}$ and change the denominator to x as follows.

$\underset{x\to 0}{lim} \dfrac{\sin 3x}{\sin 2x} = \underset{x\to 0}{lim} \left(\dfrac{\sin 3x}{x} \text{ X } \dfrac{x}{\sin 2x}\right)$ (When x cancels out it gives the original expression)

$= \underset{x\to 0}{lim} \left(\dfrac{\sin 3x}{x} \text{ X } \left(\dfrac{\sin 2x}{x}\right)^{-1}\right)$ (Note that the inverse of $\dfrac{x}{\sin 2x}$ has been taken)

$= \underset{x\to 0}{lim} \dfrac{\sin 3x}{x} \text{ X } \left(\underset{x\to 0}{lim} \dfrac{\sin 2x}{x}\right)^{-1}$

$= 3 \text{ x } 2^{-1}$ (Note that $\underset{x\to 0}{lim} \dfrac{\sin 3x}{x} = 3$ and $\underset{x\to 0}{lim} \dfrac{\sin 2x}{x} = 2$)

$= 3 \text{ x } \dfrac{1}{2}$

$= \dfrac{3}{2}$

This example shows that: $\underset{x\to 0}{lim} \dfrac{\sin a\,x}{\sin b\,x} = \dfrac{a}{b}$

6. Evaluate $\underset{x\to 0}{lim} \dfrac{\sin 2x}{\sin 5x}$

Solution

$\underset{x\to 0}{lim} \dfrac{\sin 2x}{\sin 5x}$

From the rule established in example 5 above, we can see that the solution to this problem is $\dfrac{2}{5}$.

Hence, $\underset{x\to 0}{lim} \dfrac{\sin 2x}{\sin 5x} = \dfrac{2}{5}$ (Since $\underset{x\to 0}{lim} \dfrac{\sin a\,x}{\sin b\,x} = \dfrac{a}{b}$)

Exercise 26

1. Evaluate $\underset{x\to 0}{lim}\ 2x^3 - 4x^2 + x + 9$

2. Evaluate $\underset{x\to 0}{lim}\ \dfrac{5x^2 + x - 8}{x^2 - 2x + 5}$

3. Evaluate $\underset{x\to 5}{lim}\ \dfrac{x^2 - x - 20}{x - 5}$

4. Evaluate $\underset{x\to 7}{lim}\ \dfrac{x^2 - 49}{x - 7}$

5. Evaluate $\lim_{x \to 0} (5x - 1)(3x + 7)(3x + 2)$

6. Evaluate $\lim_{x \to \infty} \dfrac{2x^3 + x^2 - 9x - 5}{5x^3 + 2x^2 - 7x + 2}$

7. Evaluate $\lim_{x \to \infty} \dfrac{x^4 - 5x^2 - 2x - 6}{3x^4 + 2x + 3}$

8. Evaluate $\lim_{x \to 4} \dfrac{x^2 - 25}{x^2 + 2x - 15}$

9. Evaluate $\lim_{x \to 8} \dfrac{2x^2 - 17x + 8}{8 - x}$

10. Evaluate $\lim_{m \to 0} \dfrac{(5+m)^2 - 25}{m}$

11. Determine the limiting value of $\dfrac{\sqrt{x} - 2}{x - 4}$ as x tends to 4 .

12. Find the limit of $\dfrac{x^3 - 27}{x - 3}$ as $x \to 3$

13. Find the limit of $\dfrac{x^3 + 8}{x + 2}$ as $x \to -2$

14. Evaluate $\lim_{x \to 25} \dfrac{5 - \sqrt{x}}{25 - x}$

15. Evaluate $\lim_{x \to 1} \dfrac{x^3 - 1}{x^7 - 1}$

16. Determine if the function $f(x) = x^2 + 3x - 1$ is continuous at the point $x = 2$

17. Determine if the function $f(x) = \dfrac{5x - 2}{3x^2 - x - 2}$ is continuous at $x = -1$

18. Determine if the function $f(x) = \dfrac{2x^2 - 18}{x - 3}$ is continuous at $x = 3$.

19. Determine if the function $f(x) = \dfrac{x^3 + 64}{x + 4}$ is continuous at $x = -4$.

20. Test for the continuity of the function: $f(x) = \begin{cases} -x + 5 & \text{if } x > 2 \\ x^3 + 10 & \text{if } x \le 2 \end{cases}$

21. Determine the continuity of the function: $f(x) = \begin{cases} 3x^2 & \text{if } x \ge -3 \\ x + 30 & \text{if } x < -3 \end{cases}$

22. Determine if the function below is continuous or not.

$$f(x) = \begin{cases} \dfrac{x^2 - 25}{x + 5} & \text{if } x \ge -5 \\ \dfrac{x + 5}{2x + 10} & \text{if } x < -5 \end{cases}$$

23. If $f(x) = \begin{cases} \dfrac{x^3 - 27}{x - 3} \\ 1 \quad \text{if } x = 3 \end{cases}$ test for the continuity of the function at $x = 3$.

24. If $f(x) = \begin{cases} 2 - 3x & x < 1 \\ x^3 + 4 & x \ge 1 \end{cases}$ evaluate the following limits, if they exist.

(a) $\lim_{x \to -2} f(x)$

(b) $\lim_{x \to 1} f(x)$

25. Evaluate $\lim\limits_{x \to 2} (5 + |x - 2|)$ if it exists.

26. Evaluate $\lim\limits_{x \to 0} \dfrac{\sin 7x}{2x}$

27. Evaluate $\lim\limits_{x \to 0} \dfrac{\sin x}{3x}$

28. Find the value of $\lim\limits_{x \to \pi} \dfrac{\cos 2x}{\sin \frac{x}{2}}$

29. Evaluate $\lim\limits_{x \to 0} \dfrac{\cos 5x}{\sin 3x - 2}$

30. Evaluate $\lim\limits_{x \to 0} \dfrac{\sin 9x}{\sin 4x}$

CHAPTER 27
DIFFERENTIATION FROM FIRST PRINCIPLE

If y = f(x), then the gradient function of y or f(x) is given by:

$$g(x) = \frac{f(x + \Delta x) - f(x)}{\Delta x}$$

where Δx is the increment in x.

Applying limits has shown that as Δx tends to zero,

$$f'(x) = \lim_{\Delta x \to 0} \frac{f(x + \Delta x) - f(x)}{\Delta x}$$

$$\frac{dy}{dx} = \lim_{\Delta x \to 0} \frac{f(x + \Delta x) - f(x)}{\Delta x}$$

$\dfrac{dy}{dx}$ reads dee y dee x.

Note that f'(x) and $\dfrac{dy}{dx}$ can be interpreted as instantaneous rate of change of y with respect to x.

The technique of finding the derivative of a function by considering the limiting value is called differentiation from first principle.

Examples

1. Find the derivative of x from first principle.

Solution

Let, y = x

Increasing x by Δx will give an increase in y by Δy. This gives:

$$y + \Delta y = x + \Delta x$$
$$\Delta y = x + \Delta x - y$$
$$\Delta y = x + \Delta x - x \quad \text{(substitute } x \text{ for y since y} = x \text{ as given above from the question)}$$
$$\Delta y = \Delta x \quad (x \text{ cancels out})$$

Dividing both sides by Δx in order to get the derivative gives:

$$\frac{\Delta y}{\Delta x} = \frac{\Delta x}{\Delta x}$$
$$\frac{\Delta y}{\Delta x} = 1$$

Taking limits as Δx tends to zero gives:

$$\lim_{\Delta x \to 0} \frac{\Delta y}{\Delta x} = 1 \quad [\text{Recall that } \lim_{x \to 0} c = c, \text{ hence } \lim_{\Delta x \to 0} f(\Delta x) = 1 \text{ gives 1, since } \frac{\Delta y}{\Delta x} \text{ is regarded as f}(\Delta x)]$$

We now write $\displaystyle\lim_{\Delta x \to 0} \frac{\Delta y}{\Delta x}$ as $\dfrac{dy}{dx}$. This gives:

$$\frac{dy}{dx} = 1$$

2. Find from first principle, the derivative of $2x^3$.

Solution

Let, $y = 2x^3$

Increasing x by Δx will give an increase in y by Δy. This gives:

$$y + \Delta y = 2(x + \Delta x)^3$$

$$y + \Delta y = 2[x^3 + 3x^2\Delta x + 3x(\Delta x)^2 + (\Delta x)^3]$$

$$= 2x^3 + 6x^2\Delta x + 6x(\Delta x)^2 + 2(\Delta x)^3$$

$$\Delta y = 2x^3 + 6x^2\Delta x + 6x(\Delta x)^2 + 2(\Delta x)^3 - y$$

$$= 2x^3 + 6x^2\Delta x + 6x(\Delta x)^2 + 2(\Delta x)^3 - 2x^3 \quad \text{(Since } y = 2x^3)$$

$$\Delta y = 6x^2\Delta x + 6x(\Delta x)^2 + 2(\Delta x)^3 \quad \text{(Since } 2x^3 \text{ cancels out)}$$

Dividing both sides by Δx in order to get the derivative gives:

$$\frac{\Delta y}{\Delta x} = \frac{6x^2\Delta x}{\Delta x} + \frac{6x(\Delta x)^2}{\Delta x} + \frac{2(\Delta x)^3}{\Delta x}$$

$$\frac{\Delta y}{\Delta x} = 6x^2 + 6x\Delta x + 2(\Delta x)^2$$

Taking limits as Δx tends to zero gives:

$$\lim_{\Delta x \to 0} \frac{\Delta y}{\Delta x} = 6x^2 + 6x(0) + 2(0)^2 \quad \text{(Note that zero replaces } \Delta x)$$

$$\lim_{\Delta x \to 0} \frac{\Delta y}{\Delta x} = 6x^2$$

We now write $\lim_{\Delta x \to 0} \dfrac{\Delta y}{\Delta x}$ as $\dfrac{dy}{dx}$. This gives:

$$\frac{dy}{dx} = 6x^2$$

3. Find the derivative of $f(x) = \dfrac{1}{x^2}$ from first principle.

Solution

$$f(x) = \Delta x$$

Increasing x by Δx gives:

$$f(x + \Delta x) = \frac{1}{(x + \Delta x)^2}$$

$$f(x + \Delta x) = \frac{1}{x^2 + 2x\Delta x + (\Delta x)^2}$$

Subtract $f(x)$ from both sides of the equation

$$f(x + \Delta x) - f(x) = \frac{1}{x^2 + 2x\Delta x + (\Delta x)^2} - f(x)$$

$$f(x + \Delta x) - f(x) = \frac{1}{x^2 + 2x\Delta x + (\Delta x)^2} - \frac{1}{x^2} \quad \text{(Note that } f(x) = \frac{1}{x^2})$$

$$= \frac{x^2 - (x^2 + 2x\Delta x + (\Delta x)^2)}{x^2(x^2 + 2x\Delta x + (\Delta x)^2)}$$

$$= \frac{x^2 - x^2 - 2x\Delta x - (\Delta x)^2}{x^2(x^2 + 2x\Delta x + (\Delta x)^2)}$$

337

$$= \frac{-2x\Delta x - (\Delta x)^2}{x^2(x^2 + 2x\Delta x + (\Delta x)^2)}$$

$$f(x + \Delta x) - f(x) = \frac{\Delta x(-2x - \Delta x)}{x^2(x^2 + 2x\Delta x + (\Delta x)^2)}$$

Dividing both sides by Δx gives:

$$\frac{f(x + \Delta x) - f(x)}{\Delta x} = \frac{-2x - \Delta x}{x^2(x^2 + 2x\Delta x + (\Delta x)^2)} \qquad (\Delta x \text{ has cancelled out from the right hand side})$$

Taking limits as Δx tends to zero gives:

$$\lim_{\Delta x \to 0} \frac{f(x + \Delta x) - f(x)}{\Delta x} = \frac{-2x - 0}{x^2(x^2 + 2x(0) + (0)^2)}$$

Replacing $\displaystyle\lim_{\Delta x \to 0} \frac{f(x + \Delta x) - f(x)}{\Delta x}$ with f'(x) gives the derivative of f(x) as:

$$f'(x) = \frac{-2x}{x^2(x^2)}$$

$$f'(x) = \frac{-2}{x^3}$$

4. If $f(x) = 3x^2$

(a) write down and simplify the expression $\dfrac{f(x + h) - f(x)}{h}$ $(h \neq 0)$

(b) find $\displaystyle\lim_{\Delta x \to 0} \frac{f(x + h) - f(x)}{h}$

Solution

(a) Let, $f(x) = 3x^2$

Increasing x by h gives:

$$f(x + h) = 3(x + h)^2$$
$$= 3[x^2 + 2xh + h^2]$$
$$= 3x^2 + 6xh + 3h^2$$

Subtracting f(x) from both sides gives:

$$f(x + h) - f(x) = 3x^2 + 6xh + 3h^2 - 3x^2 \qquad (\text{Note that } f(x) = 3x^2 \text{ from the question})$$
$$= 6xh + 3h^2$$

Dividing both side by h gives:

$$\frac{f(x + h) - f(x)}{h} = \frac{6xh}{h} + \frac{3h^2}{h}$$

$$\frac{f(x + h) - f(x)}{h} = 6x + 3h$$

(b) $\dfrac{f(x + h) - f(x)}{h} = 6x + 3h$

Taking limits as h tends to zero gives:

$$\lim_{h \to 0} \frac{f(x + h) - f(x)}{h} = 6x + 3(0)$$

$$\lim_{h \to 0} \frac{f(x + h) - f(x)}{h} = 6x$$

5. If $f(x) = 4x^3 - 5$,

(a) evaluate $\dfrac{f(x+h) - f(x)}{h}$, where $h \neq 0$

(b) From your result in (5a) above, find the derivatives of $f(x)$ with respect to x.

Solution

(a) Let, $f(x) = 4x^3 - 5$

Increasing x by h gives:

$$f(x + h) = 4(x + h)^3 - 5$$
$$= 4[x^3 + 3x^2h + 3xh^2 + h^3] - 5$$
$$= 4x^3 + 12x^2h + 12xh^2 + 4h^3 - 5$$

Subtracting $f(x)$ from both sides gives:

$$f(x + h) - f(x) = 4x^3 + 12x^2h + 12xh^2 + 4h^3 - 5 - (4x^3 - 5) \quad (f(x) = 4x^3 - 5 \text{ from the question})$$
$$= 4x^3 + 12x^2h + 12xh^2 + 4h^3 - 5 - 4x^3 + 5$$
$$= 12x^2h + 12xh^2 + 4h^3$$

Dividing both side by h gives:

$$\frac{f(x+h) - f(x)}{h} = 12x^2 + 12xh + 4h^2$$

(b) $\dfrac{f(x+h) - f(x)}{h} = 12x^2 + 12xh + 4h^2$

Taking limits as h tends to zero gives:

$$\lim_{h \to 0} \frac{f(x+h) - f(x)}{h} = 12x^2 + 12x(0)\ 4(0)^2$$

$$\lim_{h \to 0} \frac{f(x+h) - f(x)}{h} = 12x^2$$

6. If $f(x) = \dfrac{x^2 + 1}{2x}$

(a) write down and simplify the expression $\dfrac{f(x + \Delta x) - f(\Delta x)}{\Delta x}$, where $\Delta x \neq 0$

(b) Find the derivatives of $y = \dfrac{x^2 + 1}{2x}$

Solution

$$f(x) = \frac{x^2 + 1}{2x}$$

$$f(x + \Delta x) = \frac{(x + \Delta x)^2 + 1}{2(x + \Delta x)}$$

$$f(x + \Delta x) = \frac{x^2 + 2x\Delta x + (\Delta x)^2 + 1}{2x + 2\Delta x}$$

Subtract $f(x)$ from both sides of the equation

$$f(x + \Delta x) - f(x) = \frac{x^2 + 2x\Delta x + (\Delta x)^2 + 1}{2x + 2\Delta x} - \frac{x^2 + 1}{2x}$$

339

$$f(x + \Delta x) - f(x) = \frac{2x[x^2 + 2x\Delta x + (\Delta x)^2 + 1] - (x^2 + 1)(2x + 2\Delta x)}{2x(2x + 2\Delta x)}$$

$$= \frac{2x^3 + 4x^2\Delta x + 2x(\Delta x)^2 + 2x - (2x^3 + 2x^2\Delta x + 2x + 2\Delta x)}{2x(2x + 2\Delta x)}$$

$$= \frac{2x^3 + 4x^2\Delta x + 2x(\Delta x)^2 + 2x - 2x^3 - 2x^2\Delta x - 2x - 2\Delta x}{2x(2x + 2\Delta x)}$$

$$= \frac{2x^2\Delta x + 2x(\Delta x)^2 - 2\Delta x}{2x(2x + 2\Delta x)}$$

$$f(x + \Delta x) - f(x) = \frac{\Delta x(2x^2 + 2x\Delta x - 2)}{2x(2x + 2\Delta x)} \qquad \text{(After factorizing the numerator)}$$

Dividing both sides by Δx gives:

$$\frac{f(x + \Delta x) - f(x)}{\Delta x} = \frac{2x^2 + 2x\Delta x - 2}{2x(2x + 2\Delta x)}$$

(b)Taking limits as Δx tends to zero gives:

$$\lim_{\Delta x \to 0} \frac{f(x + \Delta x) - f(x)}{\Delta x} = \frac{2x^2 + 2x(0) - 2}{2x(2x + 2(0))}$$

$$= \frac{2x^2 - 2}{2x(2x)}$$

$$= \frac{2(x^2 - 1)}{4x^2}$$

Replacing $\lim_{\Delta x \to 0} \dfrac{f(x + \Delta x) - f(x)}{\Delta x}$ with f'(x) gives the derivative of f(x) as:

$$f'(x) = \frac{x^2 - 1}{2x^2}$$

Separating into fractions gives:

$$f'(x) = \frac{x^2}{2x^2} - \frac{1}{2x^2}$$

$$f'(x) = \frac{1}{2} - \frac{1}{2x^2}$$

7. Find from first principle the derivatives with respect to x of $y = 5x^3 - x + 7$

Solution

$$y = 5x^3 - x + 7$$
$$y + \Delta y = 5(x + \Delta x)^3 - (x + \Delta x) + 7$$
$$y + \Delta y = 5[x^3 + 3x^2\Delta x + 3x(\Delta x)^2 + (\Delta x)^3] - x - \Delta x + 7$$
$$= 5x^3 + 15x^2\Delta x + 15x(\Delta x)^2 + 5(\Delta x)^3 - x - \Delta x + 7$$
$$\Delta y = 5x^3 + 15x^2\Delta x + 15x(\Delta x)^2 + 5(\Delta x)^3 - x - \Delta x + 7 - y$$
$$= 5x^3 + 15x^2\Delta x + 15x(\Delta x)^2 + 5(\Delta x)^3 - x - \Delta x + 7 - (5x^3 - x + 7) \quad \text{(Since } y = 5x^3 - x + 7\text{)}$$
$$\Delta y = 5x^3 + 15x^2\Delta x + 15x(\Delta x)^2 + 5(\Delta x)^3 - x - \Delta x + 7 - 5x^3 + x - 7$$
$$= 15x^2\Delta x + 15x(\Delta x)^2 + 5(\Delta x)^3 - \Delta x$$

Dividing both sides by Δx gives:

$$\frac{\Delta y}{\Delta x} = \frac{15x^2\Delta x}{\Delta x} + \frac{15x(\Delta x)^2}{\Delta x} + \frac{5(\Delta x)^3}{\Delta x} - \frac{\Delta x}{\Delta x}$$

$$\frac{\Delta y}{\Delta x} = 15x^2 + 15x\Delta x + 5(\Delta x)^2 - 1$$

Taking limits as Δx tends to zero gives:

$$\lim_{\Delta x \to 0} \frac{\Delta y}{\Delta x} = 15x^2 + 15x(0) + 5(0)^2 - 1$$

$$\lim_{\Delta x \to 0} \frac{\Delta y}{\Delta x} = 15x^2 - 1$$

We now replace $\lim_{\Delta x \to 0} \frac{\Delta y}{\Delta x}$ with $\frac{dy}{dx}$. This gives:

$$\frac{dy}{dx} = 15x^2 - 1$$

8. Differentiate $3x - \dfrac{1}{2x^2}$ from first principle.

Solution

$$y = 3x - \frac{1}{x^2}$$

$$y + \Delta y = 3(x + \Delta x) - \frac{1}{2(x + \Delta x)^2}$$

$$= 3x + 3\Delta x - \frac{1}{2[x^2 + 2x\Delta x + (\Delta x)^2]}$$

Subtract y from both sides of the equation

$$\Delta y = 3x + 3\Delta x - \frac{1}{2x^2 + 4x\Delta x + 2(\Delta x)^2} - y$$

$$= 3x + 3\Delta x - \frac{1}{2x^2 + 4x\Delta x + 2(\Delta x)^2} - \left(3x - \frac{1}{2x^2}\right)$$

$$= 3x + 3\Delta x - \frac{1}{2x^2 + 4x\Delta x + 2(\Delta x)^2} - 3x + \frac{1}{2x^2}$$

$$= 3\Delta x - \frac{1}{2x^2 + 4x\Delta x + 2(\Delta x)^2} + \frac{1}{2x^2}$$

Combining them into one fraction gives:

$$\Delta y = \frac{3\Delta x[2x^2(2x^2 + 4x\Delta x + 2(\Delta x)^2] - 2x^2 + 2x^2 + 4x\Delta x + 2(\Delta x)^2}{2x^2[2x^2 + 4x\Delta x + 2(\Delta x)^2]}$$

$$= \frac{3\Delta x[4x^4 + 8x^3\Delta x + 4x^2(\Delta x)^2] + 4x\Delta x + 2(\Delta x)^2}{2x^2[2x^2 + 4x\Delta x + 2(\Delta x)^2]} \quad (-2x^2 + 2x^2 \text{ has cancelled out})$$

$$= \frac{12x^4\Delta x + 24x^3(\Delta x)^2 + 12x^2(\Delta x)^3 + 4x\Delta x + 2(\Delta x)^2}{2x^2[2x^2 + 4x\Delta x + 2(\Delta x)^2]}$$

Factorizing the numerator gives:

$$= \frac{\Delta x[12x^4 + 24x^3\Delta x + 12x^2(\Delta x)^2 + 4x + 2\Delta x]}{2x^2[2x^2 + 4x\Delta x + 2(\Delta x)^2]}$$

Dividing both sides by Δx gives:

$$\frac{\Delta y}{\Delta x} = \frac{12x^4 + 24x^3\Delta x + 12x^2(\Delta x)^2 + 4x + 2\Delta x}{2x^2[2x^2 + 4x\Delta x + 2(\Delta x)^2]}$$ (Δx that is outside the bracket cancels out)

Taking limits as Δx tends to zero gives:

$$\lim_{\Delta x \to 0} \frac{\Delta y}{\Delta x} = \frac{12x^4 + 24x^3(0) + 12x^2(0)^2 + 4x + 2(0)}{2x^2[2x^2 + 4x(0) + 2(0)^2]}$$

$$= \frac{12x^4 + 4x}{2x^2[2x^2]}$$

$$\lim_{\Delta x \to 0} \frac{\Delta y}{\Delta x} = \frac{12x^4 + 4x}{4x^4}$$

$$= \frac{12x^4}{4x^4} + \frac{4x}{4x^4}$$

$$= 3 + \frac{1}{x^3}$$

Replacing $\lim_{\Delta x \to 0} \frac{\Delta y}{\Delta x}$ with $\frac{dy}{dx}$ gives the derivative as:

$$\frac{dy}{dx} = 3 + \frac{1}{x^3}$$

Exercise 27

1. Find the derivative of $2x$ from first principle.

2. Find from first principle, the derivative of x^2.

3. Find the derivative of $f(x) = \frac{1}{x^3}$ from first principle.

4. If $f(x) = 5x^2$

(a) write down and simplify the expression $\frac{f(x+h) - f(x)}{h}$ ($h \neq 0$)

(b) find $\lim_{h \to 0} \frac{f(x+h) - f(x)}{h}$

5. If $f(x) = 9x^3$,

(a) evaluate $\frac{f(x+h) - f(x)}{h}$, where $h \neq 0$

(b) From your result above , find the derivatives of $f(x)$ with respect to x.

6. If $f(x) = \frac{x^3 - 2}{x}$

(a) write down and simplify the expression $\frac{f(x + \Delta x) - f(\Delta x)}{\Delta x}$, where $\Delta x \neq 0$

(b) Find the derivatives of $y = \frac{x^3 - 2}{x}$

7. Find from first principle the derivatives with respect to x of $y = 3x^2 - 10x$

8. Differentiate $x + \dfrac{3}{x}$ from first principle.

9. From first principle, find the derivatives with respect to x of $y = 5x - 3x^2$

10. Differentiate $2x + \dfrac{x}{5}$ from first principle.

CHAPTER 28
GENERAL RULE OF DIFFERENTIATION AND COMPOSITE FUNCTIONS

The general rule for the derivative/differentiation of a function is as given below.

If $y = x^n$

then, $\dfrac{dy}{dx} = nx^{n-1}$

The rule for differentiating a composite function (or function of a function) is given as follows:

If $y = (2x + 5)^4$

then we write, $u = 2x + 5$

and express y as:

$y = u^4$

Therefore, $\dfrac{dy}{dx} = \dfrac{dy}{du} \times \dfrac{du}{dx}$

This rule is called the chain rule.

Examples

1. Find the derivatives of the following:

(a) $y = 2x^7$

(b) $y = \dfrac{3}{4} x^8$

(c) $y = 3\sqrt{x}$

(d) $y = \dfrac{10}{\sqrt[5]{x^3}}$

(e) $y = \dfrac{1}{x^{\frac{1}{4}}}$

Solution

(a) $y = 2x^7$

Solutions

(a) $y = 2x^7$

$\dfrac{dy}{dx} = 7 \times 2x^{7-1}$

This is done by multiplying the exponent (power) by the term and subtracting 1 from the exponent (power). Hence, the answer is:

$\dfrac{dy}{dx} = 14x^6$

(b) $y = \dfrac{3}{4} x^8$

Multiply the term by the exponent (power) (i.e. 8) and subtract 1 from the exponent. This gives:

$\dfrac{dy}{dx} = 8 \times \dfrac{3}{4} x^{8-1}$

$$= \frac{24}{4}x^7$$

$$\frac{dy}{dx} = 6x^7$$

(c) $y = 3\sqrt{x}$

Since $\sqrt{x} = x^{\frac{1}{2}}$, we can rewrite the expression as:

$$y = 3x^{\frac{1}{2}}$$

$$\frac{dy}{dx} = \frac{1}{2} \times 3x^{\frac{1}{2}-1}$$

$$= \frac{3}{2}x^{-\frac{1}{2}}$$

$$= \frac{3}{2} \times \frac{1}{x^{\frac{1}{2}}} \qquad \text{(Note that } x^{-\frac{1}{2}} = \frac{1}{x^{\frac{1}{2}}} \text{ from indices)}$$

$$\frac{dy}{dx} = \frac{3}{2x^{\frac{1}{2}}}$$

$$\frac{dy}{dx} = \frac{3}{2\sqrt{x}} \qquad \text{(Since } x^{\frac{1}{2}} = \sqrt{x} \text{)}$$

(d) $y = \dfrac{10}{\sqrt[5]{x^3}}$

Expressing the root in fractional form gives:

$$y = \frac{10}{(x^3)^{\frac{1}{5}}}$$

$$y = \frac{10}{x^{\frac{3}{5}}} \qquad \text{(The two exponents (powers) 3 and } \frac{1}{5} \text{ have been multiplied)}$$

Taking the denominator to the numerator changes the sign of its exponent as follows:

$$y = 10x^{-\frac{3}{5}}$$

Hence, $\dfrac{dy}{dx} = -\dfrac{3}{5} \times 10x^{-\frac{3}{5}-1}$

$$= -\frac{30}{5}x^{-\frac{8}{5}}$$

$$= -6x^{-\frac{8}{5}}$$

$$= -6 \times \frac{1}{x^{\frac{8}{5}}} \qquad \text{(Note that the inverse of a term changes the sign of its exponent)}$$

$$\frac{dy}{dx} = \frac{-6}{x^{\frac{8}{5}}}$$

Or, $\dfrac{dy}{dx} = \dfrac{-6}{\sqrt[5]{x^8}}$

(e) $y = \dfrac{1}{x^{\frac{1}{4}}}$

This can be expressed as:

$$y = x^{-\frac{1}{4}}$$

$$\dfrac{dy}{dx} = -\dfrac{1}{4}x^{-\frac{1}{4}-1}$$

$$= -\dfrac{1}{4}x^{-\frac{5}{4}}$$

$$= -\dfrac{1}{4} \times \dfrac{1}{x^{\frac{5}{4}}}$$

$$\dfrac{dy}{dx} = -\dfrac{1}{4x^{\frac{5}{4}}}$$ (Note that the inverse of a term changes the sign of its exponent)

Or, $\dfrac{dy}{dx} = \dfrac{1}{4\sqrt[4]{x^5}}$

2. Find the derivative of each of the following:

(a) $5x^3 - 7x^2 - 3x + 8$

(b) $\dfrac{3}{5}x^5 + 2x^3 - x$

(c) $\dfrac{2x^4 - 5x^3 - 4x^2 + 3}{x^2}$

(d) $\sqrt{x} + \dfrac{1}{\sqrt{x}}$

Solutions

(a) Let the expression be $y = 5x^3 - 7x^2 - 3x + 8$

Hence, $\dfrac{dy}{dx} = \dfrac{d(5x^3)}{dx} - \dfrac{d(7x^2)}{dx} - \dfrac{d(3x)}{dx} + \dfrac{d(8)}{dx}$

This means that each part should be differentiated separately.

Hence, $\dfrac{dy}{dx} = (3 \times 5x^{3-1}) - (2 \times 7x^{2-1}) - (1 \times 3x^{1-1}) + 0$

$\dfrac{dy}{dx} = 15x^2 - 14x - 3$ (Note that $x^0 = 1$)

Note that the derivative of a constant is zero as shown by the derivative of 8

(b) $y = \dfrac{3}{5}x^5 + 2x^3 - x$

$\quad \dfrac{dy}{dx} = \left(5 \times \dfrac{3}{5}x^{5-1}\right) + \left(3 \times 2x^{3-1}\right) - \left(1 \times x^{1-1}\right)$

$\qquad = 3x^4 + 6x^2 - 1$

(c) $y = \dfrac{2x^4 - 5x^3 - 4x^2 + 3}{x^2}$

Dividing each term in the numerator by the denominator in order to separate the expression into it different fractions gives:

$\quad y = \dfrac{2x^4}{x^2} - \dfrac{5x^3}{x^2} - \dfrac{4x^2}{x^2} + \dfrac{3}{x^2}$

$\qquad = 2x^2 - 5x - 4 + \dfrac{3}{x^2}$

$\quad y = 2x^2 - 5x - 4 + 3x^{-2}$

$\quad \dfrac{dy}{dx} = (2 \times 2x) - 5 - 0 + \left(-2 \times 3x^{-2-1}\right)$

$\qquad = 4x - 5 - 6x^{-3}$

$\quad \dfrac{dy}{dx} = 4x - 5 - \dfrac{6}{x^2}$

(d) $y = \sqrt{x} + \dfrac{1}{\sqrt{x}}$

$\qquad = x^{\frac{1}{2}} + \dfrac{1}{x^{\frac{1}{2}}}$

$\quad y = x^{\frac{1}{2}} + x^{-\frac{1}{2}}$

$\quad \dfrac{dy}{dx} = \dfrac{1}{2}x^{-\frac{1}{2}} - \dfrac{1}{2}x^{-\frac{3}{2}} \qquad$ (Note that $-\dfrac{1}{2} - 1 = -\dfrac{3}{2}$)

$\qquad = \dfrac{1}{2} \times \dfrac{1}{x^{\frac{1}{2}}} - \dfrac{1}{2} \times \dfrac{1}{x^{\frac{3}{2}}}$

$\qquad = \dfrac{1}{2x^{\frac{1}{2}}} - \dfrac{1}{2x^{\frac{3}{2}}}$

$\quad \dfrac{dy}{dx} = \dfrac{1}{2\sqrt{x}} - \dfrac{1}{2\sqrt{x^3}}$

3. If $y = (5x - 2)^3$, find $\dfrac{dy}{dx}$

Solution

$\quad y = (5x - 2)^3 \quad$ (This is a composite function)

347

Let us take u = $5x - 2$

If $5x - 2$ is replaced with U, then the question (i.e. y = $(5x - 2)^3$) becomes:

$$y = u^3$$

Hence, $\dfrac{dy}{du} = 3u^2$

Since u = $5x - 2$

then $\dfrac{du}{dx} = 5$

Therefore, $\dfrac{dy}{dx} = \dfrac{dy}{du} \times \dfrac{du}{dx}$ (Chain rule)

$$= 3u^2 \times 5$$

$$= 15u^2$$

Now substitute $5x - 2$ for u to obtain $\dfrac{dy}{dx}$ as follows:

$$\dfrac{dy}{dx} = 15(5x - 2)^2$$

4. If y = $\dfrac{1}{(5x^2 - 1)^4}$ find $\dfrac{dy}{dx}$

Solution

$$y = \dfrac{1}{(5x^2 - 1)^4}$$ (This is a composite function)

It can also be represented as follows:

$$y = (5x^2 - 1)^{-4}$$ (Its inverse changes the sign of its exponent)

Now, let us take u = $5x^2 - 1$

Hence, y = u^{-4} (When $5x^2 - 1$ is replaced with u in the original question)

Therefore, $\dfrac{dy}{du} = -4u^{-5}$

Since u = $5x^2 - 1$

then, $\dfrac{du}{dx} = 10x$

Therefore, $\dfrac{dy}{dx} = \dfrac{dy}{du} \times \dfrac{du}{dx}$ (Chain rule)

$$= -4u^{-5} \times 10x$$

$$\dfrac{dy}{dx} = -40xu^{-5}$$

Now substitute $5x^2 - 1$ for u. This gives:

$$\dfrac{dy}{dx} = -40x(5x^2 - 1)^{-5}$$

Or, $\dfrac{dy}{dx} = \dfrac{-40x}{(5x^2 - 1)^5}$ (Note the change in the sign of the exponent as it becomes

denominator)

5. If $y = (2x^3 + 7x)^{\frac{1}{2}}$ find $\dfrac{dy}{dx}$

Solution

$$y = (2x^3 + 7x)^{\frac{1}{2}}$$

Let $u = 2x^3 + 7x$

Hence, $y = u^{\frac{1}{2}}$

$$\frac{dy}{du} = \frac{1}{2} u^{-\frac{1}{2}}$$

Also, $u = 2x^3 + 7x$

$$\frac{du}{dx} = 6x^2 + 7$$

Therefore, $\dfrac{dy}{dx} = \dfrac{dy}{du} \times \dfrac{du}{dx}$

$$= \frac{1}{2} u^{-\frac{1}{2}} \times 6x^2 + 7$$

$$= \frac{6x^2 + 7}{2} u^{-\frac{1}{2}}$$

$$= \frac{6x^2 + 7}{2} \times \frac{1}{u^{1/2}}$$

$$\frac{dy}{dx} = \frac{6x^2 + 7}{2u^{1/2}}$$

Now, replace u with $2x^3 + 7x$. This gives:

$$\frac{dy}{dx} = \frac{6x^2 + 7}{2(2x^3 + 7x)^{1/2}}$$

Or, $\dfrac{dy}{dx} = \dfrac{6x^2 + 7}{2\sqrt{2x^3 + 7x}}$　　(Note that $(2x^3 + 7x)^{\frac{1}{2}} = \sqrt{2x^3 + 7x}$)

6. Find the derivative of $3x^2 - x + 9)^4$

Solution

$$y = (3x^2 - x + 9)^4$$

Let $u = 3x^2 - x + 9$

Hence, $y = u^4$

$$\frac{dy}{du} = 4u^3$$

$$\frac{du}{dx} = 6x - 1$$

Therefore, $\dfrac{dy}{dx} = \dfrac{dy}{du} \times \dfrac{du}{dx}$

$$= 4u^3 \times 6x - 1$$

$$= 4(6x - 1)u^3$$

$$= (24x - 4)u^3$$

$$\frac{dy}{dx} = (24x - 4)(3x^2 - x + 9)^3 \quad \text{(When u is replaced with } 3x^2 - x + 9)$$

7. Find the derivative of $\left(x - \dfrac{5}{x}\right)^4$

Solution

$$y = \left(x - \frac{5}{x}\right)^4$$

Let $u = x - \dfrac{5}{x}$

Therefore, $y = u^4$

$$\frac{dy}{du} = 4u^3$$

$$\frac{du}{dx} = \frac{d(x)}{dx} - \frac{d(5x^{-1})}{dx} \quad \text{(Note that } = \frac{5}{x} = 5x^{-1})$$

$$= 1 - (-1)5x^{-2}$$

$$= 1 + 5x^{-2}$$

$$\frac{du}{dx} = 1 + \frac{5}{x^2}$$

Therefore, $\dfrac{dy}{dx} = \dfrac{dy}{du} \times \dfrac{du}{dx}$

$$= 4u^3 \times \left(1 + \frac{5}{x^2}\right)$$

$$= \left(4 + \frac{20}{x^2}\right)u^3$$

$$\frac{dy}{dx} = \left(4 + \frac{20}{x^2}\right)\left(x - \frac{5}{x}\right)^3 \quad \text{(When u is replaced with } \left(x - \frac{5}{x}\right))$$

8. Differentiate with respect to x: $\dfrac{1}{2x^5 - 3x + 1}$

Solution

Let $y = \dfrac{1}{2x^5 - 3x + 1}$

Or, $y = (2x^5 - 3x + 1)^{-1} \quad \text{(Recall from indices that } \dfrac{1}{a} = a^{-1})$

Let us take $u = 2x^5 - 3x + 1$

Therefore, $y = u^{-1}$

$$\frac{dy}{du} = -1u^{-2}$$

$$\frac{du}{dx} = 10x^4 - 3$$

Therefore, $\dfrac{dy}{dx} = \dfrac{dy}{du} \times \dfrac{du}{dx}$

$$= -1u^{-2} \times 10x^4 - 3$$

$$= -1(10x^4 - 3)u^{-2}$$

$$= (-10x^4 + 3)u^{-2}$$

$$= \frac{3 - 10x^4}{u^2}$$

$$\frac{dy}{dx} = \frac{3 - 10x^4}{(2x^5 - 3x + 1)^2}$$

9. Differentiate with respect to x: $\sqrt{7 - 5x^3}$

Solution

Let $y = \sqrt{7 - 5x^3}$

Or, $y = (7 - 5x^3)^{\frac{1}{2}}$ (Recall from indices that $\sqrt{a} = a^{\frac{1}{2}}$)

Let $u = 7 - 5x^3$

Hence, $y = u^{\frac{1}{2}}$

$$\frac{dy}{du} = \frac{1}{2}u^{-\frac{1}{2}}$$

$$\frac{du}{dx} = -15x^2$$

Therefore, $\dfrac{dy}{dx} = \dfrac{dy}{du} \times \dfrac{du}{dx}$

$$= \frac{1}{2}u^{-\frac{1}{2}} \times (-15x^2)$$

$$= -\frac{15}{2}x^2 (u^{-\frac{1}{2}})$$

$$= -\frac{15x^2}{2u^{\frac{1}{2}}}$$

$$= -\frac{15x^2}{2\sqrt{u}}$$

$$\frac{dy}{dx} = -\frac{15x^2}{2\sqrt{7 - 5x^3}}$$

10. Find $\dfrac{dy}{dx}$ if $y = \dfrac{1}{\sqrt{2x^3 - 5}}$

Solution

$$y = \frac{1}{\sqrt{2x^3 - 5}}$$

Let $u = 2x^3 - 5$

hence, $y = \dfrac{1}{\sqrt{u}}$

$$y = \frac{1}{u^{\frac{1}{2}}}$$

$$y = u^{-\frac{1}{2}}$$

$$\frac{dy}{du} = -\frac{1}{2}u^{-\frac{3}{2}}$$ (Note that $-\frac{1}{2} - 1 = -\frac{3}{2}$)

$$\frac{du}{dx} = 6x^2$$

Therefore, $\dfrac{dy}{dx} = \dfrac{dy}{du} \times \dfrac{du}{dx}$

$$= -\frac{1}{2}u^{-\frac{3}{2}} \times (6x^2)$$

$$= -\frac{6}{2}x^2\,(u^{-\frac{3}{2}})$$

$$= -3x^2(u^{-\frac{3}{2}})$$

$$= -\frac{3x^2}{u^{\frac{3}{2}}}$$

$$= -\frac{3x^2}{\sqrt{u^3}}$$

$$\frac{dy}{dx} = -\frac{3x^2}{\sqrt{(2x^3-5)^3}}$$

Exercise 28

1. Find the derivatives of the following:

(a) $y = 8x^5$

(b) $y = \dfrac{2}{5}x^5$

(c) $y = \sqrt[3]{x}$

(d) $y = 7\sqrt[7]{x}$

(e) $y = \dfrac{1}{\sqrt[8]{x^5}}$

(f) $y = \dfrac{2}{x^{\frac{5}{2}}}$

2. Find the derivative of each of the following:

(a) $2x^5 - 3x^4 - 4x^3 + 5x^2 - 6x + 7$

(b) $x^7 + 2x^4 - \dfrac{3}{x}$

(c) $\dfrac{3x^9 - x^7 - 5x^4 + 2x^2 - 1}{x^3}$

(d) $5(\sqrt[4]{x}) + \dfrac{5}{\sqrt[3]{2x}}$

3. If $y = (2x - 5)^4$, find $\dfrac{dy}{dx}$

4. If $y = \dfrac{3}{(x^3 - 7)^2}$ find $\dfrac{dy}{dx}$

5. If $y = (2x^3 + 7x)^{\frac{1}{2}}$ find $\dfrac{dy}{dx}$

6. Find the derivative of $(7x^3 - x^2 + 3)^5$

7. Find the derivative of $\left(3x - \dfrac{2}{3x}\right)^3$

8. Differentiate with respect to x: $\quad -\dfrac{9}{3x^2 - x - 10}$

9. Differentiate with respect to x: $\quad \sqrt{1 - 2x^4}$

10. Find $\dfrac{dy}{dx}$ if $y = \dfrac{1}{\sqrt[3]{5x^3 - 1}}$

11. Find the derivative of $(x^5 - 3)^9$

12. Find the derivative of $\left(x - \dfrac{1}{5x}\right)^2$

13. Differentiate with respect to x: $\quad \dfrac{2}{x^3 - x - \frac{1}{x}}$

14. Differentiate with respect to x: $\quad \sqrt{5x - x^2}$

15. Find $\dfrac{dy}{dx}$ if $y = \dfrac{1}{\sqrt[5]{x^3 + 2}}$

CHAPTER 29
PRODUCT RULE OF DERIVATIVE

If $y = uv$ where u and v are functions of x, then:

$$\frac{dy}{dx} = u\frac{dv}{dx} + v\frac{du}{dx}$$

This is called the product rule of differentiation.

Similarly, If $y = uvw$ where u, v and w are functions of x, then:

$$\frac{dy}{dx} = \frac{du}{dx}vw + \frac{dv}{dx}uw + \frac{dw}{dx}uv$$

Examples

1. If $f(x) = (x - 3)(x + 4)$, find $f'(x)$

Solution

$$f(x) = (x - 3)(x + 4)$$

This is a product of two functions of x.

Let $u = x - 3$

and $v = x + 4$

Hence, $\dfrac{du}{dx} = 1$

$$\frac{dv}{dx} = 1$$

Therefore, the derivative of $f(x)$ is:

$$f'(x) = u\frac{dv}{dx} + v\frac{du}{dx}$$
$$= (x - 3) \times 1 + (x + 4) \times 1$$
$$= x - 3 + x + 4$$
$$f'(x) = 2x + 1$$

Note that another way to differentiate the function in example 1 above is to expand the bracket and differentiate directly.

2. Find the derivative of $y = (4x^2 + 1)(x^2 - 3)$

Solution

$$y = (4x^2 + 1)(x^2 - 3)$$

Let $u = 4x^2 + 1$

and $v = (x^2 - 3)$

Hence, $\dfrac{du}{dx} = 8x$

$$\frac{dv}{dx} = 2x$$

$$\frac{dy}{dx} = u\frac{dv}{dx} + v\frac{du}{dx}$$
$$= (4x^2 + 1)2x + (x^2 - 3)8x$$
$$= 8x^3 + 2x + 8x^3 - 24x$$
$$\frac{dy}{dx} = 16x^3 - 22x$$

3. If $y = x^2(1 + 2x)^{\frac{1}{2}}$, find $\frac{dy}{dx}$

<u>Solution</u>

$$y = x^2(1 + 2x)^{\frac{1}{2}}$$
$$u = x^2$$

and $v = (1 + 2x)^{\frac{1}{2}}$

Hence, $\frac{du}{dx} = 2x$

$$\frac{dv}{dx} = \frac{1}{2} \times 2 \times (1 + 2x)^{\frac{1}{2} - 1} \qquad \text{(Use of chain rule)}$$

$$= (1 + 2x)^{-\frac{1}{2}}$$

$$\frac{dv}{dx} = \frac{1}{(1+2x)^{\frac{1}{2}}}$$

Hence, $\frac{dy}{dx} = u\frac{dv}{dx} + v\frac{du}{dx}$

$$= x^2 \frac{1}{(1+2x)^{\frac{1}{2}}} + (1 + 2x)^{\frac{1}{2}} \times 2x$$

$$= \frac{x^2}{(1+2x)^{\frac{1}{2}}} + 2x(1 + 2x)^{\frac{1}{2}}$$

let us now simplify by using $(1 + 2x)^{\frac{1}{2}}$ as the LCM as follows:

$$= \frac{x^2 + 2x \,(1+2x)^{\frac{1}{2}}(1+2x)^{\frac{1}{2}}}{(1+2x)^{\frac{1}{2}}}$$

$$= \frac{x^2 + 2x \,(1+2x)}{(1+2x)^{\frac{1}{2}}} \qquad \text{(Note that } (1 + 2x)^{\frac{1}{2}} \times (1 + 2x)^{\frac{1}{2}} = (1 + 2x)^{\frac{1}{2}+\frac{1}{2}} = 1 + 2x)$$

$$= \frac{x^2 + 2x + 4x^2}{(1+2x)^{\frac{1}{2}}}$$

$$\frac{dy}{dx} = \frac{5x^2 + 2x}{\sqrt{1+2x}}$$

4. Find the derivative of $(2x + 3)^3(4x^2 - 1)^2$

<u>Solution</u>

$$y = (2x + 3)^3(4x^2 - 1)^2$$

Let $u = (2x + 3)^3$

and $v = (4x^2 - 1)^2$

$$\frac{du}{dx} = 3(2x + 3)^{3-1} \times 2 \quad \text{(Use of chain rule. Also note that 2 is from the derivative of } 2x + 3\text{)}$$

$$= 6(2x + 3)^2$$

$$\frac{dv}{dx} = 2(4x^2 - 1)^{2-1} \times 8x \quad \text{(Note that } 8x \text{ is from the derivative of } 4x^2 - 1\text{)}$$

$$= 16x(4x^2 - 1)$$

Hence, $\frac{dy}{dx} = u\frac{dv}{dx} + v\frac{du}{dx}$

$$= (2x + 3)^3 \times 16x(4x^2 - 1) + (4x^2 - 1)^2 \times 6(2x + 3)^2$$

Let us factorize the expression by taking out $(2x + 3)^2$ and $(4x^2 - 1)$ which are the common terms as follows:

$$\frac{dy}{dx} = (2x + 3)^2(4x^2 - 1)[(2x + 3)16x + (4x^2 - 1)6]$$

$$= (2x + 3)^2(4x^2 - 1)(32x^2 + 48x + 24x^2 - 6)$$

$$= (2x + 3)^2(4x^2 - 1)(56x^2 + 48x - 6)$$

$$= (2x + 3)^2(4x^2 - 1)\, 2(28x^2 + 24x - 3)$$

$$\frac{dy}{dx} = 2(2x + 3)^2(4x^2 - 1)(28x^2 + 24x - 3)$$

5. Differentiate: $y = x(x + 1)(x^2 - 4)$

Solution

$$y = x(x + 1)(x^2 - 4)$$

Expanding the first two brackets gives:

$$y = (x^2 + x)(x^2 - 4)$$

Hence, $u = (x^2 + x)$

and $v = (x^2 - 4)$

Hence, $\frac{du}{dx} = 2x + 1$

$$\frac{dv}{dx} = 2x$$

Therefore, $\frac{dy}{dx} = u\frac{dv}{dx} + v\frac{du}{dx}$

$$= (x^2 + x)2x + (x^2 - 4)(2x + 1)$$

$$= 2x^3 + 2x^2 + 2x^3 + x^2 - 8x - 4$$

$$\frac{dy}{dx} = 4x^3 + 3x^2 - 8x - 4$$

6. Find the derivative of $(x^2 + 3x - 2)^2 \sqrt{x}$

Solution

$$y = (x^2 + 3x - 2)^2 \sqrt{x}$$

$$u = (x^2 + 3x - 2)^2$$

$$v = \sqrt{x}$$

Or, $v = x^{\frac{1}{2}}$

$$\frac{du}{dx} = 2(x^2 + 3x - 2)^{2-1} \times 2x + 3 \quad \text{(Note that } 2x + 3 \text{ is from the derivative of } x^2 + 3x - 2\text{)}$$

$$= (4x + 6)(x^2 + 3x - 2)$$

$$\frac{dv}{dx} = \frac{1}{2} x^{-\frac{1}{2}} \quad \text{(Note that } 8x \text{ is from the derivative of } 4x^2 - 1\text{)}$$

$$= \frac{1}{2x^{\frac{1}{2}}}$$

$$\frac{dv}{dx} = \frac{1}{2\sqrt{x}}$$

Hence, $\dfrac{dy}{dx} = u\dfrac{dv}{dx} + v\dfrac{du}{dx}$

$$= (x^2 + 3x - 2)^2 \frac{1}{2\sqrt{x}} + \sqrt{x}(4x + 6)(x^2 + 3x - 2)$$

$$= (x^2 + 3x - 2)^2 \frac{1}{2\sqrt{x}} + 2\sqrt{x}(2x + 3)(x^2 + 3x - 2)$$

Factorize the expression by taking out $(x^2 + 3x - 2)$ which is the common factor. This gives:

$$\frac{dy}{dx} = (x^2 + 3x - 2)\left[\frac{x^2 + 3x - 2}{2\sqrt{x}} + 2\sqrt{x}(2x + 3) \right]$$

Simplifying the part in the bracket by taking $2\sqrt{x}$ as the LCM gives:

$$\frac{dy}{dx} = (x^2 + 3x - 2)\left[\frac{x^2 + 3x - 2 + 4x(2x + 3)}{2\sqrt{x}} \right] \quad \text{(Note that } 2\sqrt{x} \times 2\sqrt{x} = 4x\text{)}$$

$$= (x^2 + 3x - 2)\left[\frac{x^2 + 3x - 2 + 8x^2 + 12x}{2\sqrt{x}} \right]$$

$$\frac{dy}{dx} = \frac{(x^2 + 3x - 2)(9x^2 + 15x - 2)}{2\sqrt{x}}$$

7. Find the derivative of $(1 + x)(5x - 2)^{\frac{3}{2}}$

Solution

$$y = (1 + x)(5x - 2)^{\frac{3}{2}}$$

$$u = (1 + x)$$

$$v = (5x - 2)^{\frac{3}{2}}$$

$$\frac{du}{dx} = 1$$

$$\frac{dv}{dx} = \frac{3}{2}(5x - 2)^{\frac{3}{2} - 1} \times 5 \quad \text{(Note that the derivative of } 5x - 2 \text{ is 5)}$$

$$= \frac{15}{2}(5x - 2)^{\frac{1}{2}}$$

Hence, $\dfrac{dy}{dx} = u\dfrac{dv}{dx} + v\dfrac{du}{dx}$

$$= (1 + x) \times \frac{15}{2}(5x - 2)^{\frac{1}{2}} + (5x - 2)^{\frac{3}{2}} \times 1$$

$$= \frac{15}{2}(1 + x)(5x - 2)^{\frac{1}{2}} + (5x - 2)^{\frac{3}{2}}$$

Factorize by taking out $(5x - 2)^{\frac{1}{2}}$ (i.e. the lower exponent) which is the common factor gives:

$$\frac{dy}{dx} = (5x - 2)^{\frac{1}{2}}\left[\frac{15}{2}(1 + x) + 5x - 2\right] \quad \text{(Note that } \frac{(5x-2)^{\frac{3}{2}}}{(5x-2)^{\frac{1}{2}}} = (5x - 2)^{\frac{3}{2} - \frac{1}{2}} = 5x - 2\text{)}$$

$$= (5x - 2)^{\frac{1}{2}}\left[\frac{15}{2} + \frac{15x}{2} + 5x - 2\right]$$

$$= (5x - 2)^{\frac{1}{2}}\left[\frac{25x}{2} + \frac{11}{2}\right]$$

$$\frac{dy}{dx} = \sqrt{5x - 2}\left[\frac{25x + 11}{2}\right]$$

8. If y = $(1 + x)(2 - 3x)(2x - 1)$, find $\frac{dy}{dx}$ by using product rule.

Solution

$$y = (1 + x)(2 - 3x)(2x - 1)$$

This is a product of three expressions, u, v and w.

Hence, u = $(1 + x)$

$\qquad$ v = $(2 - 3x)$

and $\quad$ w = $(2x - 1)$

Therefore, $\dfrac{du}{dx} = 1$

$\dfrac{dv}{dx} = -3$

$\dfrac{dw}{dx} = 2$

Hence the formula for product rule of three terms is given by:

$$\frac{dy}{dx} = \frac{du}{dx}vw + \frac{dv}{dx}uw + \frac{dw}{dx}uv$$

$$= 1(2 - 3x)(2x - 1) + (-3)(1 + x)(2x - 1) + 2(1 + x)(2 - 3x)$$

$$= 4x - 2 - 6x^2 + 3x + (-3 - 3x)(2x - 1) + (2 + 2x)(2 - 3x)$$

$$= 7x - 2 - 6x^2 - 6x + 3 - 6x^2 + 3x + 4 - 6x + 4x - 6x^2$$

$$= -2 + 3 + 4 + 7x - 6x + 3x - 6x + 4x - 6x^2 - 6x^2 - 6x^2$$

$$\frac{dy}{dx} = 5 + 2x - 18x^2$$

9. Differentiate with respect to x: $(x^2 - 3x + 5)(2x - 7)$

Solution

$$y = (x^2 - 3x + 5)(2x - 7)$$

Let us differentiate this product by applying product rule but without the use u and v. This is done as follows:

$$\frac{dy}{dx} = (x^2 - 3x + 5)\frac{d(2x-7)}{dx} + (2x-7)\frac{d(x^2-3x+5)}{dx}$$

$$= (x^2 - 3x + 5)2 + (2x - 7)(2x - 3)$$

$$= 2x^2 - 6x + 10 + 4x^2 - 6x - 14x + 21$$

$$\frac{dy}{dx} = 6x^2 - 26x + 31$$

10. If $y = (5x^2 - 3)(2 + \frac{3}{x})$, find $\frac{dy}{dx}$

Solution

$$y = (5x^2 - 3)(2 + \frac{3}{x})$$

$$u = (5x^2 - 3)$$

$$v = (2 + \frac{3}{x})$$

$$\frac{du}{dx} = 10x$$

$$\frac{dv}{dx} = \frac{d(3x^{-1})}{dx} \qquad \text{(Note that } \frac{3}{x} = 3x^{-1}\text{)}$$

$$= -3x^{-2}$$

$$\frac{dv}{dx} = \frac{-3}{x^2}$$

Hence, $\frac{dy}{dx} = u\frac{dv}{dx} + v\frac{du}{dx}$

$$= (5x^2 - 3)\left(\frac{-3}{x^2}\right) + (2 + \frac{3}{x})10x$$

$$= -15 + \frac{9}{x^2} + 20x + 30$$

$$\frac{dy}{dx} = 15 + 20x + \frac{9}{x^2}$$

Exercise 29

1. If $f(x) = (2x - 1)(3x + 1)$, find $f'(x)$

2. Find the derivative of $y = (3x^2 - 5)(x^2 + 10)$

3. If $y = 5x(3 + x)^{\frac{1}{2}}$ find $\frac{dy}{dx}$

4. Find the derivative of $(x + 5)(x^2 - 7)^3$

5. Differentiate: $y = 2x(3x + 2)(2x^2 - 5)$

6. Find the derivative of $(3x^2 - 1)^3 \sqrt{2x}$

7. Find the derivative of $(9 - x)(x + 3)^{\frac{3}{4}}$

8. If $y = (2 - x)(5 - 3x^2)(x + 3)$, find $\dfrac{dy}{dx}$ by using product rule.

9. Differentiate with respect to x: $(3x^4 - x^2 + 2x)(x - 1)$

10. If $y = (3x^2 - x)(1 + \dfrac{1}{2x})$, find $\dfrac{dy}{dx}$

11. Find the derivative of $x^2(\sqrt{x^5})$

12. Find the derivative of $x^4(2x - 11)^{\frac{2}{3}}$

13. If $y = (7 + x)(1 - x^2)(5x^3 + 1)$, find $\dfrac{dy}{dx}$.

14. Differentiate with respect to x: $(3x^4 - x^3 + x^2 + 2x - 3)(5x + 4)$

15. If $y = (x^3 - 3x + 5)\left(\dfrac{1}{x^5}\right)$, find $\dfrac{dy}{dx}$

CHAPTER 30
QUOTIENT RULE OF DERIVATIVE

If $y = \dfrac{u}{v}$ where u and v are functions of x, then:

$$\frac{dy}{dx} = \frac{v\frac{du}{dx} - u\frac{dv}{dx}}{v^2}$$

This is called the quotient rule of differentiation.

Examples

1. If $y = \dfrac{3x^2 - 8x + 5}{5x - 2}$ find $\dfrac{dy}{dx}$

Solution

$$y = \frac{3x^2 - 8x + 5}{5x - 2}$$

This is of the form $y = \dfrac{u}{v}$. Therefore, we are going to apply product rule.

Let $u = 3x^2 - 8x + 5$

and $v = 5x - 2$

Hence, $\dfrac{du}{dx} = 6x - 8$

$\dfrac{dv}{dx} = 5$

Therefore, $\dfrac{dy}{dx} = \dfrac{v\frac{du}{dx} - u\frac{dv}{dx}}{v^2}$

$$= \frac{(5x - 2)(6x - 8) - (3x^2 - 8x + 5)(5)}{(5x - 2)^2}$$

$$= \frac{30x^2 - 40x - 12x + 16 - 15x^2 - 40x - 25}{(5x - 2)^2}$$

$$\frac{dy}{dx} = \frac{15x^2 - 12x - 9}{(5x - 2)^2}$$

2. Differentiate with respect to x, the function: $\dfrac{3x^2 - 2x}{x + 5}$

Solution

$$y = \frac{3x^2 - 2x}{x + 5}$$

Let $u = 3x^2 - 2x$

and $v = x + 5$

Hence, $\dfrac{du}{dx} = 6x - 2$

$\dfrac{dv}{dx} = 1$

Therefore, $\dfrac{dy}{dx} = \dfrac{v\frac{du}{dx} - u\frac{dv}{dx}}{v^2}$

$= \dfrac{(x+5)(6x-2) - (3x^2 - 2x)(1)}{(x+5)^2}$

$= \dfrac{6x^2 - 2x + 30x - 10 - 3x^2 + 2x}{(x+5)^2}$

$\dfrac{dy}{dx} = \dfrac{3x^2 + 30x - 10}{(x+5)^2}$

3. Find the differential coefficient of $y = \dfrac{-3}{x^2 + 5}$

Solution

$y = \dfrac{-3}{x^2 + 5}$

$u = -3$

and $v = x^2 + 5$

Hence, $\dfrac{du}{dx} = 0$ (The derivative of a constant is zero)

$\dfrac{dv}{dx} = 2x$

Therefore, $\dfrac{dy}{dx} = \dfrac{v\frac{du}{dx} - u\frac{dv}{dx}}{v^2}$

$= \dfrac{(x^2 + 5) \times 0 - (-3)2x}{(x^2 + 5)^2}$

$= \dfrac{0 - 6x}{(x^2 + 5)^2}$

$\dfrac{dy}{dx} = \dfrac{-6x}{(x^2 + 5)^2}$

4. Differentiate: $y = \dfrac{\sqrt{3-x}}{\sqrt{3+x}}$

Solution

$y = \dfrac{\sqrt{3-x}}{\sqrt{3+x}}$

$u = \sqrt{3-x}$

$= (3-x)^{\frac{1}{2}}$

and $v = \sqrt{3+x}$

$= (3+x)^{\frac{1}{2}}$

Hence, $\dfrac{du}{dx} = \dfrac{1}{2}(3-x)^{\frac{1}{2}-1} \times -1$ (Note that the derivative of $3-x$ is -1)

$= -\dfrac{1}{2}(3-x)^{-\frac{1}{2}}$

362

$$\frac{du}{dx} = \frac{-(3-x)^{-\frac{1}{2}}}{2}$$

$$\frac{dv}{dx} = \frac{1}{2}(3+x)^{\frac{1}{2}-1} \times 1 \quad \text{(Note that the derivative of } 3-x \text{ is } -1)$$

$$= \frac{1}{2}(3+x)^{-\frac{1}{2}}$$

$$\frac{dv}{dx} = \frac{(3+x)^{-\frac{1}{2}}}{2}$$

Therefore, $\dfrac{dy}{dx} = \dfrac{v\frac{du}{dx} - u\frac{dv}{dx}}{v^2}$

$$= \frac{(3+x)^{\frac{1}{2}}\left(\frac{-(3-x)^{-\frac{1}{2}}}{2}\right) - (3-x)^{\frac{1}{2}}\left(\frac{(3+x)^{-\frac{1}{2}}}{2}\right)}{((3+x)^{\frac{1}{2}})^2}$$

Taking out the common terms which are terms with positive exponent in order to factorize the expression gives:

$$\frac{dy}{dx} = \frac{\left((3+x)^{\frac{1}{2}}\right)\left((3-x)^{\frac{1}{2}}\right)\left[\frac{-(3-x)^{-1}}{2} - \frac{(3+x)^{-1}}{2}\right]}{3+x}$$

Note that in order to obtain the terms in the square bracket, we subtracted the exponents of the factors from the exponents of the original term. For example, $(3+x)^{\frac{1}{2}}\left(\frac{-(3-x)^{-\frac{1}{2}}}{2}\right)$ divided by $\left((3+x)^{\frac{1}{2}}\right)\left((3-x)^{\frac{1}{2}}\right)$ gave $\frac{-(3-x)^{-1}}{2}$ since the equal terms canceled out and $-\frac{1}{2} - \frac{1}{2} = -1$, which gave the exponent of -1. Similarly, $(3-x)^{\frac{1}{2}}\left(\frac{(3+x)^{-\frac{1}{2}}}{2}\right)$ divided by $\left((3+x)^{\frac{1}{2}}\right)\left((3-x)^{\frac{1}{2}}\right)$ gave $\frac{(3+x)^{-1}}{2}$ since the equal terms canceled out and $-\frac{1}{2} - \frac{1}{2} = -1$, which gave the exponent of -1 as the terms in the square bracket.

Let us now continue with the solution by simplifying further as follows:

$$\frac{dy}{dx} = \frac{\left((3+x)^{\frac{1}{2}}\right)\left((3-x)^{\frac{1}{2}}\right)\left[\frac{-(3-x)^{-1}}{2} - \frac{(3+x)^{-1}}{2}\right]}{3+x}$$

$$= \frac{\left((3+x)^{\frac{1}{2}}\right)\left((3-x)^{\frac{1}{2}}\right)\left[\frac{-1}{2(3-x)} - \frac{1}{2(3+x)}\right]}{3+x}$$

$$= \frac{\left((3+x)^{\frac{1}{2}}\right)\left((3-x)^{\frac{1}{2}}\right)\left[\frac{-(3+x) - (3-x)}{2(3-x)(3+x)}\right]}{3+x}$$

$$= \frac{\left((3+x)^{\frac{1}{2}}\right)\left((3-x)^{\frac{1}{2}}\right)\left[\frac{-3-x-3+x}{2(3-x)(3+x)}\right]}{3+x}$$

$$= \frac{\left((3+x)^{\frac{1}{2}}\right)\left((3-x)^{\frac{1}{2}}\right)\left[\frac{-6}{2(3-x)(3+x)}\right]}{3+x}$$

$$= \frac{\left((3+x)^{\frac{1}{2}}\right)\left((3-x)^{\frac{1}{2}}\right)\left[\frac{-3}{(3-x)(3+x)}\right]}{3+x}$$

$$= \frac{-3\left((3+x)^{\frac{1}{2}}\right)\left((3-x)^{\frac{1}{2}}\right)}{(3-x)(3+x)(3+x)}$$

$$= \frac{-3\left((3+x)^{\frac{1}{2}}\right)\left((3-x)^{\frac{1}{2}}\right)}{(3+x)^2(3-x)}$$

$$= -3(3+x)^{\frac{1}{2}-2}(3-x)^{\frac{1}{2}-1} \quad \text{(Subtraction of exponents due to the division above)}$$

$$= -3(3+x)^{-\frac{3}{2}}(3-x)^{-\frac{1}{2}}$$

$$= \frac{-3}{(3+x)^{\frac{3}{2}}(3-x)^{\frac{1}{2}}}$$

$$= \frac{-3}{\sqrt{(3+x)^3(3-x)}}$$

5. Differentiate with respect to x: $y = \dfrac{(2x^2-3)^3}{x}$

Solution

$$y = \frac{(2x^2-3)^3}{x}$$

$$u = (2x^2-3)^3$$

and $v = x$

Hence, $\dfrac{du}{dx} = 3(2x^2-3)^{3-1} \times 4x$

$$= 12x(2x^2-3)^2$$

$$\frac{dv}{dx} = 1$$

Therefore, $\dfrac{dy}{dx} = \dfrac{v\frac{du}{dx} - u\frac{dv}{dx}}{v^2}$

$$= \frac{x[12x(2x^2-3)^2] - (2x^2-3)^3 \times 1}{x^2}$$

$$= \frac{12x^2(2x^2-3)^2 - (2x^2-3)^3}{x^2}$$

Factorize the expression by taking out $(2x^2-3)^2$ which has the lower exponent. This gives:

$$\frac{dy}{dx} = \frac{(2x^2-3)^2[12x^2-(2x^2-3)]}{x^2}$$

$$= \frac{(2x^2-3)^2(12x^2-2x^2+3)}{x^2}$$

$$\frac{dy}{dx} = \frac{(2x^2-3)^2(10x^2+3)}{x^2}$$

6. Find the derivative of $\dfrac{\sqrt{(1+2x^2)^3}}{x}$

Solution

$$y = \frac{\sqrt{(1+2x^2)^3}}{x}$$

$$u = \sqrt{(1+2x^2)^3}$$

$$= [(1+2x^2)^3]^{\frac{1}{2}} \qquad \text{(When the square root sign is removed, we use an exponent of } \tfrac{1}{2})$$

$$u = (1+2x^2)^{\frac{3}{2}} \qquad \text{(After multiplying the exponents)}$$

$$v = x$$

Hence, $\dfrac{du}{dx} = \dfrac{3}{2}(1+2x^2)^{\frac{3}{2}-1} \times 4x$ (Note that the derivative of $3-x$ is -1)

$$= 6x(1+2x^2)^{\frac{1}{2}}$$

$$\frac{dv}{dx} = 1$$

Therefore, $\dfrac{dy}{dx} = \dfrac{v\frac{du}{dx} - u\frac{dv}{dx}}{v^2}$

$$= \frac{x\left(6x(1+2x^2)^{\frac{1}{2}}\right) - (1+2x^2)^{\frac{3}{2}} \times 1}{x^2}$$

Take out $(1+2x^2)^{\frac{1}{2}}$ which has the lower exponent and factorize the expression. This gives:

$$\frac{dy}{dx} = \frac{(1+2x^2)^{\frac{1}{2}}[6x^2 - (1+2x^2)]}{x^2} \qquad \text{(Note that } \frac{(1+2x^2)^{\frac{3}{2}}}{(1+2x^2)^{\frac{1}{2}}} \text{ gives } (1+2x^2) \text{ by subtracting exponents.}$$

$$= \frac{(1+2x^2)^{\frac{1}{2}}(6x^2 - 1 - 2x^2)}{x^2}$$

$$= \frac{(1+2x^2)^{\frac{1}{2}}(4x^2 - 1)}{x^2}$$

$$\frac{dy}{dx} = \frac{(\sqrt{1+2x^2})(4x^2 - 1)}{x^2}$$

7. If $y = \dfrac{(4x^3 - 3x^2 + x + 1)^{\frac{1}{2}}}{(x+1)^2}$

Solution

$$y = \dfrac{(4x^3 - 3x^2 + x + 1)^{\frac{1}{2}}}{(x+1)^2}$$

$u = (4x^3 - 3x^2 + x + 1)^{\frac{1}{2}}$ (After multiplying the exponents)

$v = (x + 1)^2$

Hence, $\dfrac{du}{dx} = \dfrac{1}{2}(4x^3 - 3x^2 + x + 1)^{\frac{1}{2} - 1} \times (12x^2 - 6x + 1)$

$$= \dfrac{(12x^2 - 6x + 1)(4x^3 - 3x^2 + x + 1)^{-\frac{1}{2}}}{2}$$

$\dfrac{dv}{dx} = 2(x+1)^{2-1}$

$\quad\quad = 2(x + 1)$

Therefore, $\dfrac{dy}{dx} = \dfrac{v\frac{du}{dx} - u\frac{dv}{dx}}{v^2}$

$$= \dfrac{\dfrac{(x+1)^2(12x^2 - 6x + 1)(4x^3 - 3x^2 + x + 1)^{-\frac{1}{2}} - (4x^3 - 3x^2 + x + 1)^{\frac{1}{2}}[2(x+1)]}{2}}{[(x+1)^2]^2}$$

Take out the common terms with lower exponents [i.e. $(x + 1)$ and $(4x^3 - 3x^2 + x + 1)^{-\frac{1}{2}}$] and factorize the expression. This gives:

$$\dfrac{dy}{dx} = \dfrac{\dfrac{(x+1)(4x^3 - 3x^2 + x + 1)^{-\frac{1}{2}}[(x+1)(12x^2 - 6x + 1) - (4x^3 - 3x^2 + x + 1)2]}{2}}{[(x+1)^2]^2}$$

Remember to subtract the exponents of the factors from the exponents of the original expression when simplifying. Simplifying further, the expression above gives:

$$\dfrac{dy}{dx} = \dfrac{(x+1)(4x^3 - 3x^2 + x + 1)^{-\frac{1}{2}}[12x^3 - 6x^2 + x + 12x^2 - 6x + 1 - (8x^3 - 6x^2 + 2x + 2)]}{2(x+1)^4}$$

$$= \dfrac{(x+1)(4x^3 - 3x^2 + x + 1)^{-\frac{1}{2}}(12x^3 - 6x^2 + x + 12x^2 - 6x + 1 - 8x^3 + 6x^2 - 2x - 2)}{2(x+1)^4}$$

$$= \dfrac{(x+1)(4x^3 - 3x^2 + x + 1)^{-\frac{1}{2}}(4x^3 + 12x^2 - 7x - 1)}{2(x+1)^4}$$

$$= \dfrac{(4x^3 - 3x^2 + x + 1)^{-\frac{1}{2}}(4x^3 + 12x^2 - 7x - 1)}{2(x+1)^3}$$

Note that $(x + 1)$ cancels out from the numerator and denominator.

Therefore, $\dfrac{dy}{dx} = \dfrac{4x^3 + 12x^2 - 7x - 1}{2(4x^3 - 3x^2 + x + 1)^{\frac{1}{2}}(x+1)^3}$

8. Determine $\dfrac{d}{dx}\left(\dfrac{3 + 2x - x^2}{\sqrt{1+x}}\right)$

Solution

Let $y = \dfrac{3 + 2x - x^2}{\sqrt{1+x}}$

Hence, $u = 3 + 2x - x^2$

$\qquad \dfrac{du}{dx} = 2 - 2x$

$\qquad v = \sqrt{1 + x}$

$\qquad\quad = (1 + x)^{\frac{1}{2}}$

$\qquad \dfrac{du}{dx} = \dfrac{1}{2}(1 + x)^{\frac{1}{2} - 1} \times 1$

$\qquad\quad = \dfrac{1}{2}(1 + x)^{-\frac{1}{2}}$

Therefore, $\dfrac{dy}{dx} = \dfrac{v\frac{du}{dx} - u\frac{dv}{dx}}{v^2}$

$$= \dfrac{(1 + x)^{\frac{1}{2}}(2 - 2x) - (3 + 2x - x^2)\frac{1}{2}(1 + x)^{-\frac{1}{2}}}{[(1+x)^{\frac{1}{2}}]^2}$$

Take out $(1 + x)^{-\frac{1}{2}}$ as the common factor since it has the lower exponent and factorize the expression. This gives:

$$\dfrac{dy}{dx} = \dfrac{(1 + x)^{-\frac{1}{2}}\left[(1+x)(2 - 2x) - (3 + 2x - x^2)\frac{1}{2}\right]}{1+x}$$

$$= \dfrac{(1 + x)^{-\frac{1}{2}}\left[\frac{2(1+x)(2 - 2x) - (3 + 2x - x^2)}{2}\right]}{1+x}$$

$$= \dfrac{(1 + x)^{-\frac{1}{2}}\left[(2 + 2x)(2 - 2x) - 3 - 2x + x^2\right]}{2(1+x)}$$

$$= \dfrac{(1 + x)^{-\frac{1}{2}}\left(4 - 4x + 4x - 4x^2 - 3 - 2x + x^2\right)}{2(1+x)}$$

$$= \frac{1 - 2x - 3x^2}{2(1+x)^{\frac{1}{2}}(1+x)} \qquad \text{[When } (1+x)^{-\frac{1}{2}} \text{ is taken to the denominator it becomes } (1+x)^{\frac{1}{2}}]$$

$$\frac{dy}{dx} = \frac{1 - 2x - 3x^2}{2(1+x)^{\frac{3}{2}}} \qquad \text{(The exponents of same terms have been added together, i.e. } \frac{1}{2} + 1 = \frac{3}{2})$$

Exercise 30

1. If $y = \dfrac{x^2 - 5x + 1}{x - 1}$ find $\dfrac{dy}{dx}$

2. Differentiate with respect to x, the function: $\dfrac{4x^2 - x}{2x + 3}$

3. Find the differential coefficient of $y = \dfrac{-7}{3x^2 + 1}$

4. Differentiate: $y = \dfrac{\sqrt{1+x}}{\sqrt{1-x}}$

5. Differentiate with respect to x: $y = \dfrac{(x^3 - 2)^2}{x^2}$

6. Find the derivative of $\dfrac{\sqrt{2 + 3x^2}}{x^3}$

7. If $y = \dfrac{(x^2 - x - 4)^{\frac{1}{3}}}{(2x+1)^2}$

8. Determine $\dfrac{d}{dx}\left(\dfrac{x - 2x^3}{\sqrt{2-x}}\right)$

9. Find the differential coefficient of $y = -\dfrac{1}{1 - 3x^2}$

10. Differentiate: $y = \dfrac{2 + x}{2 - x}$

CHAPTER 31
DERIVATIVE OF PARAMETRIC EQUATIONS

If $y = f(t)$ and $x = g(t)$ are two different functions of a common variable, t, then the two equations are called parametric equations. The variable t, is the parameter.

The derivative of a parametric equation such as the one stated above is obtained as follows:

$$\frac{dy}{dx} = \frac{\frac{dy}{dt}}{\frac{dx}{dt}}$$

Examples

1. If $y = 5 + t^2$ and $x = 3 - 2t^2$, find $\frac{dy}{dx}$

Solution

$$y = 5 + t^2$$

Hence, $\frac{dy}{dt} = 2t$

$$x = 3 - 2t^2$$

Hence, $\frac{dx}{dt} = -4t$

Therefore, $\frac{dy}{dx} = \frac{\frac{dy}{dt}}{\frac{dx}{dt}}$

$$= \frac{2t}{-4t}$$

$$= -\frac{1}{2} \qquad \text{(t cancels out)}$$

2. Find $\frac{dy}{dx}$ of the functions below which are expressed in the parametric form.

$$x = \frac{5}{t^2} \text{ and } y = 2t^5 - 3$$

Solution

$$x = \frac{5}{t^2}$$

$$x = 5t^{-2}$$

$$\frac{dx}{dt} = -10t^{-3}$$

$$y = 2t^5 - 3$$

$$\frac{dy}{dt} = 10t^4$$

Therefore, $\frac{dy}{dx} = \frac{\frac{dy}{dt}}{\frac{dx}{dt}}$

$$= \frac{10t^4}{-10t^{-3}}$$

$$= -t^{4-(-3)} \quad \text{(10 cancels out)}$$

$$= -t^{4+3}$$

$$= -t^7$$

3. If $V = \frac{4}{3}\pi r^3$ and $A = \pi r^2$, find $\frac{dA}{dV}$

Solution

$$V = \frac{4}{3}\pi r^3$$

$$\frac{dV}{dr} = 3\left(\frac{4}{3}\right)\pi r^2$$

$$= 4\pi r^2$$

$$A = \pi r^2$$

$$\frac{dA}{dr} = 2\pi r$$

$$\frac{dA}{dV} = \frac{\frac{dA}{dr}}{\frac{dV}{dr}}$$

$$= \frac{2\pi r}{4\pi r^2}$$

$$\frac{dA}{dV} = \frac{1}{2r}$$

4. Determine the derivative of the curve defined by the equations: $x = t^2 - 4t$ and $y = 2t^3 - 7t$.

Solution

$$y = 2t^3 - 7t$$

Hence, $\frac{dy}{dt} = 6t^2 - 7$

$$x = t^2 - 4t$$

Hence, $\frac{dx}{dt} = 2t - 4$

Therefore, $\frac{dy}{dx} = \frac{\frac{dy}{dt}}{\frac{dx}{dt}}$

$$= \frac{6t^2 - 7}{2t - 4}$$

5. Given that $v = u + at$ and $s = ut + \frac{1}{2}at^2$. Find $\frac{dv}{ds}$ if u and a are constants.

Solution

$$v = u + at$$

$$\frac{dv}{dt} = a$$

$$s = ut + \frac{1}{2}at^2$$

$$\frac{ds}{dt} = u + (2 \times \frac{1}{2} \times at)$$

$$= u + at$$

Hence, $\dfrac{dv}{ds} = \dfrac{\frac{dv}{dt}}{\frac{ds}{dt}}$

$$= \frac{a}{u + at}$$

6. If $y = \dfrac{t}{t-2}$ and $x = \dfrac{1}{t+1}$, find $\dfrac{dy}{dx}$

<u>Solution</u>

$$y = \frac{t}{t-2}$$

$$\frac{dy}{dt} = \frac{(t-2)(1) - t(1)}{(t-2)^2} \qquad \text{(Use of quotient rule where u = t and v = t − 2)}$$

$$= \frac{t - 2 - t}{(t-2)^2}$$

$$\frac{dy}{dt} = \frac{-2}{(t-2)^2}$$

$$x = \frac{1}{t+1}$$

$$\frac{dx}{dt} = \frac{(t+1)(0) - 1(1)}{(t+1)^2} \qquad \text{(Use of quotient rule where u = 1 and v = t + 1)}$$

$$= \frac{-1}{(t+1)^2}$$

Hence, $\dfrac{dy}{dx} = \dfrac{\frac{dy}{dt}}{\frac{dx}{dt}}$

$$= \frac{\frac{-2}{(t-2)^2}}{\frac{-1}{(t+1)^2}}$$

$$= \frac{-2}{(t-2)^2} \times \frac{(t+1)^2}{-1}$$

$$\frac{dy}{dx} = \frac{2(t+1)^2}{(t-2)^2}$$

7. The parametric equations of the motion of a stone are: $y = 12 + 3t - 2t^2$ and $x = 5t$. Find $\dfrac{dy}{dx}$.

<u>Solution</u>

$$y = 12 + 3t - 2t^2$$

$$\frac{dy}{dt} = 3 - 4t$$

$$x = 5t$$

$$\frac{dx}{dt} = 5$$

$$\frac{dy}{dx} = \frac{\frac{dy}{dt}}{\frac{dx}{dt}}$$

$$= \frac{3 - 4t}{5}$$

8. If the parametric equations of a parabola are $y = \dfrac{2mt^2}{1+t^2}$ and $x = \dfrac{2m}{1+t^2}$ where m is a constant, find $\dfrac{dy}{dx}$.

Solution

$$y = \frac{2mt^2}{1+t^2}$$

$$\frac{dy}{dt} = \frac{(1+t^2)(4mt) - 2mt^2(2t)}{(1+t^2)^2} \qquad \text{(Use of quotient rule where u = } 2mt^2 \text{ and v = } 1 + t^2)$$

$$= \frac{4mt + 4mt^3 - 4mt^3}{(1+t^2)^2}$$

$$\frac{dy}{dt} = \frac{4mt}{(1+t^2)^2}$$

$$x = \frac{2m}{1+t^2}$$

$$\frac{dx}{dt} = \frac{(1+t^2)(0) - 2m(2t)}{(1+t^2)^2} \qquad \text{(Use of quotient rule where u = 1 and v = t + 1)}$$

$$= \frac{-4mt}{(1+t^2)^2}$$

Hence, $\dfrac{dy}{dx} = \dfrac{\frac{dy}{dt}}{\frac{dx}{dt}}$

$$= \frac{\frac{4mt}{(1+t^2)^2}}{\frac{-4mt}{(1+t^2)^2}}$$

$$= \frac{4mt}{(1+t^2)^2} \quad \times \quad \frac{(1+t^2)^2}{-4mt}$$

$$\frac{dy}{dx} = -1 \qquad \text{(Same terms cancels out)}$$

Exercise 31

1. If $y = t^3 - 5$ and $x = t^2 + 1$ find $\dfrac{dy}{dx}$

2. Find $\dfrac{dy}{dx}$ of the functions below which are expressed in the parametric form:

   $x = \dfrac{1}{4 - t^3}$ and $y = 3 + t^2$

3. If $V = \dfrac{1}{3}\pi r^2$ and $A = \pi r l + \pi r^2$, find $\dfrac{dV}{dA}$

4. Find the derivative of the curve defined by the equations: $x = 5t^2 - t + 3$ and $y = t^2 - t - 1$.

5. Given that $F = \dfrac{m(v - u)}{t}$ and $s = \dfrac{(u+v)t}{2}$. Find $\dfrac{dF}{ds}$ if u, v and m are constants.

6. If $y = \dfrac{5}{t^2 - 1}$ and $x = \dfrac{3}{t^4 + 1}$, find $\dfrac{dy}{dx}$

7. The parametric equations of the motion of a stone are: $y = t - 3t^4$ and $x = t^2 + 3$. Find $\dfrac{dy}{dx}$.

8. If the parametric equations of a parabola are $y = \dfrac{at^3}{2}$ and $x = \dfrac{at}{5}$ where a is a constant, find $\dfrac{dy}{dx}$.

9. If $V = \dfrac{1}{3}\pi r^3$ and $A = 4\pi r^2$, find $\dfrac{dA}{dV}$

10. Given that $G = 2m + s^2$ and $H = m^2 - \dfrac{3}{s}$. Find $\dfrac{dG}{dH}$ if m is a constant.

CHAPTER 32
DERIVATIVE OF IMPLICIT FUNCTIONS

In the function y = f(x), y is said to be expressed explicitly in terms of x. However, in expressions such as $2xy - x^2y = 5$, the relationship between y and x is said to be implicit.

In order to differentiate implicit functions, y is differentiated just like x but with the addition of $\frac{dy}{dx}$ along with the value obtained.

Examples

1. Differentiate implicitly, the expression: $2x^2 + y^2 = 9$.

Solution

$$2x^2 + y^2 = 9$$

Follow the rule of differentiation and add $\frac{dy}{dx}$ to the value obtained whenever you differentiate y. Hence we differentiate each term in the expression above as follows:

$$\frac{d(2x^2)}{dx} + \frac{d(2y^2)}{dx} = \frac{d(9)}{dx}$$

$$4x + 2y\frac{dy}{dx} = 0 \qquad \text{(The derivative of } y^2 \text{ is 2y and the addition of } \frac{dy}{dx} \text{ to 2y gives } 2y\frac{dy}{dx}\text{)}$$

We now make $\frac{dy}{dx}$ the subject of the formula as follows:

$$2y\frac{dy}{dx} = -4x$$

$$\frac{dy}{dx} = \frac{-4x}{2y} \qquad \text{(When both sides are divided by 2y)}$$

$$\frac{dy}{dx} = \frac{-2x}{y}$$

2. If $x^3 + y^3 = 18xy$, find $\frac{dy}{dx}$

Solution

$$x^3 + y^3 = 18xy$$

$$3x^2 + 3y^2\frac{dy}{dx} = (18x \times 1\frac{dy}{dx}) + (y \times 18)$$

Note that 18xy is differentiated by using product rule where 18x is taken as u while y is take as v. Also, the derivative of y is what gave us $1\frac{dy}{dx}$. The above differentiation now simplifies to:

$$3x^2 + 3y^2\frac{dy}{dx} = 18x\frac{dy}{dx} + 18y$$

Collect terms in $\frac{dy}{dx}$ on one side in order to make $\frac{dy}{dx}$ the subject of the formula as follows:

$$3y^2\frac{dy}{dx} - 18x\frac{dy}{dx} = 18y - 3x^2$$

Factorizing the left hand side gives:

$$\frac{dy}{dx}(3y^2 - 18x) = 18y - 3x^2$$

Divide both sides by $3y^2 - 18x$. This gives:

$$\frac{dy}{dx} = \frac{18y - 3x^2}{3y^2 - 18x}$$

$$= \frac{3(6y - x^2)}{3(y^2 - 6x)}$$

$$\frac{dy}{dx} = \frac{6y - x^2}{y^2 - 6x} \qquad \text{(3 cancels out)}$$

3. Find $\frac{dy}{dx}$ given that $x^2y^2 - 3xy + 4xy^3 = 4$

<u>Solution</u>

$$x^2y^2 - 3xy + 4xy^3 = 4$$

Apply product rule to x^2y^2, $3xy$ and $4xy^3$ and differentiate appropriately as follows:

$$(x^2 \times 2y\frac{dy}{dx}) + (y^2 \times 2x) - [(3x \times 1\frac{dy}{dx}) + (y \times 3)] + (4x \times 3y^2\frac{dy}{dx}) + (y^3 \times 4) = 0$$

$$2x^2y\frac{dy}{dx} + 2xy^2 - (3x\frac{dy}{dx} + 3y) + 12xy^2\frac{dy}{dx} + 4y^3 = 0$$

$$2x^2y\frac{dy}{dx} + 2xy^2 - 3x\frac{dy}{dx} - 3y + 12xy^2\frac{dy}{dx} + 4y^3 = 0$$

$$2x^2y\frac{dy}{dx} - 3x\frac{dy}{dx} + 12xy^2\frac{dy}{dx} + 2xy^2 - 3y + 4y^3 = 0$$

$$2x^2y\frac{dy}{dx} - 3x\frac{dy}{dx} + 12xy^2\frac{dy}{dx} = 3y - 2xy^2 - 4y^3$$

Factorizing the left hand side gives:

$$\frac{dy}{dx}(2x^2y - 3x + 12xy^2) = 3y - 2xy^2 - 4y^3$$

$$\frac{dy}{dx} = \frac{3y - 2xy^2 - 4y^3}{2x^2y - 3x + 12xy^2}$$

4. Differentiate $x^4 + 6x^2y^2 - 5 = 0$ implicitly with respect to x.

<u>Solution</u>

$$x^4 + 6x^2y^2 - 5 = 0$$

We differentiate accordingly and apply product rule to $6x^2y^2$ as follows:

$$4x^3 + (6x^2 \times 2y\frac{dy}{dx}) + (y^2 \times 12x) - 0 = 0$$

$$4x^3 + 12x^2y\frac{dy}{dx} + 12xy^2 = 0$$

$$12x^2y\frac{dy}{dx} = -4x^3 - 12xy^2$$

375

$$\frac{dy}{dx} = \frac{-4x^3 - 12xy^2}{12x^2y}$$

$$= \frac{-4x(x^2 + 3y^2)}{12x^2y}$$

$$\frac{dy}{dx} = \frac{-(x^2 + 3y^2)}{3xy} \qquad \text{(4 and } x \text{ cancels out)}$$

5. Find $\frac{dy}{dx}$ if $\frac{x^2}{16} + \frac{y^2}{25} = 1$

Solution

$$\frac{x^2}{16} + \frac{y^2}{25} = 1$$

$$\frac{2x}{16} + \frac{2y}{25}\frac{dy}{dx} = 0$$

$$\frac{2y}{25}\frac{dy}{dx} = -\frac{2x}{16}$$

$$\frac{2y}{25}\frac{dy}{dx} = -\frac{x}{8}$$

$$\frac{dy}{dx} = \frac{-\frac{x}{8}}{\frac{2y}{25}}$$

$$= -\frac{x}{8} \times \frac{25}{2y}$$

$$\frac{dy}{dx} = -\frac{25x}{16y}$$

Exercise 32

1. Differentiate implicitly, the expression: $5x^3 + 3y = y^2$.

2. If $2x^2 + 3y^3 = 10$, find $\frac{dy}{dx}$

3. Find $\frac{dy}{dx}$ given that $xy - 3xy^2 + 4x^3 = 4$

4. Differentiate $2x^2 + xy^2 - 5 = x^3$ implicitly with respect to x.

5. Find $\frac{dy}{dx}$ if $\frac{x^3}{3} - \frac{y^2}{2} = 0$

6. Differentiate implicitly, the expression: $2x^2y + 5y = 1$.

7. If $2x^2y + 4y^3 = 2y$, find $\frac{dy}{dx}$

8. Find $\frac{dy}{dx}$ given that $5y^2 - 4xy = y$

9. Differentiate $x + y^2 - x^2 y^2 = 7$ implicitly with respect to x.

10. Find $\dfrac{dy}{dx}$ if $x^5 - \dfrac{3}{y} = 2$

CHAPTER 33
DERIVATIVE OF TRIGONOMETRIC FUNCTIONS

The derivatives of trigonometric functions are as given below:

If $y = \sin x$, then $\dfrac{dy}{dx} = \cos x$

If $y = \cos x$, then $\dfrac{dy}{dx} = -\sin x$

If $y = \tan x$, then $\dfrac{dy}{dx} = \sec^2 x$

If $y = \cot x$, then $\dfrac{dy}{dx} = -\text{cosec}^2 x$

If $y = \sec x$, then $\dfrac{dy}{dx} = \sec x \tan x$

If $y = \text{cosec} x$, then $\dfrac{dy}{dx} = -\cot x \, \text{cosec} x$

Example

1. Find the derivative of $\cos 5x$

<u>Solution</u>

Let $y = \cos 5x$

We have to use chain rule from composite function due to $5x$ which is a function of x.

Hence, let $u = 5x$

Therefore, $y = \cos u$ (By replacing $5x$ with u)

$\dfrac{du}{dx} = 5$

$\dfrac{dy}{du} = -\sin u$ (Recall that the derivative of $\cos x$ is $-\sin x$)

Therefore, $\dfrac{dy}{dx} = \dfrac{dy}{du} \times \dfrac{du}{dx}$ (Chain rule)

$\qquad = -\sin u \times 5$

$\qquad = -5\sin u$

$\dfrac{dy}{dx} = -5\sin 5x$ (Since u = 5x)

2. If $y = \sin \dfrac{1}{2} x$, find $\dfrac{dy}{dx}$

<u>Solution</u>

$y = \sin \dfrac{1}{2} x$

Let $u = \dfrac{1}{2} x$

Therefore, $y = \sin u$ (By replacing $\dfrac{1}{2} x$ with u)

$$\frac{du}{dx} = \frac{1}{2}$$

$$\frac{dy}{du} = \cos u \quad \text{(Recall that the derivative of } \sin x \text{ is } \cos x\text{)}$$

Therefore, $\dfrac{dy}{dx} = \dfrac{dy}{du} \times \dfrac{du}{dx}$ (Chain rule)

$$= \cos u \times \frac{1}{2}$$

$$= \frac{1}{2}\cos u$$

$$\frac{dy}{dx} = \frac{1}{2}\cos \frac{1}{2}x \quad \text{(Since } u = \frac{1}{2}x\text{)}$$

3. Find the derivative of $5\cos 3x$

<u>Solution</u>

    $y = 5\cos 3x$

Let $u = 3x$

Therefore, $y = 5\cos u$

$$\frac{du}{dx} = 3$$

$$\frac{dy}{du} = 5(-\sin u) \quad \text{(Note that the constant term i.e. 5 should be used to multiply the derivative)}$$

$$= -5\sin u$$

Therefore, $\dfrac{dy}{dx} = \dfrac{dy}{du} \times \dfrac{du}{dx}$

$$= -5\sin u \times 3$$

$$= -15\sin u$$

$$\frac{dy}{dx} = -15\sin 3x \quad \text{(Since } u = 3x\text{)}$$

4. Find the derivative of $\sin^2 x$

<u>Solution</u>

    $y = \sin^2 x$

Note that $\sin^2 x = \sin x \times \sin x$

Hence, let $u = \sin x$

Therefore, $y = u^2$ (i.e. $u \times u$ from $\sin x \times \sin x$)

$$\frac{du}{dx} = \cos x$$

$$\frac{dy}{du} = 2u$$

Therefore, $\dfrac{dy}{dx} = \dfrac{dy}{du} \times \dfrac{du}{dx}$

$$= 2u \times \cos x$$

$$= 2u\cos x$$

$$\frac{dy}{dx} = 2\sin x\cos x \qquad \text{(Since u = sinx)}$$

5. Differentiate with respect to x: y = tan2x

<u>Solution</u>

$$y = \tan 2x$$

$$\frac{dy}{dx} = \sec^2 2x \ \times \ \frac{d(2x)}{dx} \qquad \text{(Note that the derivative of tanx is sec}^2x\text{)}$$

$$= \sec^2 2x \ \times \ 2$$

$$= 2\sec^2 2x$$

6. If y = $\cos^3 6x^2$, differentiate y with respect to x.

<u>Solution</u>

$$y = \cos^3 6x^2$$

Let us solve this problem without the use of v as follows:

Let u = $\cos 6x^2$

Hence, y = u^3

$$\frac{du}{dx} = 12x(-\sin 6x^2) \qquad \text{(Note that 12}x\text{ is from the derivative of 6}x^2\text{)}$$

$$= -12x\sin 6x^2$$

$$\frac{dy}{du} = 3u^2$$

Therefore, $\dfrac{dy}{dx} = \dfrac{dy}{du} \ \times \ \dfrac{du}{dx}$

$$= 3u^2 \ \times \ (-12x\sin 6x^2)$$

$$= -36xu^2\sin 6x^2$$

$$= -36x(\cos 6x^2)^2\sin 6x^2 \qquad \text{(u has been replaced with cos6}x^2\text{)}$$

$$\frac{dy}{dx} = -36x\cos^2 6x^2\sin 6x^2$$

7. Find the derivative of y = sec3x

<u>Solution</u>

$$y = \sec 3x$$

$$\frac{dy}{dx} = \sec 3x\tan 3x \ \times \ \frac{d(3x)}{dx} \qquad \text{(Note that the derivative of secx is secxtanx)}$$

$$= \sec 3x\tan 3x \ \times \ 3$$

$$\frac{dy}{dx} = 3\sec 3x\tan 3x$$

8. Find the derivative of cosec4x^3

Solution

$$y = \mathrm{cosec}\,4x^3$$

$$\frac{dy}{dx} = -\mathrm{cosec}\,4x^3\cot 4x^3 \times \frac{d(4x^3)}{dx}$$ (Note that the derivative of cosec is −cosec cot)

$$= -\mathrm{cosec}\,4x^3\cot 4x^3 \times 12x^2$$

$$= -12x^2\mathrm{cosec}\,4x^3\cot 4x^3$$

9. Find the derivative of $\cot^2 2x^4$

Solution

$$y = \cot^2 2x^4$$

This can also be written as: $y = \cot 2x^4 \times \cot 2x^4$

Let $u = 2x^4$ (Take the function of x)

Also, let $v = \cot u$ (Take a function of u without taking the exponent)

Hence, $y = v^2$ (Since $v^2 = (\cot u)^2 = (\cot 2x^4)^2 = \cot^2 2x^4$. Hence, $y = v^2$)

$$\frac{du}{dx} = 8x^3$$

$$\frac{dv}{du} = -\mathrm{cosec}^2 u$$

$$\frac{dy}{dv} = 2v$$

Therefore, $\dfrac{dy}{dx} = \dfrac{dy}{dv} \times \dfrac{dv}{du} \times \dfrac{du}{dx}$

$$= 2v \times -\mathrm{cosec}^2 u \times 8x^3$$

$$= -16x^3\, v\, \mathrm{cosec}^2 u$$

$$= -16x^3 \cot u\, \mathrm{cosec}^2 u$$ (Since v = cot u)

Substituting in the original value of u gives:

$$\frac{dy}{dx} = -16x^3\cot 2x^4\mathrm{cosec}^2 2x^4$$

10. Find the derivative of $5x\sin 2x$

Solution

$$y = 5x\sin 2x$$

We are going to apply the product rule of differentiation.

Let $u = 5x$

and $v = \sin 2x$

$$\frac{du}{dx} = 5$$

$$\frac{dv}{dx} = 2\cos 2x$$ (Note that the differentiation of $2x$ gives 2)

Hence, $\dfrac{dy}{dx} = u\dfrac{dv}{dx} + v\dfrac{du}{dx}$

$$= (5x \times 2\cos 2x) + (\sin 2x \times 5)$$

$$= 10x\cos 2x + 5\sin 2x$$

$$\frac{dy}{dx} = 5(2x\cos 2x + \sin 2x)$$

11. Find $\frac{dy}{dx}$ if $y = \frac{1}{x}\sec x$

Solution

$$y = \frac{1}{x}\sec x$$

We apply product rule as follows:

$$u = \frac{1}{x}$$

$$= x^{-1}$$

$$v = \sec 3x$$

$$\frac{du}{dx} = -1x^{-2}$$

$$= \frac{-1}{x^2}$$

$$\frac{dv}{dx} = \sec 3x\tan 3x \times 3$$

$$= 3\sec 3x\tan 3x$$

Note that the derivative of $\sec x$ is $\sec x \tan x$ and 3 is from the derivative of $3x$

Hence, $\frac{dy}{dx} = u\frac{dv}{dx} + v\frac{du}{dx}$

$$= \frac{1}{x}(3\sec 3x\tan 3x) + \sec 3x\left(\frac{-1}{x^2}\right)$$

$$\frac{dy}{dx} = \frac{3}{x}\sec 3x\tan 3x - \frac{1}{x^2}\sec 3x$$

12. Find the derivative of $3\csc x^6$

Solution

$$y = 3\csc x^6$$

Let $u = x^6$

Hence, $y = 3\csc u$

$$\frac{du}{dx} = 6x^5$$

$$\frac{dy}{du} = -3\cot u \csc u \quad \text{(The constant term i.e. 5 should be used to multiply the derivative)}$$

Therefore, $\frac{dy}{dx} = \frac{dy}{du} \times \frac{du}{dx}$

$$= -3\cot u \csc u \times 6x^5$$

$$= -18x^5 \cot u \csc u$$

$$\frac{dy}{dx} = -18x^5 \cot x^6 \csc x^6 \quad \text{(Since } u = x^6\text{)}$$

13. If $y = \dfrac{1 + \cos 2x}{\sin 2x}$ find $\dfrac{dy}{dx}$.

Solution

$$y = \frac{1 + \cos 2x}{\sin 2x}$$

We have to apply quotient rule on this as follows:

$u = 1 + \cos 2x$

$v = \sin 2x$

$\dfrac{du}{dx} = -2\sin 2x$

$\dfrac{dv}{dx} = 2\cos 2x$

Hence, $\dfrac{dy}{dx} = \dfrac{v\frac{du}{dx} - u\frac{dv}{dx}}{v^2}$ (Quotient rule)

$$= \frac{\sin 2x(-2\sin 2x) - (1 + \cos 2x)(2\cos 2x)}{(\sin 2x)^2}$$

$$= \frac{-2\sin^2 2x - (2\cos 2x + 2\cos^2 2x)}{\sin^2 2x} \qquad \text{(Note that } (\sin 2x)^2 = \sin^2 2x)$$

$$= \frac{-2\sin^2 2x - 2\cos 2x - 2\cos^2 2x}{\sin^2 2x}$$

$$= \frac{-2\sin^2 2x - 2\cos^2 2x - 2\cos 2x}{\sin^2 2x}$$

$$= \frac{-2(\sin^2 2x + \cos^2 2x) - 2\cos 2x}{\sin^2 2x}$$

$$= \frac{-2(1) - 2\cos 2x}{\sin^2 2x} \qquad \text{(Note that } \sin^2 x + \cos^2 x = 1, \text{ hence } \sin^2 2x + \cos^2 2x = 1)$$

$$\frac{dy}{dx} = \frac{-2(1 + \cos 2x)}{\sin^2 2x}$$

14. Differentiate $y = \dfrac{\sin^2 x}{x}$

Solution

$$y = \frac{\sin^2 x}{x}$$

This is also quotient rule.

Therefore, $u = \sin^2 x$

$v = x$

Let us follow a direct and systematic way of differentiating trigonometric functions.

Hence, in order to differentiate $\sin 2x$:

First, differentiate the exponent of $\sin^2$ without changing the trigonometric term. This gives:

$2\sin^{2-1}$

$= 2\sin$

Then add the term in x. This gives:

   $2\sin x$

The next step is to differentiate sin which gives cos. Also add the term in x to obtain $\cos x$.

Finally, differentiate the term in x. Hence, differentiating x gives 1.

Now multiply the three terms obtained in the three steps above. This gives:

   $2\sin x \; \times \; \cos x \; \times \; 1$

   $= 2\sin x \cos x$

Hence, $\dfrac{du}{dx} = 2\sin x \cos x \qquad$ (As obtained above)

Since, $v = x$

Then, $\dfrac{dv}{dx} = 1$

Hence, $\dfrac{dy}{dx} = \dfrac{v\dfrac{du}{dx} - u\dfrac{dv}{dx}}{v^2}$

$$= \frac{x(2\sin x \cos x) - \sin^2 x(1)}{x^2}$$

$$= \frac{2x\sin x \cos x - \sin^2 x}{x^2}$$

$$= \frac{\sin x(2x\cos x - \sin x)}{x^2}$$

15. Differentiate with respect to x: $\dfrac{\sec 2x}{x^3 + 1}$

Solution

   $y = \dfrac{\sec 2x}{x^3 + 1}$

   $u = \sec 2x$

   $v = x^3 + 1$

Hence, $\dfrac{du}{dx} = 2\sec 2x \tan 2x$

Then, $\dfrac{dv}{dx} = 3x^2$

Hence, $\dfrac{dy}{dx} = \dfrac{v\dfrac{du}{dx} - u\dfrac{dv}{dx}}{v^2}$

$$= \frac{(x^3 + 1)(2\sec 2x \tan 2x) - \sec 2x(3x^2)}{(x^3 + 1)^2}$$

$$= \frac{(x^3 + 1)(2\sec 2x \tan 2x) - 3x^2\sec 2x}{(x^3 + 1)^2}$$

16. Find the derivative of $\dfrac{\cos \sqrt{x}}{1 + x}$

Solution

$$y = \frac{\cos \sqrt{x}}{1 + x}$$

$$u = \cos\sqrt{x}$$

$$= \cos x^{\frac{1}{2}}$$

$$v = 1 + x$$

Hence, $\dfrac{du}{dx} = \dfrac{d(x^{\frac{1}{2}})}{dx} \ \times \ \dfrac{d(\cos)}{dx}$

Note that $\dfrac{d(\cos)}{dx}$ means the derivative of cos which gives $-\sin$, and then $x^{\frac{1}{2}}$ is added to it to give

$-\sin x^{\frac{1}{2}}$. Hence we continue as follows:

$$\frac{du}{dx} = \frac{1}{2}x^{-\frac{1}{2}} \ \times \ -\sin x^{\frac{1}{2}}$$

$$= \frac{1}{2x^{\frac{1}{2}}} \ \times \ -\sin x^{\frac{1}{2}}$$

$$\frac{du}{dx} = \frac{-\sin x^{\frac{1}{2}}}{2x^{\frac{1}{2}}}$$

Also, $\dfrac{dv}{dx} = 1$

Hence, $\dfrac{dy}{dx} = \dfrac{v\frac{du}{dx} - u\frac{dv}{dx}}{v^2}$

$$= \frac{(1 + x)\left(\dfrac{-\sin x^{\frac{1}{2}}}{2x^{\frac{1}{2}}}\right) - \cos x^{\frac{1}{2}}(1)}{(1 + x)^2}$$

$$= \frac{-\sin \sqrt{x}}{2\sqrt{x}(1+x)} - \frac{\cos \sqrt{x}}{(1 + x)^2} \qquad \text{(When the fractions are separated)}$$

$$\frac{dy}{dx} = \frac{-(1+x)\sin \sqrt{x} - 2\sqrt{x}\cos \sqrt{x}}{2\sqrt{x}(1 + x)^2} \qquad \text{(When the fractions are combined)}$$

17. Find $\dfrac{dy}{dx}$ if $y = \dfrac{1 - x^2}{1 + \cos x}$

Solution

$$y = \frac{1 - x^2}{1 + \cos x}$$

$$u = 1 - x^2$$

$$v = 1 + \cos x$$

$$\frac{du}{dx} = -2x$$

$$\frac{dv}{dx} = -\sin x$$

Hence, $\dfrac{dy}{dx} = \dfrac{v\frac{du}{dx} - u\frac{dv}{dx}}{v^2}$

$$= \dfrac{(1+\cos x)(-2x) - (1-x^2)(-\sin x)}{(1+\cos x)^2}$$

$$\dfrac{dy}{dx} = \dfrac{-2x(1+\cos x) + (1-x^2)\sin x}{(1+\cos x)^2}$$

Note that the negative sign from $-\sin x$ changed the negative sign at the middle to a positive sign since negative sign multiplied by negative sign gives a positive sign.

18. Find the derivative of $8x\sin x^2$

<u>Solution</u>

$\quad y = 8x\sin x^2$

We apply product rule as follows:

$\quad u = 8x$

$\quad \dfrac{du}{dx} = 8$

$\quad v = \sin x^2$

$\quad \dfrac{dv}{dx} = \dfrac{d(\sin)}{dx} \; \text{x} \; \dfrac{d(x^2)}{dx}$

$\qquad = \cos x^2 \; \text{x} \; 2x$

$\qquad = 2x\cos x^2$

Note that $\dfrac{d(\sin)}{dx}$ gives cos which result to $\cos x^2$ when x^2 from the question is added

Hence, $\dfrac{dy}{dx} = u\dfrac{dv}{dx} + v\dfrac{du}{dx}$

$\qquad = 8x(2x\cos x^2) + \sin x^2(8)$

$\qquad = 16x^2\cos x^2 + 8\sin x^2$

$\dfrac{dy}{dx} = 8(2x^2\cos x^2 + \sin x^2)$

19. Differentiate with respect to x: $\sec 2x\sin^3 2x$

Solution

$\quad y = \sec 2x\sin^3 2x$

We apply product rule as follows:

$\quad u = \sec 2x$

$\quad \dfrac{du}{dx} = \dfrac{d(\sec)}{dx} \; \text{x} \; \dfrac{d(2x)}{dx}$

$\qquad = \sec 2x\tan 2x \; \text{x} \; 2$

$\quad \dfrac{du}{dx} = 2\sec 2x\tan 2x$

Note that $\dfrac{d(\sec)}{dx}$ gives sectan, but remember to add $2x$ after sec and tan respectively to obtain sec$2x$tan$2x$

$v = \sin^3 2x$

In order to directly differentiate a trigonometric term with exponent like this (i.e. $\sin^3$) we have three solutions to multiply as follows:

First solution: consider only the exponent and differentiate $\sin^3$. This gives:

$$\frac{d(\sin^3)}{dx} = 3\sin^2$$

We now add $2x$ from the question to obtain $3\sin^2 2x$

Second solution: $\dfrac{d(\sin)}{dx} = \cos$ which gives $\cos 2x$

Third solution: $\dfrac{d(2x)}{dx} = 2$

Multiply theses three solutions to give the derivative of $v = \sin^3 2x$ as follows:

$$\frac{dv}{dx} = 3\sin^2 2x \times \cos 2x \times 2$$

$$= 6\sin^2 2x \cos 2x$$

Now, $\dfrac{dy}{dx} = u\dfrac{dv}{dx} + v\dfrac{du}{dx}$

$= \sec 2x(6\sin^2 2x\cos 2x) + \sin^3 2x(2\sec 2x\tan 2x)$

$= \dfrac{1}{\cos 2x}(6\sin^2 2x\cos 2x) + \sin^3 2x(2\dfrac{1}{\cos 2x}\dfrac{\sin 2x}{\cos 2x})$ (Note that $\sec 2x = \dfrac{1}{\cos 2x}$ and $\tan 2x = \dfrac{\sin 2x}{\cos 2x}$)

$= 6\sin^2 2x + 2\sin^2 2x(\dfrac{\sin 2x}{\cos 2x}\dfrac{\sin 2x}{\cos 2x})$

Note that one $\sin 2x$ has been taken out of $\sin^3 2x$ and placed inside the bracket.

$= 6\sin^2 2x + 2\sin^2 2x(\tan 2x\tan x)$ (Since $\dfrac{\sin 2x}{\cos 2x} = \tan 2x$)

$= 6\sin^2 2x + 2\sin^2 2x(\tan^2 2x)$

$\dfrac{dy}{dx} = 2\sin^2 2x(3 + \tan^2 2x)$ (After factorization)

20. Differentiate with respect to x: $\dfrac{\sec x - \tan x}{\sec x + \tan x}$

Solution

$y = \dfrac{\sec x - \tan x}{\sec x + \tan x}$

$u = \sec x - \tan x$

$\dfrac{du}{dx} = \sec x\tan x - \sec^2 x$

$v = \sec x + \tan x$

$\dfrac{dv}{dx} = \sec x\tan x + \sec^2 x$

$$\frac{dy}{dx} = \frac{v\frac{du}{dx} - u\frac{dv}{dx}}{v^2}$$

$$= \frac{(\sec x + \tan x)(\sec x \tan x - \sec^2 x) - (\sec x - \tan x)(\sec x \tan x + \sec^2 x)}{(\sec x + \tan x)^2}$$

Expanding bracket in the numerator gives:

$$\frac{dy}{dx} = \frac{\sec^2 x \tan x - \sec^3 x + \sec x \tan^2 x - \sec^2 x \tan x - [\sec^2 x \tan x + \sec^3 x - \sec x \tan^2 x - \sec^2 x \tan x]}{(\sec x + \tan x)^2}$$

$$= \frac{\sec^2 x \tan x - \sec^3 x + \sec x \tan^2 x - \sec^2 x \tan x - \sec^2 x \tan x - \sec^3 x + \sec x \tan^2 x + \sec^2 x \tan x]}{(\sec x + \tan x)^2}$$

$$= \frac{2\sec x \tan^2 x - 2\sec^3 x}{(\sec x + \tan x)^2} \qquad \text{(Note that all the } \sec^2 x \tan x \text{ have cancelled out)}$$

$$= \frac{2\sec x(\tan^2 x - \sec^2 x)}{(\sec x + \tan x)^2}$$

$$= \frac{2\sec x(-1)}{(\sec x + \tan x)^2} \qquad \text{(Note that } \tan^2 x - \sec^2 x = -1\text{)}$$

$$= \frac{-2\sec x}{(\sec x + \tan x)^2}$$

$$\frac{dy}{dx} = -\frac{2\sec x}{(\sec x + \tan x)^2}$$

21. Find the derivative of $\sqrt{\dfrac{\cos 2x}{1 + \sin 2x}}$

Solution

$$y = \sqrt{\frac{\cos 2x}{1 + \sin 2x}}$$

This can also be written as:

$$y = \frac{(\cos 2x)^{\frac{1}{2}}}{(1 + \sin 2x)^{\frac{1}{2}}}$$

$$u = (\cos 2x)^{\frac{1}{2}}$$

Using the chain rule, we obtain $\dfrac{du}{dx}$ as follows:

$$\frac{du}{dx} = \frac{1}{2}(\cos 2x)^{\frac{1}{2} - 1} \times \frac{d(\cos 2x)}{dx}$$

$$= \frac{1}{2}(\cos 2x)^{-\frac{1}{2}} \times -2\sin 2x$$

$$= \frac{-2\sin 2x}{2(\cos 2x)^{\frac{1}{2}}}$$

$$= \frac{-\sin 2x}{(\cos 2x)^{\frac{1}{2}}}$$

$$v = (1 + \sin 2x)^{\frac{1}{2}}$$

$$\frac{dv}{dx} = \frac{1}{2}(1+\sin 2x)^{\frac{1}{2}-1} \times \frac{d(1+\sin 2x)}{dx}$$

$$= \frac{1}{2}(1+\sin 2x)^{-\frac{1}{2}} \times 2\cos 2x$$

$$= \frac{2\cos 2x}{2(1+\sin 2x)^{\frac{1}{2}}}$$

$$= \frac{\cos 2x}{(1+\sin 2x)^{\frac{1}{2}}}$$

$$\frac{dy}{dx} = \frac{v\frac{du}{dx} - u\frac{dv}{dx}}{v^2}$$

$$= \frac{(1+\sin 2x)^{\frac{1}{2}}\left(\dfrac{-\sin 2x}{(\cos 2x)^{\frac{1}{2}}}\right) - (\cos 2x)^{\frac{1}{2}}\left(\dfrac{\cos 2x}{(1+\sin 2x)^{\frac{1}{2}}}\right)}{[(1+\sin 2x)^{\frac{1}{2}}]^2}$$

$$= \frac{\left(\dfrac{-\sin 2x(1+\sin 2x)^{\frac{1}{2}}}{(\cos 2x)^{\frac{1}{2}}}\right) - \left(\dfrac{\cos 2x(\cos 2x)^{\frac{1}{2}}}{(1+\sin 2x)^{\frac{1}{2}}}\right)}{1+\sin 2x}$$

$$= \frac{\left(\dfrac{-\sin 2x(1+\sin 2x) - \cos 2x(\cos 2x)}{(\cos 2x)^{\frac{1}{2}}(1+\sin 2x)^{\frac{1}{2}}}\right)}{1+\sin 2x}$$

Note that $(\cos 2x)^{\frac{1}{2}} \times (\cos 2x)^{\frac{1}{2}} = \cos 2x$ (Since the exponents are added). Similarly:

$(1+\sin 2x)^{\frac{1}{2}} \times (1+\sin 2x)^{\frac{1}{2}} = 1+\sin 2x$. Simplifying the above expression further, gives:

$$\frac{dy}{dx} = \frac{-\sin 2x(1+\sin 2x) - \cos^2 2x}{(\cos 2x)^{\frac{1}{2}}(1+\sin 2x)^{\frac{1}{2}}(1+\sin 2x)}$$

$$= \frac{-\sin 2x - \sin^2 2x - \cos^2 2x}{(\cos 2x)^{\frac{1}{2}}(1+\sin 2x)^{\frac{1}{2}}(1+\sin 2x)}$$

$$= \frac{-\sin 2x - (\sin^2 2x + \cos^2 2x)}{(\cos 2x)^{\frac{1}{2}}(1+\sin 2x)^{\frac{1}{2}}(1+\sin 2x)}$$

$$= \frac{-\sin 2x - 1}{(\cos 2x)^{\frac{1}{2}}(1+\sin 2x)^{\frac{1}{2}}(1+\sin 2x)} \qquad \text{(Note that } \sin^2 2x + \cos^2 2x = 1)$$

$$= \frac{-(1+\sin 2x)}{(\cos 2x)^{\frac{1}{2}}(1+\sin 2x)^{\frac{1}{2}}(1+\sin 2x)}$$

$$= \frac{-1}{(\cos 2x)^{\frac{1}{2}}(1+\sin 2x)^{\frac{1}{2}}} \qquad \text{(Note that } 1+\sin 2x \text{ cancels out)}$$

22. Differentiate with respect to x: $\sin x - 2x\cos x$

Solution

$$y = \sin x - 2x\cos x$$

Treat $2x\cos x$ using product rule.

$$\frac{dy}{dx} = \frac{d(\sin x)}{dx} - \left(2x\,\frac{d(\cos x)}{dx} + \cos x\,\frac{d(2x)}{dx}\right)$$
$$= \cos x - [2x(-\sin x) + \cos x(2)]$$
$$= \cos x + 2x\sin x - 2\cos x$$
$$= 2x\sin x - \cos x$$

23. Find the derivative of $\cos^5 3x^4$

Solution

$$y = \cos^5 3x^4$$

A direct way of differentiating this problem is applied as follows:

$$\frac{dy}{dx} = \frac{d(\cos^5)}{dx} \; \text{x} \; \frac{d(\cos)}{dx} \; \text{x} \; \frac{d(3x^4)}{dx}$$
$$= 5\cos^4 3x^4 \; \text{x} \; (-\sin 3x^4) \; \text{x} \; 12x^3$$

Note that the derivative of $\cos^5$ gives $5\cos^4$ (do not change the trigonometric term, i.e. cos) and then the addition of $3x^4$ from the question gives $5\cos^4 3x^4$.

Similarly the derivative of cos gives $-\sin$ and the addition of $3x^4$ from the question gives $-\sin 3x^4$.

Hence, multiplying the terms above gives:

$$\frac{dy}{dx} = -60x^3\cos^4 3x^4 \sin 3x^4$$

24. Differentiate with respect to x: $\sin^8 15x^6$

Solution

$$y = \sin^8 15x^6$$

We can also differentiate this problem directly as follows:

$$\frac{dy}{dx} = \frac{d(\sin^8)}{dx} \; \text{x} \; \frac{d(\sin)}{dx} \; \text{x} \; \frac{d(15x^6)}{dx}$$
$$= 8\sin^7 15x^6 \; \text{x} \; \cos 15x^6 \; \text{x} \; 90x^5$$

Note that the derivative of $\sin^8$ gives $8\sin^7$ (in this case do not change the trigonometric term, i.e. sin) and then the addition of $15x^6$ from the question gives $8\sin^7 15x^6$.

Similarly the derivative of sin gives cos and the addition of $15x^6$ from the question gives $\cos 15x^6$. Hence, we continue as follows:

$$\frac{dy}{dx} = 8\sin^7 15x^6 \; \text{x} \; \cos 15x^6 \; \text{x} \; 90x^5$$
$$= 720x^5\sin^7 15x^6 \cos 15x^6$$

Exercise 33

1. Find the derivative of $\tan 2x$

2. If $y = \cos\dfrac{1}{5}x$, find $\dfrac{dy}{dx}$

3. Find the derivative of $10\cos 5x$

4. Find the derivative of $\sin^3 x$

5. Differentiate with respect to x: $y = \tan^2 x$

6. If $y = \sin^4 3x^5$, differentiate y with respect to x.

7. Find the derivative of $y = \operatorname{cosec} 6x^2$

8. Find the derivative of $\sec 2x^5$

9. Find the derivative of $\tan 3x^2$

10. Find the derivative of $x^2 \cos 3x$

11. Find $\dfrac{dy}{dx}$ if $y = \dfrac{3}{x^3}\sin 2x$

12. Find the derivative of $12\sec x^4$

13. If $y = \dfrac{2 - \sin 5x}{\tan x}$ find $\dfrac{dy}{dx}$.

14. Differentiate $y = \dfrac{\sin^3 x}{2x}$

15. Differentiate with respect to x: $\dfrac{\cot 2x}{x + 3}$

16. Find the derivative of $\dfrac{\sin \sqrt[3]{3}}{5x}$

17. Find $\dfrac{dy}{dx}$ if $y = \dfrac{x^2 - 3}{\sec 2x}$

18. Find the derivative of $3x^2 \cos 3x^2$

19. Differentiate with respect to x: $\cot x \cos^2 x$

20. Differentiate with respect to x: $\dfrac{\sin x - \cos x}{\sin x + \cos x}$

21. Find the derivative of $\dfrac{\sec 4x}{\cos 4x - 2}$

22. Differentiate with respect to x: $\cos 3x - x^2 \sec x$

23. Find the derivative of $\sin^9 x^3$

24. Differentiate with respect to x: $\sin 6x^{\frac{1}{2}}$

25. Find $\dfrac{dy}{dx}$ if $y = \dfrac{1}{x^2}\sin x^3$

26. Find the derivative of $\cos^3 2x^5$

27. If $y = \dfrac{\cos 10x}{\sin 2x}$ find $\dfrac{dy}{dx}$.

28. Differentiate $y = \dfrac{3\sin x^4}{2x}$

391

29. Differentiate with respect to x: $\dfrac{\tan 5x}{2x - 1}$

30. Find the derivative of $\dfrac{\tan^2 x}{2x}$

CHAPTER 34
DERIVATIVE OF INVERSE FUNCTIONS

If the derivative of a function is given by $\dfrac{dy}{dx}$, then the derivative of the inverse function is given

by: $\quad \dfrac{1}{\frac{dy}{dx}} = \dfrac{dx}{dy}$

Or, $\quad \dfrac{dy}{dx} = \dfrac{1}{\frac{dx}{dy}}$

Examples

1. Find $\dfrac{dx}{dy}$ if $y = \sqrt[3]{x}$

Solution

Method 1

$y = \sqrt[3]{x}$

Or $y = x^{\frac{1}{3}}$

Let us make x the subject of the formula. The inverse of $\dfrac{1}{3}$ is 3. Hence raise both sides to the

exponent 3 as follows:

$y^3 = (x^{\frac{1}{3}})^3$

$y^3 = x^1 \qquad$ (Note that 1 was obtained from $\dfrac{1}{3}$ x 3)

Hence, $x = y^3$

Therefore, $\dfrac{dx}{dy} = 3y^2$

Method 2

$y = \sqrt[3]{x}$

Or $y = x^{\frac{1}{3}}$

$\dfrac{dy}{dx} = \dfrac{1}{3} x^{\frac{1}{3} - 1}$

$\qquad = \dfrac{1}{3} x^{-\frac{2}{3}}$

$\dfrac{dy}{dx} = \dfrac{1}{3x^{\frac{2}{3}}}$

Hence, $\dfrac{dx}{dy} = \dfrac{1}{\frac{dy}{dx}}$

$\qquad = \dfrac{3x^{\frac{2}{3}}}{1} \qquad$ (This means the inverse of $\dfrac{1}{3x^{\frac{2}{3}}}$)

393

$$= 3x^{\frac{2}{3}}$$

$$= 3(\sqrt[3]{x})^2 \qquad \text{[Recall from indices that } x^{\frac{a}{b}} = (\sqrt[b]{x})^a]$$

$$\frac{dx}{dy} = 3y^2 \qquad \text{(Since } y = \sqrt[3]{x})$$

2. If $y = \sqrt[5]{2x - 3}$ find $\dfrac{dx}{dy}$

Solution

$$y = \sqrt[5]{2x - 3}$$

Or $y = (2x - 3)^{\frac{1}{5}}$

Let us make x the subject of the formula. The inverse of $\dfrac{1}{5}$ is 5. Hence raise both sides to the exponent 5 as follows:

$$y^5 = [(2x - 3)^{\frac{1}{5}}]^5$$

$$y^5 = 2x - 3 \qquad \text{(Note that } \tfrac{1}{5} \text{ x } 5 = 1, \text{ which cancels the fractional exponent)}$$

Hence, $2x = y^5 + 3$

$$x = \frac{y^5 + 3}{2}$$

$$x = \frac{y^5}{2} + \frac{3}{2} \qquad \text{(When we separate into fractions by dividing each part by the denominator)}$$

Therefore, $\dfrac{dx}{dy} = \dfrac{5y^4}{2}$

3. If $y = x^3 - 5$. Find $\dfrac{dx}{dy}$

Solution

$$y = x^3 - 5$$
$$y + 5 = x^3$$
$$x^3 = y + 5$$

$$x = (y + 5)^{\frac{1}{3}} \qquad \text{(By raising both sides to an exponent of the inverse of 3 which is } \tfrac{1}{3})$$

$$\frac{dx}{dy} = \frac{1}{3}(y + 5)^{\frac{1}{3} - 1} \text{ x } 1$$

Note that the 1 was obtained from the derivative of y + 5 since chain rule was used.

$$\frac{dx}{dy} = \frac{1}{3}(y + 5)^{-\frac{2}{3}}$$

$$\frac{dx}{dy} = \frac{1}{3(y+5)^{\frac{2}{3}}}$$

Recall from indices that $x^{\frac{a}{b}} = (\sqrt[b]{x})^a$. Applying this rule gives:

$$\frac{dx}{dy} = \frac{1}{3[\sqrt[3]{(y+5)^2}]}$$

4. If $y = \frac{1}{2}x^4 + 3$, find $\frac{dx}{dy}$

Solution

$$y = \frac{1}{2}x^4 + 3$$

$$y - 3 = \frac{1}{2}x^4$$

$$2(y - 3) = x^4$$

$$x^4 = 2y - 6$$

$x = (2y - 6)^{\frac{1}{4}}$ (This is obtained by raising both sided to an exponent of the inverse of 4, i.e. $\frac{1}{4}$)

Hence we use chain rule to determine $\frac{dx}{dy}$ as follows:

$$\frac{dx}{dy} = \frac{1}{4}(2y - 6)^{\frac{1}{4}-1} \times 2$$ (Note that 2 is from the derivative of $2y - 6$)

$$\frac{dx}{dy} = \frac{1}{2}(2y - 6)^{-\frac{3}{4}}$$

$$\frac{dx}{dy} = \frac{1}{2(2y - 6)^{\frac{3}{4}}}$$

Hence, $\frac{dx}{dy} = \frac{1}{2[\sqrt[4]{(2y-6)^3}]}$ [This is obtained from the law of indices given by: $x^{\frac{a}{b}} = (\sqrt[b]{x})^a$]

5. If $y = \frac{x + 2}{x}$, find $\frac{dx}{dy}$.

Solution

$$y = \frac{x + 2}{x}$$

$xy = x + 2$ (When we cross multiply)

$xy - x = 2$

Factorizing the left hand side gives:

$x(y - 1) = 2$

$$x = \frac{2}{y - 1}$$

$x = 2(y - 1)^{-1}$ (Take note of the use of negative exponent when the denominator goes up)

Hence we use chain rule to determine $\frac{dx}{dy}$ as follows:

$$\frac{dx}{dy} = -1 \times 2(y - 1)^{-1-1} \times 1$$ (Note that 1 is from the derivative of $y - 1$)

$$\frac{dx}{dy} = -2(y-1)^{-2}$$

$$\frac{dx}{dy} = \frac{-2}{(y-1)^2}$$

6. Find $\frac{dx}{dy}$ if $y = \frac{1}{x+2}$

Solution

$$y = \frac{1}{x+2}$$

$$x + 2 = \frac{1}{y}$$

$$x = \frac{1}{y} - 2$$

$$x = y^{-1} - 2$$

$$\frac{dx}{dy} = -1 \times y^{-1-1}$$

$$= -y^{-2}$$

$$\frac{dx}{dy} = \frac{-1}{y^2}$$

7. Find $\frac{dx}{dy}$ if $y = x^{\frac{2}{3}}$

Solution

$$y = x^{\frac{2}{3}}$$

Raise both sides to the exponent $\frac{3}{2}$ i.e. the inverse of $\frac{2}{3}$. This gives:

$$y^{\frac{3}{2}} = (x^{\frac{2}{3}})^{\frac{3}{2}}$$

$$y^{\frac{3}{2}} = x \quad \text{(Note that } \frac{2}{3} \times \frac{3}{2} = 1, \text{ and } x^1 = x)$$

$$x = y^{\frac{3}{2}}$$

$$\frac{dx}{dy} = \frac{3}{2} \times y^{\frac{3}{2}-1}$$

$$= \frac{3}{2} y^{\frac{1}{2}}$$

$$\frac{dx}{dy} = \frac{3\sqrt{y}}{2}$$

8. If $y = (5x + 7)^{\frac{3}{10}}$ find $\frac{dx}{dy}$.

Solution

$$y = (5x + 7)^{\frac{3}{10}}$$

Raise both sides to the exponent $\frac{10}{3}$ i.e. the inverse of $\frac{3}{10}$. This gives:

$$y^{\frac{10}{3}} = [(5x+7)^{\frac{3}{10}}]^{\frac{10}{3}}$$

$$y^{\frac{10}{3}} = 5x + 7 \quad \text{(Note that } \frac{3}{10} \times \frac{10}{3} = 1)$$

$$y^{\frac{10}{3}} - 7 = 5x$$

$$x = \frac{y^{\frac{10}{3}} - 7}{5}$$

$$= \frac{y^{\frac{10}{3}}}{5} - \frac{7}{5}$$

$$x = \frac{1}{5} y^{\frac{10}{3}} - \frac{7}{5}$$

$$\frac{dx}{dy} = \frac{10}{3} \times \frac{1}{5} y^{\frac{10}{3} - 1}$$

$$= \frac{2}{3} y^{\frac{7}{3}}$$

$$\frac{dx}{dy} = \frac{2\sqrt[3]{y^7}}{3}$$

Exercise 34

1. Find $\frac{dx}{dy}$ if $y = 7x^3$

2. If $y = \sqrt[3]{1+x^2}$ find $\frac{dx}{dy}$

3. If $y = 2x^5 - 3$. Find $\frac{dx}{dy}$

4. If $y = \frac{2}{3}x^3 - 9$, find $\frac{dx}{dy}$

5. If $y = \frac{2x+1}{x}$, find $\frac{dx}{dy}$.

6. Find $\frac{dx}{dy}$ if $y = \frac{3}{x^2 - 5}$

7. Find $\frac{dx}{dy}$ if $y = 2x^{\frac{1}{4}}$

8. If $y = (x-3)^{\frac{1}{5}}$ find $\frac{dx}{dy}$.

9. If $y = \dfrac{4x+5}{x}$, find $\dfrac{dx}{dy}$.

10. Find $\dfrac{dx}{dy}$ if $y = \dfrac{1}{x^3+8}$

CHAPTER 35
DERIVATIVES OF INVERSE TRIGONOMETRIC FUNCTIONS

Recall that if $\sin x = y$, then $x = \sin^{-1}y$. This is referred to as inverse trigonometric function. The derivatives of inverse trigonometric functions are given below.

If $y = \sin^{-1}x$, then $\dfrac{dy}{dx} = \dfrac{1}{\sqrt{1-x^2}}$

If $y = \cos^{-1}x$, then $\dfrac{dy}{dx} = \dfrac{-1}{\sqrt{1-x^2}}$

If $y = \tan^{-1}x$, then $\dfrac{dy}{dx} = \dfrac{1}{1+x^2}$

If $y = \cot^{-1}x$, then $\dfrac{dy}{dx} = \dfrac{-1}{1+x^2}$

If $y = \sec^{-1}x$, then $\dfrac{dy}{dx} = \dfrac{1}{x\sqrt{x^2-1}}$

If $y = \operatorname{cosec}^{-1}x$, then $\dfrac{dy}{dx} = \dfrac{-1}{x\sqrt{x^2-1}}$

Note that $\sin^{-1}x$ can also be written as $\arcsin x$. Other inverse function can be written in a similar way.

Examples

1. If $y = \cos^{-1}x$ find the $\dfrac{dy}{dx}$.

Solution

$\quad y = \cos^{-1}x$

$\dfrac{dy}{dx} = \dfrac{-1}{\sqrt{1-x^2}}$

2. Find the derivative of $\sin^{-1}3x$

Solution

$\quad y = \sin^{-1}3x$

Let us use the chain rule to solve this problem

Let $u = 3x$

Hence, $y = \sin^{-1}u$

$\quad \dfrac{du}{dx} = 3$

$\quad \dfrac{dy}{du} = \dfrac{1}{\sqrt{1-u^2}}$

Therefore, $\dfrac{dy}{dx} = \dfrac{dy}{du} \times \dfrac{du}{dx}$ (Chain rule)

$$= \frac{1}{\sqrt{1-u^2}} \times 3$$

$$= \frac{3}{\sqrt{1-u^2}}$$

$$= \frac{3}{\sqrt{1-(3x)^2}} \qquad \text{(Since } u = 3x\text{)}$$

$$\frac{dy}{dx} = \frac{3}{\sqrt{1-9x^2}}$$

3. Find the derivative of $\cot^{-1}x^2$

Solution

$$y = \cot^{-1}x^2$$

Let $u = x^2$

Hence, $y = \cot^{-1}u$

$$\frac{du}{dx} = 2x$$

$$\frac{dy}{du} = \frac{-1}{1+u^2}$$

Therefore, $\dfrac{dy}{dx} = \dfrac{dy}{du} \times \dfrac{du}{dx}$

$$= \frac{-1}{1+u^2} \times 2x$$

$$= \frac{-2x}{1+u^2}$$

$$= \frac{-2x}{1+(x^2)^2} \qquad \text{(since } u = x^2\text{)}$$

$$\frac{dy}{dx} = \frac{-2x}{1+x^4}$$

4. If $y = \sec^{-1}2x^3$, find $y = \dfrac{dy}{dx}$

Solution

$$y = \sec^{-1}2x^3$$

Let $u = 2x^3$

Hence, $y = \sec^{-1}u$

$$\frac{du}{dx} = 6x^2$$

$$\frac{dy}{du} = \frac{1}{u\sqrt{u^2-1}}$$

Therefore, $\dfrac{dy}{dx} = \dfrac{dy}{du} \times \dfrac{du}{dx}$

$$= \frac{1}{u\sqrt{u^2-1}} \times 6x^2$$

$$= \frac{6x^2}{u\sqrt{u^2-1}}$$

$$= \frac{6x^2}{2x^3\sqrt{(2x^3)^2-1}} \qquad \text{(since } u = 2x^3\text{)}$$

$$\frac{dy}{dx} = \frac{3}{x\sqrt{4x^6-1}} \qquad \text{(Note that } \frac{6x^2}{2x^3} = \frac{3}{x}\text{)}$$

5. Find the derivative $x^2\tan^{-1}x$

<u>Solution</u>

$$y = x^2\tan^{-1}x$$

we apply the product rule as follows:

$$u = x^2$$

$$v = \tan^{-1}x$$

$$\frac{du}{dx} = 2x$$

$$\frac{dv}{dx} = \frac{1}{1+x^2}$$

$$\frac{dy}{dx} = u\frac{dv}{dx} + v\frac{du}{dx} \qquad \text{(product rule)}$$

$$= x^2\left(\frac{1}{1+x^2}\right) + \tan^{-1}x(2x)$$

$$\frac{dy}{dx} = \frac{x^2}{1+x^2} + 2x\tan^{-1}x$$

6. If $y = 3x - 1\,\cosec^{-1}x^3$, find $\dfrac{dy}{dx}$.

<u>Solution</u>

$$y = 3x - 1\,\cosec^{-1}x^3$$

$$u = 3x - 1$$

$$\frac{du}{dx} = 3$$

$$v = \cosec^{-1}x^3$$

$$\frac{dv}{dx} = \frac{-1}{x^3\sqrt{(x^3)^2-1}} \times 3x^2 \qquad \text{(From use of chain rule. Note that } 3x^2 \text{ is from the derivative of } x^3\text{)}$$

$$= \frac{-3x^2}{x^3\sqrt{x^6-1}}$$

$$\frac{dv}{dx} = \frac{-3}{x\sqrt{x^6-1}}$$

$$\frac{dy}{dx} = u\frac{dv}{dx} + v\frac{du}{dx} \qquad \text{(product rule)}$$

$$= 3x - 1 \left(\frac{-3}{x\sqrt{x^6 - 1}} \right) + \text{cosec}^{-1}x^3(3)$$

$$\frac{dy}{dx} = \frac{-3(3x - 1)}{x\sqrt{x^6 - 1}} + 3\text{cosec}^{-1}x^3$$

7. Find the derivative of $\cos^{-1}(5x - 3)$

Solution

$$y = \cos^{-1}(5x - 3)$$

Let $u = 5x - 3$

Hence, $y = \cos^{-1}u$

$$\frac{du}{dx} = 5$$

$$\frac{dy}{du} = \frac{-1}{\sqrt{1 - u^2}}$$

Therefore, $\dfrac{dy}{dx} = \dfrac{dy}{du} \times \dfrac{du}{dx}$

$$= \frac{-1}{\sqrt{1 - u^2}} \times 5$$

$$= \frac{-5}{\sqrt{1 - u^2}}$$

$$= \frac{-5}{\sqrt{1 - (5x - 3)^2}} \qquad \text{(Since } u = 5x - 3\text{)}$$

Expanding the bracket gives:

$$\frac{dy}{dx} = \frac{-5}{\sqrt{1 - (25x^2 - 15x - 15x + 9)}}$$

$$= \frac{-5}{\sqrt{1 - 25x^2 + 15x + 15x - 9}}$$

$$\frac{dy}{dx} = \frac{-5}{\sqrt{-25x^2 + 30x - 8}}$$

8. If $y = \tan^{-1}\left(\dfrac{1}{m}\right)$, find $\dfrac{dy}{dm}$

Solution

$$y = \tan^{-1}\left(\frac{1}{m}\right)$$

Let $u = \dfrac{1}{m} = m^{-1}$

Hence, $y = \tan^{-1}u$

$$\frac{du}{dm} = -1m^{-2}$$

$$\frac{du}{dm} = \frac{-1}{m^2}$$

$$\frac{dy}{du} = \frac{1}{1 + u^2}$$

Therefore, $\dfrac{dy}{dm} = \dfrac{dy}{du} \times \dfrac{du}{dm}$

$= \dfrac{1}{1+u^2} \times \dfrac{-1}{m^2}$

$= \dfrac{1}{1+\left(\frac{1}{m}\right)^2} \times \dfrac{-1}{m^2}$

$= \dfrac{-1}{m^2\left[1+\left(\frac{1}{m}\right)^2\right]}$

$= \dfrac{-1}{m^2\left(1+\frac{1}{m^2}\right)}$

$\dfrac{dy}{dm} = \dfrac{-1}{m^2+1}$

9. If $y = \sin 3x$, find $\dfrac{dx}{dy}$.

Solution

$y = \sin 3x$

$\sin^{-1}y = 3x$ (Recall that if $\sin a = b$, then $a = \sin^{-1}b$)

$x = \dfrac{\sin^{-1}y}{3}$

$x = \dfrac{1}{3}\sin^{-1}y$

Hence, $\dfrac{dx}{dy} = \dfrac{1}{3}\dfrac{1}{\sqrt{1-y^2}}$

$\dfrac{dx}{dy} = \dfrac{1}{3\sqrt{1-y^2}}$

Note that this example asked us to find $\dfrac{dx}{dy}$ and not $\dfrac{dy}{dx}$

10. If $y = (\cos 5x)^2$ find $\dfrac{dx}{dy}$.

Solution

$y = (\cos 5x)^2$

This can also be written as $\cos^2 5x$. Hence:

$y = \cos^2 5x$.

Or, $\cos^2 5x = y$

taking the square root of both sides gives:

$\cos 5x = \sqrt{y}$

$5x = \cos^{-1}\sqrt{y}$

$$x = \frac{\cos^{-1}\sqrt{y}}{5}$$

$$= \frac{1}{5}\cos^{-1}\sqrt{y}$$

$$x = \frac{1}{5}\cos^{-1}y^{\frac{1}{2}}$$

Hence, $\dfrac{dx}{dy} = \dfrac{1}{5} \cdot \dfrac{-1}{\sqrt{1-(y^{\frac{1}{2}})^2}} \times \dfrac{1}{2}y^{-\frac{1}{2}}$ (Note that $\dfrac{1}{2}y^{-\frac{1}{2}}$ is from the derivative of $y^{\frac{1}{2}}$)

$$= \frac{-1}{5\sqrt{1-y}} \times \frac{1}{2y^{\frac{1}{2}}}$$

$$= \frac{-1}{10y^{\frac{1}{2}}\sqrt{1-y}}$$

$$= \frac{-1}{10\sqrt{y}\sqrt{1-y}}$$

$$\frac{dx}{dy} = \frac{-1}{10\sqrt{y(1-y)}}$$

Exercise 35

1. If $y = \sin^{-1}2x$ find the $\dfrac{dy}{dx}$.

2. Find the derivative of $\cos^{-1}x$

3. Find the derivative of $\cot^{-1}3x^2$

4. If $y = \cot^{-1}x^4$, find $y = \dfrac{dy}{dx}$

5. Find the derivative $3x\sec^{-1}x$

6. If $y = x^2 + 3\tan^{-1}x$, find $\dfrac{dy}{dx}$.

7. Find the derivative of $\tan^{-1}(x + 5)$

8. If $y = \cos^{-1}\left(\dfrac{1}{x}\right)$, find $\dfrac{dy}{dx}$

9. If $y = \cos5x$, find $\dfrac{dx}{dy}$.

10. If $y = (\sec x)^2$ find $\dfrac{dx}{dy}$.

11. Find the derivative $5x\tan^{-1}3x$

12. If $y = x^2 + 4\sin^{-1}x^5$, find $\dfrac{dy}{dx}$.

13. Find the derivative of $\csc^{-1}x^3$

14. If $y = \tan 2x$, find $\dfrac{dx}{dy}$.

15. If $y = \sin^3 5x$, find $\dfrac{dx}{dy}$.

CHAPTER 36
DERIVATIVES OF HYPERBOLIC FUNCTIONS

Hyperbolic functions are functions in calculus which are expressed by the combination of the exponential functions e^x and e^{-x}. The derivatives of the six main hyperbolic functions are as given below.

1. If $y = \sinh x$, then $\dfrac{dy}{dx} = \cosh x$

2. If $y = \cosh x$, then $\dfrac{dy}{dx} = \sinh x$

3. If $y = \tanh x$, then $\dfrac{dy}{dx} = \text{sech}^2 x$

4. If $y = \coth x$, then $\dfrac{dy}{dx} = -\text{cosech}^2 x$

5. If $y = \text{sech} x$, then $\dfrac{dy}{dx} = -\text{sech} x \tanh x$

6. If $y = \text{cosech} x$, then $\dfrac{dy}{dx} = -\text{cosech} x \coth x$

Note that $\sinh x = \dfrac{e^x - e^{-x}}{2}$ and $\cosh x = \dfrac{e^x + e^{-x}}{2}$

Examples

1. If $y = \cosh x - 5\sinh x$, find $\dfrac{dy}{dx}$.

Solution

$$y = \cosh x - 5\sinh x$$

$$\frac{dy}{dx} = \sinh x - 5\cosh x$$

2. Find the derivative of $2x^3 \coth x$

Solution

$$y = 2x^3 \coth x$$

Using product rule gives $\dfrac{dy}{dx}$ as follows:

$$\frac{dy}{dx} = 2x^3\left[\frac{d(\coth x)}{dx}\right] + \coth x\left[\frac{d(2x^3)}{dx}\right]$$

$$= 2x^3(-\text{cosech}^2 x) + \coth x(6x^2)$$

$$= -2x^3\text{cosech}^2 x + 6x^2\coth x$$

$$\frac{dy}{dx} = -2x^2(x\text{cosech}^2 x - 3\coth x)$$

Take note of the change in sign of the term in the bracket. This is due to the negative sign outside the bracket. Expanding the bracket gives the original expression that was factorized.

3. If $y = \dfrac{\cosh x}{x^2 + 1}$ find $\dfrac{dy}{dx}$.

Solution

Let us use product rule to obtained $\dfrac{dy}{dx}$ as follows:

$$\frac{dy}{dx} = \frac{(x^2 + 1)\left[\frac{d(\cosh x)}{dx}\right] - \cosh x\left[\frac{d(x^2 + 1)}{dx}\right]}{(x^2 + 1)^2}$$

$$= \frac{(x^2 + 1)(\sinh x) - \cosh x(2x)}{(x^2 + 1)^2}$$

$$\frac{dy}{dx} = \frac{\sinh x(x^2 + 1) - 2x\cosh x}{(x^2 + 1)^2}$$

4. Find the derivative of $(\sinh 3x)^2$

Solution

$y = (\sinh 3x)^2$ (This can also be written as $\sinh^2 3x$)

Let $u = \sinh 3x$

Hence, $y = u^2$

$\dfrac{du}{dx} = 3\cosh 3x$ (Note that 3 is from the derivative of $3x$)

$\dfrac{dy}{du} = 2u$

$\dfrac{dy}{dx} = \dfrac{dy}{du} \times \dfrac{du}{dx}$

$\quad = 2u \times 3\cosh 3x$

$\quad = 6u\cosh 3x$

$\quad = 6\sinh 3x \cosh 3x$ (u has been replaced with $\sinh 3x$)

5. Find the derivative of $\sinh^4 2x^3$

Solution

$y = \sinh^4 2x^3$

Let $u = \sinh 2x^3$

Hence, $y = u^4$

$\dfrac{du}{dx} = 6x^2\cosh 2x^3$ (Note that $6x^2$ is from the derivative of $2x^3$)

$\dfrac{dy}{du} = 4u^3$

$\dfrac{dy}{dx} = \dfrac{dy}{du} \times \dfrac{du}{dx}$

$\quad = 4u^3 \times 6x^2\cosh 2x^3$

$\quad = 24x^2 u^3\cosh 2x^3$

$\dfrac{dy}{dx} = 24x^2\sinh^3 2x^3\cosh 2x^3$ (u has been replaced with $\sinh 2x^3$)

6. Find the derivative of $\text{cosech}4x^3$

Solution

$$y = \text{cosech}4x^3$$

$$\frac{dy}{dx} = -\text{cosech}4x^3\text{coth}4x^3 \times \frac{d(4x^3)}{dx}$$ (Note that the derivative of cosech is −cosechcoth)

$$= -\text{cosech}4x^3\text{coth}4x^3 \times 12x^2$$

$$= -12x^2\text{cosech}4x^3\text{coth}4x^3$$

7. Find the derivative of $\text{coth}^2 2x^4$

Solution

$$y = \text{coth}^2 2x^4$$

Let $u = \text{coth}2x^4$

Hence, $y = u^2$

$$\frac{du}{dx} = -\text{cosech}^2 2x^4 \times \frac{d(2x^4)}{dx}$$

$$= -\text{cosech}^2 2x^4 \times 8x^3$$

$$\frac{du}{dx} = -8x^3\text{cosech}^2 2x^4$$

$$\frac{dy}{du} = 2u$$

Therefore, $\dfrac{dy}{dx} = \dfrac{dy}{du} \times \dfrac{du}{dx}$

$$= 2u \times (-8x^3\text{cosech}^2 2x^4)$$

$$= -16x^3 u \text{ cosech}^2 2x^4$$

$$\frac{dy}{dx} = -16x^3\text{coth}2x^4\text{cosech}^2 2x^4$$ (Since $u = \text{coth}2x^4$)

8. Find the derivative of $5x\sinh2x$

Solution

$$y = 5x\sinh2x$$

We apply the product rule of differentiation as follows:

Let $u = 5x$

and $v = \sinh2x$

$$\frac{du}{dx} = 5$$

$$\frac{dv}{dx} = 2\cosh2x$$ (Note that the differentiation of $2x$ gives 2)

Hence, $\dfrac{dy}{dx} = u\dfrac{dv}{dx} + v\dfrac{du}{dx}$

$$= (5x \times 2\cosh2x) + (\sinh2x \times 5)$$

$$= 10x\cosh2x + 5\sinh2x$$

$$\frac{dy}{dx} = 5(2x\cosh 2x + \sinh 2x)$$

9. If $y = \dfrac{1 + \cosh 2x}{\sinh 2x}$ find $\dfrac{dy}{dx}$.

Solution

$$y = \frac{1 + \cosh 2x}{\sinh 2x}$$

We apply quotient rule as follows:

$u = 1 + \cosh 2x$

$v = \sinh 2x$

$\dfrac{du}{dx} = 2\sinh 2x$

$\dfrac{dv}{dx} = 2\cosh 2x$

Hence, $\dfrac{dy}{dx} = \dfrac{v\frac{du}{dx} - u\frac{dv}{dx}}{v^2}$ (Quotient rule)

$$= \frac{\sinh 2x(2\sinh 2x) - (1 + \cosh 2x)(2\cosh 2x)}{(\sinh 2x)^2}$$

$$= \frac{2\sin^2 2x - (2\cosh 2x + 2\cosh^2 2x)}{\sinh^2 2x} \qquad \text{(Note that } (\sinh 2x)^2 = \sinh^2 2x)$$

$$= \frac{2\sinh^2 2x - 2\cosh 2x - 2\cosh^2 2x}{\sinh^2 2x}$$

$$= \frac{2\sinh^2 2x - 2\cosh^2 2x - 2\cosh 2x}{\sinh^2 2x}$$

$$= \frac{-2(\cosh^2 2x - \sinh^2 2x) - 2\cosh 2x}{\sinh^2 2x}$$

$$= \frac{-2(1) - 2\cosh 2x}{\sinh^2 2x} \qquad \text{(Note that } \cosh^2 2x - \sinh^2 2x = 1)$$

$$\frac{dy}{dx} = \frac{-2(1 + \cosh 2x)}{\sinh^2 2x}$$

Or, $\dfrac{dy}{dx} = \dfrac{-2(1 + \cosh 2x)}{\cosh^2 2x - 1}$ (Note that since $\cosh^2 2x - \sinh^2 2x = 1$, then $\cosh^2 2x - 1 = \sinh^2 2x$)

$$= \frac{-2(1 + \cosh 2x)}{(\cosh 2x + 1)(\cosh 2x - 1)}$$

Note that from difference of two squares we have: $a^2 - b^2 = (a + b)(a - b)$.

Hence, $\cosh^2 2x - 1$ is also a difference of two squares since 1 is also 1^2. Therefore, $\cosh^2 2x - 1 = (\cosh 2x + 1)(\cosh 2x - 1)$ as represented above. Hence the expression above becomes:

$$\frac{dy}{dx} = \frac{-2}{(\cosh 2x - 1)}$$

10. Differentiate $y = \dfrac{\sinh^2 x}{x}$

Solution

$$y = \frac{\sinh^2 x}{x}$$

We use quotient rule as follows:

Therefore, $u = \sinh^2 x$

$\quad v = x$

Hence, $\dfrac{du}{dx} = 2\sinh x \cosh x$

Since, $v = x$

Then, $\dfrac{dv}{dx} = 1$

Hence, $\dfrac{dy}{dx} = \dfrac{v\frac{du}{dx} - u\frac{dv}{dx}}{v^2}$

$$= \frac{x(2\sinh x \cosh x) - \sinh^2 x(1)}{x^2}$$

$$= \frac{2x\sinh x \cosh x - \sin^2 x}{x^2}$$

$$= \frac{\sinh x(2x\cosh x - \sinh x)}{x^2}$$

Exercise 36

1. If $y = \sinh 3x - 2\cosh x$, find $\dfrac{dy}{dx}$.

2. Find the derivative of $5x^2 \text{sech} x$

3. If $y = \dfrac{\sinh 2x}{3x^2}$ find $\dfrac{dy}{dx}$.

4. Find the derivative of $\cosh^2 3x$

5. Find the derivative of $\cosh^3 4x^5$

6. Find the derivative of $\coth 2x^5$

7. Find the derivative of $(\text{sech} 2x^2)^3$

8. Find the derivative of $x^2 \cosh 5x$

9. If $y = \dfrac{\tanh 5x}{\coth 3x}$ find $\dfrac{dy}{dx}$.

10. Differentiate $y = \dfrac{2\cosh^3 x}{3x}$

CHAPTER 37
DERIVATIVE OF LOGARITHMIC FUNCTIONS

If $y = \log_a x$, then $\dfrac{dy}{dx} = \dfrac{1}{x}\log_a e$, where a is any base.

If $y = \log_e x$, then $\dfrac{dy}{dx} = \dfrac{1}{x}$

Note that $\log_e x$ can also be represented as $\ln x$ and the value of e is 2.718 (to 3 decimal places)

Examples

1. Find the derivative of $\log_a 2x$.

Solution

$\quad y = \log_a 2x$

We will use chain rule since we have a function that is not just x but $2x$.

Let $u = 2x$

Therefore, $y = \log_a u$

$\dfrac{du}{dx} = 2$

$\dfrac{dy}{du} = \dfrac{1}{u}\log_a e$

$\dfrac{dy}{dx} = \dfrac{dy}{du} \times \dfrac{du}{dx}$

$\quad = \dfrac{1}{u}\log_a e \times 2$

$\quad = \dfrac{2}{u}\log_a e$

$\quad = \dfrac{2}{2x}\log_a e \qquad$ (since $u = 2x$)

$\dfrac{dy}{dx} = \dfrac{1}{x}\log_a e$

2. Find the derivative of $\log_a(5x - 1)$

Solution

$\quad y = \log_a(5x - 1)$

Let $u = 5x - 1$

Therefore, $y = \log_a u$

$\dfrac{du}{dx} = 5$

$\dfrac{dy}{du} = \dfrac{1}{u}\log_a e$

$\dfrac{dy}{dx} = \dfrac{dy}{du} \times \dfrac{du}{dx}$

$$= \frac{1}{u} \log_a e \; \times \; 5$$

$$= \frac{5}{u} \log_a e$$

$$\frac{dy}{dx} = \frac{5}{5x-1} \log_a e \qquad (\text{since } u = 5x - 1)$$

3. Differentiate $\log_a(4x-3)^2$ with respect to x

<u>Solution</u>

$$y = \log_a(4x-3)^2$$

Let $u = (4x-3)^2$

Therefore, $y = \log_a u$

$$\frac{du}{dx} = 2(4x-3)^{2-1} \; \times \; 4 \qquad (\text{Note that 4 is from the derivative of } 4x-3)$$

$$\frac{du}{dx} = 8(4x-3)$$

$$\frac{dy}{du} = \frac{1}{u} \log_a e$$

$$\frac{dy}{dx} = \frac{dy}{du} \; \times \; \frac{du}{dx}$$

$$= \frac{1}{u} \log_a e \; \times \; 8(4x-3)$$

$$= \frac{8(4x-3)}{u} \log_a e$$

$$= \frac{8(4x-3)}{(4x-3)^2} \log_a e \qquad [\text{since } u = (4x-3)^2]$$

$$\frac{dy}{dx} = \frac{8}{4x-3} \log_a e \qquad (\text{One } 4x-3 \text{ cancels out})$$

4. If $y = \log_a\sqrt{1+2x}$, find $\dfrac{dy}{dx}$

<u>Solution</u>

$$y = \log_a\sqrt{1+2x}$$

Let $u = \sqrt{1+2x} = (1+2x)^{\frac{1}{2}}$

Therefore, $y = \log_a u$

$$\frac{du}{dx} = \frac{1}{2}(1+2x)^{\frac{1}{2}-1} \; \times \; 2$$

$$= (1+2x)^{-\frac{1}{2}} \qquad (\text{Note that } \frac{1}{2} \times 2 = 1)$$

$$\frac{du}{dx} = \frac{1}{(1+2x)^{\frac{1}{2}}}$$

$$\frac{dy}{du} = \frac{1}{u} \log_a e$$

$$\frac{dy}{dx} = \frac{dy}{du} \times \frac{du}{dx}$$

$$= \frac{1}{u} \log_a e \times \frac{1}{(1+2x)^{\frac{1}{2}}}$$

$$= \frac{1}{(1+2x)^{\frac{1}{2}}} \log_a e \times \frac{1}{(1+2x)^{\frac{1}{2}}} \qquad \text{[Note that u has been replaced with } (1+2x)^{\frac{1}{2}}\text{]}$$

$$\frac{dy}{dx} = \frac{1}{1+2x} \log_a e \qquad \text{[Note that } (1+2x)^{\frac{1}{2}} \times (1+2x)^{\frac{1}{2}} = (1+2x)^1 \text{ by adding exponents]}$$

5. Find $\frac{dy}{dx}$ given that $y = \log_a \frac{1-3x}{1+3x}$

Solution

$$y = \log_a \frac{1-3x}{1+3x}$$

Or, $y = \log_a(1-3x) - \log_a(1+3x) \qquad \left(\text{Recall that } \log_x\left(\frac{a}{b}\right) = \log_x a - \log_x b\right)$

$$\frac{dy}{dx} = \frac{-3}{1-3x} \log_a e - \frac{3}{1+3x} \log_a e$$

$$= -\log_a e \left(\frac{3}{1-3x} + \frac{3}{1+3x}\right)$$

$$= -\log_a e \left[\frac{3(1+3x) + 3(1-3x)}{(1-3x)(1+3x)}\right]$$

$$= -\log_a e \left[\frac{3+9x+3-9x}{1-9x^2}\right] \qquad \text{[Note that } (1-3x)(1+3x) = 1-9x^2\text{]}$$

$$= -\log_a e \left(\frac{6}{1-9x^2}\right)$$

$$\frac{dy}{dx} = \frac{-6}{1-9x^2} \log_a e$$

6. If $y = \log_{10}(x^2 - 2)$, find $\frac{dy}{dx}$.

Solution

$$y = \log_{10}(x^2 - 2)$$

The value 10 represents a in other examples. So we are going to differentiate y like the examples above except that we will write 10 wherever 'a' should be.

$$y = \log_{10}(x^2 - 2)$$
$$u = x^2 - 2$$

Hence, $y = \log_{10} u$

$$\frac{du}{dx} = 2x$$

$$\frac{dy}{du} = \frac{1}{u} \log_{10} e \qquad \text{(Just like differentiating } \log_a u\text{)}$$

$$\frac{dy}{dx} = \frac{dy}{du} \times \frac{du}{dx}$$

$$= \frac{1}{u} \log_{10} e \times 2x$$

$$= \frac{2x}{u} \log_{10} e$$

$$\frac{dy}{dx} = \frac{2x}{x^2 - 2} \log_{10} e$$

7. Find $\frac{dy}{dx}$ if $y = \log_{10} \frac{1}{x}$

<u>Solution</u>

$$y = \log_{10} \frac{1}{x}$$

Let us solve this question directly without using $u = \frac{1}{x}$ as follows:

$$\frac{dy}{dx} = \frac{\frac{d\left(\frac{1}{x}\right)}{dx}}{\frac{1}{x}} \times \log_{10} e$$

$$= \frac{\frac{d(x^{-1})}{dx}}{x^{-1}} \times \log_{10} e$$

$$= \frac{-x^{-2}}{x^{-1}} \times \log_{10} e \qquad \text{(Note that the derivative of } x^{-1} \text{ is } -x^{-2}\text{)}$$

$$= -x^{-2} \times x^1 \times \log_{10} e \qquad \left(\text{Since } \frac{1}{x^{-1}} = x^1\right)$$

$$= -x^{-1} \times \log_{10} e \qquad \text{(After adding the exponents of } x\text{)}$$

$$\frac{dy}{dx} = -\frac{1}{x} \log_{10} e$$

8. Find the derivative of $\log_e(2 - x^3)$

<u>Solution</u>

$$y = \log_e(2 - x^3)$$

Note that the base here is 'e' and not 'a'.

Let $u = 2 - x^3$

Hence, $y = \log_e u$

$$\frac{du}{dx} = -3x^2$$

$$\frac{dy}{du} = \frac{1}{u} \qquad \left(\text{Recall that the derivative of } \log_e x = \frac{1}{x}\right)$$

$$\frac{dy}{dx} = \frac{dy}{du} \times \frac{du}{dx}$$

$$= \frac{1}{u} \times -3x^2$$

$$= \frac{-3x^2}{u}$$

$$\frac{dy}{dx} = \frac{-3x^2}{2 - x^3}$$

9. Find the derivative of $(\log_e 5x)^2$

Solution

$y = (\log_e 5x)^2$

Let $u = \log_e 5x$

Hence, $y = u^2$

$$\frac{du}{dx} = \frac{\frac{d(5x)}{dx}}{5x}$$

$$= \frac{5}{5x}$$

$$\frac{du}{dx} = \frac{1}{x}$$

$$\frac{dy}{du} = 2u$$

$$\frac{dy}{dx} = \frac{dy}{du} \times \frac{du}{dx}$$

$$= 2u \times \frac{1}{x}$$

$$= \frac{2u}{x}$$

$$\frac{dy}{dx} = \frac{2}{x} \log_e 5x \quad \text{(Since } u = \log_e 5x\text{)}$$

10. If $y = \ln\sqrt{3x^2 - 4}$, find $\frac{dy}{dx}$.

Solution

$y = \ln\sqrt{3x^2 - 4}$

Note that $\ln\sqrt{3x^2 - 4}$ is also the same as $\log_e\sqrt{3x^2 - 4}$

Hence, $y = \log_e\sqrt{3x^2 - 4}$

Or, $y = \log_e(3x^2 - 4)^{\frac{1}{2}}$

Let $u = (3x^2 - 4)^{\frac{1}{2}}$

Hence, $y = \log_e u$

$$\frac{du}{dx} = \frac{1}{2}(3x^2 - 4)^{\frac{1}{2}-1} \times 6x \quad \text{(Note that } 6x \text{ is from the derivative of } 3x^2 - 4\text{)}$$

$$= \frac{6x}{2}(3x^2 - 4)^{-\frac{1}{2}}$$

$$\frac{du}{dx} = \frac{3x}{(3x^2 - 4)^{\frac{1}{2}}}$$

$$\frac{dy}{du} = \frac{1}{u}$$

415

$$\frac{dy}{dx} = \frac{dy}{du} \times \frac{du}{dx}$$

$$= \frac{1}{u} \times \frac{3x}{(3x^2 - 4)^{\frac{1}{2}}}$$

$$= \frac{1}{(3x^2 - 4)^{\frac{1}{2}}} \times \frac{3x}{(3x^2 - 4)^{\frac{1}{2}}}$$

$$\frac{dy}{dx} = \frac{3x}{3x^2 - 4} \quad \text{(Note that } (3x^2 - 4)^{\frac{1}{2}} \times (3x^2 - 4)^{\frac{1}{2}} = 3x^2 - 4, \text{ after adding their exponents)}$$

11. Find $\frac{dy}{dx}$ if $y = \sqrt[3]{x}\ \ln 2x$.

Solution

$y = \sqrt[3]{x}\ \ln 2x$

We are going to apply product rule here.

$$u = \sqrt[3]{x} = x^{\frac{1}{3}}$$

$v = \ln 2x$ (Note that $\ln 2x$ is the same as $\log_e 2x$)

$$\frac{du}{dx} = \frac{1}{3} x^{\frac{1}{3} - 1}$$

$$= \frac{1}{3} x^{-\frac{2}{3}}$$

$$\frac{dv}{dx} = \frac{2}{2x}$$

$$= \frac{1}{x}$$

Hence, $\frac{dy}{dx} = u\frac{dv}{dx} + v\frac{du}{dx}$ (Product rule)

$$= x^{\frac{1}{3}} \left(\frac{1}{x}\right) + \ln 2x \left(\frac{1}{3} x^{-\frac{2}{3}}\right)$$

$$= x^{\frac{1}{3}} (x^{-1}) + \frac{1}{3} x^{-\frac{2}{3}} \ln 2x$$

$$= x^{-\frac{2}{3}} \left(1 + \frac{\ln 2x}{3}\right)$$

$$= x^{-\frac{2}{3}} \left(\frac{3 + \ln 2x}{3}\right)$$

$$\frac{dy}{dx} = \frac{1}{x^{\frac{2}{3}}} \left(\frac{3 + \ln 2x}{3}\right)$$

Or, $\frac{dy}{dx} = \frac{1}{\sqrt[3]{x^2}} \left(\frac{3 + \ln 2x}{3}\right)$

12. Find $\frac{dy}{dx}$ given that $y = \ln(1 + 2x)^2$

Solution

$y = \ln(1 + 2x)^2$ [Also $y = \log_e(1 + 2x)^2$]

Let $u = (1 + 2x)^2$

Hence, $y = \ln u$

$$\frac{du}{dx} = 2(1 + 2x) \times \frac{d(2x)}{dx}$$

$$= 2(1 + 2x) \times 2x$$

$$\frac{du}{dx} = 4(1 + 2x)$$

$$\frac{dy}{du} = \frac{1}{u}$$

$$\frac{dy}{dx} = \frac{1}{u} \times 4(1 + 2x)$$

$$= \frac{4(1+2x)}{u}$$

$$= \frac{4(1+2x)}{(1+2x)^2}$$

$$= \frac{4}{1+2x}$$

13. Differentiate with respect to x: $5x\ln(3x^2 - 2)$

Solution

$y = 5x\ln(3x^2 - 2)$

By using product rule:

$u = 5x$

$v = \ln(3x^2 - 2)$

$$\frac{du}{dx} = 5$$

$$\frac{dv}{dx} = \frac{6x}{3x^2 - 2}$$

Hence, $\frac{dy}{dx} = u\frac{dv}{dx} + v\frac{du}{dx}$

$$= 5x\left(\frac{6x}{3x^2 - 2}\right) + \ln(3x^2 - 2)(5)$$

$$\frac{dy}{dx} = \frac{30x^2}{3x^2 - 2} + 5\ln(3x^2 - 2)$$

14. Find the derivative of $\frac{\ln x}{x}$

Solution

$$y = \frac{\ln x}{x}$$

We are going to use quotient rule as follows:

$u = \ln x$

$v = x$

417

$$\frac{du}{dx} = \frac{1}{x}$$

$$\frac{dv}{dx} = 1$$

Hence, $\dfrac{dy}{dx} = \dfrac{v\dfrac{du}{dx} - u\dfrac{dv}{dx}}{v^2}$

$$= \frac{x\left(\frac{1}{x}\right) - \ln x \,(1)}{x^2}$$

$$\frac{dy}{dx} = \frac{1 - \ln x}{x^2}$$

Exercise 37

1. Find the derivative of $\log_a 7x$.

2. Find the derivative of $\log_a(x^3 + 5)$

3. Differentiate $\log_a(2x^3 - 5)^6$ with respect to x

4. If $y = \log_a x^{\frac{2}{3}}$, find $\dfrac{dy}{dx}$

5. Find $\dfrac{dy}{dx}$ given that $y = \log_a \dfrac{1+x^2}{1-x^2}$

6. If $y = \log_5(5x^3 - 1)$, find $\dfrac{dy}{dx}$.

7. Find $\dfrac{dy}{dx}$ if $y = \log_2 \dfrac{1}{x^2}$

8. Find the derivative of $\log_e(x^2 + 3)$

9. Find the derivative of $(\log_e x)^3$

10. If $y = \ln\sqrt{1 - 2x^5}$, find $\dfrac{dy}{dx}$.

11. Find $\dfrac{dy}{dx}$ if $y = x^4 \ln x^2$

12. Find $\dfrac{dy}{dx}$ given that $y = \ln(3 - 7x)^4$

13. Differentiate with respect to x: $3x^2\ln(4x^3 + 1)$

14. Find the derivative of $\dfrac{\ln x^2}{x^2}$

15. Find the derivative of $\log_e(1 - 5x^2)^3$

CHAPTER 38
DERIVATIVE OF EXPONENTIAL FUNCTIONS

If $y = a^x$, then $\dfrac{dy}{dx} = a^x \log_e a$, where 'a' is any number.

If $y = e^x$, then $\dfrac{dy}{dx} = e^x$

Examples

1. Differentiate with respect to x: a^{5x}

<u>Solution</u>

$y = a^{5x}$

Let $u = 5x$

$y = a^u$

$\dfrac{du}{dx} = 5$

$\dfrac{dy}{du} = a^u \log_e a$

Hence, $\dfrac{dy}{dx} = \dfrac{dy}{du} \ \times \ \dfrac{du}{dx}$

$= a^u \log_e a \ \times \ 5$

$= 5a^u \log_e a$

$\dfrac{dy}{dx} = 5a^{5x} \log_e a \qquad$ (since $u = 5x$)

2. Find the derivative of $a^{x^2 - 3x + 4}$

<u>Solution</u>

$y = a^{x^2 - 3x + 4}$

Let us solve this example directly without using $u = x^2 - 3x + 4$

$\dfrac{dy}{dx} = \dfrac{d(x^2 - 3x + 4)}{dx} \ \times \ a^{x^2 - 3x + 4} \ \times \log_e a$

$= (2x - 3)(a^{x^2 - 3x + 4} \ \times \ \log_e a)$

3. If $y = 3x^2 a^{5x}$, find $\dfrac{dy}{dx}$.

<u>Solution</u>

$y = 3x^2 a^{5x}$

We apply product rule as follows:

$u = 3x^2$

$v = a^{5x}$

$$\frac{du}{dx} = 6x$$

$$\frac{dv}{dx} = 5a^{5x}\log_e a$$

Hence, $\dfrac{dy}{dx} = u\dfrac{dv}{dx} + v\dfrac{du}{dx}$

$$= 3x^2(5a^{5x}\log_e a) + a^{5x}(6x)$$

$$= 15x^2 a^{5x}\log_e a + 6xa^{5x}$$

Factorizing the expression above gives:

$$\frac{dy}{dx} = 3xa^{5x}(5x\log_e a + 2)$$

4. Find the derivative of $\dfrac{e^x + e^{-x}}{e^x - e^{-x}}$

Solution

$$y = \frac{e^x + e^{-x}}{e^x - e^{-x}}$$

$$u = e^x + e^{-x}$$

$$v = e^x - e^{-x}$$

$$\frac{du}{dx} = e^x - e^{-x} \quad \text{(The derivative of } e^{-x} = -1 \times e^{-x} = -e^{-x}. \text{ The value } -1 \text{ is from the derivative of } -x)$$

$$\frac{dv}{dx} = e^x + e^{-x}$$

Hence, $\dfrac{dy}{dx} = \dfrac{v\dfrac{du}{dx} - u\dfrac{dv}{dx}}{v^2}$ (Quotient rule)

$$= \frac{(e^x - e^{-x})(e^x - e^{-x}) - (e^x + e^{-x})(e^x + e^{-x})}{(e^x - e^{-x})^2}$$

Expanding the numerator and denominator gives:

$$\frac{dy}{dx} = \frac{(e^x)(e^x) - (e^x)(e^{-x}) - (e^{-x})(e^x) + (e^{-x})(e^{-x}) - [(e^x)(e^x) + (e^x)(e^{-x}) + (e^{-x})(e^x) + (e^{-x})(e^{-x})]}{(e^x - e^{-x})^2}$$

$$= \frac{e^{2x} - 1 - 1 + e^{-2x} - (e^{2x} + 1 + 1 + e^{-2x})}{(e^x - e^{-x})^2} \quad \text{(Note that exponents have been added, and } e^0 =$$

1)

$$= \frac{e^{2x} - 2 + e^{-2x} - e^{2x} - 1 - 1 - e^{-2x})}{(e^x - e^{-x})^2}$$

$$\frac{dy}{dx} = \frac{-4}{(e^x - e^{-x})^2} \quad (e^{2x} \text{ and } e^{-2x} \text{ cancel out)}$$

5. If $y = e^{\sqrt{x}}$ find $\dfrac{dy}{dx}$.

Solution

$$y = e^{\sqrt{x}} = e^{x^{\frac{1}{2}}}$$

$$\frac{dy}{dx} = \frac{d(x)^{\frac{1}{2}}}{dx} \times e^{x^{\frac{1}{2}}}$$

$$= \frac{1}{2}x^{-\frac{1}{2}} \times e^{x^{\frac{1}{2}}}$$

$$= \frac{1}{2}\left(\frac{1}{x^{\frac{1}{2}}}\right) \times e^{x^{\frac{1}{2}}}$$

$$= \frac{e^{x^{\frac{1}{2}}}}{2x^{\frac{1}{2}}}$$

$$\frac{dy}{dx} = \frac{e^{\sqrt{x}}}{2\sqrt{x}}$$

6. Find $\frac{dy}{dx}$ if $y = x^{\frac{1}{2}} e^{x^{\frac{1}{2}}}$

Solution

$$y = x^{\frac{1}{2}} e^{x^{\frac{1}{2}}}$$

From product rule:

$$\frac{dy}{dx} = x^{\frac{1}{2}} \frac{d(e^{x^{\frac{1}{2}}})}{dx} + e^{x^{\frac{1}{2}}} \frac{d(x^{\frac{1}{2}})}{dx}$$

$$= x^{\frac{1}{2}}\left(\frac{e^{x^{\frac{1}{2}}}}{2x^{\frac{1}{2}}}\right) + e^{x^{\frac{1}{2}}}\left(\frac{1}{2}x^{-\frac{1}{2}}\right) \quad \text{(Note that the derivative of } e^{x^{\frac{1}{2}}} \text{ is } \frac{e^{x^{\frac{1}{2}}}}{2x^{\frac{1}{2}}} \text{ from example 5)}$$

$$= \frac{e^{x^{\frac{1}{2}}}}{2} + \frac{e^{x^{\frac{1}{2}}}}{2x^{\frac{1}{2}}}$$

$$= \frac{x^{\frac{1}{2}}e^{x^{\frac{1}{2}}} + e^{x^{\frac{1}{2}}}}{2x^{\frac{1}{2}}}$$

$$\frac{dy}{dx} = \frac{e^{x^{\frac{1}{2}}}(x^{\frac{1}{2}} + 1)}{2x^{\frac{1}{2}}}$$

7. Find the derivative of $a^{2x} - a^{-2x}$

Solution

$$y = a^{2x} - a^{-2x}$$

$$\frac{dy}{dx} = \frac{d(2x)}{dx}(a^{2x}\log_e a) - \left[\frac{d(-2x)}{dx}(a^{-2x}\log_e a)\right]$$

$$= 2a^{2x}\log_e a - (-2a^{-2x}\log_e a)$$

$$= 2a^{2x}\log_e a + 2a^{-2x}\log_e a$$

$$= 2\log_e a(a^{2x} + a^{-2x})$$

8. Find the derivative of $e^{2x}\log_e 3x$

<u>Solution</u>

$$y = e^{2x}\log_e 3x$$

By product rule:

$$u = e^{2x}$$

$$v = \log_e 3x \quad (\text{or } v = \ln 3x)$$

$$\frac{du}{dx} = 2e^{2x}$$

$$\frac{dv}{dx} = \frac{3}{3x}$$

$$= \frac{1}{x}$$

$$\frac{dy}{dx} = e^{2x}\left(\frac{1}{x}\right) + \log_e 3x(2e^{2x})$$

$$= \frac{e^{2x}}{x} + 2e^{2x}\log_e 3x$$

$$\frac{dy}{dx} = e^{2x}(\frac{1}{x} + 2\log_e 3x)$$

9. Differentiate with respect to x: $2\sqrt{xe^x}$

<u>Solution</u>

$$y = 2\sqrt{xe^x}$$

Or, $y = 2\sqrt{x}\,\sqrt{e^x}$

From product rule:

$$u = 2\sqrt{x} = 2x^{\frac{1}{2}}$$

$$v = \sqrt{e^x} = (e^x)^{\frac{1}{2}}$$

$$v = e^{\frac{1}{2}x} \quad \text{(When the exponents are multiplied)}$$

$$\frac{du}{dx} = \frac{1}{2} \times 2x^{-\frac{1}{2}}$$

$$= x^{-\frac{1}{2}}$$

$$\frac{dv}{dx} = \frac{1}{2}e^{\frac{1}{2}x} \quad \text{(Note that } \frac{1}{2} \text{ is from the derivative of } \frac{1}{2}x\text{)}$$

$$\frac{dy}{dx} = u\frac{dv}{dx} + v\frac{du}{dx}$$

$$= 2x^{\frac{1}{2}}(\frac{1}{2}e^{\frac{1}{2}x}) + e^{\frac{1}{2}x}(x^{-\frac{1}{2}})$$

$$= (\frac{1}{2} \times 2x^{\frac{1}{2}} \times e^{\frac{1}{2}x}) \times \frac{e^{\frac{1}{2}x}}{x^{\frac{1}{2}}}$$

$$= x^{\frac{1}{2}}e^{\frac{1}{2}x} + \frac{e^{\frac{1}{2}x}}{x^{\frac{1}{2}}}$$

$$= \frac{x\,e^{\frac{1}{2}x} + e^{\frac{1}{2}x}}{x^{\frac{1}{2}}}$$

$$= \frac{e^{\frac{1}{2}x}(x+1)}{x^{\frac{1}{2}}}$$

$$= \frac{e^{\frac{x}{2}}(x+1)}{x^{\frac{1}{2}}}$$

$$\frac{dy}{dx} = \frac{\sqrt{e^x}(x+1)}{\sqrt{x}}$$

10. Find the derivative of $2x\log_{10}x$

Solution

$y = 2x\log_{10}x$ (Note that this is similar to $2x\log_a x$)

$u = 2x$

$v = \log_{10}x$

$\dfrac{du}{dx} = 2$

$\dfrac{dv}{dx} = \dfrac{1}{x}\log_{10}e$ (Recall that the derivative of $\log_a x$ is $\dfrac{1}{x}\log_a e$)

$\dfrac{dy}{dx} = u\dfrac{dv}{dx} + v\dfrac{du}{dx}$

$\qquad = 2x\left(\dfrac{1}{x}\log_{10}e\right) + \log_{10}x(2)$

$\qquad = 2\log_{10}e + 2\log_{10}x$

$\dfrac{dy}{dx} = 2(\log_{10}e + \log_{10}x)$

Or, $\dfrac{dy}{dx} = 2\left(\dfrac{\log_e e}{\log_e 10} + \log_{10}x\right)$

Note that the rule of change of base in logarithm that has been applied above is given by:

If $\log_a b$ is to be converted to a new base, e, then: $\log_a b = \dfrac{\log_e b}{\log_e a}$

$\dfrac{dy}{dx} = 2\left(\dfrac{1}{\log_e 10} + \log_{10}x\right)$ (Recall that $\log_x x = 1$, or $\log_e e = 1$)

11. Find the derivative of 10^{2x}

Solution

$y = 10^{2x}$ (This is similar to a^{2x})

$\dfrac{dy}{dx} = \dfrac{d(2x)}{dx} \times 10^{2x}\log_e 10$

$$= 2 \text{ x } 10^{2x} \log_e 10$$

$$= (10^{2x})2\log_e 10$$

$$= 10^{2x}\log_e 10^2 \quad \text{(Recall that } a\log_x y = \log_x y^a\text{)}$$

$$\frac{dy}{dx} = 10^{2x}\log_e 100$$

12. If $y = e^{\ln x}$ find $\dfrac{dy}{dx}$

<u>Solution</u>

$$y = e^{\ln x} \quad \text{(Or } y = e^{\log_e x} \text{ since } \ln x = \log_e x\text{)}$$

Let, $u = \ln x$

$$y = e^u$$

$$\frac{du}{dx} = \frac{1}{x}$$

$$\frac{dy}{du} = e^u$$

$$\frac{dy}{dx} = e^u \text{ x } \frac{1}{x}$$

$$= \frac{e^u}{x}$$

$$= \frac{e^{\ln x}}{x} \quad \text{(Since } u = \ln x\text{)}$$

$$= \frac{e^{\log_e x}}{x} \quad \text{(Since } \ln x = \log_e x\text{)}$$

$$= \frac{x}{x} \quad \text{(Recall the identity that } m^{\log_m n} = n\text{)}$$

$$\frac{dy}{dx} = 1$$

This example shows that calculus works, since the derivative of x is 1, and the derivative of $e^{\log_e x} = 1$ because $e^{\log_e x} = x$. Hence, we could have solved this example by stating that:

$$\frac{d(e^{\ln x})}{dx} = \frac{d(x)}{dx} = 1$$

13. What is the derivative of $\dfrac{e^{\frac{1}{x}}}{x^2}$

<u>Solution</u>

$$y = \frac{e^{\frac{1}{x}}}{x^2}$$

From quotient rule:

$$u = e^{\frac{1}{x}}$$

$$v = x^2$$

$$\frac{du}{dx} = \frac{d\left(\frac{1}{x}\right)}{dx} \times e^{\frac{1}{x}}$$

$$= \frac{d(x^{-1})}{dx} \times e^{\frac{1}{x}}$$

$$= -1x^{-2} \times e^{\frac{1}{x}}$$

$$= \frac{-1}{x^2} \times e^{\frac{1}{x}}$$

$$\frac{du}{dx} = \frac{-e^{\frac{1}{x}}}{x^2}$$

$$\frac{du}{dx} = 2x$$

$$\frac{dy}{dx} = \frac{v\frac{du}{dx} - u\frac{dv}{dx}}{v^2} \qquad \text{(Quotient rule)}$$

$$= \frac{x^2\left(\frac{-e^{\frac{1}{x}}}{x^2}\right) - e^{\frac{1}{x}}(2x)}{(x^2)^2}$$

$$= \frac{-e^{\frac{1}{x}} - 2x\,e^{\frac{1}{x}}}{x^4} \qquad \text{(Note that } x^2 \text{ has cancelled out)}$$

14. Find the derivative of $e^{-5x} + 5e$

<u>Solution</u>

$$y = e^{-5x} + 5e$$

$$\frac{dy}{dx} = \frac{d(-5x)}{dx} \times e^{-5x} + \frac{d(5e)}{dx}$$

$$= -5e^{-5x} + 0 \qquad \text{(Note that 5e is a constant and the derivative of a constant is zero)}$$

$$\frac{dy}{dx} = -5e^{-5x}$$

15. If $y = e^{(2x+3)^2}$ find $\frac{dy}{dx}$.

<u>Solution</u>

$$y = e^{(2x+3)^2}$$

Let $u = (2x + 3)^2$

Hence, $y = e^u$

$$\frac{du}{dx} = 2(2x + 3)^{2-1} \times \frac{d(2x)}{dx} \qquad \text{(Using chain rule)}$$

$$= 2(2x + 3) \times 2$$

$$= 4(2x + 3)$$

$$\frac{dy}{du} = e^u$$

$$\frac{dy}{dx} = \frac{dy}{du} \times \frac{du}{dx}$$

$$= e^u \times 4(2x + 3)$$

$$= e^{(2x+3)^2} \times 4(2x + 3) \qquad \text{(Note that } u = (2x + 3)^2)$$

$$\frac{dy}{dx} = 4(2x + 3)e^{(2x+3)^2}$$

16. Differentiate with respect to θ, $5^{2\theta}$

Solution

$$y = 5^{2\theta} \qquad \text{(This is like } y = a^{2\theta})$$

$$\frac{dy}{d\theta} = \frac{d(2\theta)}{d\theta} \times 5^{2\theta}\log_e 5$$

$$= 2 \times 5^{2\theta}\log_e 5$$

$$= (5^{2\theta})2\log_e 5$$

$$= 5^{2\theta}\log_e 5^2 \qquad \text{(Note that } 2\log_e 5 = \log_e 5^2)$$

$$\frac{dy}{d\theta} = 5^{2\theta}\log_e 25$$

Exercise 38

1. Differentiate with respect to x: $5a^{2x}$

2. Find the derivative of $a^{5x^3 - x}$

3. If $y = x^4 a^{3x}$, find $\frac{dy}{dx}$.

4. Find the derivative of $\frac{3 + e^{-x}}{e^x}$

5. If $y = 2e^{\sqrt{3x}}$ find $\frac{dy}{dx}$.

6. Find $\frac{dy}{dx}$ if $y = 6x^3 \, e^{x^2}$

7. Find the derivative of $a^{-5x} + a^{5x}$

8. Find the derivative of $3e^{4x}\log_e 2x^2$

9. Differentiate with respect to x: $3x^2\sqrt{e^{5x}}$

10. Find the derivative of $x^4 \log_{10} 7x$

11. Find the derivative of 6^{3x}

12. Find the derivative of $e^{x^2 - 3x}$

13. What is the derivative of $\frac{2e^{\frac{3}{x}}}{10x^5}$

14. Find the derivative of $1 - 2e^{-3x}$

15. If $y = e^{(x^2+x)^3}$ find $\dfrac{dy}{dx}$.

16. Differentiate with respect to x, 2^{x^2}

17. Find the derivative of $\dfrac{a^x - a^{-x}}{a^{-x}}$

18. Find the derivative of $e^x \ln 10 x^3$

19. Differentiate with respect to x: $e^{2x}\sqrt{5x}$

20. Find the derivative of $2x^5 \log_{10} 5x^4$

CHAPTER 39
LOGARITHMIC DIFFERENTIATION

A function of x could be raised to an exponent which is also a function of x. An example is x^x. The differentiation of such a function is called logarithmic differentiation.

This type of function is differentiated by first taking the logarithm of both sides of the equation, and then differentiating implicitly before making $\frac{dy}{dx}$ the subject of the formula (i.e. solving for $\frac{dy}{dx}$). Finally, y (i.e. the function) is substituted into the final answer.

Examples

1. Find the derivative of x^x.

Solution

$$y = x^x$$

This is a function of x raised to an exponent which is also a function of x.

Hence, take the logarithm to base e of both sides as follows:

$$\log_e y = \log_e x^x$$
$$\log_e y = x\log_e x \quad \text{(Note that } \log_m n^b = b\log_m n\text{)}$$

Now differentiate y implicitly and differentiate the right hand side using product rule. This gives:

$$\frac{1}{y}\frac{dy}{dx} = x\frac{1}{x} + \log_e x(1) \quad \text{(Use of product rule on the right hand side, with u = } x \text{ and v = } \log_e x\text{)}$$

$$\frac{1}{y}\frac{dy}{dx} = 1 + \log_e x$$

Multiply both sides by y to make $\frac{dy}{dx}$ the subject of the formula as follows:

$$\frac{dy}{dx} = y(1 + \log_e x)$$
$$\frac{dy}{dx} = x^x(1 + \log_e x) \quad \text{(Since } y = x^x\text{)}$$

2. Find the derivative of $x^{\ln x}$.

Solution

$$y = x^{\ln x}$$

Taking the logarithm of both sides gives:

$$\log_e y = \log_e x^{\ln x}$$
$$\log_e y = \ln x\log_e x$$

Differentiate y implicitly and treat the right hand side using product rule as follows:

$$\frac{1}{y}\frac{dy}{dx} = \ln x\left(\frac{1}{x}\right) + \log_e x\left(\frac{1}{x}\right)$$

428

$$\frac{1}{y}\frac{dy}{dx} = \left(\frac{\ln x}{x}\right) + \left(\frac{\ln x}{x}\right) \qquad \text{(Note that } \log_e x \text{ can also be written as } \ln x)$$

$$\frac{1}{y}\frac{dy}{dx} = \frac{2\ln x}{x}$$

Multiply both sides by y. This gives:

$$\frac{dy}{dx} = y\left(\frac{2\ln x}{x}\right)$$

$$\frac{dy}{dx} = x^{\ln x}\left(\frac{2\ln x}{x}\right) \qquad \text{(Since } y = x^{\ln x})$$

Or, $\dfrac{dy}{dx} = \dfrac{2}{x}x^{\ln x}\ln x$

3. If $y = \ln(2 + x)^x$ find $\dfrac{dy}{dx}$.

Solution

$$y = \ln(2 + x)^x$$

In this case we are not going to take the logarithm of both sides since there is already logarithm on the right hand side. Hence, we proceed as follows:

$$y = \ln(2 + x)^x$$

$$y = x\log_e(2 + x) \qquad \text{[Note that } \ln(2 + x) \text{ can also be written as } \log_e(2 + x)]$$

Using product rule gives:

$$\frac{dy}{dx} = x\left(\frac{1}{2 + x}\right) + \log_e(2 + x)(1)$$

$$\frac{dy}{dx} = \frac{x}{2 + x} + \log_e(2 + x)$$

4. Find $\dfrac{dy}{dx}$ if $y = (x^2 - 3)^{5x}$

Solution

$$y = (x^2 - 3)^{5x}$$

Taking the natural logarithm of both sides gives:

$$\log_e y = \log_e(x^2 - 3)^{5x}$$

$$\log_e y = 5x\log_e(x^2 - 3)$$

Using implicit differentiation for the left hand side and product rule for the right hand side gives:

$$\frac{1}{y}\frac{dy}{dx} = 5x\left(\frac{2x}{x^2 - 3}\right) + \log_e(x^2 - 3)(5)$$

$$\frac{1}{y}\frac{dy}{dx} = \frac{10x^2}{x^2 - 3} + 5\log_e(x^2 - 3)$$

Multiply both sides by y to obtain:

$$\frac{dy}{dx} = y\left[\frac{10x^2}{x^2 - 3} + 5\log_e(x^2 - 3)\right]$$

$$\frac{dy}{dx} = (x^2 - 3)^{5x}\left[\frac{10x^2}{x^2 - 3} + 5\log_e(x^2 - 3)\right] \qquad \text{[Note that } y = (x^2 - 3)^{5x}\text{]}$$

5. If $y = \dfrac{(x^2 - 5)(3x - 1)^2}{x^7(2x^3 - 3)}$

Solution

This is not a function of x raised to an exponent which is also a function of x. This function can be differentiated by the use of product and quotient rule. However, the use of these rules will be a nightmare. Hence, we have to apply logarithmic differentiation. This is done as follows:

$$y = \frac{(x^2 - 5)(3x - 1)^2}{x^7(2x^3 - 3)}$$

Taking the logarithm of both sides gives:

$$\log_e y = \log_e\left[\frac{(x^2 - 5)(3x - 1)^2}{x^7(2x^3 - 3)}\right]$$

In order to continue, we have to recall the theory of logarithm as follows:

1. $\log_b(cd) = \log_b c + \log_b d$

2. $\log_b\left(\dfrac{c}{d}\right) = \log_b c - \log_b d$

3. $\log_b c^d = d\log_b c$

Therefore, we continue the logarithmic differentiation as follows:

$$\log_e y = \log_e\left[\frac{(x^2 - 5)(3x - 1)^2}{x^7(2x^3 - 3)}\right]$$

Applying the theory above gives:

$$\log_e y = \log_e[(x^2 - 5)(3x - 1)^2] - \log_e[x^7(2x^3 - 3)]$$
$$\log_e y = \log_e(x^2 - 5) + \log_e(3x - 1)^2 - [\log_e x^7 + \log_e(2x^3 - 3)]$$
$$\log_e y = \log_e(x^2 - 5) + 2\log_e(3x - 1) - 7\log_e x - \log_e(2x^3 - 3)$$

Differentiating each term accordingly gives:

$$\frac{1}{y}\frac{dy}{dx} = \frac{2x}{x^2 - 5} + \frac{2 \times 3}{3x - 1} - \frac{7}{x} - \frac{6x}{2x^3 - 3}$$

Note that each function of x is differentiated first, and the value obtained is then divided by the original function. Hence, the expression above simplifies to gives:

$$\frac{1}{y}\frac{dy}{dx} = \frac{2x}{x^2 - 5} + \frac{6}{3x - 1} - \frac{7}{x} - \frac{6x}{2x^3 - 3}$$

Multiply both sides by y to obtain $\dfrac{dy}{dx}$ as follows:

$$\frac{dy}{dx} = y\left(\frac{2x}{x^2 - 5} + \frac{6}{3x - 1} - \frac{7}{x} - \frac{6x}{2x^3 - 3}\right)$$

Now replace y with its original value from the question as follows:

$$\frac{dy}{dx} = \left[\frac{(x^2 - 5)(3x - 1)^2}{x^7(2x^3 - 3)}\right]\left(\frac{2x}{x^2 - 5} + \frac{6}{3x - 1} - \frac{7}{x} - \frac{6x}{2x^3 - 3}\right)$$

6. If $y = \dfrac{(2x-1)(x+4)(6x-5)}{(3-x)^2}$

Solution

$$y = \dfrac{(2x-1)(x+4)(6x-5)}{(3-x)^2}$$

This is a complex function where product rule and quotient rule would be difficult to apply.

Hence, we apply logarithmic differentiation as follows:

$$\log_e y = \log_e\left[\dfrac{(2x-1)(x+4)(6x-5)}{(3-x)^2}\right]$$

Applying the theory of logarithm gives:

$$\log_e y = \log_e[(2x-1)(x+4)(6x-5)] - \log_e(3-x)^2$$

$$\log_e y = \log_e(2x-1) + \log_e(x+4) + \log_e(6x-5) - 2\log_e(3-x)$$

Differentiating accordingly gives:

$$\frac{1}{y}\frac{dy}{dx} = \frac{2}{2x-1} + \frac{1}{x+4} + \frac{6}{6x-5} - \frac{2(-1)}{3-x}$$

$$\frac{1}{y}\frac{dy}{dx} = \frac{2}{2x-1} + \frac{1}{x+4} + \frac{6}{6x-5} + \frac{2}{3-x}$$

Multiply both sides by y. This gives:

$$\frac{dy}{dx} = y\left(\frac{2}{2x-1} + \frac{1}{x+4} + \frac{6}{6x-5} + \frac{2}{3-x}\right)$$

Finally replace y with its original expression as follows:

$$\frac{dy}{dx} = \left[\frac{(2x-1)(x+4)(6x-5)}{(3-x)^2}\right]\left(\frac{2}{2x-1} + \frac{1}{x+4} + \frac{6}{6x-5} + \frac{2}{3-x}\right)$$

7. Find the derivative of x^{x^2}

Solution

$$y = x^{x^2}$$

$$\log_e y = \log_e x^{x^2}$$

$$\log_e y = x^2\log_e x$$

$$\frac{1}{y}\frac{dy}{dx} = x^2\left(\frac{1}{x}\right) + \log_e x(2x)$$

$$\frac{1}{y}\frac{dy}{dx} = x + 2x\log_e x$$

$$\frac{1}{y}\frac{dy}{dx} = x(1 + 2\log_e x)$$

$$\frac{dy}{dx} = yx(1 + 2\log_e x)$$

$$= x^{x^2}x(1 + 2\log_e x) \qquad \text{(y has been replaced with } x^{x^2}\text{)}$$

$$\frac{dy}{dx} = x^{x^2+1}(1 + 2\log_e x)$$

Note that the exponents of x^{x^2} and x were added to obtain x^{x^2+1} since x is also x^1

8. If $y = e^{e^x}$ find $\dfrac{dy}{dx}$.

Solution

$y = e^{e^x}$

$\log_e y = \log_e e^{e^x}$

$\log_e y = e^x \log_e e$

$\log_e y = e^x$ (Recall that logarithm of a number to the same base is 1. Hence, $\log_e e = 1$)

$\dfrac{1}{y}\dfrac{dy}{dx} = e^x$ (Note that the derivative of e^x is e^x)

$\dfrac{dy}{dx} = ye^x$

$= e^{e^x} e^x$

$\dfrac{dy}{dx} = e^{e^x + x}$ (Their exponents have been added)

9. Find the derivative of 5^{2^x}

Solution

$y = 5^{2^x}$

$\log_e y = \log_e 5^{2^x}$

$\log_e y = 2^x \log_e 5$

Differentiate the left hand side implicitly. On the right hand side, differentiate 2^x and take $\log_e 5$ as a constant. This gives:

$\dfrac{1}{y}\dfrac{dy}{dx} = 2^x \log_e 2 \log_e 5$

Note that 2^x is treated in a similar way as a^x, and recall that the derivative of a^x is $a^x \log_e a$. Hence the derivative of 2^x is $2^x \log_e 2$, and $\log_e 5$ multiplies it since it is a constant. Therefore:

$\dfrac{dy}{dx} = y(2^x \log_e 2 \log_e 5)$

$= 5^{2^x}(2^x \log_e 2 \log_e 5)$

$\dfrac{dy}{dx} = 5^{2^x}(2^x \log_e 2)(\log_e 5)$

10. If $y = x^{\ln(2x^2 - 5)}$, find $\dfrac{dy}{dx}$.

Solution

$y = x^{\ln(2x^2 - 5)}$

$\log_e y = \log_e x^{\ln(2x^2 - 5)}$

$\log_e y = \ln(2x^2 - 5)\log_e x$

$\dfrac{1}{y}\dfrac{dy}{dx} = \ln(2x^2 - 5)\left(\dfrac{1}{x}\right) + \log_e x\left(\dfrac{4x}{2x^2 - 5}\right)$ (Use of product rule)

$$\frac{1}{y}\frac{dy}{dx} = \frac{1}{x}\ln(2x^2 - 5) + \left(\frac{4x}{2x^2-5}\right)\log_e x$$

$$\frac{dy}{dx} = y\left[\frac{1}{x}\ln(2x^2 - 5) + \left(\frac{4x}{2x^2-5}\right)\log_e x\right]$$

$$\frac{dy}{dx} = x^{\ln(2x^2-5)}\left[\frac{\ln(2x^2-5)}{x} + \left(\frac{4x}{2x^2-5}\right)\log_e x\right]$$

Exercise 39

1. Find the derivative of x^{2x}

2. Find the derivative of $3x^{\ln 2x}$.

3. If $y = \ln(1 - 4x^2)^x$ find $\frac{dy}{dx}$.

4. Find $\frac{dy}{dx}$ if $y = (6x^2 - 5)^{3x}$

5. If $y = \dfrac{(2x^3 + 1)(x - 2)^3}{3x^2(x^3 - 1)^2}$

6. If $y = \dfrac{(2x - 1)(x^2 - 2)}{(1 - x)(x - 3)^2}$

7. Find the derivative of $5x^{3x^4}$

8. If $y = e^{e^{3x}}$ find $\frac{dy}{dx}$.

9. Find the derivative of 2^{e^x}

10. If $y = x^{\ln(x^2 - 4x)}$, find $\frac{dy}{dx}$.

11. If $y = \ln(10 + 2x^2)^x$ find $\frac{dy}{dx}$.

12. Find $\frac{dy}{dx}$ if $y = (3x^3 - 8)^{2x}$

13. If $y = \dfrac{(x + 3)^2(x - 3)^2}{(x^3 - 1)}$

14. If $y = 5x^{(\ln x)^2}$, find $\frac{dy}{dx}$.

15. Find the derivative of $10x^{3x^2}$

433

CHAPTER 40
DERIVATIVE OF ONE FUNCTION WITH RESPECT TO ANOTHER

We can differentiate one function with respect to another as illustrated by the examples shown below.

Examples

1. Differentiate x^{12} with respect to x^7.

<u>Solution</u>

Let $u = x^{12}$

And $v = x^7$

$$\frac{du}{dx} = 12x^{11}$$

$$\frac{dv}{dx} = 7x^6$$

Differentiating x^{12} with respect to x^7 means differentiating u (i.e. x^{12}) with respect to v (i.e. x^7).

This means $\frac{du}{dv}$. This follows the parametric equation rule given by:

$$\frac{du}{dv} = \frac{\frac{du}{dx}}{\frac{dv}{dx}}$$

$$= \frac{12x^{11}}{7x^6}$$

$$= \frac{12}{7}x^5$$

2. Differentiate $2e^x$ with respect to $\ln 2x$.

<u>Solution</u>

The differentiation of $2e^x$ with respect to $\ln 2x$ simply means:

$$\frac{\text{The derivative of } 2e^x}{\text{The derivative of } \ln 2x}$$

Hence, $\frac{d(2e^x)}{dx} = 2e^x$

And, $\frac{d(\ln 2x)}{dx} = \frac{2}{2x} = \frac{1}{x}$

Therefore, the differentiation of $2e^x$ with respect to $\ln 2x$ is given by:

$$\frac{2e^x}{\frac{1}{x}}$$

$$= 2xe^x$$

3. Differntiate $\sin 5x$ with respect to $\cos x$.

Solution

Hence, $\dfrac{d(\sin 5x)}{dx} = 5\cos 5x$

And, $\dfrac{d(\cos x)}{dx} = -\sin x$

Therefore, the differentiation of $\sin 5x$ with respect to $\cos x$ is given by:

$$= \dfrac{\text{The derivative} \quad \text{of } \sin 5x}{\text{The derivative} \quad \text{of } \cos x}$$

$$= \dfrac{5\cos 5x}{-\sin x}$$

$$= -\dfrac{5\cos 5x}{\sin x}$$

4. Find the derivative of $2x^2 - 5$ with respect to $4x - 1$

Solution

This is obtained as follows:

$$\dfrac{\dfrac{d(2x^2 - 5)}{dx}}{\dfrac{d(4x - 1)}{dx}}$$

$$= \dfrac{4x}{4}$$

$$= x$$

Exercise 40

1. Differentiate x^3 with respect to x^5.

2. Differentiate e^{5x} with respect to $\ln x^2$

3. Differentiate $\cos x^2$ with respect to $\sin x$.

4. Find the derivative of $4x^3 - 5x^2$ with respect to $(x - 1)^2$

5. Differentiate $7x^4$ with respect to $3x^2$.

6. Differentiate $2a^x$ with respect to e^x.

7. Differentiate $\tan^2 x$ with respect to $\cos 5x$.

8. Find the derivative of $\ln(x - 1)$ with respect to e^{x-1}

9. Differentiate $\log_e x^5$ with respect to e^{5x}.

10. Differentiate $\ln(e^x - e^{-x})$ with respect to $\ln x^x$

CHAPTER 41
HIGHER DERIVATIVES (SUCCESSIVE DIFFERENTIATION)

If f(x) is differentiated it gives the first derivative denoted by f'(x) or $\dfrac{dy}{dx}$. If we differentiate f'(x),

it gives the second derivative denoted by f''(x) or $\dfrac{d^2y}{dx^2}$ (read as, dee two y dee x squared). Other

higher derivatives such as $\dfrac{d^3y}{dx^3}, \dfrac{d^4y}{dx^4}$ etc can also be obtained depending on the function.

Examples

1. Find the first, second and third derivatives of $2x^5 - 3x^4 + 5x^2 - 6$

Solution

$$y = 2x^5 - 3x^4 + 5x^2 - 6$$

$$\frac{dy}{dx} = 10x^4 - 12x^3 + 10x$$

$\dfrac{d^2y}{dx^2}$ is obtained by differentiating $\dfrac{dy}{dx}$, which means to differentiate $10x^4 - 12x^3 + 10x$.

Hence, $\dfrac{d^2y}{dx^2} = 40x^3 - 36x^2 + 10$

$\dfrac{d^3y}{dx^3}$ is obtained by differentiating $40x^3 - 36x^2 + 10$ as follows:

$$\frac{d^3y}{dx^3} = 120x^2 - 72x$$

In summary, the first derivative $\left(\dfrac{dy}{dx}\right)$ is $10x^4 - 12x^3 + 10x$, the second derivative $\left(\dfrac{d^2y}{dx^2}\right)$ is $40x^3 -$

$36x^2 + 10$, while the third derivative $\left(\dfrac{d^3y}{dx^3}\right)$ is $120x^2 - 72x$.

2. If $y = \ln x^2$, find $\dfrac{d^2y}{dx^2}$.

Solution

$$y = \ln x^2$$

$$\frac{dy}{dx} = \frac{2x}{x^2}$$

$$= \frac{2}{x}$$

$$\frac{d^2y}{dx^2} = \frac{d\left(\frac{2}{x}\right)}{dx}$$

$$= \frac{d(2x^{-1})}{dx}$$

$$= -2x^{-2}$$

$$\frac{d^2y}{dx^2} = \frac{-2}{x^2}$$

3. Find $\frac{d^3y}{dx^3}$ given that $y = e^{x^3}$

Solution

$$y = e^{x^3}$$

$$\frac{dy}{dx} = 3x^2 e^{x^3} \qquad \text{(Note that } 3x^2 \text{ is from the derivative of } x^3\text{)}$$

We now use product rule to obtain $\frac{d^2y}{dx^2}$ as follows:

$$\frac{d^2y}{dx^2} = 3x^2 \left[\frac{d\left(e^{x^3}\right)}{dx}\right] + e^{x^3} \left[\frac{d\left(3x^2\right)}{dx}\right]$$

$$= 3x^2 \left(3x^2 e^{x^3}\right) + e^{x^3}(6x)$$

$$= 9x^4 e^{x^3} + 6x e^{x^3}$$

$$\frac{d^2y}{dx^2} = e^{x^3}\left(9x^4 + 6x\right)$$

Finally, let us also use product rule to obtain $\frac{d^3y}{dx^3}$ as follows:

$$\frac{d^3y}{dx^3} = e^{x^3}\left[\frac{d\left(9x^4 + 6x\right)}{dx}\right] + \left(9x^4 + 6x\right)\left[\frac{d\left(e^{x^3}\right)}{dx}\right]$$

$$= e^{x^3}\left(36x^3 + 6\right) + \left(9x^4 + 6x\right)\left(3x^2 e^{x^3}\right)$$

$$= 36x^3 e^{x^3} + 6e^{x^3} + 27x^6 e^{x^3} + 18x^3 e^{x^3}$$

$$= 27x^6 e^{x^3} + 36x^3 e^{x^3} + 18x^3 e^{x^3} + 6e^{x^3}$$

$$= 27x^6 e^{x^3} + 54x^3 e^{x^3} + 6e^{x^3}$$

$$\frac{d^3y}{dx^3} = 3e^{x^3}\left(9x^6 + 18x^3 + 2\right)$$

4. If $y = \sin 2x^3$, find the third derivative of y.

Solution

$$y = \sin 2x^3$$

$$\frac{dy}{dx} = \cos 2x^3 \frac{d\left(2x^3\right)}{dx} \qquad \text{(Recall that the derivative of } \sin x \text{ is } \cos x\text{)}$$

$$= \cos 2x^3 \left(6x^2\right)$$

$$\frac{dy}{dx} = 6x^2 \cos 2x^3$$

We now use product rule to obtain $\frac{d^2y}{dx^2}$ as follows:

$$\frac{d^2y}{dx^2} = 6x^2\left[\frac{d\left(\cos 2x^3\right)}{dx}\right] + \cos 2x^3\left[\frac{d\left(6x^2\right)}{dx}\right]$$

437

$$= 6x^2\left[-\sin 2x^3 \frac{d(2x^3)}{dx}\right] + \cos 2x^3(12x)$$

$$= 6x^2\left[-\sin 2x^3\,(6x^2)\right] + 12x\cos 2x^3$$

$$\frac{d^2y}{dx^2} = -36x^4\sin 2x^3 + 12x\cos 2x^3$$

Again, the use of product rule gives us $\dfrac{d^3y}{dx^3}$ as follows:

$$\frac{d^3y}{dx^3} = -36x^4\left[\frac{d(\sin 2x^3)}{dx}\right] + \sin 2x^3\left[\frac{d(-36x^4)}{dx}\right] + 12x\left[\frac{d(\cos 2x^3)}{dx}\right] + \cos 2x^3\left[\frac{d(12x)}{dx}\right]$$

Note that the derivative of $\sin 2x^3$ is $6x^2\cos 2x^3$ as obtained from $\dfrac{dy}{dx}$. Similarly, the differentiation of $\cos 2x^3$ will give us $-6x^2\sin 2x^3$ (since the derivative of $\sin x$ and $\cos x$ differs only by sign and the interchanging of sin with cos or cos with sin).

We now replace the derivative of $\sin 2x^3$ and $\cos 2x^3$ with $6x^2\cos 2x^3$ and $-6x^2\sin 2x^3$ respectively in the expression above. This gives:

$$\frac{d^3y}{dx^3} = -36x^4\,(6x^2\cos 2x^3) + \sin 2x^3(-144x^3) + 12x\,(-6x^2\sin 2x^3) + \cos 2x^3(12)$$

$$= -216x^6\cos 2x^3 - 144x^3\sin 2x^3 - 72x^3\sin 2x^3 + 12\cos 2x^3$$

$$\frac{d^3y}{dx^3} = -216x^6\cos 2x^3 - 216x^3\sin 2x^3 + 12\cos 2x^3$$

Or, $\dfrac{d^3y}{dx^3} = -12(18x^6\cos 2x^3 + 18x^3\sin 2x^3 - \cos 2x^3)$

5. Find $\dfrac{d^2y}{dx^2}$ given that $y = a^{x^2} - 5$

Solution

$$y = a^{x^2} - 5$$

$$\frac{dy}{dx} = a^{x^2}\left[\frac{d(x^2)}{dx}\right]\log_e a$$

$$= a^{x^2}(2x)\log_e a$$

$$\frac{dy}{dx} = 2x\,a^{x^2}\log_e a$$

We now use product rule to find $\dfrac{d^2y}{dx^2}$ as follows:

$$\frac{d^2y}{dx^2} = 2x\left[\frac{d\left(a^{x^2}\log_e a\right)}{dx}\right] + a^{x^2}\log_e a\left[\frac{d(2x)}{dx}\right]$$

$$= 2x\left[\frac{\log_e a\,d\left(a^{x^2}\right)}{dx}\right] + a^{x^2}\log_e a(2)$$

$$= 2x[\log_e a(2x a^{x^2}\log_e a)] + 2a^{x^2}\log_e a$$

$$= 4x^2 a^{x^2}(\log_e a)^2 + 2a^{x^2}\log_e a$$

$$\frac{d^2y}{dx^2} = 2a^{x^2}\log_e a(2x^2\log_e a + 1)$$

6. Find the second derivative of $y = \dfrac{\cos x}{x^3}$

Solution

$$y = \frac{\cos x}{x^3}$$

We use quotient rule as follows:

$$\frac{dy}{dx} = \frac{x^3\frac{d(\cos x)}{dx} - \cos x\frac{d(x^3)}{dx}}{(x^3)^2}$$

$$= \frac{x^3(-\sin x) - \cos x(3x^2)}{x^6}$$

$$\frac{dy}{dx} = \frac{-x^3\sin x - 3x^2\cos x}{x^6}$$

Divide each part by x^6 to separate into fractions as follows:

$$\frac{dy}{dx} = \frac{-x^3\sin x}{x^6} - \frac{3x^2\cos x}{x^6}$$

$$\frac{dy}{dx} = \frac{-\sin x}{x^3} - \frac{3\cos x}{x^4}$$

We now apply quotient rule again as follows:

$$\frac{d^2y}{dx^2} = \frac{x^3\frac{d(-\sin x)}{dx} - (-\sin x)\frac{d(x^3)}{dx}}{(x^3)^2} - \left[\frac{x^4\frac{d(3\cos x)}{dx} - 3\cos x\frac{d(x^4)}{dx}}{(x^4)^2}\right]$$

$$= \frac{x^3(-\cos x) + \sin x(3x^2)}{x^6} - \left[\frac{x^4(-3\sin x) - 3\cos x(4x^3)}{x^8}\right]$$

$$= \frac{-x^3\cos x + 3x^2\sin x}{x^6} + \frac{3x^4\sin x + 12x^3\cos x}{x^8}$$

Separating into fractions gives:

$$\frac{d^2y}{dx^2} = \frac{-x^3\cos x}{x^6} + \frac{3x^2\sin x}{x^6} + \frac{3x^4\sin x}{x^8} + \frac{12x^3\cos x}{x^8}$$

$$= \frac{-\cos x}{x^3} + \frac{3\sin x}{x^4} + \frac{3\sin x}{x^4} + \frac{12\cos x}{x^5}$$

Combining the fractions again by using x^5 as LCM gives:

$$\frac{d^2y}{dx^2} = \frac{-x^2\cos x + 3x\sin x + 3x\sin x + 12\cos x}{x^5}$$

$$\frac{d^2y}{dx^2} = \frac{6x\sin x + 12\cos x - x^2\cos x}{x^5}$$

7. Find the third derivative of $y = \log_e(1 + 2x)^2$

Solution

$$y = \log_e(1 + 2x)^2$$

$$\frac{dy}{dx} = \frac{2(1+2x)^{2-1} \times \frac{d(2x)}{dx}}{(1+2x)^2}$$

[Note the use of chain rule in finding the derivative of $(1 + 2x)^2$]

$$= \frac{2(1+2x) \times 2}{(1+2x)^2}$$

$$= \frac{4(1+2x)}{(1+2x)^2}$$

$$= \frac{4}{1+2x} \qquad (1 + 2x \text{ cancels out})$$

$$\frac{dy}{dx} = 4(1 + 2x)^{-1}$$

Using chain rule gives $\frac{d^2y}{dx^2}$ as follows:

$$\frac{d^2y}{dx^2} = -1 \times 4(1 + 2x)^{-1-1} \times \frac{d(2x)}{dx}$$

$$= -4(1 + 2x)^{-2} \times 2$$

$$= -8(1 + 2x)^{-2}$$

Use of chain rule again gives $\frac{d^3y}{dx^3}$ as follows:

$$\frac{d^3y}{dx^3} = -2 \times -8(1 + 2x)^{-2-1} \times \frac{d(2x)}{dx}$$

$$= 16(1 + 2x)^{-3} \times 2$$

$$= 32(1 + 2x)^{-3}$$

$$\frac{d^3y}{dx^3} = \frac{32}{(1+2x)^3}$$

8. If y = sec2x, find $\frac{d^3y}{dx^3}$.

Solution

$$y = \sec 2x$$

$$\frac{dy}{dx} = 2\sec 2x \tan 2x$$

We now use product rule to obtain $\frac{d^2y}{dx^2}$ as follows:

$$\frac{d^2y}{dx^2} = 2\sec 2x \frac{d(\tan 2x)}{dx} + \tan 2x \frac{d(2\sec 2x)}{dx}$$

$$= 2\sec 2x(2\sec^2 2x) + \tan 2x(2 \times 2 \sec 2x \tan 2x)$$

$$\frac{d^2y}{dx^2} = 4\sec^3 2x + 4\sec 2x \tan^2 2x$$

We now use chain rule for $4\sec^3 2x$ and product rule for $4\sec 2x \tan^2 2x$ to obtain $\frac{d^3y}{dx^3}$ as follows:

$$\frac{d^3y}{dx^3} = 3 \times 4 \sec^2 2x \frac{d(\sec 2x)}{dx} + 4\sec 2x \frac{d(\tan^2 2x)}{dx} + \tan^2 2x \frac{d(4\sec 2x)}{dx}$$

$$= 12\sec^2 2x(2\sec 2x \tan 2x) + 4\sec 2x \left[2\tan 2x \frac{d(\tan 2x)}{dx}\right] + \tan^2 2x[4(2\sec 2x \tan 2x)]$$

440

$$= 24\sec^3 2x\tan 2x + 8\sec 2x\tan 2x(2\sec^2 2x) + 8\sec 2x\tan^3 2x$$
$$= 24\sec^3 2x\tan 2x + 16\sec^3 2x\tan 2x + 8\sec 2x\tan^3 2x$$
$$\frac{d^3y}{dx^3} = 40\sec^3 2x\tan 2x + 8\sec 2x\tan^3 2x$$

9. Given that y = 2cos5x + 3sin5x, prove that $\frac{d^2y}{dx^2}$ + 25y = 0

Solution

$$y = 2\cos 5x + 3\sin 5x$$

$$\frac{dy}{dx} = 2 \times 5(-\sin 5x) + 3 \times 5(\cos 5x)$$

$$= -10\sin 5x + 15\cos 5x$$

$$\frac{d^2y}{dx^2} = -10 \times 5(\cos 5x) + 15 \times 5(-\sin 5x)$$

$$= -50\cos 5x - 75\sin 5x$$

We now obtain 25y as follows:

$$y = 2\cos 5x + 3\sin 5x$$

$$25y = 25(2\cos 5x + 3\sin 5x)$$

$$25y = 50\cos 5x + 75\sin 5x$$

We now simplify $\frac{d^2y}{dx^2}$ + 25y as follows:

$$\frac{d^2y}{dx^2} + 25y = -50\cos 5x - 75\sin 5x + 50\cos 5x + 75\sin 5x$$

$$= 0 \quad \text{(Equal terms with opposite signs cancel out)}$$

Therefore, $\frac{d^2y}{dx^2}$ + 25y = 0   (As proven above)

10. If $y = x + \sqrt{4 + x^2}$, show that $(4 + x^2)\frac{d^2y}{dx^2} + x\frac{dy}{dx} - y = 0$

Solution

$$y = x + \sqrt{4 + x^2}$$

$$y = x + (4 + x^2)^{\frac{1}{2}}$$

$$\frac{dy}{dx} = 1 + \frac{1}{2}(4 + x^2)^{\frac{1}{2} - 1} \times \frac{d(x^2)}{dx}$$

$$= 1 + \frac{1}{2}(4 + x^2)^{-\frac{1}{2}} \times 2x$$

$$\frac{dy}{dx} = 1 + x(4 + x^2)^{-\frac{1}{2}}$$

Using product rule, we obtain $\frac{d^2y}{dx^2}$ as follows:

$$\frac{d^2y}{dx^2} = 0 + x\left[\frac{d(4+x^2)^{-\frac{1}{2}}}{dx}\right] + (4+x^2)^{-\frac{1}{2}}\frac{d(x)}{dx}$$

$$= x\left[-\frac{1}{2}(4+x^2)^{-\frac{1}{2}-1}\frac{d(x^2)}{dx}\right] + (4+x^2)^{-\frac{1}{2}}(1)$$

$$= x\left[-\frac{1}{2}(4+x^2)^{-\frac{3}{2}} \times 2x\right] + (4+x^2)^{-\frac{1}{2}}$$

$$= -x^2(4+x^2)^{-\frac{3}{2}} + (4+x^2)^{-\frac{1}{2}}$$

$$\frac{d^2y}{dx^2} = \frac{-x^2}{(4+x^2)^{\frac{3}{2}}} + \frac{1}{(4+x^2)^{\frac{1}{2}}}$$

Let us now simplify $(4+x^2)\dfrac{d^2y}{dx^2} + x\dfrac{dy}{dx} - y$ as follows:

$$(4+x^2)\left[\frac{-x^2}{(4+x^2)^{\frac{3}{2}}} + \frac{1}{(4+x^2)^{\frac{1}{2}}}\right] + x[1 + x(4+x^2)^{-\frac{1}{2}}] - [x + (4+x^2)^{\frac{1}{2}}]$$

Expanding the brackets gives:

$$-x^2(4+x^2)^{1-\frac{3}{2}} + (4+x^2)^{1-\frac{1}{2}} + x + x^2(4+x^2)^{-\frac{1}{2}} - x - (4+x^2)^{\frac{1}{2}}$$

$$= -x^2(4+x^2)^{-\frac{1}{2}} + (4+x^2)^{\frac{1}{2}} + x + x^2(4+x^2)^{-\frac{1}{2}} - x - (4+x^2)^{\frac{1}{2}}$$

$$= 0 \quad \text{(Note that equal terms with opposite signs cancel out one another to give zero)}$$

Therefore, $(4+x^2)\dfrac{d^2y}{dx^2} + x\dfrac{dy}{dx} - y = 0$ (As proven above)

Exercise 41

1. Find the first, second and third derivatives of $x^4 - 2x^3 - 7x^2 + 1$

2. If $y = \ln 2x^3$, find $\dfrac{d^2y}{dx^2}$.

3. Find $\dfrac{d^2y}{dx^2}$ given that $y = e^{5x^4}$

4. If $y = \sin^2 x^5$, find the second derivative of y.

5. Find $\dfrac{d^3y}{dx^3}$ given that $y = \ln(4x - 10)$

6. Find the second derivative of $y = \dfrac{\sin x}{x}$

7. Find the third derivative of $y = (\ln x)^2$

8. If $y = \cot x$, find $\dfrac{d^3y}{dx^3}$.

9. Given that $y = \sin^2 x + \cos 2x$, find $\dfrac{d^2y}{dx^2} - y$

10. If $y = 2e^{2x} + 5e^{-x}$ evaluate $\dfrac{d^2y}{dx^2} - \dfrac{dy}{dx} - 2y$

11. Given that $y = \cos^2 x + 2\sin x^2$ find $\dfrac{d^2y}{dx^2} - \dfrac{dy}{dx}$

12. Find the second derivative of $y = \dfrac{\tan x}{3x}$

13. If $y = e^{-x}\cos 2x$, find the second derivative of y.

14. If $y = 2\cos 3x + 5\sin 3x$, evaluate $\dfrac{d^2y}{dx^2} + 9y$

15. Find the second derivative of $y = \dfrac{\sin 3x}{3x}$

CHAPTER 42
MISCELLANEOUS PROBLEMS ON DIFFERENTIAL CALCULUS

This chapter covers worked examples on general problems involving combination of topics treated in the various previous chapters. More challenging problems would also be covered here.

Examples

1. If $y = \dfrac{1}{(2x^3 - 5)^4}$ find $\dfrac{dy}{dx}$.

Solution

$$y = \frac{1}{(2x^3 - 5)^4}$$

This can be written as:

$$y = (2x^3 - 5)^{-4}$$

We now use chain rule to differentiate it as follows:

$$\frac{dy}{dx} = -4(2x^3 - 5)^{-4-1} \times \frac{d(2x^3 - 5)}{dx}$$

$$= -4(2x^3 - 5)^{-5} \times 6x^2$$

$$= -24x^2(2x^3 - 5)^{-5}$$

$$\frac{dy}{dx} = \frac{-24x^2}{(2x^3 - 5)^5}$$

2. Given that $f(x) = 5x^4 - 3x^3 - 7x^2 + 9$, find $f'(2)$

Solution

$$f(x) = 5x^4 - 3x^3 - 7x^2 + 9$$

We differentiate $f(x)$ to obtain $f'(x)$ as follows:

$$f'(x) = 20x^3 - 9x^2 - 14x$$

In order to find $f'(2)$, we simply substitute 2 for x in $f'(x)$ as follows:

$$f'(x) = 20x^3 - 9x^2 - 14x$$

$$f'(2) = 20(2^3) - 9(2^2) - 14(2)$$

$$= 20(8) - 9(4) - 28$$

$$= 160 - 36 - 28$$

$$f'(2) = 96$$

3. A function $f(x)$ is given by $f(x) = \dfrac{\sqrt{4 + 3x^2}}{x^3}$. Find:

(a) the derivative of $f(x)$

(b) the gradient of $f(x)$ at the point $(2, \frac{1}{2})$

<u>Solution</u>

(a) $f(x) = \dfrac{\sqrt{4 + 3x^2}}{x^3}$

We apply quotient rule to differentiate f(x) as follows:

$f'(x) = \dfrac{x^3\left[\dfrac{d(\sqrt{4+3x^2})}{dx}\right] - \sqrt{4+3x^2}\left[\dfrac{d(x^3)}{dx}\right]}{(x^3)^2}$

$= \dfrac{x^3\left[\dfrac{d(4+3x^2)^{\frac{1}{2}}}{dx}\right] - (4+3x^2)^{\frac{1}{2}}(3x^2)}{x^6}$

$= \dfrac{x^3\left[\frac{1}{2}(4+3x^2)^{\frac{1}{2}-1} \ \text{x} \ \dfrac{d(3x^2)}{dx}\right] - 3x^2(4+3x^2)^{\frac{1}{2}}}{x^6}$

$= \dfrac{x^3\left[\frac{1}{2}(4+3x^2)^{-\frac{1}{2}} \ \text{x} \ 6x\right] - 3x^2(4+3x^2)^{\frac{1}{2}}}{x^6}$

$= \dfrac{x^3\left[3x\,(4+3x^2)^{-\frac{1}{2}}\right] - 3x^2(4+3x^2)^{\frac{1}{2}}}{x^6}$

$= \dfrac{3x^4(4+3x^2)^{-\frac{1}{2}} - 3x^2(4+3x^2)^{\frac{1}{2}}}{x^6}$

Let us factorize the expression above by taking $3x^2(4 + 3x^2)^{-\frac{1}{2}}$ as the common factor. This gives:

$f'(x) = \dfrac{3x^2(4+3x^2)^{-\frac{1}{2}}[x^2 - (4+3x^2)]}{x^6}$

Note that $(4 + 3x^2)^{\frac{1}{2}} \div (4 + 3x^2)^{-\frac{1}{2}} = (4 + 3x^2)^{\frac{1}{2}-(-\frac{1}{2})} = 4 + 3x^2$ as the exponent becomes 1. Recall that exponents are subtracted during division.

$= \dfrac{3x^2(4+3x^2)^{-\frac{1}{2}}(x^2 - 4 - 3x^2)}{x^6}$

$= \dfrac{3x^2(4+3x^2)^{-\frac{1}{2}}(-2x^2 - 4)}{x^6}$

$= \dfrac{-3x^2(4+3x^2)^{-\frac{1}{2}}(2x^2 + 4)}{x^6}$

$= \dfrac{-3(4+3x^2)^{-\frac{1}{2}}(2x^2 + 4)}{x^4}$ (Note that x^2 has cancelled out of x^6)

$f'(x) = \dfrac{-3(2x^2 + 4)}{x^4(4+3x^2)^{\frac{1}{2}}}$

(b) The gradient of f(x) at the point $(2, \frac{1}{2})$ is obtained by substituting 2 for x in f'(x). Note that

the point $(2, \frac{1}{2})$ is at $x = 2$ and $y = \frac{1}{2}$. We ignore $y = \frac{1}{2}$ since y is not in the expression for f'(x).

Hence, $\quad f'(x) = \dfrac{-3(2x^2 + 4)}{x^4(4 + 3x^2)^{\frac{1}{2}}}$

At $(2, \frac{1}{2})$, $\quad f'(2) = \dfrac{-3(2(2)^2 + 4)}{(2)^4(4 + 3(2)^2)^{\frac{1}{2}}}$

$\quad = \dfrac{-3(8 + 4)}{16(4 + 12)^{\frac{1}{2}}}$

$\quad = \dfrac{-3(12)}{16(16)^{\frac{1}{2}}}$

$\quad = \dfrac{-36}{16 \text{ x } 4}$ $\quad$ (Note that $(16)^{\frac{1}{2}} = \sqrt{16} = 4$)

$f'(2) = \dfrac{-9}{16}$ $\quad$ (In its lowest term)

Therefore the gradient of f(x) at the point $(2, \frac{1}{2})$ is $\dfrac{-9}{16}$

4. If $x^2y^2 - 3xy + 4xy^3 = 4$, find:

(a) the derivative of the expression

(b) the gradient at $(-1, 2)$

Solution

(a) $\quad x^2y^2 - 3xy + 4xy^3 = 4$

The use of implicit differentiation combined with product rule gives us the derivative as follows:

$$x^2\frac{d(y^2)}{dx} + y^2\frac{d(x^2)}{dx} - \left[3x\frac{d(y)}{dx} + y\frac{d(3x)}{dx}\right] + 4x\frac{d(y^3)}{dx} + y^3(4) = \frac{d(4)}{dx}$$

$$x^2\left(2y\frac{dy}{dx}\right) + y^2(2x) - \left[3x\frac{dy}{dx} + y(3)\right] + 4x\left(3y^2\frac{dy}{dx}\right) + 4y^3 = 0$$

$$2x^2y\frac{dy}{dx} + 2xy^2 - 3x\frac{dy}{dx} - 3y + 12xy^2\frac{dy}{dx} + 4y^3 = 0$$

Collect terms in $\dfrac{dy}{dx}$ on one side of the equation. This gives:

$$2x^2y\frac{dy}{dx} - 3x\frac{dy}{dx} + 12xy^2\frac{dy}{dx} = 3y - 2xy^2 - 4y^3$$

Factorizing the left hand side gives:

$$\frac{dy}{dx}(2x^2y - 3x + 12xy^2) = 3y - 2xy^2 - 4y^3$$

Divide both sides by the terms in the bracket. This gives:

$$\frac{dy}{dx} = \frac{3y - 2xy^2 - 4y^3}{12x^2y - 3x + 12xy^2}$$

(b) At the point (−1, 2) the gradient of the expression is obtained by simply substituting −1 for x and 2 for y in the expression for the derivative. This is done as follows:

$$\frac{dy}{dx} = \frac{3y - 2xy^2 - 4y^3}{12x^2y - 3x + 12xy^2}$$

$$= \frac{3(2) - 2(-1)(2)^2 - 4(2)^3}{12(-1)^2(2) - 3(-1) + 12(-1)(2)^2}$$

$$= \frac{6 + 8 - 32}{24 + 3 - 48}$$

$$= \frac{-18}{-21}$$

$$= \frac{6}{7}$$

5. Find the derivative of $(x + 3y^2)^3 = 7$

Solution

$(x + 3y^2)^3 = 7$

We differentiate implicitly and apply chain rule as follows:

$$3(x + 3y^2)^{3-1} \times \frac{d(x+3y^2)}{dx} = 0$$

$$3(x + 3y^2)^2\left(1 + 6y\frac{dy}{dx}\right) = 0$$

Dividing both sides by $3(x + 3y^2)^2$ gives:

$$1 + 6y\frac{dy}{dx} = 0 \qquad \text{(Note that } \frac{0}{3(x + 3y^2)^2} = 0\text{)}$$

$$6y\frac{dy}{dx} = -1$$

$$\frac{dy}{dx} = -\frac{1}{6y}$$

6. If $y = (3x^2 - 2x + 5)(2x - 3)$, find $\frac{dy}{dx}$.

Solution

$y = (3x^2 - 2x + 5)(2x - 3)$

We differentiate the expression by applying product rule as follows:

$$\frac{dy}{dx} = (3x^2 - 2x + 5)(2) + (2x - 3)(6x - 2)$$

$$= 6x^2 - 4x + 10 + 12x^2 - 4x - 18x + 6$$

$$\frac{dy}{dx} = 18x^2 - 26x + 16$$

7. Find the derivative of $a^{1 + \tan x}$

Solution

$y = a^{1 + \tan x}$

448

Let $u = 1 + \tan x$

Hence, $y = a^u$

$$\frac{du}{dx} = \sec^2 x$$

$$\frac{dy}{du} = a^u \log_e a$$

Hence, $\dfrac{dy}{dx} = \dfrac{dy}{du} \times \dfrac{du}{dx}$

$$= a^u \log_e a \times \sec^2 x$$

$$\frac{dy}{dx} = a^{1+\tan x} \sec^2 x \log_e a$$

8. Find the derivative of $\dfrac{\log_e x}{1+\cos x}$

Solution

$$y = \frac{\log_e x}{1+\cos x}$$

We apply product rule as follows:

$$\frac{dy}{dx} = \frac{(1+\cos x)\frac{d(\log_e x)}{dx} - \log_e x \frac{d(1+\cos x)}{dx}}{(1+\cos x)^2}$$

$$= \frac{(1+\cos x)\frac{1}{x} - \log_e x(-\sin x)}{(1+\cos x)^2}$$

$$= \frac{\frac{1+\cos x}{x} + \sin x \log_e x}{(1+\cos x)^2}$$

$$= \frac{\frac{1+\cos x + x \sin x \log_e x}{x}}{(1+\cos x)^2}$$

$$\frac{dy}{dx} = \frac{1+\cos x + x \sin x \log_e x}{x(1+\cos x)^2}$$

9. Differentiate with respect to x: $\ln(\cos x + \sin x)$

Solution

$$y = \ln(\cos x + \sin x)$$

$$\frac{dy}{dx} = \frac{\frac{d(\cos x + \sin x)}{dx}}{\cos x + \sin x}$$

$$= \frac{-\sin x + \cos x}{\cos x + \sin x}$$

$$\frac{dy}{dx} = \frac{\cos x - \sin x}{\cos x + \sin x}$$

10. If $y = e^{\sin 2x + \cos x}$ find $\dfrac{dy}{dx}$.

Solution

$$y = e^{\sin 2x + \cos x}$$

$$\frac{dy}{dx} = \frac{d(\sin 2x + \cos x)}{dx} \times e^{\sin 2x + \cos x}$$

$$= (2\cos 2x - \sin x)(e^{\sin 2x + \cos x})$$

11. Given that $y = e^x - e^{-x}$ show that $\dfrac{d^3y}{dx^3} + \dfrac{d^2y}{dx^2} + \dfrac{dy}{dx} + y = 4e^x$

Solution

$$y = e^x - e^{-x}$$

$$\frac{dy}{dx} = e^x - (-e^{-x}) \quad \text{(Note that } \frac{d(e^{-x})}{dx} = \frac{d(-x)}{dx} \times e^{-x} = -1 \times e^{-x} = -e^{-x})$$

$$= e^x + e^{-x}$$

$$\frac{d^2y}{dx^2} = e^x - e^{-x}$$

$$\frac{d^3y}{dx^3} = e^x + e^{-x}$$

Let us now substitute corresponding term into $\dfrac{d^3y}{dx^3} + \dfrac{d^2y}{dx^2} + \dfrac{dy}{dx} + y$ as follows:

$$(e^x + e^{-x}) + (e^x - e^{-x}) + (e^x + e^{-x}) + (e^x - e^{-x})$$

Collecting like terms together gives:

$$e^x + e^x + e^x + e^x + e^{-x} - e^{-x} + e^{-x} - e^{-x}$$

$$= 4e^x \quad \text{(Note that } e^{-x} \text{ cancels out each other)}$$

Hence, $\dfrac{d^3y}{dx^3} + \dfrac{d^2y}{dx^2} + \dfrac{dy}{dx} + y = 4e^x$ (As proven above)

12. Differentiate with respect to x: $\ln\left(\dfrac{1 - 3x^2}{1 + 3x^2}\right)^{\frac{1}{2}}$

Solution

$$y = \ln\left(\frac{1 - 3x^2}{1 + 3x^2}\right)^{\frac{1}{2}}$$

$$= \log_e\left(\frac{1 - 3x^2}{1 + 3x^2}\right)^{\frac{1}{2}} \quad \text{(Note that "ln" is } \log_e)$$

$$= \log_e\left[\frac{(1 - 3x^2)^{\frac{1}{2}}}{(1 + 3x^2)^{\frac{1}{2}}}\right]$$

$$= \log_e(1 - 3x^2)^{\frac{1}{2}} - \log_e(1 + 3x^2)^{\frac{1}{2}}$$

$$y = \frac{1}{2}\log_e(1 - 3x^2) - \frac{1}{2}\log_e(1 + 3x^2)$$

$$\frac{dy}{dx} = \frac{1}{2}\frac{\frac{d(1-3x^2)}{dx}}{1-3x^2} - \frac{1}{2}\frac{\frac{d(1+3x^2)}{dx}}{1+3x^2}$$

$$= \frac{1}{2}\frac{-6x}{1-3x^2} - \frac{1}{2}\frac{6x}{1+3x^2}$$

$$= \frac{-3x}{1-3x^2} - \frac{3x}{1+3x^2} \qquad \text{(Note that } \frac{1}{2} \text{ reduces } 6x \text{ to } 3x)$$

$$= \frac{-3x(1+3x^2) - 3x(1-3x^2)}{(1-3x^2)(1+3x^2)}$$

$$= \frac{-3x - 9x^3 - 3x + 9x^3)}{(1-3x^2)(1+3x^2)}$$

$$= \frac{-6x}{1 + 3x^2 - 3x^2 - 9x^4}$$

$$\frac{dy}{dx} = \frac{-6x}{1 - 9x^4}$$

13. If $y = \dfrac{x}{\sqrt{9 - x^2}}$ show that: $(9 - x^2)\dfrac{d^2y}{dx^2} = 3x\dfrac{dy}{dx}$

Solution

$$y = \frac{x}{\sqrt{9 - x^2}}$$

Applying quotient rule gives $\dfrac{dy}{dx}$ as follows:

$$\frac{dy}{dx} = \frac{\sqrt{9 - x^2}(1) - x\frac{d(\sqrt{9-x^2})}{dx}}{(\sqrt{9 - x^2})^2}$$

$$= \frac{(9 - x^2)^{\frac{1}{2}}(1) - x\frac{d[(9-x^2)^{\frac{1}{2}}]}{dx}}{[(9 - x^2)^{\frac{1}{2}}]^2}$$

$$= \frac{(9 - x^2)^{\frac{1}{2}} - x\left[\frac{1}{2}(9-x^2)^{\frac{1}{2} - 1}\right] \times (-2x)}{9 - x^2} \qquad \text{(Note that } -2x \text{ is from the derivative of } -x^2)$$

$$= \frac{(9 - x^2)^{\frac{1}{2}} - x\left[-x(9 - x^2)^{-\frac{1}{2}}\right]}{9 - x^2}$$

$$= \frac{(9 - x^2)^{\frac{1}{2}} + x^2(9 - x^2)^{-\frac{1}{2}}}{9 - x^2}$$

$$= \frac{(9 - x^2)^{\frac{1}{2}} + \frac{x^2}{(9 - x^2)^{\frac{1}{2}}}}{9 - x^2}$$

$$= \frac{\dfrac{9 - x^2 + x^2}{(9 - x^2)^{\frac{1}{2}}}}{9 - x^2}$$

$$= \frac{9}{(9 - x^2)^{\frac{1}{2}}(9 - x^2)}$$

$$= \frac{9}{(9 - x^2)^{\frac{3}{2}}}$$

$$\frac{dy}{dx} = 9(9 - x^2)^{-\frac{3}{2}}$$

Applying chain rule gives $\dfrac{d^2 y}{dx^2}$ as follows:

$$\frac{d^2 y}{dx^2} = \frac{-3}{2} \times 9(9 - x^2)^{-\frac{3}{2} - 1} \times -2x \qquad \text{(Note that } -2x \text{ is from the derivative of } -x^2)$$

$$= \frac{-27}{2} \times -2x(9 - x^2)^{-\frac{5}{2}}$$

$$= 27x(9 - x^2)^{-\frac{5}{2}}$$

$$\frac{d^2 y}{dx^2} = \frac{27x}{(9 - x^2)^{\frac{5}{2}}}$$

Let us now show that $(9 - x^2)\dfrac{d^2 y}{dx^2} = 3x\dfrac{dy}{dx}$

We simplify $(9 - x^2)\dfrac{d^2 y}{dx^2}$ as follows:

$$(9 - x^2)\frac{27x}{(9 - x^2)^{\frac{5}{2}}}$$

$$= \frac{27x(9 - x^2)}{(9 - x^2)^{\frac{5}{2}}}$$

$$= 27x(9 - x^2)^{1 - \frac{5}{2}}$$

$$= 27x(9 - x^2)^{-\frac{3}{2}}$$

$$= \frac{27x}{(9 - x^2)^{\frac{3}{2}}}$$

Hence $(9 - x^2)\dfrac{d^2 y}{dx^2}$ gives us $\dfrac{27x}{(9 - x^2)^{\frac{3}{2}}}$

Let us now simplify $3x\dfrac{dy}{dx}$ as follows:

$$3x[9(9 - x^2)^{-\frac{3}{2}}] \qquad \text{(Note that } \frac{dy}{dx} = 9(9 - x^2)^{-\frac{3}{2}} \text{ as obtained above)}$$

$$= 27x(9 - x^2)^{-\frac{3}{2}}$$

$$= \frac{27x}{(9 - x^2)^{\frac{3}{2}}}$$

Hence $3x\dfrac{dy}{dx}$ also gives us $\dfrac{27x}{(9-x^2)^{\frac{3}{2}}}$

Therefore, $(9-x^2)\dfrac{d^2y}{dx^2} = 3x\dfrac{dy}{dx}$ as both sides give $\dfrac{27x}{(9-x^2)^{\frac{3}{2}}}$

14. Given that $y = \dfrac{x}{x-1}$, show that: $(x-1)\dfrac{d^2y}{dx^2} + 2\dfrac{dy}{dx} = 0$

Solution

$$y = \dfrac{x}{x-1}$$

Using quotient rule gives $\dfrac{dy}{dx}$ as follows:

$$\dfrac{dy}{dx} = \dfrac{(x-1)(1) - x(1)}{(x-1)^2}$$

$$= \dfrac{x-1-x}{(x-1)^2}$$

$$\dfrac{dy}{dx} = \dfrac{-1}{(x-1)^2}$$

Using chain rule gives $\dfrac{d^2y}{dx^2}$ as follows:

$$\dfrac{dy}{dx} = \dfrac{-1}{(x-1)^2}$$

$$= -1(x-1)^{-2}$$

$$\dfrac{d^2y}{dx^2} = -2 \times -1(x-1)^{-2-1} \times \dfrac{d(x)}{dx}$$

$$= 2(x-1)^{-3} \times 1$$

$$\dfrac{d^2y}{dx^2} = \dfrac{2}{(x-1)^3}$$

From the question, let us now simplify $(x-1)\dfrac{d^2y}{dx^2} + 2\dfrac{dy}{dx}$ as follows:

$$(x-1)\dfrac{d^2y}{dx^2} + 2\dfrac{dy}{dx}$$

$$= (x-1)\dfrac{2}{(x-1)^3} + (2)\dfrac{-1}{(x-1)^2}$$

$$= \dfrac{2}{(x-1)^2} - \dfrac{2}{(x-1)^2}$$

$$= 0$$

This shows that $(x-1)\dfrac{d^2y}{dx^2} + 2\dfrac{dy}{dx}$ is equal to zero.

15. Find, with respect to x, the derivative of $\left(x - \dfrac{5}{x}\right)^3$

Solution

$$y = \left(x - \frac{5}{x}\right)^3$$

This can also be written as:

$$y = (x - 5x^{-1})^3$$

$$\frac{dy}{dx} = 3(x - 5x^{-1})^{3-1} \times \frac{d(x - 5x^{-1})}{dx} \qquad \text{(By use of chain rule)}$$

$$= 3(x - 5x^{-1})^2 \times 1 - (-1 \times 5x^{-1-1})$$

$$= 3(x - 5x^{-1})^2 \times 1 - (-5x^{-2})$$

$$= 3(x - 5x^{-1})^2 \times (1 + 5x^{-2})$$

$$= 3(x - 5x^{-1})^2(1 + 5x^{-2})$$

$$\frac{dy}{dx} = 3\left(x - \frac{5}{x}\right)^2 \left(1 + \frac{5}{x^2}\right)$$

16. Given that $\exp(2x^2 + 2y^2 - 16) = x + y$, find:

(a) $\dfrac{dy}{dx}$

(b) $\dfrac{dy}{dx}$ at $\left(\dfrac{1}{2} \ \dfrac{1}{2}\right)$

Solution

$$\exp(2x^2 + 2y^2 - 16) = x + y$$

This can also be written as

$$e^{2x^2 + 2y^2 - 16} = x + y \qquad \text{(Note that } \exp x = e^x\text{)}$$

We now differentiate implicitly as follows:

$$\frac{d(2x^2 + 2y^2 - 16)}{dx} \times (e^{2x^2 + 2y^2 - 16}) = \frac{d(x)}{dx} + \frac{d(y)}{dx}$$

$$\left(4x + 4y\frac{dy}{dx}\right)(e^{2x^2 + 2y^2 - 16}) = 1 + \frac{dy}{dx}$$

Expanding the bracket gives:

$$4x(e^{2x^2 + 2y^2 - 16}) + 4y\frac{dy}{dx}(e^{2x^2 + 2y^2 - 16}) = 1 + \frac{dy}{dx}$$

Collecting terms in $\dfrac{dy}{dx}$ on the left hand side gives:

$$4y\frac{dy}{dx}(e^{2x^2 + 2y^2 - 16}) - \frac{dy}{dx} = 1 - 4x(e^{2x^2 + 2y^2 - 16})$$

Factorizing the left hand side gives:

$$\frac{dy}{dx}[4y(e^{2x^2 + 2y^2 - 16}) - 1] = 1 - 4x(e^{2x^2 + 2y^2 - 16})$$

Hence, $\dfrac{dy}{dx} = \dfrac{1 - 4x(e^{2x^2 + 2y^2 - 16})}{4y(e^{2x^2 + 2y^2 - 16}) - 1}$

Or, $\dfrac{dy}{dx} = \dfrac{-[4x(e^{2x^2 + 2y^2 - 16}) - 1]}{4y(e^{2x^2 + 2y^2 - 16}) - 1}$

(b) In order to find $\dfrac{dy}{dx}$ at $\left(\dfrac{1}{2}\ \dfrac{1}{2}\right)$, we simply substitute $x = \dfrac{1}{2}$ and $y = \dfrac{1}{2}$ into the expression for $\dfrac{dy}{dx}$ as follows:

$$\frac{dy}{dx} = \frac{-[4x(e^{2x^2 + 2y^2 - 16}) - 1]}{4y(e^{2x^2 + 2y^2 - 16}) - 1}$$

Since $x = \dfrac{1}{2}$ and $y = \dfrac{1}{2}$, this simplifies to give:

$$\frac{dy}{dx} \text{ at } \left(\frac{1}{2}\ \frac{1}{2}\right) = \frac{-[4\left(\frac{1}{2}\right)(e^{2x^2 + 2y^2 - 16}) - 1]}{4\left(\frac{1}{2}\right)(e^{2x^2 + 2y^2 - 16}) - 1}$$

$$= \frac{-[2(e^{2x^2 + 2y^2 - 16}) - 1]}{2(e^{2x^2 + 2y^2 - 16}) - 1}$$

The numerator cancels out the denominator to give:

$$\frac{dy}{dx} = \frac{-1}{1}$$

$$\frac{dy}{dx} = -1$$

17. If $y = x^3 - 2x^2 + 5$, show that: $x\dfrac{dy}{dx} - 3y - 2x^2 + 15 = 0$

Solution

$$y = x^3 - 2x^2 + 5$$

$$\frac{dy}{dx} = 3x^2 - 4x$$

Let us now simplify $x\dfrac{dy}{dx} - 3y - 2x^2 + 15$ as follows:

$$x\frac{dy}{dx} - 3y - 2x^2 + 15$$
$$= x(3x^2 - 4x) - 3y - 2x^2 + 15$$
$$= x(3x^2 - 4x) - 3(x^3 - 2x^2 + 5) - 2x^2 + 15 \quad \text{(Note that } x^3 - 2x^2 + 5 \text{ has been substituted for y)}$$
$$= 3x^3 - 4x^2 - 3x^3 + 6x^2 - 15 - 2x^2 + 15$$
$$= 0$$

Therefore, $x\dfrac{dy}{dx} - 3y - 2x^2 + 15 = 0$ as proven above.

18. Determine $\dfrac{d^2}{dx^2}\left(x\sin\dfrac{1}{x}\right)$

Solution

This means the second derivative of $x\sin\dfrac{1}{x}$

Let $y = x\sin\dfrac{1}{x}$

Or, $y = x\sin x^{-1}$

455

We now apply product rule to obtain $\dfrac{dy}{dx}$ as follows:

$$\dfrac{dy}{dx} = x\left[\dfrac{d(x^{-1})}{dx}\cos x^{-1}\right] + \sin x^{-1}\left[\dfrac{d(x)}{dx}\right]$$

$$= x(-x^{-2}\cos x^{-1}) + \sin x^{-1}(1)$$

$$= x(\dfrac{-1}{x^2}\cos x^{-1}) + \sin x^{-1}$$

$$= \dfrac{-x}{x^2}\cos x^{-1} + \sin x^{-1}$$

$$= \dfrac{-\cos x^{-1}}{x} + \sin x^{-1}$$

$$\dfrac{dy}{dx} = -\dfrac{1}{x}\cos\dfrac{1}{x} + \sin\dfrac{1}{x}$$

Or, $\dfrac{dy}{dx} = -x^{-1}\cos x^{-1} + \sin x^{-1}$

We now obtain $\dfrac{d^2y}{dx^2}$ as follows:

$$\dfrac{d^2y}{dx^2} = -x^{-1}\left[\dfrac{d(\cos x^{-1})}{dx}\right] + \cos x^{-1}\left[\dfrac{d(-x^{-1})}{dx}\right] + \dfrac{d(\sin x^{-1})}{dx}$$

$$= -x^{-1}\left[\dfrac{d(x^{-1})}{dx}(-\sin x^{-1})\right] + \cos x^{-1}(x^{-2}) + \left[\dfrac{d(x^{-1})}{dx}\cos x^{-1}\right]$$

$$= -x^{-1}[-x^{-2}(-\sin x^{-1})] + \cos x^{-1}(x^{-2}) - x^{-2}(\cos x^{-1})$$

$$= -x^{-1}(x^{-2}\sin x^{-1}) + x^{-2}\cos x^{-1} - x^{-2}\cos x^{-1}$$

$$= -x^{-1}(x^{-2}\sin x^{-1}) \quad \text{(Note that } x^{-2}\cos x^{-1} \text{ cancels out)}$$

$$= -x^{-3}\sin x^{-1}$$

$$\dfrac{d^2y}{dx^2} = -\dfrac{1}{x^3}\sin\dfrac{1}{x}$$

19. Given that $y = \dfrac{e^x + e^{-x}}{e^x - e^{-x}}$

(a) find $\dfrac{d^2y}{dx^2}$

(b) show that $\dfrac{d^2y}{dx^2} + 2y\dfrac{dy}{dx} = 0$

Solution

(a) $y = \dfrac{e^x + e^{-x}}{e^x - e^{-x}}$

By using quotient rule we obtain $\dfrac{dy}{dx}$ as follows:

$$\dfrac{dy}{dx} = \dfrac{e^x - e^{-x}\left[\dfrac{d(e^x + e^{-x})}{dx}\right] - \left(e^x + e^{-x}\left[\dfrac{d(e^x - e^{-x})}{dx}\right]\right)}{(e^x - e^{-x})^2}$$

$$= \dfrac{(e^x - e^{-x})(e^x - e^{-x}) - (e^x + e^{-x})(e^x + e^{-x})}{(e^x - e^{-x})^2}$$

$$= \frac{(e^x - e^{-x})^2 - (e^x + e^{-x})^2}{(e^x - e^{-x})^2}$$

Separating into fractions gives:

$$= \frac{(e^x - e^{-x})^2}{(e^x - e^{-x})^2} - \frac{(e^x + e^{-x})^2}{(e^x - e^{-x})^2}$$

$$\frac{dy}{dx} = 1 - \frac{(e^x + e^{-x})^2}{(e^x - e^{-x})^2}$$

We now apply quotient and chain rules to obtain $\dfrac{d^2y}{dx^2}$ as follows:

$$\frac{d^2y}{dx^2} = 0 - \frac{(e^x - e^{-x})^2[2(e^x + e^{-x})(e^x - e^{-x})] - (e^x + e^{-x})^2[2(e^x - e^{-x})(e^x + e^{-x})]}{[(e^x - e^{-x})^2]^2}$$

$$= -\left[\frac{2(e^x - e^{-x})^3(e^x + e^{-x}) - 2(e^x + e^{-x})^3(e^x - e^{-x})}{(e^x - e^{-x})^4}\right]$$

$$= \frac{-2(e^x - e^{-x})^3(e^x + e^{-x}) + 2(e^x + e^{-x})^3(e^x - e^{-x})}{(e^x - e^{-x})^4}$$

$$= \frac{2(e^x + e^{-x})^3(e^x - e^{-x}) - 2(e^x - e^{-x})^3(e^x + e^{-x})}{(e^x - e^{-x})^4} \quad \text{(After rearranging the numerator)}$$

Separating into fractions gives:

$$= \frac{2(e^x + e^{-x})^3(e^x - e^{-x})}{(e^x - e^{-x})^4} - \frac{2(e^x - e^{-x})^3(e^x + e^{-x})}{(e^x - e^{-x})^4}$$

$$\frac{d^2y}{dx^2} = \frac{2(e^x + e^{-x})^3}{(e^x - e^{-x})^3} - \frac{2(e^x + e^{-x})}{e^x - e^{-x}}$$

(b) Let us simplify $\dfrac{d^2y}{dx^2} + 2y\dfrac{dy}{dx}$ as follows:

$$\frac{d^2y}{dx^2} + 2y\frac{dy}{dx}$$

$$= \frac{2(e^x + e^{-x})^3}{(e^x - e^{-x})^3} - \frac{2(e^x + e^{-x})}{e^x - e^{-x}} + 2\left(\frac{e^x + e^{-x}}{e^x - e^{-x}}\right)\left[1 - \frac{(e^x + e^{-x})^2}{(e^x - e^{-x})^2}\right]$$

Expanding bracket gives:

$$= \frac{2(e^x + e^{-x})^3}{(e^x - e^{-x})^3} - \frac{2(e^x + e^{-x})}{e^x - e^{-x}} + 2\left(\frac{e^x + e^{-x}}{e^x - e^{-x}}\right) - \frac{2(e^x + e^{-x})^3}{(e^x - e^{-x})^3}$$

$= 0$ (Since equal terms with opposite signs cancel out each other)

Therefore, $\dfrac{d^2y}{dx^2} + 2y\dfrac{dy}{dx}$ gives zero as proven above.

Or, $\dfrac{d^2y}{dx^2} + 2y\dfrac{dy}{dx} = 0$

20. Find the derivative of $\ln(\tan 2x)$

Solution

$y = \ln(\tan 2x)$

$$\frac{dy}{dx} = \frac{\frac{d(\tan 2x)}{dx}}{\tan 2x}$$

$$= \frac{2\sec^2 2x}{\tan 2x}$$

Or, $\dfrac{dy}{dx} = \dfrac{2\left(\frac{1}{\cos 2x}\right)\left(\frac{1}{\cos 2x}\right)}{\left(\frac{\sin 2x}{\cos 2x}\right)}$ (Note that $\sec 2x = \dfrac{1}{\cos 2x}$ and $\tan 2x = \dfrac{\sin 2x}{\cos 2x}$)

$$= \left(\frac{2}{\cos 2x}\right)\left(\frac{1}{\cos 2x}\right) \text{ x } \frac{\cos 2x}{\sin 2x}$$

$$\frac{dy}{dx} = \frac{2}{\sin 2x \cos 2x} \qquad \text{(Since } \cos 2x \text{ cancels } \cos 2x\text{)}$$

21. Given that $y = x^3 + 3x^2$, determine $2\dfrac{dy}{dx} - x\dfrac{d^2y}{dx^2}$

Solution

$$y = x^3 + 3x^2$$

$$\frac{dy}{dx} = 3x^2 + 6x$$

$$\frac{d^2y}{dx^2} = 6x + 6$$

Hence, $2\dfrac{dy}{dx} - x\dfrac{d^2y}{dx^2}$ is simplified as follows:

$$2(3x^2 + 6x) - x(6x + 6)$$

$$= 6x^2 + 12x - 6x^2 - 6x$$

$$= 6x$$

Therefore, $2\dfrac{dy}{dx} - x\dfrac{d^2y}{dx^2} = 6x$

22. Find $\dfrac{dy}{dx}$ if $y = 3(3x + \sqrt{x})^2$

Solution

$$y = 3(3x + \sqrt{x})^2$$

Applying chain rule gives:

$$\frac{dy}{dx} = 3 \text{ x } 2\,(3x + x^{\frac{1}{2}})^{2-1} \text{ x } \frac{d(3x + x^{\frac{1}{2}})}{dx} \qquad \text{(Note that } \sqrt{x} = x^{\frac{1}{2}})$$

$$= 6(3x + x^{\frac{1}{2}}) \text{ x } \left(3 + \frac{1}{2}x^{-\frac{1}{2}}\right)$$

$$= 6(3x + x^{\frac{1}{2}})\left(3 + \frac{1}{2x^{\frac{1}{2}}}\right)$$

$$= 6(3x + \sqrt{x})\left(3 + \frac{1}{2\sqrt{x}}\right)$$

Expanding the bracket gives:

$$= 6\left(9x + \frac{3x}{2\sqrt{x}} + 3\sqrt{x} + \frac{1}{2}\right)$$

$$= 6\left(9x + \frac{3}{2}\sqrt{x} + 3\sqrt{x} + \frac{1}{2}\right) \qquad \text{(Note that } \frac{3x}{2\sqrt{x}} = \frac{3}{2}x^{1-\frac{1}{2}} = \frac{3}{2}x^{\frac{1}{2}} = 3\sqrt{x})$$

$$= 6\left(9x + \frac{9}{2}\sqrt{x} + \frac{1}{2}\right)$$

$$\frac{dy}{dx} = 54x + 27\sqrt{x} + 3$$

23. If $y = \dfrac{5x^2 + 7}{x^4}$,

(a) find $\dfrac{d^2y}{dx^2}$

(b) show that $x^2\dfrac{d^2y}{dx^2} + 7x\dfrac{dy}{dx} + 8y = 0$

Solution

(a) $y = \dfrac{5x^2 + 7}{x^4}$

$$= \frac{5x^2}{x^4} + \frac{7}{x^4}$$

$$= \frac{5}{x^2} + \frac{7}{x^4}$$

$$y = 5x^{-2} + 7x^{-4}$$

$$\frac{dy}{dx} = -10x^{-3} - 28x^{-5}$$

Similarly,

$$\frac{d^2y}{dx^2} = 30x^{-4} + 140x^{-6}$$

(b) Let us now simplify $x^2\dfrac{d^2y}{dx^2} + 7x\dfrac{dy}{dx} + 8y$ by substituting appropriately as follows:

$$x^2(30x^{-4} + 140x^{-6}) + 7x(-10x^{-3} - 28x^{-5}) + 8(5x^{-2} + 7x^{-4})$$

$$= 30x^{-2} + 140x^{-4} - 70x^{-2} - 196x^{-4} + 40x^{-2} + 56x^{-4}$$

$$= 30x^{-2} + 40x^{-2} - 70x^{-2} + 140x^{-4} + 56x^{-4} - 196x^{-4}$$

$$= 0$$

Therefore $x^2\dfrac{d^2y}{dx^2} + 7x\dfrac{dy}{dx} + 8y$ gives zero as obtained above.

Or, $x^2\dfrac{d^2y}{dx^2} + 7x\dfrac{dy}{dx} + 8y = 0$

24. If $y = (2x + 5)^4 + \dfrac{x - 1}{2x - 1}$ find $\dfrac{dy}{dx}$.

Solution

$$(2x + 5)^4 + \frac{x - 1}{2x - 1}$$

We now use chain rule and quotient rule as follows:

$$\frac{dy}{dx} = 4(2x+5)^{4-1} \times \frac{d(2x+5)}{dx} + \frac{(2x-1)\frac{d(x-1)}{dx} - \left[(x-1)\frac{d(2x-1)}{dx}\right]}{(2x-1)^2}$$

$$= 4(2x+5)^3(2) + \frac{(2x-1)(1) - [(x-1)(2)]}{(2x-1)^2}$$

$$= 8(2x+5)^3 + \frac{(2x-1) - 2(x-1)}{(2x-1)^2}$$

$$= 8(2x+5)^3 + \frac{2x-1-2x+2}{(2x-1)^2}$$

$$\frac{dy}{dx} = 8(2x+5)^3 + \frac{1}{(2x-1)^2}$$

Exercise 42

1. If $y = x^2(3x^4 - 5)^3$ find $\frac{dy}{dx}$.

2. Given that $f(x) = 2x^5 - x^4 + 2x^3 + x^2 - 3x + 4$ find $f'(-2)$

3. A function $f(x)$ is given by $f(x) = \frac{\sqrt{(3x^2-1)^3}}{x^2}$. Find:

(a) the derivative of $f(x)$

(b) the gradient of $f(x)$ at the point $(1, -2)$

4. If $3xy^3 - y^2 - 4x^3 = 5y$, find:

(a) the derivative of the expression

(b) the gradient at $(-1, 1)$

5. Find the derivative of $(2x^2 + y^2)^2 = 0$

6. If $y = (x^2 + 5)^3(x^3 - 1)$, find $\frac{dy}{dx}$.

7. Find the derivative of $e^{\sin x + \tan x}$

8. Find the derivative of $\frac{x^2}{\sin^2 x}$

9. Differentiate with respect to x: $\ln(\sin^2 x + \cos 3x)$

10. If $y = a^{\cos 5x}$ find $\frac{dy}{dx}$.

11. Given that $y = x^2 e^{-3x}$ evaluate $\frac{d^2y}{dx^2} + \frac{dy}{dx} - y$

12. Differentiate with respect to x: $\sin\left(\frac{1-2x^3}{x^2}\right)$

13. If $y = \frac{2x-1}{x^2}$, find $x^4 \frac{d^2y}{dx^2} - 3(2x-1)\frac{dy}{dx}$

14. Given that $y = 5e^{-2x} + 3e^x$, evaluate $\dfrac{d^2y}{dx^2} + \dfrac{dy}{dx} - 2y$

15. Find the derivative of $\left(\dfrac{1}{x^2} - \dfrac{2}{x}\right)^5$

16. Given that $e^{x^3 - y^3} = xy$, find:

(a) $\dfrac{dy}{dx}$

(b) $\dfrac{dy}{dx}$ at $(1, -1)$

17. If $y = \sin(\sin x)$, evaluate $\dfrac{d^2y}{dx^2} + \tan x \, \dfrac{dy}{dx} + y\cos^2 x$

18. Determine $\dfrac{d^2}{dx^2}\left(x^2 \cos \dfrac{1}{x^2}\right)$

19. Given that $y = \dfrac{1 + e^{-x}}{1 - e^{-x}}$

(a) find $\dfrac{d^2y}{dx^2}$

(b) evaluate $\dfrac{d^2y}{dx^2} - \dfrac{dy}{dx} + y$

20. Find the derivative of $\ln(\sec^2 x)$

21. Given that $y = x + \tan x$, determine $\cos^2 x \dfrac{d^2y}{dx^2} - 2y + 2x$

22. Find $\dfrac{dy}{dx}$ if $y = \left(3x + \dfrac{\sqrt{5x}}{2}\right)^3$

23. If $y = \dfrac{x^3 + 2x^2 - 5x - 3}{x^2}$, find $\dfrac{d^2y}{dx^2}$

24. If $y = (x^3 + 1)^3 + \dfrac{5x - 2}{x^2 + 1}$ find $\dfrac{dy}{dx}$.

25. Given that $y = 2\cos 5x + 7\sin 5x$, determine $\dfrac{d^2y}{dx^2} + 25y$

CHAPTER 43
COLLECTION AND TABULATION OF DATA

When a large volume of data is obtained, it is necessary to present such data in frequency table. Sometimes the tally system which involves the use of vertical and horizontal strokes is applied.

Examples

1. A die is rolled 50 times and the following data is obtained. Represent the data in a frequency table.

4	6	4	3	5	3	1	4	6	5	6	4	2
6	4	5	6	2	1	6	4	3	4	6	1	5
1	3	6	2	2	4	3	4	5	3	4	1	2
3	1	2	1	5	3	4	3	4	2	5		

Solutions

The data which ranges from 1 to 6 is summarized as shown on the table below.

Number on die	Frequency
1	7
2	7
3	9
4	12
5	7
6	8

The data can also be represented on a horizontal table as shown below.

Number in die	1	2	3	4	5	6
Frequency	7	7	9	12	7	8

2. The scores of 40 students in a physics test are presented below. Prepare a frequency distribution table for the data.

64	66	68	63	70	63	67	64	70	69
66	64	65	70	62	70	66	69	67	64
61	63	67	62	68	64	63	69	70	63
63	61	68	67	68	63	61	67	69	68

Solution

The data which ranges from 61 to 70 is summarized as shown on the table below.

Score	61	62	63	64	65	66	67	68	69	70
Frequency	3	2	7	5	1	3	5	5	4	5

Exercise 43

1. The marks obtained in an examination by 40 students in a class are as shown below. Represent the data in a frequency table.

71	74	74	70	70	72	74	74	65	69
66	68	65	73	66	72	69	69	67	65
71	73	67	68	68	69	70	69	71	65
67	67	68	72	74	73	67	67	69	68

2. The score of 20 students in a test are as shown below. Represent the score in a frequency table.

6	6	8	9	5	6	7	5	6	9
8	6	7	5	9	5	9	6	6	5

3. The number of seeds in a sample of 40 cocoa pods are as shown below. Represent the information using a frequency table.

28	22	28	28	27	29	20	20	20	24
21	25	25	20	22	20	26	29	30	24
21	23	27	22	28	30	23	29	20	23
23	21	28	27	28	23	21	27	30	28

4. The ages of 30 students in a senior high school is represented below. Show the data using a frequency table.

14	15	12	13	10	13	11	14	10	12
15	11	15	10	12	10	11	13	14	14
15	13	12	11	14	14	13	12	10	15

CHAPTER 44
MEAN, MEDIAN AND MODE OF UNGROUPED DATA

The mean, median and mode are averages of sets of statistical data. They are called "measures of central tendency".

Mean

The mean of a set of data is obtained by adding all the data and then dividing the result by the total number in the data set.

For an ungrouped data given in a frequency table, the mean can be calculated by using the formula:

$$\bar{x} = \frac{\sum fx}{\sum f}$$

Where $\bar{x}$ is the mean, $\sum$ is a symbol representing summation, x is each number in the data, and f is the frequency.

Median

The median of a data is the middle number when the data is arranged in an increasing or decreasing order of size. For an odd number of data, the position of the middle number is obtained by the expression:

Median = number in the $(\frac{N+1}{2})$th position

where N is the total number of data.

If there is an even number of data, the median is the average of the two middle numbers. In such a case the positions of the two middle numbers is obtained by the expression:

$$\text{Median} = \frac{\text{Number in the } (\frac{N}{2})\text{th position} + \text{Number in the } (\frac{N+2}{2})\text{th position}}{2}$$

However, for large data in a frequency table which has an even number of data, the median is given by:

$$\text{Median} = \frac{\text{Number in the } (\frac{\sum f}{2})\text{th position} + \text{Number in the } (\frac{\sum f +2}{2})\text{th position}}{2}$$

Where $\sum f$ is the total frequency of the data.

Note that for an odd number of data, only one number will be at the middle. However, for an even number of data, two numbers will be at the middle. The average of the two numbers gives the median of the data.

Mode

The mode is the most occurring number in a set of data. It is the number with the highest frequency. If a set of data has two modes, we say it is bimodal.

Examples

1. Find the mean, median and mode of the data below:

2, 5, 0, 3, 1, 6, 9, 7, 3

Solution

There are 9 numbers in the data. So, the mean is obtained as follows:

$$\text{Mean} = \frac{2+5+0+3+1+6+9+7+3}{9}$$

$$= \frac{36}{9}$$

$$= 4$$

∴   The mean is 4

In order to calculate the median, first arrange the numbers in ascending order as follows:

0, 1, 2, 3, 3, 5, 6, 7, 9

By inspection, the number that is at the middle of the data is 3

∴   The median is 3

Or, since there are 9 numbers in the data and 9 is an odd number, then the position of the middle number is obtained as follows:

$$\text{Median} = \text{number in the } (\frac{N+1}{2})\text{th position}$$

$$= \text{number in the } (\frac{9+1}{2})\text{th position}$$

$$= \text{number in the } (\frac{10}{2})\text{th position}$$

$$= \text{number in the } 5^{th} \text{ position.}$$

$$= 3 \text{ (Since 3 is in the } 5^{th} \text{ position in the data arranged above)}$$

∴ The median is 3

The most occurring number in the data is 3 since it appears twice while every other number appears once.

∴ The mode is 3.

2. Find: a. the mean;

 b. the median;

 c. the mode of the data below.

6, 7, 10, 5, 11, 5, 9, 7, 10, 13, 5, 8, 7, 5, 12

Solutions

a.   There are 15 numbers in the data. So, the mean is obtained as follows:

$$\text{Mean} = \frac{6+7+10+5+11+5+9+7+10+13+5+8+7+5+12}{15}$$

$$= \frac{120}{15}$$

$$= 8$$

∴   The mean is 8

b. In order to calculate the median, first arrange the numbers in ascending order as follows:

5, 5, 5, 5, 6, 7, 7, 7, 8, 9, 10, 10, 11, 12, 13

By inspection, the number that is at the middle of the data is 7

∴   The median is 7

Or, since there are 15 numbers in the data and 15 is an odd number, then the position of the middle number is obtained as follows:

$$\text{Median} = \text{number in the } (\frac{N+1}{2})\text{th position}$$

$$= \text{number in the } (\frac{15+1}{2})\text{th position}$$

$$= \text{number in the } (\frac{16}{2})\text{th position}$$

$$= \text{number in the } 8^{\text{th}} \text{ position.}$$

$$= 7 \quad \text{(Since 7 is in the } 8^{\text{th}} \text{ position of the data)}$$

∴   The median is 7

c. The most occurring number in the data is 5. It occurs four times.

∴   The mode is 5.

3. Find the mean, median and mode of the data below:

11, 14, 10, 16, 18, 12, 11, 15, 10, 11, 15, 13

Solutions

There are 12 numbers in the data. So, the mean is obtained as follows:

$$\text{Mean} = \frac{11+14+10+16+18+12+11+15+10+11+15+13}{12}$$

$$= \frac{156}{12}$$

$$= 13$$

∴ The mean is 13

In order to calculate the median, first arrange the numbers in ascending order as follows:

10, 10, 11, 11, 11, 12, 13, 14, 15, 15, 16, 18

By inspection, the two numbers that are at the middle of the data are 12 and 13. So, we take their average.

$$\text{median} = \frac{12+13}{2}$$

$$= \frac{25}{2}$$

$$= 12.5$$

∴ The median is 12.5

Or, since there are 12 numbers in the data and 12 is an even number, then the positions of the two middle numbers and their average are obtained as follows:

$$\text{Median} = \frac{\text{Number in the } \left(\frac{N}{2}\right)\text{th position} + \text{Number in the } \left(\frac{N+2}{2}\right)\text{th position}}{2}$$

$$= \frac{\text{Number in the } \left(\frac{12}{2}\right)\text{th position} + \text{Number in the } \left(\frac{12+2}{2}\right)\text{th position}}{2}$$

$$= \frac{\text{Number in the 6th position} + \text{Number in the } \left(\frac{14}{2}\right)\text{th position}}{2}$$

$$= \frac{12 + 13}{2} \quad \text{(Note that 12 is in the 6}^{\text{th}} \text{ position while 13 is in the 7}^{\text{th}} \text{ position)}$$

$$= \frac{25}{2}$$

$$= 12.5$$

∴ The median is 12.5

The most occurring number in the data is 11 since it appears three times.

∴ The mode is 11.

4. Find the mean, median and mode of the data below:

50, 56, 58, 52, 55, 59, 51, 55

Solutions

There are 8 numbers in the data. So, the mean is obtained as follows:

$$\text{Mean} = \frac{50 + 56 + 58 + 52 + 55 + 59 + 51 + 55}{8}$$

$$= \frac{436}{8}$$

$$= 54.5$$

∴ The mean is 54.5

In order to calculate the median, first arrange the numbers in ascending order as follows:

50, 51, 52, 55, 55, 56, 58, 59.

By inspection, the two numbers that are at the middle of the data are 55 and 55. So, we take their average.

$$\text{median} = \frac{55+55}{2}$$
$$= \frac{110}{2}$$
$$= 55$$

∴   The median is 55

Or, since there are 8 numbers in the data and 8 is an even number, then the positions of the two middle numbers and their average are obtained as follows:

$$\text{Median} = \frac{\text{Number in the }\left(\frac{N}{2}\right)\text{th position} + \text{Number in the }\left(\frac{N+2}{2}\right)\text{th position}}{2}$$

$$= \frac{\text{Number in the }\left(\frac{8}{2}\right)\text{th position} + \text{Number in the }\left(\frac{8+2}{2}\right)\text{th position}}{2}$$

$$= \frac{\text{Number in the 4th position} + \text{Number in the }\left(\frac{10}{2}\right)\text{th position}}{2}$$

$$= \frac{\text{Number in the 4th position} + \text{Number in the 5th position}}{2}$$

$$= \frac{55 + 55}{2} \quad \text{(Note that 55 is in the 4}^{th}\text{ and the 5}^{th}\text{ position)}$$

$$= \frac{110}{2}$$

$$= 55$$

∴   The median is 55

The most occurring number in the data is 55.

∴   The mode is 55.

5. The table below shows the marks of 50 students in a test.

Mark	1	2	3	4	5	6	7
No of Student	8	16	10	5	3	6	2

Calculate: a. the mean b. the median c. the mode of the marks

Solutions

a. Note that the number of students is also the frequency. Presenting the table as shown below allows for easy calculation of the mean.

Mark (x)	No of student (f)	Fx
1	8	8
2	16	32
3	10	30
4	5	20
5	3	15
6	6	36
7	2	14
Total:	$\sum f = 50$	$\sum fx = 155$

Note that the column fx is obtained by multiplying the values of numbers in the column f by numbers in the column x.

Mean, $\bar{x} = \dfrac{\sum fx}{\sum f}$

$= \dfrac{155}{50}$

$= 3.1$

b. Since there are 50 students, i.e. the total frequency is 50, and 50 is an even number, then the positions of the two middle marks and their average are obtained as follows:

$$\text{Median} = \frac{\text{Number in the } \left(\frac{\sum f}{2}\right)\text{th position} + \text{Number in the } \left(\frac{\sum f + 2}{2}\right)\text{th position}}{2}$$

$$= \frac{\text{Number in the } \left(\frac{50}{2}\right)\text{th position} + \text{Number in the } \left(\frac{50+2}{2}\right)\text{th position}}{2}$$

$$= \frac{\text{Number in the 25th position} + \text{Number in the } \left(\frac{52}{2}\right)\text{th position}}{2}$$

$$= \frac{\text{Number in the 25th position} + \text{Number in the 26th position}}{2}$$

$$= \frac{3+3}{2} \quad \text{(Note that 3 is the mark in the 25}^{\text{th}} \text{ position and in the 26}^{\text{th}} \text{ position)}$$

$$= \frac{6}{2}$$

$$= 3$$

∴   The median is 3

Use the frequency (number of students) to locate the marks in the 25th and 26th position as follows:

The first frequency of 8 shows that mark 1 occupies the position of 1st to 8th. Adding the second frequency of 16 to the first frequency gives, 8 + 16 = 24. This shows that after the 8th position occupied by the mark 1, the positions 9th to 24th is occupied by the mark 2. Adding the third frequency of 10 to the previous sum of frequencies gives, 10 + 24 = 34. This shows that after the 24th position occupied by the mark 2, the positions 25th to 34th is occupied by the mark 3. Hence, 3 is the mark in the 25th and 26th position which are at the middle of the data.

c. The mode is the mark that has the highest frequency. From the table, the mark 2 has the highest frequency of 16. So, the mode is 2.

∴ The mode is 2.

Note that the mode is not the frequency itself, but that particular mark that has the highest frequency. Avoid the mistake of taking 16 (frequency) as the mode.

6. The table below shows the ages of 30 students in a school.

Age	10	11	12	13	14	15
No of Student	1	4	3	7	9	6

Calculate: a. the mean b. the median c. the mode of the ages

Solutions

a. Using the number of students as the frequency, the table can be presented for easy calculation of the mean as follows:

Age (x)	No of student (f)	Fx
10	1	10
11	4	44
12	3	36
13	7	91
14	9	126
15	6	90
Total:	$\sum f = 30$	$\sum fx = 397$

Note that the column fx is obtained by multiplying the values of numbers in the column f by numbers in the column x.

$$\text{Mean, } \bar{x} = \frac{\Sigma fx}{\Sigma f}$$

$$= \frac{397}{30}$$

$$= 13.3$$

b.   Since there are 30 students, i.e. the total frequency is 30, and 30 is an even number, then the positions of the two middle ages and their average are obtained as follows:

$$\text{Median} = \frac{\text{Age in the }\left(\frac{\Sigma f}{2}\right)\text{th position } + \text{Age in the }\left(\frac{\Sigma f+2}{2}\right)\text{th position}}{2}$$

$$= \frac{\text{Age in the }\left(\frac{30}{2}\right)\text{th position } + \text{Age in the }\left(\frac{30+2}{2}\right)\text{th position}}{2}$$

$$= \frac{\text{Age in the 15th position } + \text{Age in the }\left(\frac{32}{2}\right)\text{th position}}{2}$$

$$= \frac{\text{Age in the 15th position } + \text{Age in the 16th position}}{2}$$

$$= \frac{13 + 14}{2}$$

$$= \frac{27}{2}$$

$$= 13.5$$

∴   The median is 13.5

Note that 13 is the age in the 15th position while 14 is the age in the 16th position.

The frequency (number of students) was used to locate the ages in the 15th and 16th position as follows:

The first frequency of 1 shows that age 10 occupies the 1st position. Adding the second frequency of 4 to the first frequency gives, 1 + 4 = 5. This shows that after the 1th position occupied by the age 10, the positions 2nd to 5th is occupied by the age 11. Adding the third frequency of 3 to the previous sum of frequencies gives, 3 + 5 = 8. This shows that after the 5th position occupied by the age 11, the positions 6th to 8th is occupied by the age 12. Adding the fourth frequency of 7 to the previous sum of frequencies gives, 7 + 8 = 15. This shows that after the 8th position occupied by the age 12, the positions 9th to 15th is occupied by the age 13. Adding the fifth frequency of 9 to the previous sum of frequencies gives, 9 + 15 = 24. This shows that after the 15th position occupied by the age 13, the positions 16th to 24th is occupied by the age 14. Hence, 13 is the age in the 15th position while 14 is the age in the 16th position.

c.   The mode is the age that has the highest frequency. From the table, the age 14 has the highest frequency of 9.

∴ The mode is 14.

Note that 9 is the frequency. It should not be taken as the mode.

Range

Range is the difference between the highest and lowest values in a given set of data. It is a measure of dispersion or variation.

Examples

1. Find the range of the following set of numbers: 4, 8, 2, 5, 8, 3, 6, 4, 9, 2, 5

Solution

The highest number in the data set is 9, while the lowest number is 2.

∴   Range = Highest number – Lowest number

= 9 – 2 = 7

Range = 7

2. The monthly salaries of five workers in a company are: $845, $1205, $694, $626 and $864. What is the range of the salaries?

Solution

Range = Highest salary – Lowest salary

= 1205 – 626

Range = $579

Exercise 44

1. Find the mean, median and mode of the data below:

1, 6, 10, 4, 1, 2, 5, 2, 3, 2, 8

2. Find: a. the mean;

b. the median;

c. the mode of the data below.

20, 24, 21, 25, 22, 25, 28, 26, 20, 23, 25, 27 and 26

3. Find the mean, median and mode of the data below:

101, 105, 120, 116, 109, 112, 118, 115, 105 and 111

4. Find the mean, median and mode of the data below:

0, 6, 8, 2, 5, 9, 1, 5, 4, 7, 5, 2, 3, 3

5. The table below shows the marks of 50 students in a test.

Mark	3	4	5	6	7	8	9
No of Student	8	16	10	5	3	6	2

Calculate: a. the mean b. the median c. the mode of the marks

6. The table below shows the ages of 30 students in a school.

Age	10	11	12	13	14	15
No of Student	1	4	3	7	9	6

Calculate: a. the mean b. the median c. the mode of the ages

7. Find the range of the following set of data

a. 12, 17, 21, 15, 19, 13, 11, 16, 22, 12, 13

b. 231kg, 258kg, 213kg, 243kg, 216kg, 271kg, 262kg, 219kg, 238kg, 231kg.

CHAPTER 45
COLLECTION AND TABULATION OF GROUPED DATA

Statistical data containing numerous values is easier to work with when the values are grouped into class intervals.

Examples

1. The data below gives the marks of 30 students in an exam.

43	45	50	47	51	58	52	47	42	54
61	50	45	55	57	41	46	49	51	50
59	44	53	57	49	40	48	52	51	58

Taking class intervals 40 – 44, 45 – 49, ..., construct a frequency distribution for the data.

Solution

The data is summarized as shown on the table below. Note that the highest value in the given data falls within the range 60 – 64.

Class interval	40 – 44	45 - 49	50 - 54	55 - 59	60 - 64
Frequency	5	8	10	6	1

2. The data below gives the ages of lecturers in a university.

34	62	54	41	51	63	31	44	48	50
33	45	59	55	47	31	39	55	60	40
63	45	53	55	36	58	61	34	34	43
47	35	43	51	35	48	42	51	36	31

Taking class intervals 31 – 35, 36 – 40, ..., construct a frequency table for the data.

Solution

The data is summarized as shown on the table below. Note that the highest value in the given data falls within the range 61 – 65.

Age	31 – 35	36 - 40	41 - 45	46 - 50	51 - 55	56 - 60	61 - 65
Frequency	9	4	7	5	8	3	4

Terms Used in Grouped Data

The table below will be used to explain the terms used in grouped data.

Class interval	Frequency
8 – 14	3
15 – 21	5
22 – 28	8
29 – 35	18

1. Class limit: The end numbers in each class interval are called the class limits. 8 is the lower class limit, while 14 is the upper class limit of the first class interval.

2. Class boundaries: The class boundary for the second class interval is 14.5 – 21.5. The lower class boundary is 14.5 which is obtained by subtracting 0.5 from 15 (the lower class limit). The upper class boundary is 21.5 which is obtained by adding 0.5 to 21 (the upper class limit). Other class boundaries are obtained in a similar way.

3. Class width: For each class interval the difference between the upper class boundary and the lower class boundary gives the class width. From the table above the class width for the third class interval is 28.5 – 21.5 = 7

4. Class mid-value: This is half of the sum of the lower and upper class limit of a given class interval. The class-value of the first class interval is given by: $\dfrac{8+14}{2} = \dfrac{22}{2} = 11$.

Examples

1. Copy and complete the table below.

Class interval	Frequency	Class boundary	Class width	Class mid-value
55 – 59	3			
60 – 64	2			
65 – 69	5			
70 – 74	4			
75 – 79	1			

Solution

The completed table is as shown below

Class interval	Frequency	Class boundary	Class width	Class mid-value
55 – 59	3	54.5 – 59.5	5	57
60 – 64	2	59.5 – 64.5	5	62
65 – 69	5	64.5 – 69.5	5	67
70 – 74	4	69.5 – 74.5	5	72
75 – 79	1	74.5 – 79.5	5	77

Note that the class boundaries are obtained by subtracting and adding 0.5 to the lower and upper class limits respectively. This 0.5 is obtained by finding the difference between the lower class limit of one class and the upper class limit of the previous class and dividing the result by 2. This gives, for example, $\frac{60-59}{2} = \frac{1}{2} = 0.5$.

The class width is the difference between the upper and lower class boundaries. The class mid-values are obtained by taking the mean of the upper and lower class limits.

2. Copy and complete the table below.

Class interval	Frequency	Class boundary	Class width	Class mid-value
0 – 19	2			
20 – 39	8			
40 – 59	3			
60 – 79	1			
80 – 99	4			

Solution
The completed table is as shown below

Class interval	Frequency	Class boundary	Class width	Class mid-value
0 – 19	2	-0.5 – 19.5	20	9.5
20 – 39	8	19.5 – 39.5	20	29.5
40 – 59	3	39.5 – 59.5	20	49.5
60 – 79	1	59.5 – 79.5	20	69.5
80 – 99	4	79.5 – 99.5	20	89.5

Exercise 45

1. The data below gives the scores of 50 students in an exam.

43	45	50	47	51	58	52	47	42	54
61	50	45	55	57	41	46	49	51	50
59	44	53	57	49	40	48	52	51	58
48	54	43	54	61	60	49	57	45	42
56	45	57	61	54	62	44	47	46	62

Taking class intervals 40 – 44, 45 – 49, ..., construct a frequency distribution for the data.

2. The data below shows the weights in kg of students in a school.

24	32	44	51	31	23	51	34	48	40
53	45	29	35	27	51	29	35	50	30
43	55	53	35	26	28	41	44	54	43
27	45	33	41	55	28	32	51	26	39

Taking class intervals 21 – 25, 26 – 30, ..., construct a frequency table for the data.

3. Copy and complete the table below.

Class interval	Frequency	Class boundary	Class width	Class mid-value
5 – 9	2			
10 – 14	5			
15 – 19	5			
20 – 24	7			
25 – 29	1			

4. Copy and complete the table below.

Class interval	Frequency	Class boundary	Class width	Class mid-value
1 – 20	1			
21 – 40	4			
41 – 60	7			
61 – 80	3			
81 – 100	5			

5. Copy and complete the table below.

Class interval	Frequency	Class boundary	Class width	Class mid-value
0 – 90	2			
100 – 190	4			
200 – 290	1			
300 – 390	7			
400 – 490	1			

CHAPTER 46
MEAN, MEDIAN AND MODE OF GROUPED DATA

Mean

The mean of a grouped data can be calculated by substituting the class mid value as the values of x in the formula given by:

$$\text{Mean } \bar{x} = \frac{\sum fx}{\sum f}$$

Median

The median of a grouped data can be estimated by:

$$\text{Median} = L + C\left(\frac{\frac{\sum f}{2} - CF_{bm}}{F_m}\right)$$

Where, $\frac{\sum f}{2}$ determines the median class

L = Lower class boundary of the median class

CF_{bm} = Cumulative frequency before the median class

F_m = Frequency of the median class

C = Class width

Mode

The mode of a grouped data can be calculated as follows:

$$\text{Mode} = L + C\left(\frac{\Delta_1}{\Delta_1 + \Delta_2}\right)$$

Where, L = Lower class boundary of modal class

C = Class width

Δ_1 = Difference between the frequency of the modal class and the frequency before it

Δ_2 = Difference between the frequency of the modal class and the frequency after it

Examples

1. The following table shows the frequency distribution of ages, in years of 50 people at a bus stop.

Ages	10 - 19	20 - 29	30 - 39	40 - 49	50 - 59	60 - 69
Number of people	6	12	16	9	5	2

Calculate: a. the mean

b. the median

c. the mode of the distribution

Solution

Ages	Number of people (f)	Cumulative frequency	Class mid-value(x)	fx	Class boundary	Class width
10 – 19	6	6	14.5	87	9.5-19.5	10
20 – 29	12	6+12=18	24.5	294	19.5-29.5	10
30 – 39	16	18+16=34	34.5	552	29.5-39.5	10
40 – 49	9	34+9=43	44.5	400.5	39.5-49.5	10
50 – 59	5	43+5=48	54.5	272.5	49.5-59.5	10
60 – 69	2	48+2=50	64.5	129	59.5-69.5	10
	$\sum f = 50$			$\sum f = 1735$		

a. Mean $\bar{x} = \dfrac{\sum fx}{\sum f}$

$= \dfrac{1735}{50} = 34.7$

b. Median $= L + C\left(\dfrac{\frac{\sum f}{2} - CF_{bm}}{F_m}\right)$

$\dfrac{\sum f}{2} = \dfrac{50}{2} = 25$

This shows that the median class falls in the 25th position

∴ Median class $= 30 - 39$

Lower class boundary of median class, $L = 29.5$

Class width, $C = 10$

Cumulative frequency before the median class, $CF_{bm} = 18$

Frequency of the median class, $F_m = 16$

∴ Median $= L + C\left(\dfrac{\frac{\sum f}{2} - CF_{bm}}{F_m}\right)$

$= 29.5 + 10\left(\dfrac{25 - 18}{16}\right)$

$= 29.5 + 10\left(\dfrac{7}{16}\right)$

$= 29.5 + 4.375 = 33.875$

∴ Median $= 33.9$ (To 1 d.p)

c. Mode $= L + C\left(\dfrac{\Delta_1}{\Delta_1 + \Delta_2}\right)$

Modal class = 30 – 39

Lower class boundary of modal class, L = 29.5

Class width = 10

Δ_1 = Modal class frequency – frequency before it

    = 16 - 12 = 4

Δ_2 = Modal class frequency – frequency after it

    = 16 - 9 = 7

$\therefore$ Mode = L + C$\left(\dfrac{\Delta_1}{\Delta_1 + \Delta_2}\right)$

      = 29.5 + 10$\left(\dfrac{4}{4+7}\right)$

      = 29.5 + 10$\left(\dfrac{4}{11}\right)$

      = 29.5 + 3.64 = 33.14

$\therefore$ Mode = 33.1 (To 1 d.p)

2. The data below is the weight of students in a high school.

Weight	31 - 35	36 - 40	41 - 45	46 - 50	51 - 55	56 - 60	61 - 65
Number of student	2	9	7	5	8	3	6

Determine: a. the mean

b. the median

c. the mode of the weights

Solution

Weight	Number of student (f)	Cumulative frequency	Class mid-value(x)	fx	Class boundary	Class width
31 – 35	2	2	33	66	30.5-35.5	5
36 – 40	9	2+9=11	38	349	35.5-40.5	5
41 – 45	7	11+7=18	43	301	40.5-45.5	5
46 – 50	5	18+5=23	48	240	45.5-50.5	5
51 – 55	8	23+8=31	53	424	50.5-55.5	5
56 – 60	3	31+3=34	58	174	55.5-60.5	5
61 – 65	6	34+6=40	63	378	60.5-65.5	5
	$\sum f = 40$			$\sum fx = 1932$		

a. Mean $\bar{x} = \dfrac{\sum fx}{\sum f}$

$$= \frac{1932}{40}$$

$$= 48.3$$

b. Median $= L + C\left(\dfrac{\frac{\Sigma f}{2} - CF_{bm}}{F_m}\right)$

$$\frac{\Sigma f}{2} = \frac{40}{2} = 20$$

This shows that the median class falls in the 20th position

∴ Median class $= 46 - 50$

Lower class boundary of median class, $L = 45.5$

Class width, $C = 5$

Cumulative frequency before the median class, $CF_{bm} = 18$

Frequency of the median class, $F_m = 5$

∴ Median $= L + C\left(\dfrac{\frac{\Sigma f}{2} - CF_{bm}}{F_m}\right)$

$$= 45.5 + 5\left(\frac{20 - 18}{5}\right)$$

$$= 45.5 + 5\left(\frac{2}{5}\right)$$

$$= 45.5 + 2 = 47.5$$

∴ Median $= 47.5$

c. Mode $= L + C\left(\dfrac{\Delta_1}{\Delta_1 + \Delta_2}\right)$

Modal class $= 36 - 40$

Lower class boundary of modal class, $L = 35.5$

Class width $= 5$

Δ_1 = Modal class frequency – frequency before it

$\quad = 9 - 2 = 7$

Δ_2 = Modal class frequency – frequency after it

$\quad = 9 - 7 = 2$

∴ Mode $= L + C\left(\dfrac{\Delta_1}{\Delta_1 + \Delta_2}\right)$

$$= 35.5 + 5\left(\frac{7}{7 + 2}\right)$$

$$= 35.5 + 5\left(\frac{7}{9}\right)$$

$$= 35.5 + 3.89 = 39.39$$

∴ Mode $= 39.4$ (To 1 d.p)

1. The following table shows the weights of 30 people at a company.

Weight	60 - 64	65 - 69	70 - 74	75 - 79	80 - 84	85 - 89
Number of people	1	12	7	5	3	2

Calculate: a. the mean

b. the median

c. the mode of the distribution

2. The data below is the load distribution in tones, a chain can support.

Load	83 - 85	86 - 88	89 - 91	92 - 94	95 - 97
Number of chain	2	8	5	14	1

Determine: a. the mean

b. the median

c. the mode of the weights

3. The data below is the ages, in years, of 50 people at a party.

Ages	1 - 20	21 - 40	41 - 60	61 - 80	81-100
Number of people	3	21	17	7	2

Determine: a. the mean

b. the median

c. the mode of the weights

CHAPTER 47
MEAN DEVIATION

The mean deviation of a set of data is the mean of the absolute deviation of the values from the mean of the group. The mean deviation for data not given in a frequency table is given by:

$$\text{Mean deviation} = \frac{\Sigma |x - \bar{x}|}{N}$$ where x is each value in the data, $\bar{x}$, is the mean and N is the number of values in the data.

For data given in a frequency table, the mean deviation is given by:

$$\text{Mean deviation} = \frac{\Sigma f|x - \bar{x}|}{\Sigma f}$$

Examples

1. Calculate the mean deviation of the following data: 2, 4, 1, 3, 0

Solution

Let us first calculate the mean of the data.

$$\text{Mean, } \bar{x} = \frac{2 + 4 + 1 + 3 + 0}{5}$$

$$= \frac{10}{5}$$

$$= 2$$

$$\therefore \bar{x} = 2$$

The deviation from the mean $(x - \bar{x})$ is now tabulated as follows.

Data (x)	$x - \bar{x}$ ($\bar{x} = 2$)	$|x - \bar{x}|$
2	0	0
4	2	2
1	-1	1
3	1	1
0	-2	2
		$\Sigma|x - \bar{x}| = 6$

$$\therefore \quad \text{Mean deviation} = \frac{\Sigma |x - \bar{x}|}{N} = \frac{6}{5} = 1.2$$

2. Calculate the mean deviation of the following data: 6, 2, 5, 8, 3, 6, 4, 5, 7, 4.

Solution

Let us first calculate the mean of the data.

Mean, $\bar{X} = \dfrac{6 + 2 + 5 + 8 + 3 + 6 + 4 + 5 + 7 + 4}{10}$

$= \dfrac{50}{10}$

$= 5$

$\therefore \bar{x} = 5$

The deviation from the mean $(x - \bar{x})$ is now tabulated as follows.

Data (x)	$x - \bar{x}$ ($\bar{x} = 5$)	$|x - \bar{x}|$
6	1	1
2	-3	3
5	0	0
8	3	3
3	-2	2
6	1	1
4	-1	1
5	0	0
7	2	2
4	-1	1
		$\sum|x - \bar{x}| = 14$

$\therefore$ Mean deviation $= \dfrac{\sum|x - \bar{x}|}{N} = \dfrac{14}{10} = 1.4$

3. The marks obtained by 40 students in a mathematics test are as shown below. Calculate the mean deviation of the data.

Marks	31 - 40	41 - 50	51 - 60	61 - 70	71 - 80	81 - 90	91- 100
Number of student	1	2	8	11	8	6	4

Solution

The table below summarizes the determination of the mean and the values needed for the mean deviation. Note that the mean used on the table has been calculated below the table.

Mark	mid-value x	$x - \bar{x}$ $\bar{x} = 69.75$	$\lvert x - \bar{x}\rvert$	No of student, f	fx	$f\lvert x - \bar{x}\rvert$
31 – 40	35.5	-34.25	34.25	1	35.5	34.25
41 – 50	45.5	-24.25	24.25	2	91	48.50
51 – 60	55.5	-14.24	14.24	8	444	114
61 – 70	65.5	-4.25	4.25	11	720.5	46.75
71 – 80	75.5	5.75	5.75	8	604	46
81 – 90	85.5	15.75	15.75	6	513	94.5
91 – 100	95.5	25.75	25.75	4	382	103
				$\sum f = 40$	$\sum fx = 2790$	

Mean, $\bar{x} = \dfrac{\sum fx}{\sum f} = \dfrac{2790}{40} = 69.75$

Using the values from the table above, $\sum f\lvert x - \bar{x}\rvert = 34.25 + 48.50 + 114 + 46.75 + 46 + 94.5 + 103 = 487$

∴ Mean deviation $= \dfrac{\sum f\lvert x - \bar{x}\rvert}{\sum f}$

$= \dfrac{487}{40}$

∴ Mean deviation $= 12.2$

4. The ages of 50 people in a hospital are as shown below. Calculate the mean deviation of the ages.

Age	1 - 5	6 - 10	11 – 15	16 - 20	21 - 25	26 - 30
Number of people	6	9	14	10	4	7

Solution

The table below summarizes the determination of the mean and the values needed for the mean deviation. The mean has been calculated below the table.

Age	mid-value x	$x - \bar{x}$ $\bar{x} = 14.8$	$\lvert x - \bar{x}\rvert$	No of people, f	fx	$f\lvert x - \bar{x}\rvert$
1 – 5	3	-11.8	11.8	6	18	70.8
6 – 10	8	-6.8	6.8	9	72	61.2
11 – 15	13	-1.8	1.8	14	182	25.2
16 – 20	18	3.2	3.2	10	180	32
21 – 25	23	8.2	8.2	4	92	32.8
26 – 30	28	13.2	13.2	7	196	92.4
				$\sum f = 50$	$\sum fx = 740$	

Mean, $\bar{x} = \dfrac{\Sigma fx}{\Sigma f} = \dfrac{740}{50} = 14.8$

Using the values from the table above, $\Sigma f|x - \bar{x}| = 70.8 + 61.2 + 25.2 + 32 + 32.8 + 92.4 = 314.4$

$\therefore$ Mean deviation $= \dfrac{\Sigma f|x - \bar{x}|}{\Sigma f}$

$= \dfrac{314.4}{50}$

$\therefore$ Mean deviation $= 6.29$

5. The table below shows the number of cars owned by some political public office holders.

Number of cars	1	2	3	4	5	6
Number of politicians	9	15	11	7	3	5

Calculate the mean deviation of the data.

Solution

The table below summarises the calculations of the mean and the mean deviation.

Cars X	$x - \bar{x}$ $\bar{x} = 2.9$	$|x - \bar{x}|$	No of politicians, f	Fx	$f|x - \bar{x}|$
1	-1.9	1.9	9	9	17.1
2	-0.9	0.9	15	30	13.5
3	0.1	0.1	11	33	1.1
4	1.1	1.1	7	28	7.7
5	2.1	2.1	3	15	6.3
6	3.1	3.1	5	30	15.5
			$\Sigma f = 50$	$\Sigma fx = 145$	

Mean, $\bar{x} = \dfrac{\Sigma fx}{\Sigma f} = \dfrac{145}{50} = 2.9$

Using the values from the table above, $\Sigma f|x - \bar{x}| = 17.1 + 13.5 + 1.1 + 7.7 + 6.3 + 15.5 = 61.2$

$\therefore$ Mean deviation $= \dfrac{\Sigma f|x - \bar{x}|}{\Sigma f}$

$= \dfrac{61.2}{50}$

$\therefore$ Mean deviation $= 1.224$

Exercise 47

1. Calculate the mean deviation of the following data: 0, 5, 7, 4, 5, 3

2. Calculate the mean deviation of the following data: 4, 6, 5, 9, 9, 5, 2, 4, 8, 6, 8

3. Calculate the mean deviation of the following data: 1, 3, 1, 4, 6

4. The marks obtained by 30 students in a physics test are as shown below. Calculate the mean deviation of the data.

Marks	0 - 9	10 - 19	20 - 29	30 - 39	40 - 49	50 – 59	60- 69
Number of student	4	1	5	8	3	2	7

5. The ages of 100 people in a village are as shown below. Calculate the mean deviation of the ages.

Age	11 - 20	21 - 30	31 – 40	41 - 50	51 - 60	61 - 70
Number of people	12	9	15	24	29	11

6. The number of employees in 50 enterprises are as shown below. Calculate the mean deviation of the data.

Marks	0 - 4	5 - 9	10 - 14	15 - 19	20 - 24	25 – 29	30- 34
Number of student	2	11	15	3	4	2	13

7. The table below shows the number of farms owned by some people in a city.

Number of farms	2	4	6	8	10	12
Number of people	3	5	10	6	8	8

Calculate the mean deviation of the data.

8. A die is rolled 50 times and the following data is obtained.

2	5	4	3	5	3	1	4	6	5	6	4	2
6	1	5	6	2	1	6	4	3	4	3	1	6
1	3	6	4	2	4	3	4	5	3	4	1	2
3	1	2	2	5	6	4	3	4	6	5		

a. Present the data in a frequency table

b. Calculate the mean deviation of the data.

CHAPTER 48
VARIANCE AND STANDARD DEVIATION

Variance is the mean of the squares of the deviations from the mean. Standard deviation is the positive square root of the variance.

Variance, standard deviation and mean deviation are also regarded as measures of dispersion or variation.

The variance of data not given on a frequency table is given by:

$$\text{Variance} = \frac{\sum(x - \bar{x})^2}{N}$$

For data given on a frequency table, the variance is given by:

$$\text{Variance} = \frac{\sum f(x - \bar{x})^2}{\sum f}$$

Standard deviation is the square root of variance.

Examples

1. Calculate the variance and standard deviation of the following data: 4, 2, 1, 5.

Solution

Let us first calculate the mean of the data.

$$\text{Mean, } \bar{x} = \frac{4 + 2 + 1 + 5}{4} = \frac{12}{4} = 3$$

We now present the deviation from the mean as follows.

Data x	$x - \bar{x}$ ($\bar{x} = 3$)	$(x - \bar{x})^2$
4	1	1
2	-1	1
1	-2	4
5	2	4
		$\sum(x - \bar{x})^2 = 10$

$$\text{Variance} = \frac{\sum(x - \bar{x})^2}{N} = = \frac{10}{4} = 2.5$$

∴   Standard deviation = $\sqrt{2.5}$ = 1.58

2. Calculate the variance and standard deviation of the data below:
   2, 5, 3, 2, 6, 5, 7, 2.

Solution

Let us first calculate the mean of the data as follows:

$$\text{Mean, } \bar{x} = \frac{2 + 5 + 3 + 2 + 6 + 5 + 7 + 2}{8}$$

$$= \frac{32}{8}$$

$$= 4$$

The deviation from the mean is as presented below.

Data x	$x - \bar{x}$ ($\bar{x} = 4$)	$(x - \bar{x})^2$
2	-2	4
5	1	1
3	-1	1
2	-2	4
6	2	4
5	1	1
7	3	9
2	-2	4
		$\sum(x - \bar{x})^2 = 28$

$$\text{Variance} = \frac{\sum(x - \bar{x})^2}{N} = \frac{28}{8} = 3.5$$

$$\therefore \quad \text{Standard deviation} = \sqrt{3.5} = 1.87$$

3. The distances in Km, from school to the homes of 30 students are as shown below.
Calculate:

a. the variance

b. the standard deviation of the data

Distance (Km)	0 - 4	5 - 9	10 - 14	15 - 19	20 - 24	25 - 29
Number of students	2	10	8	6	3	1

Solutions

The working is set out as shown on the table below

Distance	mid-value x	No of student f	fx	$x - \bar{x}$ $\bar{x} = 12.2$	$(x - \bar{x})^2$	$f(x - \bar{x})^2$
0 – 4	2	2	4	-10.2	104.04	208.08
5 – 9	7	10	70	-5.2	27.04	270.4
10 – 14	12	8	96	-0.2	0.04	0.32
15 – 19	17	6	102	4.8	23.04	138.24
20 – 24	22	3	66	9.8	96.04	288.12
25 – 29	27	1	27	14.8	219.04	219.04
		$\sum f = 30$	$\sum fx = 365$			

Mean, $\bar{x} = \dfrac{\sum fx}{\sum f} = \dfrac{365}{30} = 12.2$

a. Using the values from the table above, $\sum f(x - \bar{x})^2 = 208.08 + 270.4 + 0.32 + 138.24 + 288.12 + 219.04 = 1124.2$

$\therefore$ Variance $= \dfrac{\sum f(x - \bar{x})^2}{\sum f}$

$\qquad = \dfrac{1124.2}{30}$

$\therefore$ Variance $= 37.5$

b. Standard deviation $= \sqrt{\text{Variance}}$

$\qquad\qquad = \sqrt{37.5}$

$\therefore$ Standard deviation $= 6.1$

4. The projected population in millions, of 20 states in a country are as shown below. Calculate:

a. the variance

b. the standard deviation of the data

Population	1 - 5	6 - 10	11 – 15	16 - 20	21 - 25	26 - 30
Number of state	1	8	5	3	2	1

The working is set out as shown on the table below

Popula-tion	mid-value x	No of states f	fx	$x - \bar{x}$ $\bar{x} = 13$	$(x - \bar{x})^2$	$f(x - \bar{x})^2$
1 – 5	3	1	3	-10	100	100
6 – 10	8	8	64	-5	25	200
11 – 15	13	5	65	0	0	0
16 – 20	18	3	54	5	25	75
21 – 25	23	2	46	10	100	200
26 – 30	28	1	28	15	225	225
		$\sum f = 20$	$\sum fx = 260$			

Mean, $\bar{x} = \dfrac{\sum fx}{\sum f} = \dfrac{260}{20} = 13$

a. Using the values from the table above, $\sum f(x - \bar{x})^2 = 100 + 200 + 0 + 75 + 200 + 225 = 800$

$\therefore$ Variance $= \dfrac{\sum f(x - \bar{x})^2}{\sum f}$

$\qquad = \dfrac{800}{20}$

$\therefore$ Variance = 40

b. Standard deviation $= \sqrt{\text{Variance}}$

$\qquad\qquad\qquad = \sqrt{40}$

$\therefore$ Standard deviation = 6.3

5. The scores obtained by 100 students in a test are as shown below. Calculate:

a. the variance

b. the standard deviation of the scores

Scores	2	3	4	5	6	7
Number of student	10	22	18	30	12	8

Solutions

The working is set out as shown on the table below

Scores x	No of students f	fx	x - x̄ x̄ = 4.4	(x - x̄)²	f(x - x̄)²
2	12	24	-2.4	5.76	69.12
3	18	54	-1.4	1.96	35.28
4	22	88	-0.4	0.16	3.52
5	24	120	0.6	0.36	8.64
6	14	84	1.6	2.56	35.84
7	10	70	2.6	6.76	67.6
	$\sum f = 100$	$\sum fx = 440$			

Mean, $\bar{x} = \dfrac{\sum fx}{\sum f} = \dfrac{440}{100} = 4.4$

a. Using the values from the table above, $\sum f(x - \bar{x})^2 = 69.12 + 35.28 + 3.52 + 8.64 + 35.84 + 67.6$
$= 220$

$\therefore$ Variance $= \dfrac{\sum f(x - \bar{x})^2}{\sum f}$

$= \dfrac{220}{100}$

$\therefore$ Variance $= 2.2$

b. Standard deviation $= \sqrt{\text{Variance}}$

$= \sqrt{2.2}$

$\therefore$ Standard deviation $= 1.48$

Exercise 48

1. Calculate the variance and standard deviation of the following data: 3, 5, 4, 7, 6.

2. Calculate the variance and standard deviation of the data below:
   1, 0, 4, 3, 5, 8, 6, 4, 7, 2.

3. The scores of 50 students in a test are as shown below. Calculate:
a. the variance
b. the standard deviation of the data

Scores	0 – 9	10 - 19	20 – 29	30 - 39	40 – 49	50 - 59
Number of students	5	12	6	18	5	4

4. The ages of employees in an organization are as shown below. Calculate:
a. the variance
b. the standard deviation of the data

Age	20 - 24	25 - 29	30 - 34	35 - 39	40 - 44
Number of empoyees	8	6	3	1	2

5. The scores obtained by 40 students in a test are as shown below. Calculate:
a. the variance
b. the standard deviation of the scores

Scores	5	6	7	8	9	10
Number of student	1	2	4	12	20	1

CHAPTER 49
QUARTILES AND PERCENTILES BY INTERPOLATION METHOD

When a distribution is divide into four equal parts, it is called a quartile. When it is divided into hundred equal parts, such a division is called percentile.

The first quartile is also called lower quartile, and it is denoted by Q_1.

The second quartile is also called median, and it is denoted by Q_2.

The third quartile is also called upper quartile, and it is denoted by Q_3.

The lower quartile for a grouped data is calculated as follows:

$$Q_1 = L_1 + C\left(\frac{\frac{\Sigma f}{4} - CF_{bQ_1}}{F_{Q_1}}\right)$$

Where, $\frac{\Sigma f}{4}$ determines the lower quartile class

L_1 = Lower class boundary of the lower quartile class

CF_{bQ_1} = Cumulative frequency before the lower quartile class

F_{Q_1} = Frequency of the lower quartile class

C = Class width

The median is calculated as follows:

$$Q_2 = L_2 + C\left(\frac{\frac{\Sigma f}{2} - CF_{bm}}{F_m}\right)$$

Where, $\frac{\Sigma f}{2}$ determines the median class

L_2 = Lower class boundary of the median class

CF_{bm} = Cumulative frequency before the median class

F_m = Frequency of the median class

C = Class width

The upper quartile is calculated as follows:

$$Q_3 = L_3 + C\left(\frac{\frac{3\Sigma f}{4} - CF_{bQ_3}}{F_{Q_3}}\right)$$

Where, $\frac{3\Sigma f}{4}$ determines the upper quartile class

L_3 = Lower class boundary of the upper quartile class

CF_{bQ_3} = Cumulative frequency before the upper quartile class

F_{Q_3} = Frequency of the upper quartile class

C = Class width

The interquartile range is given by:

Interquartile range = $Q_3 - Q_1$

The semi-interquartile range is also called quartile deviation, and it is given by:

Semi-interquartile range = $\dfrac{Q_3 - Q_1}{2}$

The percentile is calculated as follows:

$$P_N = L_N + C\left(\dfrac{\frac{N\Sigma f}{100} - CF_{bP_N}}{F_{P_N}}\right)$$

Where P_N is the N percentile, and $\dfrac{N\Sigma f}{100}$ determines the N percentile class

L_N = Lower class boundary of the N percentile class

CF_{bP_N} = Cumulative frequency before the N percentile class

F_{P_N} = Frequency of the N percentile class

C = Class width

Examples

1. The following is the record of marks of 40 students in an examination:

64 84 91 58 43 86 73 33 76 80 57 33 53 29 40 27 72 19 51
67 37 14 18 92 13 45 61 39 23 22 22 41 27 51 63 47 19 35
39 76

Using class interval 11 – 20, 21 – 30, ..., prepare a frequency table for the distribution.
Hence calculate the:

a. median

b. lower quartile

c. upper quartile

d. interquartile range

e. quartile deviation/semi-interquartile range

f. 40th percentile

g. 85th percentile

Solutions

The frequency table is as shown below.

Class interval	Frequency
11 – 20	5
21 – 30	6
31 – 40	7
41 – 50	4
51 – 60	5
61 – 70	4
71 – 80	5
81 – 90	2
91 – 100	2

a. In order to calculate the median, a table of the class boundaries and cumulative frequency has to be drawn as shown below.

Class interval	Class boundary	Frequency	Cumulative frequency	Class width
11 – 20	10.5 – 20.5	5	5	10
21 – 30	20.5 – 30.5	6	11	10
31 – 40	30.5 – 40.5	7	18	10
41 – 50	40.5 – 50.5	4	22	10
51 – 60	50.5 – 60.5	5	27	10
61 – 70	60.5 – 70.5	4	31	10
71 – 80	70.5 – 80.5	5	36	10
81 – 90	80.5 – 90.5	2	38	10
91 - 100	90.5 – 100.5	2	40	10

The median is calculated as follows:

$$Q_2 = L_2 + C\left(\frac{\frac{\Sigma f}{2} - CF_{bm}}{F_m}\right)$$

$\frac{\Sigma f}{2} = \frac{40}{2} = 20$. This shows that the median class is at the 20th position. This is the class, 41 – 50. This position is obtained by counting the frequency to get to the 20th position. 5 + 6 + 7 = 18. This shows that the 18th position is occupied by the class 31 – 40. After this class, the next frequency is 4. When 4 positions are added to 18 positions, it gives 22. This means that these 4 positions are the 19th, 20th, 21st and 22nd positions. These 4 positions are occupied by the class 41 – 50 as shown on the table. Hence the class in the 20th position is 41 – 50. You can also look at the cumulative frequency and see where the 20th position class falls.

L_2 = Lower class boundary of the median class = 40.5

CF_{bm} = Cumulative frequency before the median class = 18

F_{m} = Frequency of the median class = 4

C = Class width = 10. The class limit is the difference between an upper class boundary and a lower class boundary. For example, 20.5 − 10.5 = 10.

$$\therefore \quad Q_2 = L_2 + C\left(\frac{\frac{\Sigma f}{2} - CF_{bm}}{F_m}\right)$$

$$= 40.5 + 10\left(\frac{\frac{40}{2} - 18}{4}\right)$$

$$= 40.5 + 10\left(\frac{20 - 18}{4}\right)$$

$$= 40.5 + 10\left(\frac{2}{4}\right)$$

$$= 40.5 + \left(\frac{10 \times 2}{4}\right)$$

$$= 40.5 + 5$$

$$Q_2 = 45.5$$

b. The lower quartile is calculated as follows:

$$Q_1 = L_1 + C\left(\frac{\frac{\Sigma f}{4} - CF_{bQ1}}{F_{Q1}}\right)$$

$\frac{\Sigma f}{4} = \frac{40}{4} = 10$. Hence the lower quartile class is at the 10th position. This class is, 21 − 30.

L_1 = Lower class boundary of the lower quartile class = 20.5

CF_{bQ_1} = Cumulative frequency before the lower quartile class = 5

F_{Q_1} = Frequency of the lower quartile class = 6

C = Class width = 10

$$\therefore \quad Q_1 = L_1 + C\left(\frac{\frac{\Sigma f}{4} - CF_{bQ1}}{F_{Q1}}\right)$$

$$= 20.5 + 10\left(\frac{\frac{40}{4} - 5}{6}\right)$$

$$= 20.5 + 10\left(\frac{10 - 5}{6}\right)$$

$$= 20.5 + 10\left(\frac{5}{6}\right)$$

$$= 20.5 + \left(\frac{10 \times 5}{6}\right)$$

$$= 20.5 + 8.3$$

$$Q_1 = 28.8$$

c. The upper quartile is calculated as follows:

$$Q_3 = L_3 + C\left(\frac{\frac{3\Sigma f}{4} - CF_{bQ3}}{F_{Q3}}\right)$$

$\frac{3\Sigma f}{4} = \frac{3 \times 40}{4} = 30$. Hence the upper quartile class is at the 30th position. This class is, 61 − 70.

L_3 = Lower class boundary of the upper quartile class = 60.5

CF_{bQ_3} = Cumulative frequency before the upper quartile class = 27

F_{Q_3} = Frequency of the upper quartile class = 4

C = Class width = 10

$$\therefore \quad Q_3 = L_3 + C\left(\frac{\frac{3\Sigma f}{4} - CF_{bQ_3}}{F_{Q_3}}\right)$$

$$= 60.5 + 10\left(\frac{\frac{3 \times 40}{4} - 27}{4}\right)$$

$$= 60.5 + 10\left(\frac{30 - 27}{4}\right)$$

$$= 60.5 + 10\left(\frac{3}{4}\right)$$

$$= 60.5 + \left(\frac{10 \times 3}{4}\right)$$

$$= 60.5 + 7.5$$

$$Q_3 = 68$$

d. Interquartile range = $Q_3 - Q_1$

$$= 68 - 28.8$$

$$= 39.2$$

e. Quartile deviation/semi-interquartile range is given by:

$$Q = \frac{Q_3 - Q_1}{2}$$

$$= \frac{68 - 28.8}{2}$$

$$= \frac{39.2}{2}$$

$$Q = 19.6$$

f. The 40th percentile is calculated as follows:

$$P_N = L_N + C\left(\frac{\frac{N\Sigma f}{100} - CF_{bP_N}}{F_{P_N}}\right)$$

$$P_N = P_{40}$$

$\frac{N\Sigma f}{100} = \frac{40 \times 40}{100}$ = 16. Hence the 40th percentile class is at the 16th position. This class is: 31 - 40

$L_N = L_{40}$ = Lower class boundary of the 40th percentile class = 30.5

$CF_{bP_N} = CF_{bP_{40}}$ = Cumulative frequency before the 40th percentile class = 11

$F_{P_N} = F_{P_{40}}$ = Frequency of the 40th percentile class = 7

C = Class width = 10

Hence, $P_{40} = L_{40} + C\left(\dfrac{\frac{40\sum f}{100} - CF_{bP40}}{F_{P40}}\right)$

$= 30.5 + 10\left(\dfrac{\frac{40 \times 40}{100} - 11}{7}\right)$

$= 30.5 + 10\left(\dfrac{16 - 11}{7}\right)$

$= 30.5 + 10\left(\dfrac{5}{7}\right)$

$= 30.5 + \left(\dfrac{10 \times 5}{7}\right)$

$= 30.5 + 7.1$

$P_{40} = 37.6$

g. The 85th percentile is calculated as follows:

$P_N = L_N + C\left(\dfrac{\frac{N\sum f}{100} - CF_{bPN}}{F_{PN}}\right)$

$P_N = P_{85}$

$\dfrac{N\sum f}{100} = \dfrac{85 \times 40}{100} = 34$. Hence the 85th percentile class is at the 34th position. This class is: 71 - 80

$L_N = L_{85}$ = Lower class boundary of the 85th percentile class = 70.5

$CF_{bP_N} = CF_{bP_{85}}$ = Cumulative frequency before the 85th percentile class = 31

$F_{P_N} = F_{P_{85}}$ = Frequency of the 85th percentile class = 5

C = Class width = 10

Hence, $P_{85} = L_{85} + C\left(\dfrac{\frac{85 \times 40}{100} - CF_{bP85}}{F_{P85}}\right)$

$= 70.5 + 10\left(\dfrac{\frac{85 \times 40}{100} - 31}{5}\right)$

$= 70.5 + 10\left(\dfrac{34 - 31}{5}\right)$

$= 70.5 + 10\left(\dfrac{3}{5}\right)$

$= 70.5 + \left(\dfrac{10 \times 3}{5}\right)$

$= 70.5 + 6$

$P_{85} = 76.5$

2. The table below shows the distribution of marks scored by students in an examination.

Class interval	Frequency
60 – 64	2
65 – 69	4
70 – 74	7
75 – 79	13
80 – 84	10
85 – 89	8
90 – 94	5
95 – 99	1

From the data, calculate:

a. median

b. lower quartile

c. upper quartile

d. interquartile range

e. semi-interquartile range

f. 70th percentile

g. the pass mark if 25% of the students passed

h. the pass mark if it was later agreed that only 40% of the students should fail.

Solution

a. In order to calculate the median, a table of the class boundaries and cumulative frequency has to be drawn as shown below.

Class interval	Class boundary	Frequency	Cumulative frequency	Class width
60 – 64	59.5 – 64.5	2	2	5
65 – 69	64.5 – 69.5	4	6	5
70 – 74	69.5 – 74.5	7	13	5
75 – 79	74.5 – 79.5	13	26	5
80 – 84	79.5 – 84.5	10	36	5
85 – 89	84.5 – 89.5	8	44	5
90 – 94	89.5 – 94.5	5	49	5
95 – 99	94.5 – 99.5	1	50	5

The median is calculated as follows:

$$Q_2 = L_2 + C\left(\frac{\frac{\sum f}{2} - CF_{bm}}{F_m}\right)$$

$\frac{\sum f}{2} = \frac{50}{2} = 25$. This shows that the median class is at the 25th position. This is the class, 75 – 79.

This is obtained by looking at the cumulative frequency to see where the 25th position class falls.

L_2 = Lower class boundary of the median class = 74.5

CF_{bm} = Cumulative frequency before the median class = 13

F_m = Frequency of the median class. This is also 13

C = Class width = 5

$$\therefore \quad Q_2 = L_2 + C\left(\frac{\frac{\Sigma f}{2} - CF_{bm}}{F_m}\right)$$

$$= 74.5 + 5\left(\frac{\frac{50}{2} - 13}{13}\right)$$

$$= 74.5 + 5\left(\frac{25 - 13}{13}\right)$$

$$= 74.5 + 5\left(\frac{12}{13}\right)$$

$$= 74.5 + \left(\frac{5 \times 12}{13}\right)$$

$$= 74.5 + 4.6$$

$$Q_2 = 79.1$$

b. The lower quartile is calculated as follows:

$$Q_1 = L_1 + C\left(\frac{\frac{\Sigma f}{4} - CF_{bQ_1}}{F_{Q_1}}\right)$$

$\frac{\Sigma f}{4} = \frac{50}{4}$ = 12.5. Hence the lower quartile class is at the 12th or 13th position. This class is, 70 – 74.

L_1 = Lower class boundary of the lower quartile class = 69.5

CF_{bQ_1} = Cumulative frequency before the lower quartile class = 6

F_{Q_1} = Frequency of the lower quartile class = 7

C = Class width = 5

$$\therefore \quad Q_1 = L_1 + C\left(\frac{\frac{\Sigma f}{4} - CF_{bQ_1}}{F_{Q_1}}\right)$$

$$= 69.5 + 5\left(\frac{\frac{50}{4} - 6}{7}\right)$$

$$= 69.5 + 5\left(\frac{12.5 - 6}{7}\right)$$

$$= 69.5 + 5\left(\frac{6.5}{7}\right)$$

$$= 69.5 + \left(\frac{5 \times 6.5}{7}\right)$$

$$= 69.5 + 4.6$$

$$Q_1 = 74.1$$

c. The upper quartile is calculated as follows:

$$Q_3 = L_3 + C\left(\frac{\frac{3\Sigma f}{4} - CF_{bQ3}}{F_{Q3}}\right)$$

$\frac{3\Sigma f}{4} = \frac{3 \times 50}{4} = 37.5$. Hence the upper quartile class is at the 37th or 38th position. This class is, 85 − 89.

L_3 = Lower class boundary of the upper quartile class = 84.5

CF_{bQ_3} = Cumulative frequency before the upper quartile class = 36

F_{Q_3} = Frequency of the upper quartile class = 8

C = Class width = 5

$$\therefore \quad Q_3 = L_3 + C\left(\frac{\frac{3\Sigma f}{4} - CF_{bQ3}}{F_{Q3}}\right)$$

$$= 84.5 + 5\left(\frac{\frac{3 \times 50}{4} - 36}{8}\right)$$

$$= 84.5 + 5\left(\frac{37.5 - 36}{8}\right)$$

$$= 84.5 + 5\left(\frac{1.5}{8}\right)$$

$$= 84.5 + 0.9$$

$$Q_3 = 85.4$$

d. Interquartile range = $Q_3 - Q_1$

$$= 85.4 - 74.1$$

$$= 11.3$$

e. Semi-interquartile range, $Q = \frac{Q_3 - Q_1}{2}$

$$= \frac{85.4 - 74.1}{2}$$

$$= \frac{11.3}{2}$$

$$Q = 5.65$$

f. The 70th percentile is calculated as follows:

$$P_N = L_N + C\left(\frac{\frac{N\Sigma f}{100} - CF_{bPN}}{F_{PN}}\right)$$

$P_N = P_{70}$

$\frac{N\Sigma f}{100} = \frac{70 \times 50}{100} = 35$. Hence the 70th percentile class is at the 35th position. This class is: 80 - 84

$L_N = L_{70}$ = Lower class boundary of the 70th percentile class = 79.5

$CF_{bP_N} = CF_{bP_{70}}$ = Cumulative frequency before the 70th percentile class = 26

$F_{P_N} = F_{P_{70}}$ = Frequency of the 70th percentile class = 10

C = Class width = 5

Hence, $P_{70} = L_{70} + C\left(\dfrac{\frac{70\sum f}{100} - CF_{bP_{70}}}{F_{P_{70}}}\right)$

$= 79.5 + 5\left(\dfrac{\frac{70 \times 50}{100} - 26}{10}\right)$

$= 79.5 + 5\left(\dfrac{35 - 26}{10}\right)$

$= 79.5 + 5\left(\dfrac{9}{10}\right)$

$= 79.5 + 4.5$

$P_{70} = 84$

g. If 25% of the students passed, then the first 75% (i.e. 100 – 25) of the students failed. This means that the pass mark is at the 75th percentile.

Note that the pass mark is always at the failure percentile.

Hence the 75th percentile is calculated as follows:

$P_N = L_N + C\left(\dfrac{\frac{N\sum f}{100} - CF_{bP_N}}{F_{P_N}}\right)$

$P_N = P_{75}$

$\dfrac{N\sum f}{100} = \dfrac{75 \times 50}{100} = 37.5$. Hence the 75th percentile class is at the 37.5th position. This class is: 85 - 89

$L_N = L_{75}$ = Lower class boundary of the 75th percentile class = 84.5

$CF_{bP_N} = CF_{bP_{75}}$ = Cumulative frequency before the 75th percentile class = 36

$F_{P_N} = F_{P_{75}}$ = Frequency of the 75th percentile class = 8

C = Class width = 5

Hence, $P_{75} = L_{75} + C\left(\dfrac{\frac{75\sum f}{100} - CF_{bP_{75}}}{F_{P_{75}}}\right)$

$= 84.5 + 5\left(\dfrac{\frac{75 \times 50}{100} - 36}{8}\right)$

$= 84.5 + 5\left(\dfrac{37.5 - 36}{8}\right)$

$= 84.5 + 5\left(\dfrac{1.5}{8}\right)$

$= 84.5 + 0.9$

$P_{75} = 85.4$

Hence the pass mark is 85.4

h. If 40% of the students should fail, then the pass mark is at the 40th percentile.
Hence the 40th percentile is calculated as follows:

$$P_N = L_N + C\left(\frac{\frac{N\Sigma f}{100} - CF_{bP_N}}{F_{P_N}}\right)$$

$P_N = P_{40}$

$\frac{N\Sigma f}{100} = \frac{40 \times 50}{100} = 20$. Hence the 40th percentile class is at the 20th position. This class is: 75 - 79

$L_N = L_{40}$ = Lower class boundary of the 40th percentile class = 74.5

$CF_{bP_N} = CF_{bP_{40}}$ = Cumulative frequency before the 40th percentile class = 13

$F_{P_N} = F_{P_{40}}$ = Frequency of the 40th percentile class = 13

C = Class width = 5

Hence, $P_{40} = L_{40} + C\left(\frac{\frac{40\Sigma f}{100} - CF_{bP_{40}}}{F_{P_{40}}}\right)$

$$= 74.5 + 5\left(\frac{\frac{40 \times 50}{100} - 13}{13}\right)$$

$$= 74.5 + 5\left(\frac{20 - 13}{13}\right)$$

$$= 74.5 + 5\left(\frac{7}{13}\right)$$

$$= 74.5 + 2.7$$

$P_{40} = 77.2$

Hence the pass mark is 77.2

3. The table below shows the masses of some items sold in a supermarket.

Mass	1.5 – 1.9	2.0 – 2.4	2.5 – 2.9	3.0 – 3.4	3.5 – 3.9	4.0 – 4.5
Number of Items	5	12	6	18	5	4

From the table given above, estimate:

a. median

b. lower quartile

c. upper quartile

d. the 55th percentile

Solution

a. In order to calculate the median, a table of the class boundaries and cumulative frequency has to be drawn as shown below.

Class interval	Class boundary	Frequency	Cumulative frequency	Class width
1.5 – 1.9	1.45 – 1.95	5	5	0.5
2.0 – 2.4	1.95 – 2.45	12	17	0.5
2.5 – 2.9	2.45 – 2.95	6	23	0.5
3.0 – 3.4	2.95 – 3.45	18	41	0.5
3.5 – 3.9	3.45 – 3.95	5	46	0.5
4.0 – 4.4	3.95 – 4.45	4	50	0.5

Note that in computing the class boundaries, a difference between an upper class limit and a lower class limit, such as, 2.0 – 1.9 = 0.1, is first determined and then divided by 2 to give, 0.1/2 = 0.05. It is this 0.05 that is added and subtracted from the class limit values to obtain the class boundary values. This is the method applied in obtaining the class boundaries of any given grouped data.

The median is calculated as follows:

$$Q_2 = L_2 + C\left(\frac{\frac{\Sigma f}{2} - CF_{bm}}{F_m}\right)$$

$\frac{\Sigma f}{2} = \frac{50}{2} = 25$. Hence the median class is at the 25th position. This is the class, 3.0 – 3.4. This is obtained by looking at the cumulative frequency to see where the 25th position class falls.

L_2 = Lower class boundary of the median class = 2.95

CF_{bm} = Cumulative frequency before the median class = 23

F_m = Frequency of the median class = 18

C = Class width = 0.5, i.e. 1.95 – 1.45 = 0.5.

$$\therefore \quad Q_2 = L_2 + C\left(\frac{\frac{\Sigma f}{2} - CF_{bm}}{F_m}\right)$$

$$= 2.95 + 0.5\left(\frac{\frac{50}{2} - 23}{18}\right)$$

$$= 2.95 + 0.5\left(\frac{25 - 23}{18}\right)$$

$$= 2.95 + 0.5\left(\frac{2}{18}\right)$$

$$= 2.95 + 0.06$$

$$Q_2 = 3.01$$

b. The lower quartile is calculated as follows:

$$Q_1 = L_1 + C\left(\frac{\frac{\Sigma f}{4} - CF_{bQ1}}{F_{Q1}}\right)$$

$\frac{\Sigma f}{4} = \frac{50}{4} = 12.5$. This shows that the lower quartile class is at the 12th or 13th position. This class is, 2.0 – 2.4.

L_1 = Lower class boundary of the lower quartile class = 1.95

CF_{bQ_1} = Cumulative frequency before the lower quartile class = 5

F_{Q_1} = Frequency of the lower quartile class = 12

C = Class width = 0.5

$$\therefore \quad Q_1 = L_1 + C\left(\frac{\frac{\Sigma f}{4} - CF_{bQ1}}{F_{Q1}}\right)$$

$$= 1.95 + 0.5\left(\frac{\frac{50}{4} - 5}{12}\right)$$

$$= 1.95 + 0.5\left(\frac{12.5 - 5}{12}\right)$$

$$= 1.95 + 0.5\left(\frac{7.5}{12}\right)$$

$$= 1.95 + 0.31$$

$$Q_1 = 2.26$$

c. The upper quartile is calculated as follows:

$$Q_3 = L_3 + C\left(\frac{\frac{3\Sigma f}{4} - CF_{bQ3}}{F_{Q3}}\right)$$

$\frac{3\Sigma f}{4} = \frac{3 \times 50}{4} = 37.5$. This shows that the upper quartile class is at the 37th or 38th position. This class is, 3.0 – 3.4.

L_3 = Lower class boundary of the upper quartile class = 2.95

CF_{bQ_3} = Cumulative frequency before the upper quartile class = 23

F_{Q_3} = Frequency of the upper quartile class = 18

C = Class width = 0.5

$$\therefore \quad Q_3 = L_3 + C\left(\frac{\frac{3\Sigma f}{4} - CF_{bQ3}}{F_{Q3}}\right)$$

$$= 2.95 + 0.5\left(\frac{\frac{3 \times 50}{4} - 23}{18}\right)$$

$$= 2.95 + 0.5\left(\frac{37.5 - 23}{18}\right)$$

$$= 2.95 + 0.5\left(\frac{14.5}{18}\right)$$

$$= 2.95 + 0.4$$

$Q_3 = 3.35$

d. The 55th percentile is calculated as follows:

$$P_N = L_N + C\left(\frac{\frac{N\Sigma f}{100} - CF_{bP_N}}{F_{P_N}}\right)$$

$P_N = P_{55}$

$\frac{N\Sigma f}{100} = \frac{55 \times 50}{100}$ = 27.5. Hence the 55th percentile class is at the 27th and 28th position. This class is: 3.0 – 3.4

$L_N = L_{55}$ = Lower class boundary of the 55th percentile class = 2.95

$CF_{bP_N} = CF_{bP_{55}}$ = Cumulative frequency before the 55th percentile class = 23

$F_{P_N} = F_{P_{55}}$ = Frequency of the 55th percentile class = 18

C = Class width = 0.5

Hence, $P_{55} = L_{55} + C\left(\frac{\frac{55\Sigma f}{100} - CF_{bP_{55}}}{F_{P_{55}}}\right)$

$= 2.95 + 0.5\left(\frac{\frac{55 \times 50}{100} - 23}{18}\right)$

$= 2.95 + 0.5\left(\frac{27.5 - 23}{18}\right)$

$= 2.95 + 0.5\left(\frac{4.5}{18}\right)$

$= 2.95 + 0.13$

$P_{55} = 3.08$

Exercise 49

1. The following is the record of marks of 40 students in an examination:

34 74 92 58 46 76 73 23 66 70 57 43 53 39
50 37 82 29 54 77 67 19 18 96 15 55 41 29
33 52 22 81 77 81 58 27 20 55 49 96

Using class interval 11 – 20, 21 – 30, ..., prepare a frequency table for the distribution. Hence calculate the:

a. median

b. lower quartile

c. upper quartile

d. interquartile range

e. quartile deviation/semi-interquartile range

f. 30th percentile

g. 68th percentile

2. The table below shows the distribution of marks scored by students in an examination.

Class interval	Frequency
10 – 14	1
15 – 19	3
20 – 24	8
25 – 29	11
30 – 34	7
35 – 39	9
40 – 44	10
45 – 49	5

From the data, calculate:

a. median

b. lower quartile

c. upper quartile

d. interquartile range

e. semi-interquartile range

f. 80th percentile

g. the pass mark if 35% of the students passed

h. the pass mark if 15% the students should fail.

3. The table below shows the height of some flowers sold in a farm.

Mass	0.5 – 0.9	1.0 – 1.4	1.5 – 1.9	2.0 – 2.4	2.5 – 2.9	3.0 – 3.4
No of Items	4	15	12	9	7	3

From the table given above, estimate:

a. median

b. lower quartile

c. upper quartile

d. the 45th percentile

e. pass mark if 90% of the students passed

4. The table below shows the distribution of marks scored by students in an test.

Class interval	Frequency
0 – 4	1
5 – 9	4
10 – 14	7
15 – 19	5
20 – 24	9
25 – 29	1
30 – 34	2
35 – 39	1

From the data, calculate:

a. median

b. lower quartile

c. upper quartile

d. interquartile range

e. semi-interquartile range

f. 60th percentile

g. the pass mark if 10% of the students passed

5. The table below shows the weight in gram of some seeds found in some cocoa pods.

Mass	0 – 0.4	0.5 – 0.9	1.0 – 1.4	1.5 – 1.9	2.0 – 2.4	2.5 – 2.9
No of Items	9	21	16	22	28	4

From the table given above, estimate:

a. median

b. lower quartile

c. upper quartile

d. the 25th percentile

e. pass mark if 68% of the students passed

CHAPTER 50
THE BASIC THEORY OF PROBABILITY

Probability is the likelihood of an event happening. Mathematically probability is given by:

$$\text{Probability} = \frac{\text{number of required outcome}}{\text{number of total or possible outcome}}$$

If the probability of an event happening is x, then the probability that it will not happen will be given by: $\quad 1-x$

Probability must lie between the values of 0 and 1. If an event cannot happen, then its probably is 0. If an event is certain to happen, then its probability is 1.

Mutually Exclusive Events

When there is no member/element common between two or more similar events, then we say they are mutually exclusive events. For example the event of odd numbers or even numbers are mutually exclusive. They are disjoint sets.

Addition Law of Probability

If two events are mutually exclusive, then the probability of one or the other happening is the sum of their individual probabilities.

Independent Events

When a die is thrown, and a coin is tossed, these two events have no effect on each other. Such events are called independent events

Product law of probability

If two events are independent, then the probability of both events happening is the product is the product (multiplication) of their individual probabilities.

CHAPTER 51
PROBABILITY ON SIMPLE EVENTS

Examples

1. The table below give the number of students in each age group in a class.

Age (Years)	12	13	14	15	16	17
number of students	6	3	10	4	2	5

If a student is chosen at random, find the probability that the student is:

(a) 13 years old

(b) 15 years old or less

(c) at least 16 years old

(d) most 13 years old

(e) not 17 years old

<u>Solution</u>

(a) Pr. (13 years old) $= \dfrac{\text{Number of students who are 13 years old}}{\text{Total number of students}}$

$= \dfrac{3}{30}$

$= \dfrac{1}{10}$ (when $\dfrac{3}{30}$ is express in its lowest term, it gives $\dfrac{1}{10}$)

(b) Pr. (15 years or less) $= \dfrac{\text{Students who are 15 years and below}}{\text{Total number of students}}$

$= \dfrac{4+10+3+6}{30}$

$= \dfrac{23}{30}$

(c) Pr. (At least 16 years old) $= \dfrac{\text{Students who are 16 years and above}}{\text{Total number of students}}$

$= \dfrac{2+5}{30}$

$= \dfrac{7}{30}$

(d) Pr. (At most 13 years) $= \dfrac{\text{Students who are 13 years and below}}{\text{Total number of students}}$

$= \dfrac{3+6}{30}$

$= \dfrac{9}{30}$

$$= \frac{3}{10} \quad \text{(When expressed in its lowest term)}$$

(e) Pr. (17 years old) $= \dfrac{\text{number of students who are 17 years old}}{\text{total number of students}}$

$$= \frac{5}{30}$$

$$= \frac{1}{6}$$

Therefore, Pr. (Not 17 years old) = 1 - Pr. (17 years old)

$$= 1 - \frac{1}{6}$$

$$= \frac{5}{6}$$

2. The probability that a seed will germinate is $\frac{2}{5}$. What is the probability that it will not germinate?

Solution

Pr. (It will germinate) $= \frac{2}{5}$

Pr. (It will not germinate) $= 1 - \frac{3}{5}$

$$= \frac{2}{5}$$

3. A letter is chosen at random from the alphabet. Find the probability that it is one of the letters of the word: PROBABILITY.

Solution

In this case a letter should not be counted more than once. Avoiding repetition, the word can now be written as:

PROBALITY (i.e. 9 letters). Note that there are 26 letters of the alphabet.

Therefore, Pr. (one letter from PROBABILITY) $= \frac{9}{26}$

4. The probability that a boy gains admission into a higher institution is $\frac{3}{7}$. What is the probability that he does not gain admission into the institution?

Solution

Pr. (He gains admission) $= \frac{3}{7}$

Pr. (He does not gain admissions) = $1 - \dfrac{3}{7}$

$\qquad = \dfrac{4}{7}$

5. Out of every 100 cars, 4 develop mechanical fault within 6 months of purchase. What is the probability of buying a car which will not develop a mechanical fault within 6 months of purchase?

Solution

Total number of cars is 100. Number of cars with fault within 6 months is 4. Number of cars without fault within an months of purchase is going is 96, (i.e. 100 - 4 = 96).

Therefore, Pr. (Buying a car that will not develop fault) = $\dfrac{\text{Number of cars without fault}}{\text{Total number of cars}}$

$\qquad = \dfrac{96}{100}$

$\qquad = \dfrac{24}{25}$ (In its lowest term after equal division by 4)

6. In Mr. Smith's extended family, the number of males is 16, while the number of females is 14. Find the probability that Mr. Smith has:

(a) a male child

(b) a female child

Solution

(a) Total number of family members = 16 + 14 = 30

Therefore, Pr. (a male child) = $\dfrac{\text{Family members who are males}}{\text{Total number of family members}}$

$\qquad = \dfrac{16}{30}$

$\qquad = \dfrac{8}{15}$

(b) Pr. (a female child) = $\dfrac{\text{Family members who are females}}{\text{Total number of family members}}$

$\qquad = \dfrac{14}{30}$

$\qquad = \dfrac{7}{15}$

7. A survey shows that 36% of all women take size 8 shoes. What is the probability that Khan's grandmother takes size 8 shoes?

Solution

Pr. (Khan's grandmother takes size 8 shoes) = $\dfrac{36}{100}$ (Note that the total percentage is always 100%)

$\qquad\qquad = \dfrac{9}{25}$ (In its lowest term)

8. In a secondary school, 46 out of every 50 students are at least 130cm tall. What is the probability that a student chosen at random from the school is less than 130cm tall?

Solution

Total number of students for the sample = 50

Number of students who are at least 130cm tall = 46

Number of students who are less than 130cm tall = 50 - 46 = 4

Therefore, Pr. (a student less than 130cm tall) = $\dfrac{\text{Number of students less than 130 cm tall}}{\text{Total number of students in the sample}}$

$\qquad = \dfrac{4}{50}$

$\qquad = \dfrac{2}{25}$

9. A number is chosen at random between 1 and 16, both inclusive. What is the probability that it is:

(a) even

(b) prime

(c) odd or prime

(d) divisible by 4

(e) a perfect square or a perfect cube

Solution

(a) Total numbers in all from 1 to 16 = 16

The even numbers are 2, 4, 6, 8, 10, 12, 14, 16

Therefore the number of even numbers is 8

Hence Pr. (even number selected) = $\dfrac{\text{Number of even numbers}}{\text{Total numbers in all}}$

$\qquad = \dfrac{8}{16}$

$\qquad = \dfrac{1}{2}$

(b) The prime numbers are 2, 3, 5, 7, 11, 13

Therefore the number of prime numbers is 6

Hence Pr. (prime number selected) = $\dfrac{\text{Number of prime numbers}}{\text{Total numbers in all}}$

$= \dfrac{6}{16}$

$= \dfrac{3}{8}$

(c) The odd numbers are 1, 3, 5, 7, 9, 11, 13, 15

The prime numbers are 1, 3, 5, 7, 11, 13

Since OR in probability means addition, then we add all the odd and prime numbers together, but we must not count any number twice. This gives 1, 3, 5, 7, 9, 11, 13, 15, which is a total of 8 numbers.

Hence Pr. (odd or prime number selected) = $\dfrac{\text{Number of odd and even numbers}}{\text{Total numbers in all}}$

$= \dfrac{8}{16}$

$= \dfrac{1}{8}$

(d) The numbers divisible by 4 are 4, 8, 12, 16

This gives a total of 4 numbers

Hence Pr. (a number divisible by 4) = $\dfrac{\text{The four numbers divisible by 4}}{\text{Total numbers in all}}$

$= \dfrac{4}{16}$

$= \dfrac{1}{4}$

(e) The perfect square numbers are 1, 4, 9, 16

The perfect cube numbers are 1, 8

Since OR in probability means addition, then we add all the set of values above without counting any number twice. This gives 1, 4, 8, 9, 15, which is a total of 5 numbers.

Hence Pr. (perfect square or perfect cube selected) = $\dfrac{15}{16}$

10. A letter is chosen at random from the alphabet. Find the probability that it is:

(a) T

(b) E or P

(c) not B or G

(d) either D, J, N, U, W or Y

(e) one of the letters of the word REJECTED

Solution

(a) There are 26 letters of the alphabet, out of which there is 1 T.

Therefore, Pr. (T) = $\dfrac{\text{Number of Ts}}{\text{Total numbers of alphabets}}$

$$= \dfrac{1}{26}$$

(b) Pr. (E or P) = $\dfrac{\text{Number of Es and Ps}}{\text{Total numbers of alphabets}}$

$$= \dfrac{6}{26}$$

$$= \dfrac{1}{13}$$

(c) Pr. (B or G) = $\dfrac{2}{26}$ = $\dfrac{1}{13}$

Therefore, Pr. (not B or G) = 1 - Pr. (B or G)

$$= 1 - \left(\dfrac{6}{13}\right)$$

$$= \dfrac{12}{13}$$

(d) The letters D, J, N, U, W and Y makes a total of 6 letters.

Pr. (D, J, N, U, W or Y) = $\dfrac{6}{26}$

$$= \dfrac{3}{13}$$

(e) Writing the letters of the word REJECTED without repeating a letter gives REJCTD. This gives a total of 6 letters

Therefore Pr. (one of the letters of REJECTED) = $\dfrac{6}{26}$

$$= \dfrac{3}{13}$$

11. A letter is selected at random from the word PROBABILITY. What is the probability of selecting the letter B.

Solution

In this case the total letters of the word PROBABILITY gives 11. The repeated letters should be counted more than once since this is not a case of letter from the alphabet. In the 26 alphabet each letter appears once, that is why they are counted once. But in PROBABILITY (or other words that might be given) some letters appear more than once, hence they should be counted as many times as they appear.

In PROBABILITY, B appears 2 times.

Therefore, Pr. (selecting B) = $\dfrac{2}{11}$

Exercise 51

1. The table below give the number of students in each mark group in a class.

Mark	5	6	7	8	9	10
Number of students	3	6	2	4	1	4

If a student is chosen at random, find the probability that the student scored:

(a) 7 marks

(b) 6 marks or less

(c) at least 9 marks

(d) at most 8 marks

(e) 5 or 8 maks

2. The probability that a seed will germinate is $\frac{3}{4}$. What is the probability that it will not germinate?

3. A letter is chosen at random from the alphabet. Find the probability that it is one of the letters of the word: MATHEMATICS.

4. The probability that a man wins an election is $\frac{3}{5}$. What is the probability that he does not win.

5. Out of every 10 bulbs, 2 do not last long. What is the probability that a bulb will last long when lit?

6. In family, the number of males is 3, while the number of females is 2. Find the probability that another child born into the family is:

(a) a male child

(b) a female child

7. A survey shows that 44% of all women take size 7 shoes. What is the probability that a mother of two takes size 7 shoes?

8. In a secondary school, 30 out of every 100 students are at least 160cm tall. What is the probability that a student chosen at random from the school is less than 160cm tall?

9. A number is chosen at random between 1 and 20, both inclusive. What is the probability that it is:

(a) prime

(b) odd

(c) even or prime

(d) divisible by 3

(e) a number less than 10 or a perfect cube

10. A letter is chosen at random from the alphabet. Find the probability that it is:

(a) F

(b) M or Q or Y

(c) in the word COME

(d) either in the word BUT or in REMOVE

(e) one of the letters of the word SURPRISED

11. A letter is selected at random from the word RESPIRATION. What is the probability of selecting the letter I.

CHAPTER 52
PROBABILITY ON PACK OF PLAYING CARDS

A pack of playing cards contains 52 cards of 4 types. There are 13 clubs, 13 diamonds, 13 hearts and 13 spades. Each of the set of 13 cards contains Ace (A), 2, 3, 4, 5, 6, 7, 8, 9, 10, Jack (J), Queen (Q), and King (K). This means that out of the 52 cards, each card is four in number, i.e. Aces are 4 in number, 1s are 4 in number, 2s are 4 in number, 3s are 4 in number, 4s are 4 in number, 5s are 4 in number, 6s are 4 in number, 7s are 4 in number, 8s are 4 in number, 9s are 4 in number, 10s are 4 in number, Jacks are 4 in number, Queens are 4 in number, and Kings are 4 in number. Clubs and spades are black, diamonds and hearts are red. This means that there are 26 black cards and 26 red cards. This also means that out of the 4 Aces cards, 2 are black and 2 are red. Out of the four cards that are 1, two are black and two are red, out of the four cards that are 2, two are black and two are red, and so on.

Examples

1. A card is picked at random from a pack of playing cards. Find the probability of picking a spade. <u>Solution</u>

There are 13 spades in a pack of playing cards.

Therefore, Pr. (picking a spade) $= \dfrac{\text{Number of Spades}}{\text{Total numbers of cards}}$

$$= \frac{13}{52}$$

$$= \frac{1}{4} \quad \text{(In its lowest term)}$$

2. A card is picked at random from a pack of playing cards. Find the probability of picking a red card.

<u>Solution</u>

There are 26 red cards in a pack of playing cards.

Therefore, Pr. (picking a red card) $= \dfrac{\text{Number of red cards}}{\text{Total numbers of cards}}$

$$= \frac{26}{52}$$

$$= \frac{1}{2} \quad \text{(In its lowest term)}$$

3. A card is picked at random from a pack of playing cards. Find the probability of picking a red 5.

Solution

There are two red 5 cards in a pack of playing cards.

Therefore, Pr. (picking a red 5) = $\dfrac{\text{Number of red 5}}{\text{Total numbers of cards}}$

$= \dfrac{2}{52}$

$= \dfrac{1}{26}$ (In its lowest term)

4. A card is picked at random from a pack of playing cards. Find the probability of picking a 3.

Solution

There are 4 cards that are 3 in a pack of playing cards.

Therefore, Pr. (picking a 3) = $\dfrac{\text{Number of cards that are 3}}{\text{Total numbers of cards}}$

$= \dfrac{4}{52}$

$= \dfrac{1}{13}$ (In its lowest term)

5. A card is picked at random from a pack of playing cards. Find the probability of picking a black Ace.

Solution

There are 2 cards that are black ace in a pack of playing cards.

Therefore, Pr. (picking a black ace) = $\dfrac{\text{Number of cards that are black ace}}{\text{Total numbers of cards}}$

$= \dfrac{2}{52}$

$= \dfrac{1}{26}$ (In its lowest term)

6. A card is picked at random from a pack of playing cards. Find the probability of picking a card that is not a Jack.

Solution

(a) There are 4 cards that are Jacks.

Therefore, Pr. (picking a Jack) = $\dfrac{\text{Number of jacks}}{\text{Total numbers of cards}}$

$$= \frac{4}{52}$$

$$= \frac{1}{13}$$

Hence, Pr. (picking a card that in not a Jack) = 1 - Pr. (picking a Jack)

$$= 1 - \frac{1}{13}$$

$$= \frac{12}{13}$$

7. A card is picked at random from a pack of playing cards. Find the probability of picking

(a) a black or red card

(b) a 2 or a 5

(c) either a heart or the king of spades

(d) a club or a red Queen

(e) a diamond or a 9

(f) a 6 or a black card

Solution

(a) There are 26 black cards and 26 red card

Since or in probability means plus, then we have to add the numbers. This gives a total of: 26 + 26 = 52

Therefore, Pr. (picking a black or red card) = $\dfrac{\text{Number of black and red cards}}{\text{Total numbers of cards}}$

$$= \frac{52}{52}$$

$$= 1$$

(b) There are 4 cards that are 2, and 4 cards that are 5. This gives a total of 8 cards.

Therefore, Pr. (picking a 2 or a 5) = $\dfrac{8}{52}$

$$= \frac{2}{13}$$

(c) There are 13 cards that are Hearts, and 1 king that is a spade. This gives a total of 14 cards.

Therefore, Pr. (picking either a heart or the king of spades) = $\dfrac{14}{52}$

$$= \frac{7}{26}$$

(d) There are 13 cards that are club, and 2 cards that are red Queen, (i.e. the Queen of hearts and the queen of diamond). This gives a total of 15 cards.

Therefore, Pr. (picking a club or a red Queen) = $\dfrac{15}{52}$

(e) There are 13 cards that are diamonds, and 4 cards that are 9. But one of the 9 is in diamond and has already been counted among the 13 diamonds. So it must not be counted twice. Hence we count the other three 9 (each from clubs, hearts and spades). This will give a total of 16 (13 + 3) cards.

Therefore, Pr. (picking a diamond or a 9) = $\dfrac{16}{52}$

$= \dfrac{4}{13}$

(f) There are 4 cards that are 6, and 26 cards that are black. But two of the 26 black cards are among the four cards that are 6, and these two black 6 cards have already been counted among the 26 black cards. So they must not be counted twice. Hence we count the other two 6 cards that are red. This will give a total of 28 (26 + 2) cards.

Therefore, Pr. (picking a 6 or a back card) = $\dfrac{28}{52}$

$= \dfrac{7}{13}$

8. A card is picked at random from a pack of playing cards and then replaced. A second card is picked. What is the probability of picking:
(a) a 3 and a 10
(b) a queen and an ace
(c) two kings
(d) two red cards
(e) two cards of different colours
(f) two cards of the same colour

Solution

In probability problems, when two items are selected, it is important to logically analyse the situation when solving the problem. This will help you to know if addition (use of OR) is involved or multiplication (use of AND) is involved. For example, for a queen and a king to be selected, it simply means that, either the queen is selected first and then the king, or the king is selected first and then the queen. When this logical analysis is understood, then most questions in probability become easy to solve.

(a) There are four cards that are 3, and four cards that are 10

Therefore, Pr. (picking a 3) = $\dfrac{4}{52}$

$= \dfrac{1}{13}$

Similarly, Pr. (picking a 10) = $\dfrac{4}{52}$

$$= \frac{1}{13}$$

Recall that "and" in probability means multiplication.

The probability of picking a 3 and a 10 means that:

Either the first is a 3 AND the second is a 10, OR the first is a 10 AND the second is a 3.

This can be calculated by putting x in place of AND and + in place of OR in the above statement as follows:

Pr. (picking a 3) x Pr. (picking a 10) + Pr. (picking a 10) x Pr. (picking a 3)

$$= (\frac{1}{13} \times \frac{1}{13}) + (\frac{1}{13} \times \frac{1}{13})$$

$$= \frac{1}{169} + \frac{1}{169}$$

$$= \frac{2}{169}$$

Therefore, Pr. (picking a 3 and a 10) $= \frac{2}{169}$

(b) There are 4 cards that are queen, and 4 cards that are ace

Therefore, Pr. (picking a queen) $= \frac{4}{52}$

$$= \frac{1}{13}$$

Similarly, Pr. (picking an ace) $= \frac{4}{52}$

$$= \frac{1}{13}$$

The probability of picking a queen and an ace means that:

Either you first pick a queen AND then an ace, OR you first pick an ace AND then a queen.

This can be calculated by putting x in place of AND and + in place of OR in the above statement as follows:

Pr. (picking a queen) x Pr. (picking an ace) + Pr. (picking an ace) x Pr. (picking a queen)

$$= (\frac{1}{13} \times \frac{1}{13}) + (\frac{1}{13} \times \frac{1}{13})$$

$$= \frac{1}{169} + \frac{1}{169}$$

$$= \frac{1}{169}$$

Therefore, Pr. (picking a queen and an ace) $= \frac{2}{169}$

(c) There are four cards that are King

Therefore, Pr. (picking a king) $= \frac{4}{52}$

$$= \frac{1}{13}$$

The probability of picking two kings means that:

The first is a king AND the second is a king

= Pr. (picking a king) x Pr. (picking a king)

$$= \frac{1}{13} \times \frac{1}{13}$$

$$= \frac{1}{169}$$

Therefore, Pr. (picking two kings) = $\frac{1}{169}$

(d) There are 26 cards that are red

Therefore, Pr. (picking a red card) = $\frac{26}{52}$

$$= \frac{1}{2}$$

The probability of picking two red cards means that:

The first is a red card AND the second is a red card

= Pr. (picking a red card) x Pr. (picking a red card)

$$= \frac{1}{2} \times \frac{1}{2}$$

$$= \frac{1}{4}$$

Therefore, Pr. (picking two red cards) = $\frac{1}{4}$

(e) There are two colours of cards, red and black.

Therefore, Pr. (picking a red card) = $\frac{1}{2}$ (i.e from $\frac{26}{52}$ since there are 26 red cards)

Similarly, Pr. (picking a black card) = $\frac{1}{2}$ (i.e from $\frac{26}{52}$ since there are also 26 black cards)

The probability of picking two cards of different colours means that:

Either the first is a black card AND the second is a red card, OR the first is a red card AND the second is a black card.

This can be calculated by putting x in place of AND and + in place of OR in the above statement as follows:

Pr. (picking a black card) x Pr. (picking a red card) + Pr. (picking a red card) x Pr. (picking a black card)

$$= (\frac{1}{2} \times \frac{1}{2}) + (\frac{1}{2} \times \frac{1}{2})$$

$$= \frac{1}{4} + \frac{1}{4}$$

$$= \frac{2}{4}$$

$$= \frac{1}{2}$$

Therefore, Pr. (picking two cards of different colours) = $\frac{1}{2}$

(f) Pr. (picking two cards of the same colours) = 1 - Pr. (picking two cards of different colours)

$$= 1 - \frac{1}{2}$$

$$= \frac{1}{2}$$

Note that this can also be solved by using the logical process which is:

Either the first is red AND the second is red OR the first is black AND the second is black. This will also give $\frac{1}{2}$

9. Two cards are picked at random one after the other without replacement from a pack of playing cards. What is the probability of picking:

(a) a 5 and a 7

(b) a king and a jack

(c) two aces

(d) two diamond cards

(e) two black cards

(f) a red and a black card

(g) two cards of the same colours

Solution

This problem involves picking a card without replacement. This means that when one card is picked out, the total number of cards remaining in the pack become reduced to 51. That number of that particular type of card also reduces by 1.

(a) There are four cards that are 5. There are also four cards that are 7.

Hence the probability of picking a 5 and a 7 means that:

Either first picking a 5 AND then a 7, OR first picking a 7 AND then a 5.

Now, let us calculate each of the probabilities as follows:

Pr. (first card is a 5) = $\frac{4}{52}$ (There are four cards that are 5)

$\quad = \frac{1}{13}$ (In its lowest term)

We now have 51 cards left in the pack.

Therefore, Pr. (second card is a 7) = $\frac{4}{51}$ (There are four cards that are 7, and a total of 51 cards remaining in the pack)

Or,

Pr. (first card is a 7) = $\frac{4}{52}$ (There are four cards that are 7)

$\quad = \frac{1}{13}$ (In its lowest term)

We now have 51 cards left in the pack.

Therefore, Pr. (second card is a 5) = $\frac{4}{51}$ (There are four cards that are 5, and a total of 51 cards remaining in the pack)

Hence the probability of picking a 5 and a 7 means that:

Either first picking a 5 AND then a 7, OR first picking a 7 AND then a 5. Which is computed as:

Pr. (picking a 5 and a 7) = Pr. (first card is a 5) x Pr. (second card is a 7) + Pr. (first card is a 7) x Pr. (second card is a 5)

$$= (\frac{1}{13} \times \frac{4}{51}) + (\frac{1}{13} \times \frac{4}{51})$$

$$= \frac{4}{663} + \frac{4}{663}$$

$$= \frac{8}{663}$$

(b) There are four cards that are kings. There are also four cards that are jacks.

Now, let us calculate each of the probabilities as follows:

Pr. (first card is a king) = $\frac{4}{52}$ (There are four cards that are kings)

$$= \frac{1}{13}$$ (In its lowest term)

We now have 51 cards left in the pack.

Therefore, Pr. (second card is a jack) = $\frac{4}{51}$ (There are four cards that are jack, and a total of 51 cards remaining in the pack)

Or,

Pr. (first card is a jack) = $\frac{4}{52}$ (There are four cards that are jack)

$$= \frac{1}{13}$$ (In its lowest term)

We now have 51 cards left in the pack.

Therefore, Pr. (second card is a king) = $\frac{4}{51}$ (There are four cards that are king, and a total of 51 cards remaining in the pack)

Hence the probability of picking a king and a jack means that:

Either first picking a king AND then a jack, OR first picking a jack AND then a king. This is computed as:

Pr. (picking a king and a queen) = Pr. (first card is a king) x Pr. (second card is a jack) + Pr. (first card is a jack) x Pr. (second card is a king)

$$= (\frac{1}{13} \times \frac{4}{51}) + (\frac{1}{13} \times \frac{4}{51})$$

$$= \frac{4}{663} + \frac{4}{663}$$

$$= \frac{8}{663}$$

(c) There are 4 cards that are aces.

Hence the probability of picking two aces means that:

The first is an ace, and the second is an ace.

Now, let us calculate each of the probabilities as follows:

Pr. (first card is an ace) = $\frac{4}{52}$ (There are 4 cards that are aces)

$\quad = \frac{1}{13}$ (In its lowest term)

We now have 3 aces left in the pack, and a total of 51 cards left in the pack.

Therefore, Pr. (second card is an ace) = $\frac{3}{51}$

Hence the probability of picking two aces is given by:

Pr. (picking two aces) = Pr. (first card is an ace) x Pr. (second card is an ace)

$\quad = \frac{1}{13} \times \frac{3}{51}$

$\quad = \frac{3}{663}$

(d) There are 13 cards that are diamonds.

Hence the probability of picking two diamonds means that:

The first is a diamond, and the second is a diamond.

Now, let us calculate each of the probabilities as follows:

Pr. (first card is a diamond) = $\frac{13}{52}$ (There are 13 cards that are diamonds)

$\quad = \frac{1}{4}$ (In its lowest term)

We now have 12 diamonds left in the pack, and a total of 51 cards left in the pack.

Therefore, Pr. (second card is a diamond) = $\frac{1}{13}$

$\quad = \frac{4}{17}$ (In its lowest term)

Hence the probability of picking two diamonds is given by:

Pr. (picking two diamonds) = Pr. (first card is a diamond) x Pr. (second card is a diamond)

$\quad = \frac{1}{4} \times \frac{4}{17}$

$\quad = \frac{4}{68}$

$\quad = \frac{1}{17}$ (In its lowest term)

(e) There are 26 black cards.

Hence the probability of picking two black cards means that:

The first is a black card, and the second is a black card.

Now, let us calculate each of the probabilities as follows:

Pr. (first card is a black card) = $\dfrac{26}{52}$

$\qquad = \dfrac{1}{2}$ (In its lowest term)

We now have 25 black cards left in the pack, and a total of 51 cards left in the pack.

Therefore, Pr. (second card is a black card) = $\dfrac{25}{51}$

Hence the probability of picking two black cards is given by:

Pr. (picking two black cards) = Pr. (first card is a black card) x Pr. (second card is a black card)

$\qquad = \dfrac{1}{2} \times \dfrac{25}{51}$

$\qquad = \dfrac{25}{102}$

(f) The logical explanation for this situation is that:

Either the first card is red AND the second is black OR the first card is black and the second is red.

There are 26 red cards and also 26 black cards.

Now, let us calculate each of the probabilities as follows:

Pr. (first card is a red card) = $\dfrac{26}{52}$

$\qquad = \dfrac{1}{2}$ (In its lowest term)

We now have 51 cards left in the pack.

Therefore, Pr. (second card is a black card) = $\dfrac{26}{51}$ (There are 26 black cards, and a total of 51 cards remaining in the pack)

Or,

Pr. (first card is a black card) = $\dfrac{26}{52}$

$\qquad = \dfrac{1}{2}$ (In its lowest term)

We now have 51 cards left in the pack.

Therefore, Pr. (second card is a red card) = $\dfrac{26}{51}$ (There are 26 red cards, and a total of 51 cards remaining in the pack)

Hence the probability of picking a red card and a black card means that:

Either first picking a red card AND then a black card, OR first picking a black card AND then a red card. This is computed as:

Pr. (picking a red and black cards) = Pr. (first card is a red card) x Pr. (second card is a black card) + Pr. (first card is a black card) x Pr. (second card is a red card)

$$= (\frac{1}{2} \times \frac{26}{51}) + (\frac{1}{2} \times \frac{26}{51})$$

$$= \frac{26}{102} + \frac{26}{102}$$

$$= \frac{52}{102}$$

$$= \frac{26}{51}$$

(g) The logical explanation for this situation is that:

Either the first card is red AND the second is red OR the first card is black and the second is black.

There are 26 red cards and also 26 black cards.

Now, let us calculate each of the probabilities as follows:

Pr. (first card is a red card) $= \frac{26}{52}$

$= \frac{1}{2}$ (In its lowest term)

We now have 25 red cards left and a total of 51 cards left in the pack.

Therefore, Pr. (second card is a red card) $= \frac{25}{51}$

Or,

Pr. (first card is a black card) $= \frac{26}{102}$

$= \frac{1}{2}$ (In its lowest term)

We now have 25 black cards left and a total of 51 cards left in the pack.

Therefore, Pr. (second card is a black card) $= \frac{25}{51}$

Hence the probability of picking two cards of the same colour means that:

Either picking a red card AND then another red card, OR picking a black card AND then another black card. This is computed as:

Pr. (picking two cards of the same colour) = Pr. (first card is a red card) x Pr. (second card is a red card) + Pr. (first card is a black card) x Pr. (second card is a black card)

$$= (\frac{1}{2} \times \frac{25}{51}) + (\frac{1}{2} \times \frac{25}{51})$$

$$= \frac{25}{102} + \frac{25}{102}$$

$$= \frac{50}{102}$$

$$= \frac{25}{51}$$

Alternatively, this question can also be solved as follows:

Recall that question (f) above gives the probability of picking a red and a black card. This also means the probability of picking two cards of different colours.

Hence the probability of picking two cards of different colours as given in (f) above = $\dfrac{26}{51}$

Therefore, Pr. (picking two cards of the same colour) = 1 - Pr. (picking two cards of different colours)   (Note that they are opposite statements)

$$= 1 - \dfrac{26}{51}$$

$$= \dfrac{51-26}{51}$$

$$= \dfrac{25}{51} \quad \text{(As obtained before)}$$

10. If three cards are picked from a pack of playing cards with replacement, what is the probability if getting:
(a) at least two clubs
(b) at most two clubs

Solution
I am going to be using a special type of tree diagram without actually drawing the diagram. Now, the total outcome in a selection of three items involving two events (i.e. a club or not a club) is given by:
$$2^n,$$
where n is the number of selection made.
In the question, n = 3, since three cards were picked.
Hence total outcome = 2^3 = 2 x 2 x 2

$$= 8.$$

Now, in order to write out the outcomes, let us use the letter C to represent a club and letter N to represent not a club. Note that in tree diagrams like this, only two letters should be used in writing the outcomes since the question involves the picking of only one type of item (club). Hence the outcome is written as follows:

(CCC), (CCN), (CNC), (CNN), (NCC), (NCN), (NNC), (NNN)

Note that there are 8 ways of arranging the two letters in the brackets. There is no fast rule in carrying out the arrangement. You just have to make sure that no two brackets have the same arrangement of the letters. Also make sure the number of brackets is complete.

(a) In order to determine the probability of getting at least two clubs, we need to compute the probabilities of the brackets that contain at least 2 clubs. They are, (CCC), (CCN), (CNC), and (NCC). Note that at least two, means two and above, (i.e. two and three clubs in this case). Hence the probability of getting at least two clubs = (CCC) or (CCN) or (CNC) or (NCC)
Now, let us compute each of the probabilities.

There are 13 clubs in a pack of cards, and there are 39 cards that are not club. Note that this is a case of with replacement, which means that the total number of cards in the pack is always complete. Hence:

(CCC) = Pr. (first card is a club) x Pr. (second card is a club) x Pr. (third card is a club)

$$= \frac{13}{52} \times \frac{13}{52} \times \frac{13}{52}$$

$$= \frac{1}{4} \times \frac{1}{4} \times \frac{1}{4}$$

$$= \frac{1}{64}$$

(CCN) = Pr. (first card is a club) x Pr. (second card is a club) x Pr. (third card is not a club)

$$= \frac{13}{52} \times \frac{13}{52} \times \frac{39}{52} \quad \text{(Note that there are 39 cards that are not club)}$$

$$= \frac{1}{4} \times \frac{1}{4} \times \frac{3}{4}$$

$$= \frac{3}{64}$$

(CNC) = Pr. (first card is a club) x Pr. (second card is not a club) x Pr. (third card is a club)

$$= \frac{13}{52} \times \frac{39}{52} \times \frac{13}{52}$$

$$= \frac{1}{4} \times \frac{3}{4} \times \frac{1}{4}$$

$$= \frac{3}{64}$$

(NCC) = Pr. (first card is not a club) x Pr. (second card is a club) x Pr. (third card is a club)

$$= \frac{39}{52} \times \frac{13}{52} \times \frac{13}{52}$$

$$= \frac{3}{4} \times \frac{1}{4} \times \frac{1}{4}$$

$$= \frac{3}{64}$$

Therefore, Pr. (getting at least two clubs) = (CCC) or (CCN) or (CNC) or (NCC)

$$= (CCC) + (CCN) + (CNC) + (NCC)$$

$$= \frac{1}{64} + \frac{3}{64} + \frac{3}{64} + \frac{3}{64}$$

$$= \frac{10}{64}$$

$$= \frac{5}{32}$$

(b) In order to determine the probability of getting at most two clubs, we need to compute the probabilities of the brackets that contain at most 2 clubs. From the outcome brackets given above, the ones that contain at most two clubs are, (CCN), (CNC), (CNN), (NCC), (NCN), (NNC), (NNN). Note that at most two, means two and below, (i.e. two, one and zero clubs in this case). Hence the probability of getting at most two clubs = (CCN) or (CNC) or (CNN) or (NCC) or (NCN) or (NNC) or (NNN)

Now, let us compute each of the probabilities. Hence:

$(CCN) = \dfrac{3}{64}$ (As calculated in (a) above)

$(CNC) = \dfrac{3}{64}$ (As calculated in (a) above)

(CNN) = Pr. (first card is a club) x Pr. (second card is not a club) x Pr. (third card is not a club)

$$= \dfrac{13}{52} \times \dfrac{39}{52} \times \dfrac{39}{52}$$

$$= \dfrac{1}{4} \times \dfrac{3}{4} \times \dfrac{3}{4})$$

$$= \dfrac{9}{64}$$

$(NCC) = \dfrac{3}{64}$ (As calculated in (a) above)

(NCN) = Pr. (first card is not a club) x Pr. (second card is a club) x Pr. (third card is not a club)

$$= \dfrac{39}{52} \times \dfrac{13}{52} \times \dfrac{39}{52}$$

$$= \dfrac{3}{4} \times \dfrac{1}{4} \times \dfrac{3}{4}$$

$$= \dfrac{9}{64}$$

(NNC) = Pr. (first card is not a club) x Pr. (second card is not a club) x Pr. (third card is a club)

$$= \dfrac{39}{52} \times \dfrac{39}{52} \times \dfrac{13}{52}$$

$$= \dfrac{3}{4} \times \dfrac{3}{4} \times \dfrac{1}{4}$$

$$= \dfrac{9}{64}$$

(NNN) = Pr. (first card is not a club) x Pr. (second card is not a club) x Pr. (third card is not a club)

$$= \dfrac{39}{52} \times \dfrac{39}{52} \times \dfrac{39}{52}$$

$$= \dfrac{3}{4} \times \dfrac{3}{4} \times \dfrac{3}{4}$$

$$= \dfrac{27}{64}$$

Therefore, Pr. (getting at most two clubs) = (CCN) or (CNC) or (CNN) or (NCC) or (NCN) or (NNC) or (NNN)

$$= (CCN) + (CNC) + (CNN) + (NCC) + (NCN) + (NNC) + (NNN)$$

$$= \dfrac{3}{64} + \dfrac{3}{64} + \dfrac{9}{64} + \dfrac{3}{64} + \dfrac{9}{64} + \dfrac{9}{64} + \dfrac{27}{64}$$

$$= \dfrac{63}{64}$$

Alternatively, a shorter method of solving this problem is as follows:

Pr. (getting at most two clubs) = 1 - Pr. (getting three clubs)   (This is because the only item not in (b) is CCC. All 8 outcomes in brackets make a total probability of 1, but 7 of the outcomes are in (b), hence 1 minus the outcome not in (b) gives (b). This can also be expressed in fraction as, (b) = $\frac{7}{8}$ = $\frac{8}{8}$ - $\frac{1}{8}$). Therefore:

Pr. (getting at most two clubs) = 1 - Pr. (getting three clubs)

= 1 - CCC

= 1 - $\frac{1}{64}$ (Note that CCC computed in (a) = $\frac{1}{64}$)

= $\frac{63}{64}$ (As obtained before)

11. If three cards are chosen from a pack of playing cards without replacement, what is the probability of getting:

(a) at least two diamonds

(b) at most one diamond?

Solution

In order to write out the outcomes, let us use the letter D to represent a diamond and letter N to represent not a diamond.

Hence the outcomes are written as follows:

(DDD), (DDN), (DND), (DNN), (NDD), (NDN), (NND), (NNN)

(a) In order to determine the probability of getting at least two diamonds, we need to compute the probabilities of the brackets that contain at least 2 diamonds. They are, (DDD), (DDN), (DND), and (NDD).

Hence the probability of getting at least two diamonds = (DDD) or (DDN) or (DND) or (NDD)

Now, let us compute each of the probabilities.

There are 13 diamonds in a pack of cards, and there are 39 cards that are not diamonds. Note that this is a case of without replacement, which means that after each selection, both the total number of cards left and the number of the particular card picked, are reduced by 1. Hence:

(DDD) = Pr. (first card is a diamond) x Pr. (second card is a diamond) x Pr. (third card is a diamond)

= $\frac{13}{52}$ x $\frac{12}{51}$ x $\frac{11}{50}$ (Note that the number of diamond and the total number of card left, keep reducing by 1 after each selection)

= $\frac{1}{4}$ x $\frac{12}{51}$ x $\frac{11}{50}$

= $\frac{132}{10200}$

= $\frac{11}{850}$ (In its lowest term, after equal division by 12)

(DDN) = Pr. (first card is a diamond) x Pr. (second card is a diamond) x Pr. (third card is not a diamond)

$$= \frac{13}{52} \times \frac{12}{51} \times \frac{39}{50} \quad \text{(Note that there are 39 cards that are not diamond)}$$

$$= \frac{1}{4} \times \frac{4}{17} \times \frac{39}{50}$$

$$= \frac{156}{3400}$$

$$= \frac{39}{850} \quad \text{(After equal division by 4)}$$

(DND) = Pr. (first card is a diamond) x Pr. (second card is not a diamond) x Pr. (third card is a diamond)

$$= \frac{13}{52} \times \frac{39}{51} \times \frac{12}{50}$$

$$= \frac{1}{4} \times \frac{13}{17} \times \frac{6}{25}$$

$$= \frac{78}{1700}$$

$$= \frac{39}{850}$$

(NDD) = Pr. (first card is not a diamond) x Pr. (second card is a diamond) x Pr. (third card is a diamond)

$$= \frac{39}{52} \times \frac{13}{51} \times \frac{12}{50}$$

$$= \frac{3}{4} \times \frac{13}{51} \times \frac{6}{25}$$

$$= \frac{234}{5100}$$

$$= \frac{39}{850}$$

Therefore, Pr. (getting at least two diamonds) = (DDD) or (DDN) or (DND) or (NDD)

$$= (DDD) + (DDN) + (DND) + (NDD)$$

$$= \frac{11}{850} + \frac{39}{850} + \frac{39}{850} + \frac{39}{850}$$

$$= \frac{128}{850}$$

$$= \frac{64}{425}$$

(b) In order to determine the probability of getting at most one diamond, we need to compute the probabilities of the brackets that contain at most one diamond. From the outcome brackets given above, the ones that contain at most one diamond are, (DNN), (NDN), (NND), (NNN). Note that at most one, means one and below, (i.e. one and zero diamond in this case). Hence the probability of getting at most one diamond = (DNN) + (NDN) + (NND) + (NNN) Now, let us compute each of the probabilities. Hence:

(DNN) = Pr. (first card is a diamond) x Pr. (second card is not a diamond) x Pr. (third card is not a diamond)

$$= \frac{13}{52} \times \frac{39}{51} \times \frac{38}{50}$$

$$= \frac{1}{4} \times \frac{13}{17} \times \frac{19}{25}$$

$$= \frac{247}{1700}$$

(NDN) = Pr. (first card is not a diamond) x Pr. (second card is a diamond) x Pr. (third card is not a diamond)

$$= \frac{39}{52} \times \frac{13}{51} \times \frac{38}{50}$$

$$= \frac{3}{4} \times \frac{13}{51} \times \frac{19}{25}$$

$$= \frac{741}{5100}$$

$$= \frac{247}{1700} \quad \text{(In its lowest term after equal division by 3)}$$

(NND) = Pr. (first card is not a diamond) x Pr. (second card is not a diamond) x Pr. (third card is a diamond)

$$= \frac{39}{52} \times \frac{38}{51} \times \frac{13}{50}$$

$$= \frac{3}{4} \times \frac{38}{51} \times \frac{13}{50}$$

$$= \frac{1482}{10200}$$

$$= \frac{247}{1700} \quad \text{(After equal division by 6)}$$

(NNN) = Pr. (first card is not a diamond) x Pr. (second card is not a diamond) x Pr. (third card is not a diamond)

$$= \frac{39}{52} \times \frac{38}{51} \times \frac{37}{50}$$

$$= \frac{3}{4} \times \frac{38}{51} \times \frac{37}{50}$$

$$= \frac{4218}{10200}$$

$$= \frac{703}{1700} \quad \text{(After equal division by 6)}$$

Therefore, Pr. (getting at most one diamond) = (DNN) or (NDN) or (NND) or (NNN)

$$= (DNN) + (NDN) + (NND) + (NNN)$$

$$= \frac{247}{1700} + \frac{247}{1700} + \frac{247}{1700} + \frac{703}{1700}$$

$$= \frac{1444}{1700}$$

$$= \frac{361}{425}$$

Exercise 52

1. A card is picked at random from a pack of playing cards. Find the probability of picking a jack.

2. A card is picked at random from a pack of playing cards. Find the probability of picking a black 4.

3. A card is picked at random from a pack of playing cards. Find the probability of picking a red king.

4. A card is picked at random from a pack of playing cards. Find the probability of picking a either a black or red card.

5. A card is picked at random from a pack of playing cards. Find the probability of picking a black Queen.

6. A card is picked at random from a pack of playing cards. Find the probability of picking a card that is not an Ace.

7. A card is picked at random from a pack of playing cards. Find the probability of picking

(a) a queen or a king

(b) a 3 or a 9

(c) either a jack or the queen of diamonds

(d) a spade or a black 7

(e) a club or a red king

(f) a 2 or a red card

8. A card is picked at random from a pack of playing cards and then replaced. A second card is picked. What is the probability of picking:

(a) an 8 and a 5

(b) a black card and a 4

(c) two cards between 2 and 9 that have odd numbers

(d) two black cards

(e) two cards with the same number on them

(f) two cards with different number on them

9. Two cards are picked at random one after the other without replacement from a pack of playing cards. What is the probability of picking:

(a) a 4 and an ace

(b) a 2 and a 7

(c) two 8s

(d) two clubs

(e) two red cards

(f) a club and a diamond

(g) two cards that are queens

10. If three cards are picked from a pack of playing cards with replacement, what is the probability if getting:

(a) at least two 9s

(b) at most two 9s

11. If three cards are chosen from a pack of playing cards without replacement, what is the probability of getting:

(a) at least two kings

(b) at most one king?

CHAPTER 53
PROBABILITY ON TOSSING OF COINS

When a coin is tossed, the outcome can either be a head or a tail. However when two or more coins are tossed, the total outcome is obtained from 2^n, where n is the number of times the coin is tossed, or the number of coins tossed together.

Note that 'head' is the part of the coin that shows the person drawn on the coin, while the opposite side of the coin is called the 'tail'

Examples

1. A fair coin is tossed. What is the probability of getting:
(a) a head
(b) a tail

Solution

(a) There are only two possible outcomes. Head or tail.

Therefore, Pr. (getting a head) $= \dfrac{\text{Number of heads}}{\text{Total outcomes}}$

$= \dfrac{1}{2}$

(b) Pr. (getting a tail) $= \dfrac{\text{Number of tails}}{\text{Total outcomes}}$

$= \dfrac{1}{2}$

2. A coin is tossed two times. What is the probability of getting:
(a) a head and a tail
(b) at least a tail
(c) two heads
(d) two tails
(e) a head on the first toss, and a tail on the second toss.

Solution

The outcomes are written by using H for head and T for tail. The total number of outcomes will be $2^2 = 4$ (i.e. from 2^n, and n = 2 in this case)

The outcomes are: (HH), (HT), (TH), (TT).

(a) The outcomes with head and tail are (HT) and (TH). This gives 2 outcomes.

Therefore, Pr. (getting a head and tail) $= \dfrac{\text{Number of outcome with head and tail}}{\text{Total outcomes}}$

$$= \frac{2}{4}$$

$$= \frac{1}{2}$$

(b) The outcomes with at least a tail are (HT), (TH) and (TT). This gives 3 outcomes.

Therefore, Pr. (getting at least a tail) = $\dfrac{\text{Number of outcomes with at least a tail}}{\text{Total number of outcomes}}$

$$= \frac{3}{4}$$

(c) The outcome with two heads is (HH). This gives 1 outcome.

Therefore, Pr. (getting two heads) = $\dfrac{\text{Number of outcomes with heads}}{\text{Total number of outcomes}}$

$$= \frac{1}{4}$$

(d) The outcome with two tails is (TT). This gives 1 outcome.

Therefore, Pr. (getting two tails) = $\dfrac{\text{Number of outcomes with two tails}}{\text{Total number of outcomes}}$

$$= \frac{1}{4}$$

(e) The outcome with a head on the first toss, and a tail on the second toss is (HT). This gives 1 outcome

Therefore, Pr. (getting a head on the first toss, and a tail on the second toss) = $\dfrac{1}{4}$

3. A coin is tossed three times. What is the probability of getting:
(a) two heads and one tail
(b) at least one head
(c) three tails
(d) at least two heads
(e) a tail, a head and a tail

Solution

(a) The total number of outcomes will be $2^3 = 8$ (i.e. from 2^n, and n = 3 in this case)
The outcomes are: (HHH), (HTH), (HTT), (HHT), (THH), (THT), (TTH), (TTT). This gives a total of 8 outcomes

(a) The outcomes with two heads and one tail are (HTH), (HHT) and (THH). This gives 3 outcomes.
Therefore, Pr. (getting two heads and one tail) =

$$\frac{\text{Number of outcomes with two heads and one tail}}{\text{Total number of outcomes}}$$

$$= \frac{3}{8}$$

(b) The outcomes with at least one head are (HHH), (HTH), (HTT), (HHT), (THH), (THT) and (TTH). This gives 7 outcomes.

Therefore, Pr. (getting at least one head) = $\dfrac{\text{Number of outcomes with at least one head}}{\text{Total number of outcomes}}$

$$= \frac{7}{8}$$

(c) The outcome with three tails is (TTT). This gives 1 outcome.

Therefore, Pr. (getting three tails) = $\dfrac{1}{8}$

(d) The outcomes with at least two heads are (HHH), (HTH), (HHT) and (THH). This gives 4 outcomes.

Therefore, Pr. (getting at least two heads) = $\dfrac{\text{Number of outcomes with at least two heads}}{\text{Total number of outcomes}}$

$$= \frac{4}{8}$$

$$= \frac{1}{2}$$

(e) The outcome with a tail, a head and a tail is (THT). This is 1 outcome

Hence, Pr. (getting a tail, a head and a tail) = $\dfrac{1}{8}$

4. Four coins are tossed together. Find the probability of getting:
(a) two heads and two tails
(b) four tails
(c) at least three heads
(d) at least two heads
(e) one head

Solution
The total number of outcomes will be $2^4 = 16$ (i.e. from 2^n, and n = 4 in this case)
The outcomes are: (HHHH), (HHHT), (HHTT), (HTTT), (THHH), (TTHH), (TTTH), (THTH), (HTHT), (HHTH), (THHT), HTTH), (TTHT), (THTT), (HTHH), (TTTT). This gives a total of 16 outcomes.

(a) The outcomes with two heads and two tails are (HHTT), (TTHH), (THTH), (HTHT), (THHT), (HTTH). This gives 6 outcomes.

Therefore, Pr. (getting two heads and two tails) =

$$\frac{\text{Number of outcomes with two heads and two tails}}{\text{Total number of outcomes}}$$

$$= \frac{6}{16}$$

$$= \frac{3}{8}$$

(b) The outcome with four tails is (TTTT). This gives 1 outcomes.

Therefore, Pr. (getting four tails) = $\frac{1}{16}$

(c) The outcomes with at least three heads are (HHHH), (HHHT), (THHH), (HHTH), (HTHH). This gives 5 outcomes.

Therefore, Pr. (getting at least three heads) = $\frac{\text{Number of outcomes with at least three heads}}{\text{Total number of outcomes}}$

$$= \frac{5}{16}$$

(d) The outcomes with at least two heads are (HHHH), (HHHT), (HHTT), (THHH), (TTHH), (THTH), (HTHT) (HHTH), (THHT), (HTTH), (HTHH). This gives 11 outcomes.

Therefore, Pr. (getting at least two heads) = $\frac{\text{Number of outcomes with at least two heads}}{\text{Total number of outcomes}}$

$$= \frac{11}{16}$$

(e) The outcomes with one head are, (HTTT), (TTTH), (TTHT), (THTT), . This gives 4 outcomes.

Therefore, Pr. (getting one head) = $\frac{\text{Number of outcomes with one head}}{\text{Total number of outcomes}}$

$$= \frac{4}{16}$$

$$= \frac{1}{4}$$

(5) A coin is tossed five times. Find the probability of getting at least one tail.

Solution

The total number of outcomes will be $2^5 = 32$

The only outcome without a tail is (HHHHH). This is an outcome of 1

Pr. (getting no tail, i.e. all head) = $\frac{1}{32}$

Therefore, Pr. (getting at least one tail) = 1 - Pr. (getting no tail)

$$= 1 - \frac{1}{32}$$

$$= \frac{31}{32}$$

Exercise 53

1. A fair coin is tossed. What is the probability of getting:

   (a) a tail

   (b) a head

   (c) a tail or a head

2. A coin is tossed two times. What is the probability of getting:

   (a) a tail and then a head

   (b) at least a head

   (c) two tails

   (d) at least a tail

   (e) a head on the first toss, and a tail on the second toss.

3. Three coins are tossed. What is the probability of getting:

   (a) three heads

   (b) at least one tail

   (c) a head, a tail and then a head

   (d) at least one head

   (e) at least two heads

   (f) at most two tails

4. Four coins are tossed together. Find the probability of getting:

   (a) at least one head

   (b) four heads

   (c) at least two heads

   (d) at most three tails

   (e) two heads

(5) A coin is tossed five times. Find the probability of getting at least one head.

CHAPTER 54
PROBABILITY ON THROWING OF DICE

Examples

1. A fair die is thrown. Find the probability of getting:

(a) a 2

(b) a 5

(c) a 7

(d) a 4 or a 5

(e) a number less than 4

(f) an odd number

Solution

Note that a die has six faces numbered 1 to 6. That means that each number appears once.

(a) Pr. (getting a 2) = $\dfrac{\text{Number of faces having } 2}{\text{total number of faces}}$

$= \dfrac{1}{6}$

(b) Pr. (getting a 5) = $\dfrac{\text{Number of faces having } 5}{\text{total number of faces}}$

$= \dfrac{1}{6}$

(c) Pr. (getting a 7) = $\dfrac{\text{Number of faces having } 7}{\text{total number of faces}}$

$= \dfrac{0}{6}$

$= 0$ (This is a case of an impossible event)

(d) Pr. (getting a 4 or a 5) = $\dfrac{\text{Number of faces having } 4 \text{ and having } 5}{\text{total number of faces}}$

$= \dfrac{2}{6}$

$= \dfrac{1}{3}$

(e) Pr. (getting a number less than 4) = $\dfrac{\text{Number of faces having numbers less than } 4}{\text{total number of faces}}$

$= \dfrac{3}{6}$ (Note that the faces with numbers less than 4 are 3, 2, and 1. This makes a total of 3 faces)

$= \dfrac{1}{2}$

(f) Pr. (getting an odd number) = $\dfrac{\text{Number of faces having odd numbers}}{\text{total number of faces}}$

$\quad = \dfrac{3}{6}$ (Faces with odd numbers are 1, 3 and 5, i.e. three faces)

$\quad = \dfrac{1}{2}$

2. A fair die is rolled once. What is the probability of getting:

(a) a number divisible by 3

(b) a multiple of 2

(c) at least 5

(d) at most 2

(e) a prime number or an even number

(f) either a number greater that 2 or a multiple of 4

Solution

(a) Pr. (getting a number divisible by 3) = $\dfrac{\text{Number of faces having numbers divisible by 3}}{\text{total number of faces}}$

$\quad = \dfrac{2}{6}$ (Faces with numbers divisible by 3 are 3 and 6, i.e. 2 faces)

$\quad = \dfrac{1}{3}$

(b) Pr. (getting a multiple of 2) = $\dfrac{\text{Number of faces having numbers that are multiple of 2}}{\text{total number of faces}}$

$\quad = \dfrac{3}{6}$ (Faces with numbers that are multiple of 2 are 2, 4 and 6, i.e. 3 faces)

$\quad = \dfrac{1}{2}$

(c) Pr. (getting at least 5) = $\dfrac{\text{Number of faces having numbers that are at least 5}}{\text{Total number of faces}}$

$\quad = \dfrac{2}{6}$ (Faces with numbers that are at least 5 are 5 and 6, i.e. 2 faces)

$\quad = \dfrac{1}{3}$

(d) Pr. (getting at most 2) = $\dfrac{\text{Number of faces having numbers that are at least 2}}{\text{Total number of faces}}$

$\quad = \dfrac{2}{6}$ (Faces with numbers that are at most 2 are 1 and 2, i.e. 2 faces)

$\quad = \dfrac{1}{3}$

(e) Pr. (getting a prime number or an even number) =

$$\frac{\text{Number of faces having prime numbers and even numbers}}{\text{Total number of faces}}$$

$= \frac{5}{6}$ (Faces with prime numbers are 2, 3, and 5. Faces with even numbers are 2, 4, 6. This will give a total of 5 faces because 2 which is both a prime and even number should be counted once)

Therefore, Pr. (getting a prime number or an even number) $= \frac{5}{6}$

(f) Pr. (getting either a number greater that 2 or a multiple of 4) =

$$\frac{\text{Number of faces having numbers greater than 2 and numbers that are multiple of 4}}{\text{Total number of faces}} = \frac{4}{6}$$ (Faces with numbers greater than 2 are 3, 4, 5 and 6. Faces with multiple of 4 is 4. This will give a total of 4 faces because 4 which appears in both events should be counted once)

Therefore, Pr. (getting either a number greater that 2 or a multiple of 4) $= \frac{4}{6}$

$= \frac{2}{3}$

3. A die is thrown and a coin is tossed. What is the probability of getting:
(a) a 3 and a head
(b) a tail and a prime number

Solution

(a) From the die, Pr. (getting a 3) $= \frac{1}{6}$

From the coin, Pr. (getting a head) $= \frac{1}{2}$

Since AND means multiplication in probability:

Therefore, Pr. (getting a 3 and a head) = Pr. (getting a 3) x Pr. (getting a head)

$= \frac{1}{6} \times \frac{1}{2}$

$= \frac{1}{12}$

(b) (a) From the coin, Pr. (getting a tail) $= \frac{1}{2}$

From the die, Pr. (getting a prime number) $= \frac{3}{6}$ (The prime numbers are 3, i.e. 2, 3 and 5)

$= \frac{1}{2}$

Therefore, Pr. (getting a tail and a prime number) = Pr. (getting a tail) x Pr. (getting a prime number)

$$= \frac{1}{2} \times \frac{1}{2}$$

$$= \frac{1}{4}$$

4. Two fair dice are thrown at the same time. Find the probability of getting:

(a) at least one six

(b) a sum of at least 10

(c) a sum of at most 5

(d) a sum less than 3

(e) a total of seven

(f) a sum that is either a prime number or a multiple of 3

(g) a sum that is either divisible by 3 or a multiple of 2

Solution

The outcome table is as shown below. The numbers in the bracket give the outcome from the first and second die respectively. Adding the numbers in the bracket will give the respective sum that will be obtained.

Number on second die

+	1	2	3	4	5	6
1	(1,1)	(1,2)	(1,3)	(1,4)	(1,5)	(1,6)
2	(2,1)	(2,2)	(2,3)	(2,4)	(2,5)	(2,6)
3	(3,1)	(3,2)	(3,3)	(3,4)	(3,5)	(3,6)
4	(4,1)	(4,2)	(4,3)	(4,4)	(4,5)	(4,6)
5	(5,1)	(5,2)	(5,3)	(5,4)	(5,5)	(5,6)
6	(6,1)	(6,2)	(6,3)	(6,4)	(6,5)	(6,6)

Number on first die (rows labelled 1 to 6)

The outcome table above can be presented in a more direct form by adding the values in the brackets above to obtain the sum. This is as shown below. In the table below, the numbers in the brackets represent the numbers on each die. The numbers that are not in bracket are the outcomes from the sum of numbers on first and second dice.

Number on second die

+	(1)	(2)	(3)	(4)	(5)	(6)
(1)	2	3	4	5	6	7
(2)	3	4	5	6	7	8
(3)	4	5	6	7	8	9
(4)	5	6	7	8	9	10
(5)	6	7	8	9	10	11
(6)	7	8	9	10	11	12

Number on first die (rows)

Note that any of the tables above can be used to answer the questions asked above.

(a) The outcomes that can be obtained from getting at least a six are (6,1), (6,2), (6,3), (6,4), (6,5), (6,6), (1,6), (2,6), (3,6), (4,6), (5,6). They are from the first table. They are the outcomes from the 6 on the first die, and 6 on the second die respectively. The number of brackets from this outcome is 11 (when the brackets are counted). Note that the total outcomes from any of the two outcome tables above is 36. This is easily obtained from the second table by counting the numbers that are not in bracket.

Therefore, Pr. (getting at least a six) = $\dfrac{\text{Number of outcomes obtained when at least a six shows}}{\text{Total number of outcomes on the table}}$

$= \dfrac{11}{36}$

(b) A sum of at least 10 as shown on the second table above are, 10, 10, 10, 11, 11, and 12. This gives a total of 6 outcomes.

Therefore, Pr. (getting a sum of at least 10) = $\dfrac{6}{36}$ (Note that 36 is the total outcome)

$= \dfrac{1}{6}$

(c) A sum of at most 5 as shown on the second table above are, 5, 5, 5, 5, 4, 4, 4, 3, 3, and 2. This gives a total of 10 outcomes.

Therefore, Pr. (getting a sum of at most 5) = $\dfrac{10}{36}$

$= \dfrac{5}{18}$

(d) A sum less than 3 as shown on the second table above 2 only. This gives a total of 1 outcome.

Therefore, Pr. (getting a sum less than 3) = $\dfrac{1}{36}$

(e) A total of 7 as shown on the second table above appears 6 times. This gives a total of 6 outcomes.

Therefore, Pr. (getting a total of 7) = $\dfrac{6}{36}$

$\qquad\qquad = \dfrac{1}{6}$

(f) Sums which are prime numbers are 2, 3, 5, 7 and 11. Sums which are multiple of 3 are 3, 6, 9 and 12. Hence we are to count the outcomes from 2, 3, 5, 6, 7, 9, 11, and 12 (3 should be counted once). Hence, from the table, 2 appears 1 time, 3 appears 2 times, 5 appears 4 times, 6 appears 5 times, 7 appears 6 times, 9 appears 4 times, 11 appears 2 times, 12 appears 1 time. This gives a total outcome of 1 time + 2 times + 4 times + 5 times + 6 times + 4 times + 2 times + 1 time = 25. This is easier done on the table by counting all 2, 3, 5, 6, 7, 9, 11 and 12. It will also give a total of 25 outcomes.

Therefore, Pr. (getting a sum that is either a prime number or a multiple of 3) = $\dfrac{25}{36}$

(g) Sums which are divisible by 3 are 3, 6, 9 and 12. Sums which are multiples of 2 are 2, 4, 6, 8, 10 and 12. Hence we are to count the outcomes from 2, 3, 4, 6, 8, 9, 10 and 12 (6 and 12 which appear in both events should be counted once each). Hence, we go to the second table above and count all 2, 3, 4, 6, 8, 9, 10 and 12. It will give a total of 24 outcomes.

Therefore, Pr. (getting a sum that is either divisible by 3 or a multiple of 2) = $\dfrac{24}{36}$

$\qquad\qquad = \dfrac{2}{3}$

5. An unbiased die with faces numbered 1 to 6 is rolled twice. Find the probability that the product of the numbers obtained is:
(a) odd
(b) even
(c) 12
(d) prime
(e) either odd or a multiple of 5

Solution

The outcome table is as shown below. The numbers in brackets are the numbers on the die.

Number on second die

	x	(1)	(2)	(3)	(4)	(5)	(6)
	(1)	1	2	3	4	5	6
Number on	(2)	2	4	6	8	10	12
first die	(3)	3	6	9	12	15	18
	(4)	4	8	12	16	20	24
	(5)	5	10	15	20	25	30
	(6)	6	12	18	24	30	36

(a) All the odd numbers from the outcome table above are 1, 3, 3, 5, 5, 9, 15, 15, 25. This gives a total of 9 outcomes.

Therefore, Pr. (product of numbers is odd) = $\dfrac{9}{36}$ (Note that the total outcomes is 36)

$$= \dfrac{1}{4}$$

(b) Pr. (product of numbers is even) = 1 - Pr. (product of numbers is odd)

$$= 1 - \dfrac{1}{4}$$

$$= \dfrac{3}{4}$$

This can also be obtained by counting all the outcomes that are even numbers in the table above. Total even numbers is 27.

Hence, Pr. (product of numbers is even) = $\dfrac{27}{36}$

$$= \dfrac{3}{4} \quad \text{(As obtained before)}$$

(c) Pr. (product of numbers is 12) = $\dfrac{4}{36}$ (12 appears 4 times in the table)

$$= \dfrac{1}{9}$$

(d) All the prime numbers are, 2, 2, 3, 3, 5, 5. This gives a total outcomes of 6.

Therefore, Pr. (product of numbers is prime) = $\dfrac{6}{36}$

$\qquad = \dfrac{1}{6}$

(e) All products that are odd numbers are, 1, 3, 3, 5, 5, 9, 15, 15, 25. All products that are multiples of 5 are, 5, 5, 10, 10, 15, 15, 20, 20, 25, 30, 30.

They will both give a total outcome of 15. Note that, 5, 5, 15, 15, 25 are counted once under odd number. They should not be counted under multiples of 5, as this will result to double counting. Hence with this total outcome of 15,

Pr. (product of numbers is either odd or a multiple of 5) = $\dfrac{15}{36}$

$\qquad = \dfrac{5}{12}$

6. Three dice are thrown together. What is the probability of getting a total score of 10?

Solution.

If a die is thrown once, the total outcome is given by $6^1 = 6$. If two dice are thrown, the total outcome is $6^2 = 36$. Similarly, if three dice are thrown, the total outcome will be $6^3 = 216$. Now, for us to draw a table with 216 outcomes will be very tedious. So, a direct way of solving this problem will be to select the outcomes from each die that will result in a total score of 10. These outcomes are:

(6, 3, 1), (6, 2, 2), (5, 4, 1), (5, 3, 2), (4, 4, 2), (4, 3, 3)

Each of the brackets above can give us 6 outcomes. For example, the first bracket above can give us the following 6 outcomes:

(6, 3, 1): which means - First die shows 6, second die shows 3, third die shows 1

(6, 1, 3): which means - First die shows 6, second die shows 1, third die shows 3

(1, 6, 3): which means - First die shows 1, second die shows 6, third die shows 3

(1, 3, 6): which means - First die shows 1, second die shows 3, third die shows 6

(3, 1, 6): which means - First die shows 3, second die shows 1, third die shows 6

(3, 6, 1): which means - First die shows 3, second die shows 6, third die shows 1

Similarly, each of the other brackets can give us 6 outcomes.

Let us write out our outcome brackets again. They are, (6, 3, 1), (6, 2, 2), (5, 4, 1), (5, 3, 2), (4, 4, 2), (4, 3, 3)

When each of these brackets give us 6 outcomes, then we will obtain a total of 36 (i.e. 6 x 6) outcomes. Recall that our overall outcome table will give us a total of 216 (i.e. 6^3) outcomes.

Therefore, Pr. (getting a total score of 10) = $\dfrac{36}{216}$

$= \dfrac{1}{6}$

Exercise 54

1. A fair die is thrown once. Find the probability of getting:

(a) a 5

(b) a 1

(c) a 9

(d) a 2 or 3 or 6

(e) a number less than 6

(f) a prime or an even number

2. A fair die is rolled once. What is the probability of getting:

(a) a number divisible by 2

(b) a multiple of 3

(c) at least 2

(d) at most 3

(e) a perfect square or an odd number

(f) either a number greater that 5 or a multiple of 3

3. A die is thrown and a coin is tossed. What is the probability of getting:

(a) a 5 and a head

(b) a tail and a perfect cube

4. Two fair dice are thrown at the same time. Find the probability of getting:

(a) at least one four

(b) a sum of at least 6

(c) a sum of at most 10

(d) a sum less than 8

(e) a total of 12

(f) a sum that is either a perfect square or a multiple of 5

(g) a sum that is either divisible by 6 or a multiple of 4

5. An unbiased die with faces numbered 1 to 6 is rolled twice. Find the probability that the product of the numbers obtained is:

(a) prime

(b) divisible by 6

(c) 9

(d) a factor of 10

(e) either perfect cube or a multiple of 8

6. Three dice are thrown together. What is the probability of getting a total score of 11?

CHAPTER 55
MISCELLANEOUS PROBLEMS ON PROBABILITY

Examples

1. A box contains two green balls, three yellow balls and four white balls. A ball is picked at random from the box. What is the probability that it is:

(a) green

(b) yellow

(c) white

(d) blue

(e) not white

(f) either yellow or green

Solution

Total number of balls in the box = 2 + 3 + 4 = 9

(a) Pr. (that it is green) = $\dfrac{\text{Number of green balls}}{\text{Total number of balls in the box}}$

$= \dfrac{2}{9}$

(b) Pr. (that it is yellow) = $\dfrac{\text{Number of yellow balls}}{\text{Total number of balls in the box}}$

$= \dfrac{3}{9}$

$= \dfrac{1}{3}$

(c) Pr. (that it is white) = $\dfrac{\text{Number of white balls}}{\text{Total number of balls in the box}}$

$= \dfrac{4}{9}$

(d) There is no blue ball in the box.
Therefore, Pr. (that it is blue) = 0

(e) Pr. (that it is not white) = 1 - Pr. (that it is white)

$= 1 - \dfrac{4}{9}$

$= \dfrac{5}{9}$

(f) Pr. (that it is either yellow or green) = $\dfrac{\text{Number of yellow and green balls}}{\text{Total number of balls in the box}}$

$$= \frac{3+2}{9}$$

$$= \frac{5}{9}$$

Or,

Pr. (that it is either yellow or green) = Pr. (that it is yellow) + Pr. (that it is green) (Since OR means addition)

$$= \frac{1}{3} + \frac{2}{9}$$

$$= \frac{3+2}{9}$$

$$= \frac{5}{9} \quad \text{(As obtained before)}$$

2. A letter is chosen at random from the word COMPUTER. What is the probability that it is:
(a) either in the word MORE or in the word CUT
(b) either in the word COPE or in the word CUTE
(c) neither in the word ROT nor in the word CUP

Solution

(a) The total number of letters in COMPUTER is 8 letters.

In the word MORE, the number of letters is 4, while in the word CUT, the number of letters is 3. They both give a total of 7 letters.

Therefore, Pr. (that it is either in the word MORE or in the word CUT) = $\dfrac{7}{8}$

(b) In the word COPE, the number of letters is 4, while in the word CUTE, the number of letters is 4. Without counting any letter twice (i.e. C and E), the two words give a total of 6 letters (i.e. C, O, P, E, U, T).

Therefore, Pr. (that it is either in the word COPE or in the word CUTE) = $\dfrac{6}{8}$ (The total number of letters in COMPUTER is 8 letters).

$$= \frac{3}{4}$$

(c) Out of the 8 letters in COMPUTER, the letters that are neither in the word ROT nor in the word CUP are letters M and E. They are 2 letters.

Therefore, Pr. (that it is neither in the word ROT nor in the word CUP) = $\dfrac{2}{8}$

$$= \frac{1}{4}$$

(3) In a college 80% of the boys and 45% of the girls can drive a car. If a boy and a girl are chosen at random, what is the probability that:

(a) both of then can drive a car |

(b) the boy cannot drive a car and the girl can drive a car

(c) neither of them can drive a car?

(d) one of them can drive a car

Solution

The probabilities are given in percentage. Hence the total for each probability is 100%

Therefore, Pr. (a boy can drive a car) = $\dfrac{80}{100}$

$= \dfrac{4}{5}$

Pr. (a boy cannot drive a car) = $\dfrac{20}{100}$ (i.e. 100 - 80 = 20)

$= \dfrac{1}{5}$ (Can also be obtained from $1 - \dfrac{4}{5}$)

Similarly, Pr. (a girl can drive a car) = $\dfrac{45}{100}$

$= \dfrac{9}{20}$ (After equal division by 5)

Pr. (a girl cannot drive a car) = $1 - \dfrac{9}{20}$)

$= \dfrac{11}{20}$

(a) Therefore, Pr. (both of them can drive a car) = Pr. (a boy can drive a car) AND Pr. (a girl can drive a car)

$=$ Pr. (a boy can drive a car) x Pr. (a girl can drive a car)

$= \dfrac{4}{5} \times \dfrac{9}{20}$

$= \dfrac{36}{100}$

$= \dfrac{9}{25}$

(b) Pr. (the boy cannot drive a car and the girl can drive a car) = Pr. (a boy cannot drive a car) AND Pr. (a girl can drive a car)

$=$ Pr. (a boy cannot drive a car) x Pr. (a girl can drive a car)

$= \dfrac{1}{5} \times \dfrac{9}{20}$

$= \dfrac{9}{100}$

(c) Pr. (neither of them can drive a car) = Pr. (a boy cannot drive a car) AND Pr. (a girl cannot drive a car)

= Pr. (a boy cannot drive a car) x Pr. (a girl cannot drive a car)

$$= \frac{1}{5} \times \frac{11}{20}$$

$$= \frac{11}{100}$$

(d) Since we do not know which of then can drive a car, then this case is logically explained as follows:

Pr. (one of them can drive a car) = either the boy can drive a car AND the girl cannot drive a car OR the girl can drive a car AND the boy cannot drive a car.

This in now calculated as follows:

Pr. (one of them can drive a car) = Pr. (the boy can drive a car) x Pr. (the girl cannot drive a car) + Pr. (the girl can drive a car) x Pr. (the boy cannot drive a car)

$$= (\frac{4}{5} \times \frac{11}{20}) + (\frac{9}{20} \times \frac{1}{5})$$

$$= \frac{11}{25} + \frac{9}{100}$$

$$= \frac{44 + 9}{100}$$

$$= \frac{53}{100}$$

4. The probability of a seed germinating is $\frac{2}{5}$. If three of the seeds are planted, what is the probability that:

(a) none will germinate

(b) at least one will germinate

(c) at least one will not germinate

(d) only one will germinate

Solution

This is a case of selection of three items from two possible events. We are going to write our outcomes in bracket like a tree diagram method. In order to write out the outcomes, let us use the letter G to represent germinate and letter N to represent not germinate.

Hence the outcomes are written as follows:

(GGG), (GGN), (GNG), (GNN), (NGG), (NGN), (NNG), (NNN)

(a) The probability that none will germinate is given by (NNN).

From the question, the probability that a seed germinate, G = $\frac{2}{5}$. Therefore the probability that it will

not germinate, N = 1 - G = $1 - \frac{2}{5} = \frac{3}{5}$

Hence, $G = \dfrac{2}{5}$, $N = \dfrac{3}{5}$

Therefore, Pr. (that none will germinate) = (NNN)

$$= \dfrac{3}{5} \times \dfrac{3}{5} \times \dfrac{3}{5}$$

$$= \dfrac{27}{125}$$

(b) The outcomes of the probability that at least one will germinate are, (GGG), (GGN), (GNG), (GNN), (NGG), (NGN), (NNG). Hence we can compute each of the outcomes and add them together. But this will be tedious. An easier way of solving this problem is as explained below. The difference between the outcome in question (a) and (b) is (NNN). This shows that subtracting (NNN) from the total probability will give us the outcomes in question (b). Recall that the total of any probability is 1. Therefore, 1 - (NNN) = outcomes in (b)

Hence, Pr. (that at least one will germinate) = 1 - (NNN)

$$= 1 - \dfrac{27}{125} \quad \text{[Note that (NNN)} = \dfrac{27}{125} \text{ as calculated in question (a)]}$$

$$= \dfrac{108}{125}$$

(c) The outcomes of the probability that at least one will not germinate are, (GGN), (GNG), (GNN), (NGG), (NGN), (NNG), (NNN). Similar to (b) above, the difference between this outcomes of this question and the overall outcomes is (GGG).

Therefore, Pr. (that at least one will not germinate) = 1 - (GGG)

Let us calculate (GGG) as follows:

Pr. [that all three will germinate, i.e. (GGG)] $= \dfrac{2}{5} \times \dfrac{2}{5} \times \dfrac{2}{5}$

$$= \dfrac{8}{125}$$

Therefore, Pr. (that at least one will not germinate) = 1 - (GGG)

$$= 1 - \dfrac{8}{125}$$

$$= \dfrac{117}{125}$$

(d) The outcomes of the probability that only one will germinate are, (GNN), (NGN), (NNG). Hence we will calculate each of these outcomes and add them together.

(GNN) = Pr. (that the first will germinate) x Pr. (that the second will not germinate) x Pr. (that the third will not germinate)

$$= \dfrac{2}{5} \times \dfrac{3}{5} \times \dfrac{3}{5}$$

$$= \dfrac{18}{125}$$

(NGN) $= \dfrac{3}{5} \times \dfrac{2}{5} \times \dfrac{3}{5}$

$$= \frac{18}{125}$$

$$(NNG) = \frac{3}{5} \times \frac{3}{5} \times \frac{2}{5}$$

$$= \frac{18}{125}$$

Therefore, Pr. (that only one will germinate) $= \frac{18}{125} + \frac{18}{125} + \frac{18}{125}$

$$= \frac{54}{125}$$

5. When children are born, they are equally likely to be boys or girls. What is the probability that in a family of four children:

(a) three are boys and one is a girl

(b) at least two are girls

(c) two are boys and two are girls

(d) the first and second born are girls

Solution

Since children are equally likely to be boys or girls, it means that the probability of having a boy is $\frac{1}{2}$,

and the probability of having a girl is also $\frac{1}{2}$. This is similar to the case of tossing a coin (i.e. $\frac{1}{2}$ for head and $\frac{1}{2}$ for tail).

Therefore, the case of a family of four children is like when four coins are tossed. Refer to the example on tossing four coins in our previous chapter.

Let us use B for boy and G for girl to write out the total outcomes of 16 (i.e. 2^4 = 16) as shown below.

The outcomes are: (BBBB), (BBBG), (BBGG), (BGGG), (GBBB), (GGBB), (GGGB), (GBGB), (BGBG), (BBGB), (GBBG), (BGGB), (GGBG), (GBGG), (BGBB), (GGGG). This gives a total of 16 outcomes.

(a) The outcomes that the children are three boys and one girl are, (BBBG), (GBBB), (BBGB), (BGBB). This gives 4 outcomes.

Therefore, Pr. (three are boys and one is a girl) $= \frac{4}{16}$

$$= \frac{1}{4}$$

(b) The outcomes that the children are at least two girls are, (BBGG), (BGGG), (GGBB), (GGGB), (GBGB), (BGBG), (GBBG), (BGGB), (GGBG), (GBGG), (GGGG). This gives 11 outcomes.

Therefore, Pr. (at least two are girls) = $\dfrac{11}{16}$

(c) The outcomes that the children are two boys and two girls are, (BBGG), (GGBB), (GBGB), (BGBG), (GBBG), (BGGB). This gives 6 outcomes.

Therefore, Pr. (two are boys and two are girls) = $\dfrac{6}{16}$

$= \dfrac{3}{8}$

(d) The outcomes that the first and second born are girls are, (GGBB), (GGGB), (GGBG), (GGGG). This gives 4 outcomes.

Therefore, Pr. (the first and second born are girls) = $\dfrac{4}{16}$

$= \dfrac{1}{4}$

6. A bag contains three blue balls, four red balls and five white balls. Three balls are removed from the bag without replacement. What is the probability of getting:
(a) a white, blue and red balls in that order
(a) one of each colour
(c) at least two white balls

Solution
The total number of balls in the bag = 3 + 4 + 5 = 12

(a) A white, blue and red balls in that order means that the first is white, the second is blue and the third is red. This can be represented as (WBR).
Note that this is a case of without replacement. Hence after each ball is removed, the total number of ball remaining and the number of the particular ball removed are both reduced by one.

Therefore, Pr. (getting a white, blue and red balls, i.e. WBR) = $\dfrac{5}{12} \times \dfrac{3}{11} \times \dfrac{4}{10}$.   (Notice how the total balls is reduced by 1 after each ball is removed from the bag.

$= \dfrac{60}{1320}$

$= \dfrac{1}{22}$ (After equal division by 60)

(b) Let B represent blue, R represent red and W represent white. Then the outcomes for getting one of each colour are given by: (BRW), (BWR), (RBW), (RWB), (WBR), (WRB).

Let us now calculate each of them.

(BRW) = Pr. (First is blue) x Pr. (Second is red) x Pr. (Third is white)

$$= \frac{3}{12} \times \frac{4}{11} \times \frac{5}{10}$$

$$= \frac{1}{4} \times \frac{4}{11} \times \frac{1}{2}$$

$$= \frac{4}{88}$$

$$= \frac{1}{22}$$

Similarly, each of the other five outcomes, i.e. (BWR), (RBW), (RWB), (WBR), (WRB), will each give us a value of $\frac{1}{22}$ when calculated. This is because each is obtained by multiplying 3 x 4 x 5, to give the numerator, and 12 x 11 x 10, to give the denominator, which simplifies to $\frac{1}{22}$.

Therefore, Pr. (getting one of each colour) = $\frac{1}{22} + \frac{1}{22} + \frac{1}{22} + \frac{1}{22} + \frac{1}{22} + \frac{1}{22}$

$$= \frac{6}{22}$$

$$= \frac{3}{11}$$

(c) Let us write out a different outcome for this problem. Since we are concerned about one colour, we are going to use W to represent white colour, and N to represent not a white colour. This will give us 8 outcomes in brackets as usual. The outcomes are:

(WWW), (WWN), (WNW), (WNN), (NWW), (NWN), (NNW), (NNN).

The outcomes representing at least two white balls are: (WWW), (WWN), (WNW), (NWW). Number of white balls is 5. Therefore number of balls that are not white = 12 - 5 = 7, or blue + red = 3 + 4 = 7. (Blue and red ball are the balls that are not white balls).

Let us now calculate each of the outcomes above as follows:

(WWW) = Pr. (first is white) x Pr. (second is white) x Pr. (third is white)

$$= \frac{5}{12} \times \frac{4}{11} \times \frac{3}{10}$$ (Take note of the reduction in the white balls and total number of balls as each ball is removed from the bag)

$$= \frac{60}{1320}$$

$$= \frac{1}{22}$$

(WWN) = $\frac{5}{12} \times \frac{4}{11} \times \frac{7}{10}$ (Note that there are 7 balls that are not white)

$$= \frac{140}{1320}$$

$$= \frac{7}{66}$$

(WNW) = $\frac{5}{12} \times \frac{7}{11} \times \frac{4}{10}$

$$= \frac{140}{1320}$$

$$= \frac{7}{66}$$

$$(NWW) = \frac{7}{12} \times \frac{5}{11} \times \frac{4}{10}$$

$$= \frac{140}{1320}$$

$$= \frac{7}{66}$$

Therefore, Pr. (getting at least two white balls) = (WWW) or (WWN) or (WNW) or (NWW)

$$= (WWW) + (WWN) + (WNW) + (NWW)$$

$$= \frac{1}{22} + \frac{7}{66} + \frac{7}{66} + \frac{7}{66}$$

$$= \frac{3 + 7 + 7 + 7}{66}$$

$$= \frac{24}{66}$$

$$= \frac{4}{11}$$

7. A committee consist of 6 men and 4 women. A subcommittee made up of three members is randomly chosen from the committee members. What is the probability that:

(a) they are all men

(b) two of them are women?

Solution

Let us write out the outcome for this problem. Let M represent man, and W represent woman. This will give us 8 outcomes in brackets as usual. The outcomes are:

(WWW), (WWM), (WMW), (WMM), (MWW), (MWM), (MMW), (MMM).

(a) The total members in the committee are: 6 + 4 = 10.

The outcomes representing all men is (MMM)

Therefore, Pr. (they are all men, i.e. MMM) = Pr. (first is a man) x Pr. (second is a man) x Pr. (third is a man)

$$= \frac{6}{10} \times \frac{5}{9} \times \frac{4}{8}$$ (Notice the reduction in the number of men and people left, as each

member is chosen from the committee).

$$= \frac{130}{720}$$

$$= \frac{13}{72}$$

(b) The outcomes showing that two of them are women are: (WWM), (WMW), (MWW)

Let us calculate each of them as follows:

(WWM) = Pr. (the first is a woman) x Pr. (the second is a woman) x Pr. (the third is a man)

$$= \frac{4}{10} \times \frac{3}{9} \times \frac{6}{8}$$

$$= \frac{72}{720}$$

$$= \frac{1}{10}$$

$$(WMW) = \frac{4}{10} \times \frac{6}{9} \times \frac{3}{8}$$

$$= \frac{72}{720}$$

$$= \frac{1}{10}$$

$$(MWW) = \frac{6}{10} \times \frac{4}{9} \times \frac{3}{8}$$

$$= \frac{72}{720}$$

$$= \frac{1}{10}$$

Therefore, Pr. (two of them are women) = (WWM) or (WMW) or (MWW)

$$= (WWM) + (WMW) + (MWW)$$

$$= \frac{1}{10} + \frac{1}{10} + \frac{1}{10}$$

$$= \frac{3}{10}$$

8. A box contains seven blue pens and three red pens. Three pens are picked one after the other without replacement. Find the probability of picking:

(a) two blue pens

(b) at least two red pens

(c) at most two blue pens

Solution

Let B represent blue pen, and R represent red pen. The outcomes are:

(BBB), (BBR), (BRB), (BRR), (RBB), (RBR), (RRB), (RRR).

The total number of pens = 7 + 3 = 10

(a) The outcomes showing two blue pens are: (BBR), (BRB), (RBB)

Let us calculate each of them as follows:

(BBR) = Pr. (the first is a blue pen) x Pr. (the second is a blue pen) x Pr. (the third is a red pen)

$$= \frac{7}{10} \times \frac{6}{9} \times \frac{3}{8}$$

$$= \frac{126}{720}$$

$$= \frac{7}{40} \quad \text{(In its lowest term after equal division by 18)}$$

$$\text{(BRB)} = \frac{7}{10} \times \frac{3}{9} \times \frac{6}{8}$$

$$= \frac{126}{720}$$

$$= \frac{7}{40}$$

Also, $\text{(RBB)} = \frac{7}{40}$ (Similar to the once above)

Therefore, Pr. (picking two blue pens) $= \frac{7}{40} \times \frac{7}{40} \times \frac{7}{40}$

$$= \frac{21}{40}$$

(b) The outcomes representing at least two red pens are: (RRR), (RRB), (RBR), (BRR)

Let us now calculate each of the outcomes as follows:

(RRR) = Pr. (first is a red pen) x Pr. (second is a red pen) x Pr. (third is a red pen)

$$= \frac{3}{10} \times \frac{2}{9} \times \frac{1}{8} \quad \text{(Take note of the reduction in the red pens and total number of pens as}$$

each pen is picked from the box)

$$= \frac{6}{720}$$

$$= \frac{1}{120}$$

$$\text{(RRB)} = \frac{3}{10} \times \frac{2}{9} \times \frac{7}{8}$$

$$= \frac{42}{720}$$

$$= \frac{7}{120}$$

Hence, $\text{(RBR)} = \frac{7}{120}$ (This is similar to the one above)

And, $\text{(BRR)} = \frac{7}{120}$ (Same reason as above)

Therefore, Pr. (picking at least two red pens) $= \frac{1}{120} + \frac{7}{120} + \frac{7}{120} + \frac{7}{120}$

$$= \frac{1 + 7 + 7 + 7}{120}$$

$$= \frac{22}{120}$$

$$= \frac{11}{60}$$

(c) The outcomes that represent picking at most two blue pens are: (BBR), (BRB), (BRR), (RBB), (RBR), (RRB), (RRR). Note that at most two blue pens means 2, 1 or 0 blue pens.

Notice that there is only (BBB) missing from this outcome. This shows that it can be obtained by: total probability - (BBB). Which is: 1 - (BBB).

Let us calculate (BBB) as follows:

(BBB) = Pr. (first is a blue pen) x Pr. (second is a blue pen) x Pr. (third is a blue pen)

$$= \frac{7}{10} \times \frac{6}{9} \times \frac{5}{8}$$

$$= \frac{210}{720}$$

$$= \frac{7}{24} \quad \text{(After equal division by 30)}$$

Therefore, Pr. (picking at most two blue pens) = 1 - (BBB)

$$= 1 - \frac{7}{24}$$

$$= \frac{17}{24}$$

Exercise 55

1. A box contains 5 green balls, 8 yellow balls and 7 white balls. A ball is picked at random from the box. What is the probability that it is:

(a) green

(b) yellow

(c) white

(d) blue

(e) not white

(f) either yellow or green

2. A letter is chosen at random from the word NORMADIC. What is the probability that it is:

(a) either in the word MAD or in the word CORN

(b) either in the word NORM or in the word DAM

(c) neither in the word RID nor in the word CAN

(3) In a college 20% of the boys and 8% of the girls who had graduated from the college, graduated with distinction since the inception of the college. If a boy and a girl are chosen at random, what is the probability that:

(a) both of them will graduate with distinction

(b) the boy will not and the girl will graduate with distinction

(c) neither of them will graduate with distinction?

(d) one of them will graduate with distinction

4. The probability of a seed germinating is $\frac{1}{4}$. If three of the seeds are planted, what is the probability that:
(a) none will germinate
(b) at least one will germinate
(c) at least one will not germinate
(d) only one will germinate

5. When parents who are carriers of sickle cell disorder get married, they are equally likely to give birth to normal child and sick child. What is the probability that in a family of three children:
(a) two are normal and one is sick
(b) at least two are sick
(c) one is normal and two are sick
(d) the first is sick
(e) at most one is normal

6. A box contains six blue balls, three red balls and five white balls. Three balls are removed from the bag without replacement. What is the probability of getting:
(a) a white, blue and red balls in that order
(a) one of each colour
(c) at least two white balls

7. A committee consist of 4 men and 2 women. A subcommittee made up of two members is randomly chosen from the committee members. What is the probability that:
(a) they are all men
(b) one of them is a woman?

8. A bag contains 5 blue balls and seven red balls. Three balls are picked one after the other without replacement. Find the probability of picking:
(a) two blue balls
(b) at least two red balls
(c) at most two blue balls

ANSWERS TO EXERCISES

Exercise 1

1. (a) -15 (b) -11 (c) -5 (d) -12 (e) 19 (f) 6 (g) -82

2. (a) -10 (b) -22 (c) 36 (d) 30 (e) -4 (f) -6 (g) 8 (h) 4

3. (a) 12 (b) 24 (c) $\frac{71}{3}$ or $23\frac{2}{3}$ (d) $\frac{19}{8}$ or $2\frac{3}{8}$ (e) -1910

Exercise 2

1. 1000y metres   2. (38 + m)years, (38 + m + y)years   3. (b - c)years

4. (a) 100y cents (b) $\frac{n}{60}$ minutes, $\frac{n}{60 \times 60}$ 5. $(22x - x^2)cm^2$ 6. $(x - 1)(x + 10)cm^2$

7. $10x - 5y$ 8. $8x - 3y - 5$ 9. $2a - 3b$ 10. $11x^2 + 15x$ 11. $2a^2 - 7ab - 15b^2$

12. $15x^2 - 13xy + 2y^2$ 13. $16a^2 - 40a + 25$ 14. 24   15. 3   16. (a) lmn (b) 6cde

(c) $30a^3x^2y^2$ (d) 20mn (e) $(x + y)(3x - y)$ 17. (a) $\frac{3x}{5}$ (b) $\frac{4x + 3}{9x^2}$ (c) $\frac{3ax - 14b}{30bx}$

(d) $\frac{10bx + 25b + 8ab - 24a}{40ab}$ (e) $\frac{25a^3 - 15a^2 - 15a^2b - 3a + 15b}{15a^2}$ 18. (a) $\frac{-15x + 8}{30}$ (b) $\frac{3}{2(3a + 7)}$

(c) $\frac{-7a(a + 2) + 6}{(3a - 1)(a - 2)}$ 19. (a) 5 (b) 2 (c) $6a^2b^2$ (d) ab^2c^2 (e) $12x^2yz^2$ 20. $2(2m - 4n)$

21. $4(3y - 2)$ 22. $-10y(2 + 5z)$ 23. $m^2(5m^2 - 10 - m)$ 24. $(2 + 5m)(x + 3y)$ 25. $(a - 1)(3b + 7)$

26. $(5b + n)(x - y)$ 27. $(5 + y)(3 - x)$ 28. $(10 + m)(10 - m)$ 29. $(5x + 3y(5x - 3y)$

30. $7(1 + b)(1 - b)$ (31) $(6p + 2)(6p - 2)$ 32. 2600   33. 8400   34. $(x + 3)(x + 3)$

35. $(b - 5)(b + 4)$ 36. $(n - 8)(n - 6)$ 37. $(5a + 4)(a - 3)$ 38. (a) -6 (b) -5   39. (a) 23 (b) -78

(c) $\frac{6}{7}$ 40. -12   41. (a) 0 (b) -24

Exercise 3

1. (a) $-3t^4e^{12}$ (b) $64a^3b^9$ (c) $-(a^4)$ (d) g^{20} (e) $-m$ 2. (a) $\frac{1}{3a}$ (b) $\frac{1}{a}$ (c) $7x^{\frac{3}{2}}$ (d) $9x$

3. (a) $x = \frac{1}{25}$ (b) $a = \frac{1}{3}$ (c) $x = 3\frac{1}{2}$ (d) $x = 4$

Exercise 4

1. (a) $x = \frac{4}{3}$ (b) $a = \frac{5}{16}$ (c) $b = -\frac{4}{7}$ 2. (a) $x = \frac{5}{11}$ (b) $x = 1\frac{11}{20}$ (c) $\frac{113}{62}$ 3. $p = \frac{md^2}{3(md - t)}$

4. $C = \sqrt{\frac{P^2}{I^2V^2} - \frac{E^2}{I^2}}$ or $C = \frac{1}{I}\sqrt{\frac{P^2}{V^2} - E^2}$ 5. $x = \frac{R^2T - S}{b - aR^2}$ 6. $m = 0$ 7. $Y = \sqrt[3]{\frac{TC^2}{L} - B}$

or $Y = \sqrt[3]{\frac{TC^2 + BL}{L}}$ 8. $x = \sqrt{\frac{M}{2Q^2}}$ or $x = \frac{1}{Q}\sqrt{\frac{M}{2}}$

Exercise 5

1. 8   2. 4   3. Man = 100years, daughter = 20years   4. 10years   5. 150km/hr   6. 150km/hr

7. Eighteen \$10 notes and six \$1 notes   8. 15days   9. 1050litres   10. 2250km   11. 53.3km

12. 12   13. 13, 15 and 17   14. 5 and 24   15. 12   16. 15 and 25   17. 32years   18. $-\dfrac{5}{4}$

Exercise 6

1. $x = -1, y = 1$ 2. $x = 6$ 3. $m = 2, n = 2$ 4. $a = 10, b = 5$ 5. $P = 4, q = -3$ 6. $x = -2, y = 2$

7. $x = 3, y = 6$ 8. $x = \dfrac{31}{23}, y = \dfrac{3}{23}$ 9. $m = 1, n = -1$ 10. $p = -\dfrac{11}{23}, q = -\dfrac{3}{23}$ 11. $c = 3\dfrac{1}{2}, d = -1\dfrac{1}{2}$

12. $x = -\dfrac{60}{53}, y = \dfrac{250}{53}$

Exercise 7

1. Book = \$10, pen = \$5   2. Mathematics = 30, Biology = 70   3. 18.75 and 56.25   4. m = 5hours,

n = 3hours   5. $x = -2, y = 3$. Sides are 14cm, 18cm and 18cm   6. (a) $x = 6, y = 7.5$ (b) 30cm each

7. 47   8. 92   9. 35   10. $\dfrac{3}{8}$

Exercise 8

1. (a) a(7a + 9)   (b) (m + 4)(m - 4)   (c) (3p + 11)(3p - 11)   (d) 8(2b + 5)(2b - 5)   2. (a) $(3x + 5)(x - 3)$

(b) $(x + 9)(x - 8)$ (c) $(5x + 2)(x + 4)$ (d) $(10x + 6)(x - 1)$ 3. (a) Yes   (b) Yes   (c) No   (d) No

(e) No   4. (a) a = 0, or -4   (b) b = -6 or 6   (c) y = $\dfrac{7}{2}$ or $-\dfrac{7}{2}$ (d) $x = 4$ or 1   5. (a) $x = \dfrac{4}{3}$ or -1

(b) y = 2 or -3   (c) y = 2 or 18   (d) m = $\dfrac{1}{2}$ or $-\dfrac{4}{5}$ (e) $x = \dfrac{6}{7}$ or -1   6. (a) $x^2 - 3x - 40 = 0$

(b) $4x^2 + 27x - 7 = 0$ (c) $5x^2 + 32x + 12 = 0$ 7. (a) $\dfrac{15}{2}$ (b) -4   8. (a) $-\dfrac{4}{3}$ (b) $-\dfrac{11}{3}$

9. M = 5. The other root is $-\dfrac{9}{2}$ 10. K = -8. The other root is $-\dfrac{11}{3}$ 11. K = -6, P = -27

12. A = 9, B = 30   13. K = 12 or -12. The two possible equations are $m^2 - 12m + 27 = 0$ and $m^2 + 12m + 27 = 0$ 14. 64   15. $\dfrac{169}{4}$ 16. $a^2 - 6a + 9$ 17. $36m^2 + 60m + 26$ 18. $x - 4$ 19. $x + 0.6$

20. (a) y = -11 or 2   (b) $x = 3$ or 8   (c) $x = 4$ or $-\dfrac{3}{2}$ (d) $x = 1$ or $\dfrac{7}{3}$ (e) y = -1 or $-\dfrac{6}{5}$ (f) m = 2 or $-\dfrac{7}{2}$

(g) $x = 5.56$ or 1.44   21. (a) $x = 3$ or -4   (b) $x = 1$ or $\dfrac{10}{7}$ (c) $x = -\dfrac{1}{2}$ or 4   (d) $x = 2.76$ or -1.09

(e) $x = 0.52$ or -1.92   (f) n = $\dfrac{3}{2}$ twice

Exercise 9

1. -4 and -18 or 4 and 18   2. -3 and 7 or -7 and 3   3. 11years and 18years   4. 2.89cm   5. 2m

6. 11years   7. 4 years   8. -16 and -18 or 16 and 18   9. 10   10. 12years and 36years

Exercise 10

1. (a) $x = \frac{5y}{4}$ (b) $x = 12.5$ (c) $y = 12.8$ 2. (a) $h = \frac{5}{3}\sqrt{P}$ (b) $p = 144$ (c) $h = \frac{25}{6}$ 3. (a) $p = \frac{36}{q}$

(b) $q = \frac{9}{5}$ (c) $p = 7\frac{1}{5}$ 4. (a) $m = \frac{15}{\sqrt[3]{n}}$ (b) $m = 7\frac{1}{2}$ (c) $n = 24{,}146$ 5. (a) $a = \frac{8b^2}{3c}$ (b) $a = \frac{20}{3}$

(c) $b = 1.225$ 6. (a) $w = \sqrt{\frac{9h}{5L}}$ (b) $w = 2.32$ (c) $L = 0.576$ 7. (a) $x = 2 + 3y$ (b) $x = 26$

8. (a) $E = 5 + 10F$ (b) $E = 30$ (c) $F = 3\frac{1}{2}$

Exercise 11

1. (a) $(-12, 73)$ or $(2, 3)$ (b) $(-\frac{23}{9}, -\frac{10}{9})$ or $(-1, 2)$ (c) $(-\frac{13}{2}, -\frac{19}{2})$ or $(-2, 1)$ (d) $(-\frac{3}{5}, -\frac{5}{6})$ or $(\frac{1}{2}, 1)$

(e) $(\frac{23}{5}, -\frac{3}{5})$ or $(3, -1)$ (f) $(-5, 3)$ (g) $(2, 1)$ or $(\frac{14}{17}, -\frac{33}{17})$ 2. Boy = 30years old, Father = 60 years old

3. $(\frac{47}{4}, -\frac{13}{2})$ or $(5, 7)$ 4. 6years and 11years

Exercise 12

1. $x > 2$

2. $x < -1$

3. $x < 2\frac{32}{43}$

4. $x \leq -4$

5. -1, -2 and -3   6. 1, 2 and 3   7. 65, 66 and 67   8. 3, 4 and 5

9. $-1 \leq y \leq 2$

10. $-2 < y < 3$

11. $y > -1$ or $y \leq -7$

12. $4 < m < 5$

Exercise 13

1. $-5 < x < 6$ 2. $x \geq \frac{2}{3}$ or $x \leq -4$ 3. $\frac{4}{5} \leq x \leq 1$ 4. $-\frac{3}{5} < x < \frac{3}{2}$ 5. $-1.95 \leq x \leq 2.95$ 6. $-2 < x < \frac{9}{2}$

7. $x \geq \frac{6}{7}$ or $x \leq -1$ 8. $-2 \leq x \leq -\frac{7}{5}$ 9. $-3 < x < 5$ 10. $-1.5 \leq x \leq 3.7$

Exercise 14

1. 10 units   2. $x = 12$ 3. (a) $\binom{-13}{27}$ (b) $\binom{-1}{1}$ (c) $\binom{-2\frac{1}{2}}{8}$ (d) 59.1   4. (a) $p = -\frac{41}{23}$, $q = -\frac{7}{23}$

(b) 37.6   5. (a) $\binom{-11}{38}$ (b) $(\sqrt{50}, 81.9°)$ 6. $\theta = 33.7°$, $(7.2, 33.7°)$ 7. (a) $\binom{-2}{4}$, $\sqrt{20}$ (b) $\binom{-6}{-2}$, $\sqrt{40}$

(c) $\binom{6}{-3}$, 6.7   (d) $\binom{-4}{-6}$, 7.2   8. (a) $Q = (-1, 6)$, $R = (8, 9)$ (b) $\binom{1\frac{1}{2}}{2\frac{1}{2}}$ 9. 29.4 units   10. 14.3

11. (a) $\binom{12}{-8}$ (b) $\binom{-7}{9}$ (c) $\binom{-5}{8\frac{3}{4}}$ (d) 42.2   12. (a) $p = \frac{22}{7}$, $q = \frac{26}{21}$ (b) 19.6

13. (a) $PR = \binom{-4}{5}$, $|PR| = \sqrt{41}$ (b) $RQ = \binom{4}{2}$, $|RQ| = \sqrt{20}$ (c) $-5QS = \binom{15}{-10}$, $|-5QS| = \sqrt{325}$

(d) $2PQ = \binom{16}{-6}$, $|2PQ| = \sqrt{292}$ or 17.09

Exercise 15

1. $\dfrac{m+d}{m+f}$ 2. $\dfrac{x-7}{x-4}$ 3. $\dfrac{2-x}{3x-5}$ 4. $-\dfrac{(4m+n)}{(n+m)}$ 5. $\dfrac{p-r}{p+r}$ 6. $\dfrac{6a+7}{3b}$ 7. $\dfrac{13}{5(2x-y)}$ 8 $\dfrac{1+9m}{6mn}$

9. $\dfrac{5a^2+13a+4}{(a-2)(a+3)}$ 10. $\dfrac{3ab^2+4b^2-3a^3}{12a^2b^3}$ 11. $\dfrac{4+5n}{(3m-2n)(3m+2n)}$

Exercise 16

1. $m = -4$ 2. $b = -\frac{3}{2}$ or 6   3. $x = 7$ or 4.875   4. $x = -8$ 5. $x = \frac{7}{3}$ 6 (a) $a = -\frac{12}{5}$

(b) $a = 6$ or -5 7. (a) $m = -2$ (b) $m = -\frac{9}{5}$ 8 (a) $x = \frac{2}{5}$ or -3 (b) $x = -\frac{5}{2}$ or 3   9. $m = \frac{22}{5}$

10. $x = -\frac{41}{113}$ 11. 4   12. $\frac{33}{26}$

Exercise 17

1. $x = 1$, $y = 2$ 2. $a = \frac{1}{4}$, $b = -1$ 3. $c = 2$, $d = 3$ 4. $a = \frac{4}{3}$, $b = \frac{2}{3}$ 5. $p = -\frac{1}{2}$, $q = \frac{1}{2}$

6. $c = 2$, $d = -5$ 7. $x = \frac{3}{10}$, $y = \frac{2}{5}$ 8. $m = \frac{9}{17}$, $n = 9$

Exercise 18

1. $x = 9$ or -9 2. (a) $x = 2$ or $-\frac{24}{5}$ (b) $x = \frac{7}{3}$ or -3 (c) $x = -\frac{3}{4}$ or $\frac{33}{4}$ 3 (a) $x = \frac{5}{4}$ or -3

(b) No solution   4 (a) $x = \dfrac{7}{2}$ or -2 (b) $x = -5$ or -13 5 (a) $x = 2$ (b) $x = 35$ or -3

(c) $x = -\dfrac{1}{2}$, or $\dfrac{1}{2}$ 6 (a) $x = 2$ or -9 (b) $x = 0, \dfrac{9}{2}, x = \dfrac{1}{2}$ or 4 7 (a) $x = 12$ (b) $x = 1$

8 (a) No solution   (b) $x = 5$ or -2 9 (a) $x = -\dfrac{2}{3}$ or 3 (b) $x = 3.63, -17.63, 4$ or -18

10 (a) $x = \dfrac{3}{5}$ or 2 (b) $x = 1.30, -5.80, 0.84$ or -5.34

Exercise 19

1. $x \le -2$ or $x \ge \dfrac{4}{5}$ 2. $-5 < x < \dfrac{19}{3}$ 3. $-3 \le x \le 19$ 4. $m < -1$ or $m > -\dfrac{3}{7}$ 5. $x < -3$ or $x > 5$

6. $-3 < m < 2$ 7. $2 < m \le 5$ 8. $-2 < x \le -1$ 9. $x < -3$ or $x > 3$ 10. $y < 6$ or $y > -6$

11. $a \le -2.1$ or $a \ge 5.4$ 12. $y \le -3$ or $y \ge 3$ 13. $m \le -1$ or $m \ge 1$ 14. $x > -\dfrac{1}{2}$ or $x \le 2$

15. $-4 \le y \le 4$

Exercise 20

1. $x = \dfrac{16}{5}$ 2. $x = \dfrac{13}{11}$ 3. $x = \dfrac{14}{13}$ 4. $x = 0.42$ 5. $x = -41.39$ 6. $x = 0.474$ 7. $x = 2.50$ or -0.34

8. $x = 4.14$ or 0.36 9. $x = 3$ 10. $x = 2$ 11. $x = 0$ 12. $x = 1, y = -1$ 13. $x = 5, y = \dfrac{20}{3}$

14. $y = -\dfrac{4}{9}, x = \dfrac{19}{9}$ 15. $x = -\dfrac{4}{3}$ or 2

Execise 21

1 (a) $\dfrac{8}{3}$ (b) $\dfrac{5}{3}$ (c) $\dfrac{34}{9}$ (d) $\dfrac{34}{15}$ (e) $\dfrac{8}{15}$ (f) $\dfrac{34}{25}$ (g) $\dfrac{152}{27}$ 2. $4x^2 - 188x + 121 = 0$

3. $40x^2 + 89x + 40 = 0$ 4. $2x^2 + 15x - 1 = 0$ 5. $49x^2 - 189x + 9 = 0$ 6. $8x^2 + 305x - 216 = 0$

7 (a) $\dfrac{\sqrt{24}}{5}$ (b) $\dfrac{8\sqrt{24}}{25}$ 8. $5\dfrac{9}{20}$ 9. $\dfrac{13\sqrt{13}}{8}$ 10 (a) $3x^2 - 28x + 32 = 0$ (b) $3x^2 - 8x - 16 = 0$

11. $m = -9$ and $n = \dfrac{2}{3}$ or $m = -2$ and $n = 3$ 12. $P = 1$ or -23 13. $k = -\dfrac{28}{3}$ 14. $\dfrac{84}{5}$ 15 (a) $y = -\dfrac{217}{8}$

(b) $x = -\dfrac{11}{4}$ 16 (a) $y = \dfrac{513}{16}$ (b) $x = -\dfrac{15}{8}$ 17 (a) 1 and 3 or -1 and -3 (b) $k = -6$ or 10

18. $3\sqrt{3}$ 19. $K = -8$ 20. $\dfrac{\sqrt{5}}{2}$ and $\sqrt{5}$ or $-\dfrac{\sqrt{5}}{2}$ and $-\sqrt{5}$

Exercise 22

1 (a) It is not a function   (b) It is not a function   (c) It is not a function   (d) It is a function

2 (a) -5 (b) 22   (c) -1.625 (d) $81x^3 - 162x^2 + 108x - 26$ (e) $3x^3 + 27x^2 + 81x - 79$

3 (a) $-\dfrac{13}{5}$ (b) $p = \dfrac{7}{4}$ (c) $-\dfrac{153}{5}$ 4 (a) $3x - 2$ (b) $16 - 3x$ (c) $2x^2 - 5x - 3$ (d) $-\dfrac{4}{5}$

(e) $h(x) = x - 16$ (f) -14 5 (a) $2x - 4$ (b) $2x - 2$ 6 (a) $18(4x^2 + 4x + 1)$ (b) $18x^2$

(c) $36x^2 + 1$ 7 (a) Even   (b) Odd   8 (a) Even   (b) Even   (c) Neither   (d) Even

9 (a) $\dfrac{x+1}{5}$ (b) $\sqrt[3]{\dfrac{x-2}{9}}$ (c) $\sqrt[5]{\dfrac{x}{2}}$ 10 (a) $2\sqrt[3]{3} + 1$ (b) $\dfrac{1 + 2x^2}{5x^2 - 2}$ 11 (a) $\dfrac{x+1}{5 - 2x}$ (b) 0

12 (a) $2x^2 + 3x - 9$ (b) -7 13. $\dfrac{2x - 2}{5} + 7$ 14. 47   15 (a) $2x - 5$ (b) $\dfrac{x+5}{2}$ (c) -9

(d) 2   (e) $4x - 7$ 16 (a) $7x - 12$ (b) $14x^2 + 23$

Exercise 23

1 (a) $5x^2 + 5x + 6$ (b) $-2x^2 + 4x - 22$ (c) $3x^3 - 8x^2 + 6x - 11$ (d) $3x^3 + 7x^2 - 4x + 21$

(e) $9x^3 + 4x^2 + 2x + 9$ 2 (a) -82 (b) -10 3. $10x^6 + 11x^5 - 29x^4 - 2x^3 - x^2 - 2x$

4. $6x^6 - 11x^5 + 25x^4 - 5x^3 + 19x^2 - 14x - 20$ 5. $2x^6 - 3x^5 + 6x^4 - 19x^3 + 15x^2 - 30x + 45$

6. $x - 2$ 7. $2x^2 - 5x + 8$ 8. $3x^2 - 5$ 9. $2x^2 - xy + 3y^2$ 10. $5x^2 - 2xy - y^2$

11. $2x^2 + 3x + 27$ remainder 126   12. $5x^2 - 8x + 5$ remainder -9 13 (a) $x = \dfrac{15}{2}$ and 2

(b) $x = -\dfrac{3}{2}$ and $\dfrac{1}{2}$ 14. $(x+1)(x-2)(x-3)$. Hence, $x = -1, 2$ or 3   15. $x = 1, -2, 3$ or $-\dfrac{5}{2}$ 16. 11

17. $7\dfrac{1}{4}$ 18. $2x + 3$ and $2x + 1$ 19. $x = 0, -\dfrac{1}{2}$ or $\dfrac{1}{2}$ 20. $3x^2 - 5$ 21. $a = 2, b = 8$

22 (a) $m = -9, n = 10$ (b) $x^2 - x - 10$ (c) $x = 1, 3.7$ or -2.7 23. $m = \dfrac{1}{2}, n = 39$

Exercise 24

1. $\dfrac{2}{x+3} + \dfrac{3}{x-5}$ 2. $\dfrac{17}{9(x-3)} - \dfrac{16}{9(2x+3)}$ 3. $\dfrac{1}{x-4} + \dfrac{1}{x+4}$ 4. $\dfrac{1}{x} + \dfrac{1}{2(x-5)}$ 5. $\dfrac{1}{x-2} + \dfrac{9}{(x-2)^2}$

6. $\dfrac{2}{x^2} - \dfrac{5}{2x+3}$ 7. $\dfrac{2}{x+1} + \dfrac{3}{(x-2)^2}$ 8. $\dfrac{2}{3x} + \dfrac{13x - 3}{3(x^2+3)}$ 9. $\dfrac{2}{x^2+2} + \dfrac{3}{x+1}$

10. $\dfrac{3}{x-2} - \dfrac{5}{x+3} + \dfrac{2}{x-1}$ or $\dfrac{3}{x-2} + \dfrac{11 - 3x}{x^2+2x-3}$ 11. $\dfrac{2}{x+3} + \dfrac{3x - 1}{x^2-3x-3}$

12. $\dfrac{1}{x-1} + \dfrac{3}{(x-1)^2} - \dfrac{5}{(x-1)^3}$ 13. $\dfrac{1}{x} + \dfrac{5}{x^2} + \dfrac{x-2}{x^2-1}$ 14. $1 - \dfrac{1}{4(x-3)} - \dfrac{3}{4(x+1)}$

15. $x + 5 + \dfrac{17}{5(x+4)} - \dfrac{7}{5(x-1)}$ 16. 1   17. $P = -\dfrac{11}{6}, Q = -\dfrac{13}{12}$ 18. $1 - x + \dfrac{4}{x-1} + \dfrac{3x-1}{x^2+1}$

Exercise 25

1. $x = 26$ 2. $x = 7$ 3. $x = 4\dfrac{1}{2}$ or $\dfrac{1}{2}$ 4. $x = 10$ 5. $x = 1$ 6. $x = 5$ 7. $x = 5$ 8. $x = 6$ or -6

9. $x = 4$ or -2 10. $x = 1$ or 3

Exercise 26

1. 9   2. $-\dfrac{8}{5}$ 3. 9   4. 14   5. –14   6. $\dfrac{2}{5}$ 7. $\dfrac{1}{3}$ 8. –1   9. –15

10. 10   11. $\dfrac{1}{4}$ 12. 27   13. 12   14. $\dfrac{1}{10}$ 15. $\dfrac{3}{7}$ 16. Continuous

17. Continuous   18. Discontinuous   19. Not continuous (Discontinuous

20. Not continuous   21. Continuous   22. Not continuous   23. Not continuous

24(a) 8   (b) 5   26. $\dfrac{7}{2}$ 27. $\dfrac{1}{3}$ 28. 1   29. $-\dfrac{1}{2}$ 30. $2\dfrac{1}{4}$

Exercise 27

1. 2   2. $2x$ 3. $-3x^{-4}$ or $-\dfrac{3}{x^4}$ 4(a) $10x + 5h$ (b) $10x$

5(a) $27x^2 + 27xh + 9h^2$ (b) $27x^2$ 6(a) $\dfrac{2x^3 + 3x^2 \Delta x + x(\Delta x)^2 + 2}{x(x + \Delta x)}$ (b) $2x + \dfrac{2}{x^2}$

7. $6x - 10$ 8. $1 - \dfrac{3}{x^2}$ 9. $5 - 6x$ 10. $2 + \dfrac{1}{5} = 2\dfrac{1}{5}$

Exercise 28

1(a) $40x^4$ (b) $2x^4$ (c) $\dfrac{1}{3x^{\frac{2}{3}}}$ or $\dfrac{1}{3\sqrt[3]{x^2}}$ (d) $\dfrac{1}{x^{\frac{6}{7}}}$ or $\dfrac{1}{\sqrt[7]{x^6}}$ (e) $-\dfrac{5}{8x^{\frac{13}{8}}}$ or $-\dfrac{5}{8\sqrt[8]{x^{13}}}$

(f) $-\dfrac{5}{x^{\frac{7}{2}}}$ or $-\dfrac{5}{\sqrt{x^7}}$ 2. (a) $10x^4 - 12x^3 - 12x^2 + 10x - 6$ (b) $7x^6 + 8x^3 + \dfrac{3}{x^2}$

(c) $18x^5 - 4x^3 - 5 - \dfrac{2}{x^2} + \dfrac{3}{x^4}$ (d) $\dfrac{5}{4\sqrt[4]{x^3}} - \dfrac{5}{3\sqrt[3]{2x^4}}$ 3. $8(2x - 5)^3$ 4. $\dfrac{-6}{(x^3 - 7)^3}$

5. $\dfrac{6x^2 + 7}{2(2x^3 + 7x)^{\frac{1}{2}}}$ 6. $5(21x^2 - 2x)(7x^3 - x^2 + 3)^4$ 7. $\left(9 + \dfrac{2}{x^2}\right)\left(3x - \dfrac{2}{3x}\right)^2$

8. $1 + \dfrac{9(6x - 1)}{(3x^2 - x - 10)^2}$ 9. $\dfrac{-4x^3}{\sqrt{1 - 2x^4}}$ 10. $\dfrac{-5x^2}{\sqrt[3]{(5x^3 - 1)^4}}$

Exercise 29

1. $12x - 1$ 2. $12x^3 + 50x$ 3. $\dfrac{15x + 30}{2\sqrt{3 + x}}$ 4. $(7x^2 + 30 - 7)(x^2 - 7)^2$

5. $48x^3 + 24x^2 - 6x - 20$ 6. $\dfrac{\sqrt{2}\,[(3x^2 - 1)^3 + 36x^2(9x^4 - 6x^2 + 1)]}{2\sqrt{x}}$

7. $\dfrac{15 - 7x}{4(x + 3)^{\frac{1}{4}}}$ or $\dfrac{15 - 7x}{4\sqrt[4]{x + 3}}$ 8. $12x^3 + 9x^2 - 46x - 5$ 9. $15x^4 - 12x^3 - 3x^2 + 6x - 2$

10. $\dfrac{12x + 1}{2}$ 11. $\dfrac{9x^{\frac{7}{2}}}{2}$ or $\dfrac{9\sqrt{x^7}}{2}$ 12. $\dfrac{4x^3(7x - 33)}{3\sqrt[3]{2x - 11}}$

13. $-30x^5 - 175x^4 + 20x^3 + 102x^2 - 14x + 1$ 14. $75x^4 + 28x^3 + 3x^2 + 28x - 7$

15. $-\dfrac{2}{x^2} + \dfrac{12}{x^5} - \dfrac{25}{x^6}$

Exercise 30

1. $\dfrac{x^2 - 2x + 4}{(x-1)^2}$

2. $\dfrac{8x^2 + 24x - 3}{(2x+3)^2}$

3. $\dfrac{42x}{(3x^2+1)^2}$

4. $\dfrac{1}{(1-x)^{\frac{3}{2}}\sqrt{x+1}}$

5. $\dfrac{4(x^3 - 2)(x^3 + 1)}{x^3}$

6. $\dfrac{-6(x^2 + 1)}{x^4\sqrt{3x^2+2}}$

7. $\dfrac{-8x^2 + 12x + 24}{3(2x+1)^3(x^2 - x - 4)^{\frac{2}{3}}}$

8. $\dfrac{10x^3 - 2x^2 - x + 4}{2(2-x)^{\frac{3}{2}}}$

9. $\dfrac{-6x}{(1-3x^2)^2}$

10. $\dfrac{4}{(x-2)^2}$ or $\dfrac{4}{(2-x)^2}$

Exercise 31

1. $\dfrac{3t}{2}$

2. $\dfrac{2(4 - t^3)^2}{3t}$

3. $\dfrac{2r}{3(l+2r)}$

4. $\dfrac{2t - 1}{10t - 1}$

5. $\dfrac{-2m(v - u)}{(u+v)t^2}$

6. $\dfrac{5(t^4 + 1)^2}{3(t^2 - 1)^2}$

7. $\dfrac{1 - 12t^2}{2t}$

8. $\dfrac{15t^2}{2}$

9. $\dfrac{8}{r}$

10. $\dfrac{2s^3}{3}$

Exercise 32

1. $\dfrac{-15x^2}{3 - 2y}$ or $\dfrac{15x^2}{2y - 3}$

2. $\dfrac{-4x}{9y^2}$

3. $\dfrac{3y^2 + 12x^2 - y}{x - 6xy}$

4. $\dfrac{3x^2 - y^2 - 4x}{2xy}$

5. $\dfrac{x^2}{y}$

6. $\dfrac{4xy}{2x^2 + 5}$

7. $\dfrac{-2xy}{x^2 + 6y^2 - 1}$

8. $\dfrac{4y}{10y - 4x - 1}$

9. $\dfrac{2xy^2 - 1}{2y(x^2 - 1)}$

10. $\dfrac{-5x^4 y^2}{3}$

Exercise 33

1. $2\sec^2 2x$

2. $-\dfrac{1}{5}\sin\dfrac{1}{5}x$

3. $-50\sin 5x$

4. $3\sin^2 x\cos x$

5. $2\sec^2 x\tan x$

6. $60x^4\sin^3 3x^5\cos 3x^5$

7. $-12x\cot 6x^2\cosec 6x^2$

8. $10x^4\sec 2x^5\tan 2x^5$

9. $6x\sec^2 3x^2$

10. $-x(3x\sin 3x - 2\cos 3x)$

11. $\dfrac{6x\cos 2x - 9\sin 2x}{x^4}$

12. $48x^3\sec x^4\tan x^4$

13. $\dfrac{-5\tan x\cos 5x - \sec^2 x(2 - \sin 5x)}{\tan^2 x}$

14. $\dfrac{\sin^2 x(3x\cos x - \sin x)}{2x^2}$

15. $\dfrac{-(2x+6)\cosec^2 2x - \cot 2x)}{(x+3)^2}$

16. $\dfrac{-\sin\sqrt[3]{3}}{5x^2}$

17. $\dfrac{2[x - (x^2 - 3)\tan 2x]}{\sec 2x}$

18. $6x\cos 3x^2 - 18x^3\sin 3x^2$

19. $-\cos x(2\cot x\sin x + \cos x\cosec^2 x)$

20. $\dfrac{2(\sin^2 x + \sin^2 x)}{(\sin x + \cos x)^2}$ or $\dfrac{2}{(\sin x + \cos x)^2}$

21. $\dfrac{4\sec 4x\sin(4x-2) + \tan 4x\cos(4x-2)}{\cos^2(4x-2)}$

22. $-3\sin 3x - x^2\sec x\tan x - 2x\sec x$

23. $27x^2\cos x^3\sin^8 x^3$

24. $\dfrac{3\cos 6\sqrt{x}}{\sqrt{x}}$

25. $3\cos x^3 - \dfrac{2\sin x^3}{x^3}$

26. $-30x^4\cos^2 2x^5\sin 2x^2$

27. $-10\sin 2x\sin 10x - 2\cos 2x\cos 10x$

28. $6x^2\cos x^4 - \dfrac{3\sin x^4}{2x^2}$

29. $\dfrac{(10x - 5)\sec^2 5x - 2\tan 5x}{(2x-1)^2}$

30. $\dfrac{\sec^2 x\tan x}{x} - \dfrac{\tan^2 x}{2x^2}$

Exercise 34

1. $\dfrac{1}{3\sqrt[3]{7y^2}}$ 2. $\dfrac{3y^2}{2\sqrt{y^3-1}}$ 3. $\dfrac{1}{5\sqrt[5]{2(y+3)^4}}$ 4. $\dfrac{1}{\sqrt[3]{2(3y+27)^2}}$ 5. $\dfrac{-1}{(y-2)^2}$

6. $\dfrac{-3}{2y^2\sqrt{\frac{3}{y}+5}}$ 7. $2y^3$ 8. $5y^4$ 9. $\dfrac{-5}{(y-4)^2}$ 10. $\dfrac{-1}{3y^2\sqrt[3]{(\frac{1}{y}+8)^2}}$

Exercise 35

1. $\dfrac{2}{\sqrt{1-4x^2}}$ 2. $\dfrac{-1}{\sqrt{1-x^2}}$ 3. $\dfrac{-6x}{9x^4+1}$ 4. $\dfrac{-4x^3}{x^8+1}$ 5. $3\sec^{-1}x+\dfrac{3}{\sqrt{x^2-1}}$

6. $2x+\dfrac{3}{x^2+1}$ 7. $\dfrac{1}{(x+5)^2+1}$ or $\dfrac{1}{x^2+10x+26}$ 8. $\dfrac{1}{x^2\sqrt{1-\frac{1}{x^2}}}$ 9. $\dfrac{-1}{5\sqrt{1-y^2}}$

10. $\dfrac{1}{2y\sqrt{y-1}}$ 11. $5\tan^{-1}3x+\dfrac{15x}{9x^2+1}$ 12. $2x+\dfrac{20x^4}{\sqrt{1-x^{10}}}$ 13. $\dfrac{-3}{x\sqrt{x^6-1}}$

14. $\dfrac{1}{2(y^2+1)}$ 15. $\dfrac{1}{15y^{\frac{2}{3}}\sqrt{1-y^{\frac{2}{3}}}}$

Exercise 36

1. $3\cosh 3x-2\sinh x$ 2. $-5x\,\text{sech}\,x(x\tanh x-2)$ 3. $\dfrac{2(x\cosh 2x-\sinh 2x)}{3x^3}$

4. $6\cosh 3x\sinh 3x$ 5. $60x^4\cosh^2 4x^5\sinh 4x^5$ 6. $-10x^4\text{cosech}\,2x^5$

7. $-12x\,\text{sech}^3 2x^2\tanh 2x^2$ 8. $x(5x\sinh 5x+2\cosh 5x)$

9. $\dfrac{3\text{cosech}^2 3x\tanh 5x+5\coth 3x\,\text{sech}^2 5x}{\coth^2 3x}$ 10. $\dfrac{2\cosh^2 x(3x\sinh x-\cosh x)}{3x^2}$

Exercise 37

1. $\dfrac{1}{x}\log_a e$ 2. $\dfrac{3x^2}{x^3+5}\log_a e$ 3. $\dfrac{36x^2}{2x^3-5}\log_a e$ 4. $\dfrac{2}{3x}\log_a e$ 5. $\dfrac{4x}{(x^2+1)(1-x^2)}\log_a e$

or $\dfrac{-4x}{(x^2+1)(x^2-1)}\log_a e$ or $\dfrac{-4x}{(x^4-1)}\log_a e$ 6. $\dfrac{15x^2}{5x^3-1}\log_5 e$ 7. $\dfrac{-2}{x}\log_2 e$ 8. $\dfrac{2x}{x^2+3}$

9. $\dfrac{3\ln^2 x}{x}$ 10. $\dfrac{5x^4}{2x^5-1}$ 11. $2x^3(4\ln x+1)$ 12. $\dfrac{-28}{3-7x}$ or $\dfrac{28}{7x-3}$

13. $6x\ln(4x^3+1)+\dfrac{36x^4}{4x^3+1}$ 14. $\dfrac{2-4\ln x}{x^3}$ 15. $\dfrac{-30x}{1-5x^2}$ or $\dfrac{30x}{5x^2-1}$

Exercise 38

1. $10a^{2x}\log_e a$ 2. $a^{5x^2-x}\ln a(10x-1)$ 3. $a^{3x}x^3(3x\ln a+4)$

4. $-2e^{-2x}-3e^{-x}$ or $-e^{-2x}(2+3e^x)$ 5. $\dfrac{\sqrt{3}\,e^{\sqrt{3x}}}{\sqrt{x}}$ 6. $6x^2 e^{x^2}(2x^2+3)$

7. $\dfrac{5\ln a\,(a^{10x}-1)}{a^{5x}}$ 8. $\dfrac{6e^{4x}(2x\ln 2x^2+1)}{x}$ 9. $\dfrac{3x\sqrt{e^{5x}}(5x+4)}{2a}$ 10. $x^3(\log_{10}e+4\log_{10}7x)$

11. $3\ln6(6^{3x})$ 12. $2x - 3(e^{x^2 - 3x})$ 13. $\dfrac{-(5x + 3)e^{\frac{3}{x}}}{5x^7}$ 14. $6e^{-3x}$

15. $3(2x + 1)(x^2 + x)^2 e^{(x^2 + x)^3}$ 16. $2^{x^2} 2x\ln2$ 17. $2a^{2x}\ln a$ 18. $\dfrac{e^x(x\ln 10x^3 + 3)}{x}$

19. $\dfrac{\sqrt{5}(4x + 1)e^{2x}}{2\sqrt{x}}$ 20. $2x^4(4\log_{10}e + 5\log_{10}5x^4)$

Exercise 39

1. $x^{2x}(2 + 2\ln x)$ 2. $\dfrac{3x^{\ln 2x}(\ln 2x + \ln x)}{x}$ 3. $\ln(1 - 4x^2) - \dfrac{8x^2}{1 - 4x^2}$

4. $(6x^x - 5)^{3x}\left[3\ln(6x^2 - 5) + \dfrac{36x^2}{6x^2 - 5}\right]$ 5. $\dfrac{(2x^3 + 1)(x - 2)^3}{3x^2(x^3 - 1)^2}\left(\dfrac{6x^2}{2x^3 + 1} + \dfrac{3}{x - 2} - \dfrac{2}{x} - \dfrac{6x^2}{x^3 - 1}\right)$

6. $\dfrac{(2x - 1)(x^2 - 2)}{(1 - x)(x - 3)^2}\left(\dfrac{2}{2x - 1} + \dfrac{2x}{x^2 - 2} + \dfrac{1}{1 - x} - \dfrac{2}{x - 3}\right)$

7. $15x^3 x^{3x^4}(4\ln x + 1)$ or $15x^{3x^4 + 3}(4\ln x + 1)$ 8. $3e^{3x}(e^{e^{3x}})$ or $3e^{e^{3x} + 3x}$

9. $2e^x e^x \ln2$ 10. $\dfrac{x^{\ln(x^2 - 4x)}[(x - 4)\ln(x^2 - 4x) + (2x - 4)\ln x]}{x(x - 4)}$ 11. $\ln(2x^2 + 10) + \dfrac{4x^2}{2x^2 + 10}$

12. $(3x^2 - 8)^{2x}\left[2\ln(3x^3 - 8) + \dfrac{18x^3}{3x^3 - 8}\right]$ 13. $\dfrac{(x + 3)^2(x - 3)^2}{x^3 - 1}\left(\dfrac{2}{x + 3} + \dfrac{2}{x - 3} - \dfrac{3x^2}{x^3 - 1}\right)$

14. $15x^{(\ln x)^2}(\ln x)^2$ 15. $30x^{3x^2}x(2\ln x + 1)$ or $30x^{3x^2 + 1}(2\ln x + 1)$

Exercise 40

1. $\dfrac{3}{5x^2}$ 2. $\dfrac{5xe^{5x}}{2}$ 3. $\dfrac{-2x\sin x^2}{\cos x}$ 4. $\dfrac{x(6x - 5)}{x - 1}$ 5. $\dfrac{14x^2}{3}$ 6. $\dfrac{2a^x\ln a}{e^x}$

7. $\dfrac{-2\tan x \sec^2 x}{5\sin 5x}$ 8. $\dfrac{1}{(x - 1)(e^{x-1})}$ 9. $\dfrac{1}{xe^{5x}}$ 10. $\dfrac{e^x + e^{-x}}{(e^x - e^{-x})(1 + \ln x)}$

Exercise 41

1. $\dfrac{dy}{dx} = 2x(2x^2 - 3x - 7)$, $\dfrac{d^2y}{dx^2} = 2(6x^2 - 6x - 7)$, $\dfrac{d^3y}{dx^3} = 2(12x - 6)$ 2. $-\dfrac{3}{x^2}$

3. $20x^2 e^{5x^4}(20x^4 + 3)$ 4. $-50x^8\sin^2 x^5 + 40x^3\cos x^5\sin x^5 + 50x^8\cos x^5$ 5. $\dfrac{128}{(4x - 10)^2}$

6. $\dfrac{-(x^2 - 2)\sin x - 2x\cos x}{x^3}$ 7. $\dfrac{4\ln x - 6}{x^3}$ 8. $-2\csc^2 x(\csc^2 x + 2\cot^2 x)$

9. $2\cos^2 x - 3\sin^2 x - 5\cos 2x$ 10. 0

11. $-8x^2\sin x^2 + 4\cos x^2 + 2\sin^2 x - 2\cos^2 x - 4x\cos x^2 + 2\cos x\sin x$

12. $\dfrac{2[(x^2\sec^2 x + 1)\tan x - x\sec^2 x]}{3x^3}$ 13. $e^{-x}(4\sin 2x - 3\cos 2x)$ 14. 0

15. $\dfrac{-(9x^2 - 2)\sin 3x - 6x\cos 3x}{3x^3}$

Exercise 42

1. $2x(3x^4-5)^3 + 36x(3x^4-5)^3$ or $2x(3x^4-5)^2(21x^4-5)$ 2. 209

3(a) $\dfrac{\sqrt{3x^2-1}\,(3x^2+2)}{x^3}$ (b) $5\sqrt{2}$ 4(a) $\dfrac{12x^2-3y^3}{9xy^2-2y-5}$ (b) $-\dfrac{9}{16}$ 5. $-\dfrac{2}{y}$

6. $3x(x^2+5)^2(3x^3+5x-2)$ 7. $e^{\sin x + \tan x}(\sec^2 x + \cos x)$ 8. $\dfrac{2x(\sin x - x\cos x)}{\sin^3 x}$

9. $\dfrac{2\cos x \sin x - 3\sin 3x}{\sin^2 x + \cos 3x}$ 10. $-5a^{\cos 5x}\sin 5x \ln a$ 11. $5x^2 e^{-3x} - 10xe^{-3x} + 2e^{-3x}$

12. $-(2x^3+2)\cos\!\left(\dfrac{1-2x^3}{x^2}\right)$ 13. $\dfrac{4x^4-6x^3+12x^2-18x+6}{x^3}$ 14. 0

15. $5\!\left(\dfrac{2}{x^2}-\dfrac{2}{x^3}\right)\!\left(\dfrac{1}{x^2}-\dfrac{2}{x}\right)^4$ or $\dfrac{10(x-1)(1-2x)^4}{x^{11}}$ 16(a) $\dfrac{3x^2 e^{x^3-y^3}-y}{3y^2 e^{x^3-y^3}+x}$ (b) 1

17. 0   18. $\dfrac{2\sin\frac{1}{x^2}}{x^2} - \dfrac{4\cos\frac{1}{x^2}}{x^4} + 2\cos\dfrac{1}{x^2}$ 19(a) $\dfrac{2e^x(e^x+1)}{(e^x-1)^3}$

(b) $\dfrac{4(e^{2x}-e^{-x})+(1+e^{-x})(e^x-1)^3}{(e^x-1)^3(1-e^{-x})}$ 20. $2\tan x$ 21. 0

22. $3\!\left(\dfrac{\sqrt{5}}{4\sqrt{x}}+3\right)\!\left(3x+\dfrac{\sqrt{5x}}{2}\right)^2$ 23. $\dfrac{-10x-18}{x^4}$ or $\dfrac{-2(5x+9)}{x^4}$

24. $9x^2(x^3+1)^2 + \dfrac{5+4x-10x^2}{(x^2+1)^2}$ 25. 0

Exercise 43

1.

Mark	65	66	67	68	69	70	71	72	73	74
Frequency	4	2	6	5	6	3	3	2	3	5

2.

Score	5	6	7	8	9
Frequency	5	7	2	2	4

3.

No of Seed	20	21	22	23	24	25	26	27	28	29	30
Frequency/No of Pod	6	4	3	5	2	2	1	4	7	3	3

4.

Age	10	11	12	13	14	15

No of School	5	4	5	5	6	5

Exercise 44

1. Mean = 4   Median = 3   Mode = 2
2. (a) 24   (b) 25   (c) 25
3. Mean = 111.2   Median = 111.5   Mode = 105
4. Mean = 4.3   Median = 4.5   Mode = 5
5. (a) 5.1   (b) 5   (c) 4
6. (a) 13.2   (b) 13.5   (c) 14
7. (a) 11   (b) 58kg

Exercise 45

1.

Score	40 – 44	45 – 49	50 - 54	55 - 59	60 – 64
Frequency	8	14	13	9	6

2.

Weight	21 - 25	26 - 30	31 - 35	36 - 40	41 - 45	46 - 50	51 – 55
Frequency	2	9	8	2	8	2	9

3.

Class interval	Frequency	Class Boundary	Class width	Class mid-value
5 – 9	2	4.5 – 9.5	20	7
10 – 14	5	9.5 – 14.5	20	12
15 – 19	5	14.5 – 19.5	20	17
20 – 24	7	19.5 – 24.5	20	22
25 – 29	1	24.5 – 29.5	20	27

4.

Class interval	Frequency	Class Boundary	Class width	Class mid-value
1 – 20	1	0.5 – 20.5	20	10.5
21 – 40	4	20.5 – 40.5	20	30.5
41 – 60	7	40.5 – 60.5	20	50.5
61 – 80	3	60.5 – 80.5	20	70.5
81 – 100	5	80.5 – 100.5	20	90.5

5.

Class interval	Frequency	Class Boundary	Class width	Class mid-value
0 – 90	2	-5 – 95	100	45
100 – 190	4	95 – 195	100	145
200 – 290	1	195 – 295	100	245
300 – 390	7	295 – 395	100	345
400 – 490	1	395 - 495	100	445

Exercise 46

1. (a) 72.5   (b) 70.9   (c) 67.9

2. (a) 90.4   (b) 91.5   (c) 92.7

3. (a) 44.1   (b) 41.7   (c) 36.9

Exercise 47

1. 1.67   2. 1.82   3. 1.6   4. 16.27   5. 12.5   6. 9.1   7. 2.7

8. (a) The frequency table is as shown below.

No on Die	1	2	3	4	5	6
No of Times	7	7	9	11	7	9

(b) 1.41

Exercise 48

1. Variance = 2   Standard Deviation = 1.41

2. Variance = 6   Standard Deviation = 2.45

3. (a) 199.04   (b) 14.11

4. (a) 40.69   (b) 6.38

5. (a) 1.049   (b) 1.024

Exercise 49

1. The frequency table is as shown below.

Mark	11 - 20	21 – 30	31 - 40	41 – 50	51 - 60	61 - 70	71 - 80	81 - 90	91-100
No of Student	4	5	4	5	8	3	5	3	3

(a) 53   (b) 33   (c) 72.5   (d) 39.5   (e) 19.75   (f) 38   (g) 64.5

2. (a) 32.36   (b) 25.18   (c) 40.25   (d) 15.07   (e) 7.54   (f) 41.6

(g) 37.33   (h) 22.06

3. (a) 1.7   (b) 1.23   (c) 2.31   (d) 1.60   (e) 0.98

4. (a) 17.5   (b) 11.29   (c) 22.56   (d) 11.27   (e) 5.63   (f) 20.06   (g) 29.5

5. (a) 1.54   (b) 0.83   (c) 2.08   (d) 0.83   (e) 1.01

Exercise 50   There is no exercise in chapter 50.

Exercise 51

1. (a) $\dfrac{1}{10}$ (b) $\dfrac{9}{20}$ (c) $\dfrac{1}{4}$ (d) $\dfrac{3}{4}$ (e) $\dfrac{7}{20}$

2. $\dfrac{1}{4}$ 3. $\dfrac{4}{13}$ 4. $\dfrac{2}{5}$ 5. $\dfrac{4}{5}$

6. (a) $\dfrac{3}{5}$ (b) $\dfrac{2}{5}$

7. $\dfrac{11}{25}$ 8. $\dfrac{7}{10}$

9. (a) $\dfrac{2}{5}$ (b) $\dfrac{1}{2}$ (c) $\dfrac{9}{10}$ (d) $\dfrac{3}{10}$ (e) $\dfrac{9}{20}$

10. (a) $\dfrac{1}{26}$ (b) $\dfrac{3}{26}$ (c) $\dfrac{2}{13}$ (d) $\dfrac{4}{13}$ (e) $\dfrac{4}{13}$

11. $\dfrac{2}{11}$

Exercise 52

1. $\dfrac{1}{13}$ 2. $\dfrac{1}{26}$ 3. $\dfrac{1}{26}$ 4. 1   5. $\dfrac{1}{26}$ 6. $\dfrac{12}{13}$

7. (a) $\dfrac{2}{13}$ (b) $\dfrac{2}{13}$ (c) $\dfrac{5}{52}$ (d) $\dfrac{7}{26}$ (e) $\dfrac{15}{52}$ (f) $\dfrac{7}{13}$

8. (a) $\dfrac{1}{169}$ (b) $\dfrac{1}{13}$ (c) $\dfrac{6}{169}$ (d) $\dfrac{1}{4}$ (e) $\dfrac{1}{13}$ (f) $\dfrac{12}{13}$

9. (a) $\dfrac{8}{663}$ (b) $\dfrac{8}{663}$ (c) $\dfrac{1}{221}$ (d) $\dfrac{1}{17}$ (e) $\dfrac{25}{102}$ (f) $\dfrac{13}{102}$ (g) $\dfrac{1}{221}$

10. (a) $\dfrac{37}{2197}$ (b) $\dfrac{2196}{2197}$

11. (a) $\dfrac{73}{5525}$ (b) $\dfrac{5452}{5525}$

Exercise 53

1. (a) $\frac{1}{2}$ (b) $\frac{1}{2}$ (c) 1

2. (a) $\frac{1}{4}$ (b) $\frac{3}{4}$ (c) $\frac{1}{4}$ (d) $\frac{3}{4}$ (e) $\frac{1}{4}$

3. (a) $\frac{1}{8}$ (b) $\frac{7}{8}$ (c) $\frac{1}{8}$ (d) $\frac{7}{8}$ (e) $\frac{1}{2}$ (f) $\frac{7}{8}$

4. (a) $\frac{15}{16}$ (b) $\frac{1}{16}$ (c) $\frac{11}{16}$ (d) $\frac{15}{16}$ (e) $\frac{3}{8}$

5. $\frac{31}{32}$

Exercise 54

1. (a) $\frac{1}{6}$ (b) $\frac{1}{6}$ (c) $\frac{1}{6}$ (d) $\frac{1}{2}$ (e) $\frac{5}{6}$ (f) $\frac{5}{6}$

2. (a) $\frac{1}{2}$ (b) $\frac{1}{2}$ (c) $\frac{5}{6}$ (d) $\frac{1}{2}$ (e) $\frac{2}{3}$ (f) $\frac{1}{3}$

3. (a) $\frac{1}{12}$ (b) $\frac{1}{12}$

4. (a) $\frac{11}{36}$ (b) $\frac{13}{18}$ (c) $\frac{11}{12}$ (d) $\frac{7}{12}$ (e) $\frac{1}{36}$ (f) $\frac{7}{18}$ (g) $\frac{7}{18}$

5. (a) $\frac{1}{6}$ (b) $\frac{5}{12}$ (c) $\frac{1}{36}$ (d) $\frac{1}{6}$ (e) $\frac{1}{6}$

6. $\frac{1}{6}$

Exercise 55

1. (a) $\frac{1}{4}$ (b) $\frac{2}{5}$ (c) $\frac{7}{20}$ (d) 0   (e) $\frac{13}{20}$ (f) $\frac{13}{20}$

2. (a) $\frac{7}{8}$ (b) $\frac{3}{4}$ (c) $\frac{1}{4}$

3. (a) $\frac{2}{125}$ (b) $\frac{8}{125}$ (c) $\frac{92}{125}$ (d) $\frac{31}{125}$

4. (a) $\frac{27}{64}$ (b) $\frac{37}{64}$ (c) $\frac{63}{64}$ (d) $\frac{27}{64}$

5. (a) $\frac{3}{8}$ (b) $\frac{1}{2}$ (c) $\frac{3}{8}$ (d) $\frac{1}{2}$ (e) $\frac{1}{2}$

6. (a) $\frac{15}{364}$ (b) $\frac{145}{182}$ (c) $\frac{25}{91}$

7. (a) $\frac{2}{5}$ (b) $\frac{8}{15}$

8. (a) $\frac{7}{22}$ (b) $\frac{7}{11}$ (c) $\frac{21}{22}$

Please if you found this book well simplified enough for easier understanding, kindly give it a five star rating on amazon so as to encourage people to buy this book, thereby putting more money in my pocket and helping students improve on their skills on algebra and differential calculus. Thank you.

If you want to see other books written by the author, just simply search for the author's name, Kingsley Augustine on amazon.com

If you have any enquiry, suggestion or information concerning this book, please contact the author through the email below.

KINGSLEY AUGUSTINE

kingzohb2@yahoo.com